W9-ASI-656

The Earth's
Dynamic Systems

The Earth's Dynamic Systems

A Textbook in Physical Geology

by W. Kenneth Hamblin
Brigham Young University

Illustrated by
Wm. L. Chesser

BURGESS PUBLISHING COMPANY
MINNEAPOLIS, MINNESOTA

To Sally, Bill,
Lisa, Laura, and Kim

Book and cover design by Dennis Tasa

Preface

The history of the development of any scientific discipline is marked by periods of rapid advancement with major "breakthroughs" separated by longer periods of gradual growth during which new ideas are consolidated and new data are analyzed. Generally the periods of rapid advancement result from the development of new instrumentation which permits us to study something which was previously beyond our capabilities. This commonly results in the development of a new theory that unifies thinking and provides a new framework for interpretation of observational data. When a new theory is so far-reaching as to affect the entire field of study, we commonly speak of its consequences as a scientific "*revolution*." A scientific revolution does not mean that previously established facts are wrong or have been replaced but only that we have made a major step forward in our ability to understand and explain the meaning of many things that were previously matters of conjecture.

Geology is presently in such a revolution as a result of the unifying theory of plate tectonics. This theory concerns the dynamics of our planet and relates many major features, such as continents, ocean basins, earthquake belts, and vulcanism to a simple system of slow-moving material below the crust. It was developed as a result of our newly acquired ability to study the characteristics of the ocean floor, map its landforms, and measure its magnetic and seismic properties. These new data show quite clearly that the earth's crust is not rigid and fixed but is actually in motion, with the continents moving as "rafts floating on a sea" of denser rock. The continents have repeatedly split, drifted apart, collided, and have been sutured together in various patterns. With this process, they have grown larger as part of the mechanism by which earth's materials are segregated and differentiated. The sea floors, in contrast, are temporary features, opening and closing, continually being created and consumed as the crust moves.

In addition to this revolutionary thinking about the earth's crust, the space program has brought the moon and the planets well within the sphere of study of earth scientists. Detailed photographs of the surface of the moon, Mars, and Mercury are analyzed in much the same way as we analyze aerial photographs of the earth's surface, and geologists are able to make geologic and topographic maps of these planets using normal techniques of photogeology. Indeed, we have learned more about the surface of the moon and Mars in the past few years than was known about the surface of the earth only a few hundred years ago. From this new knowledge, we are able for the first time to compare planets, their surface features, their history, and their processes of evolution.

Our objective in this book is to utilize the unifying theory of plate tectonics and to consider the planet earth as a system of matter constantly changing toward a state of equilibrium. By a *system* we mean just what the dictionary defines as "regularly interacting and interdependent components forming a unified whole."

Within the earth's system we can identify two major subsystems of chemical and physical processes which produce most

v

of the major changes in our planet. One is the system of moving fluids on the earth's surface and in the atmosphere, known as the *hydrologic system*. Within this system, weathering, running water, ground water, glaciers, and wave action operate to erode the earth's surface and deposit the resulting sediment in the sea. Energy for the hydrologic system comes from the sun. The other major system involves the movement of material in the earth's interior, motion which is expressed on the surface by vulcanism, mountain chains, earthquakes, continental drift, and sea-floor spreading. Here the theory of *plate tectonics* gives us an important new insight into the internal dynamics of the earth and explains many seemingly unrelated phenomena as part of a simple mechanism of moving material resulting from radiogenic heat originating within the earth.

We express our sincere thanks to our colleagues and others who have helped by discussing problems and ideas, critically reading the manuscript, and supplying unpublished information. Those who read parts of the manuscript include: James L. Baer, Willis H. Brimhall, Myron G. Best, H. M. Davis, Thomas Hendrix, D. Hill, Wade E. Miller, Joseph R. Murphy, John Reid, James A. Rhodes, and Morris S. Petersen. We sincerely appreciate the assistance and persistence of Alice Lanyk, Dennis Tasa, and Bob Lakemacher of the Burgess Publishing Company, which extend well beyond their normal duties.

January 1975   *W. Kenneth Hamblin*

Preface

Our Approach

In studying the processes in the earth's system, we are continually confronted with the problem of scale. How can we begin to understand the nature and interrelation of things as large as rivers, mountains, continents, and ocean basins? One approach is to utilize models, a method commonly used in all branches of science and engineering to study and analyze things too large to see or handle conveniently. A model is a formulation that simulates a real-world phenomenon and has value in analyzing and predicting behavior of the real thing. Models may be physical or abstract rep-representations of the structure and function of real systems. Throughout this text, we have made extensive use of graphic models (diagrams, photographs, and maps) in an effort to provide you with an accurate image of the features and processes described. These models, especially the block diagrams, are as important as the text in providing information, and as much time should be devoted to studying them as reading the text.

The *Physiographic Map of the Surface of the Earth* at the back of the book is a semi-scale model. The map was compiled from the best sources available and represents, in perspective, the surface features of our planet. These features are the net result of processes operating in the earth's systems and should be carefully studied as you read the text. For the most effective results, you should hang this "model" on the wall of your study and identify and analyze each feature as it is discussed. Much can be gained by annotating the map as you study.

The *Shaded Relief Maps of the Moon and Mars* are essentially the same type of model as the physiographic map of the earth although they were constructed with different techniques. They are accurate models showing regional features in perspective. Continued reference to these models is absolutely essential in mastering chapters on lunar geology, the geology of Mars and Mercury, and geologic contrasts between the planets.

Organization of the Chapters

We have attempted to present the material in each chapter in a manner which will help you to recognize the essential ideas or concepts and separate them from supportive data. In each chapter, most major subdivisions are introduced by a paragraph or two under the subtitle "*Statement.*" The statement presents the major concept, principle, or theory. It is not intended as a summary or abstract of what the section contains but is an expression of the various ideas of the subject in one all-embracing point of view. It is, in effect, a "word model" of the principle in question. The "statement" may be difficult to fully comprehend the first time it is read, but further insight can be gained from the paragraphs under the subtitle "*Discussion.*" Here a more complete discussion of the ideas or principles is presented and illustrated. Evidence supporting the statement is discussed and, where pertinent to understanding the concept, a brief history of how it developed is presented. This material is designed to help you clearly understand the ideas presented in the statement.

Contents

ix

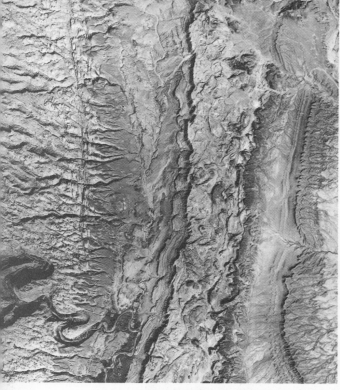
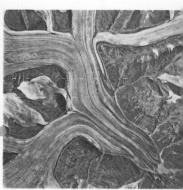

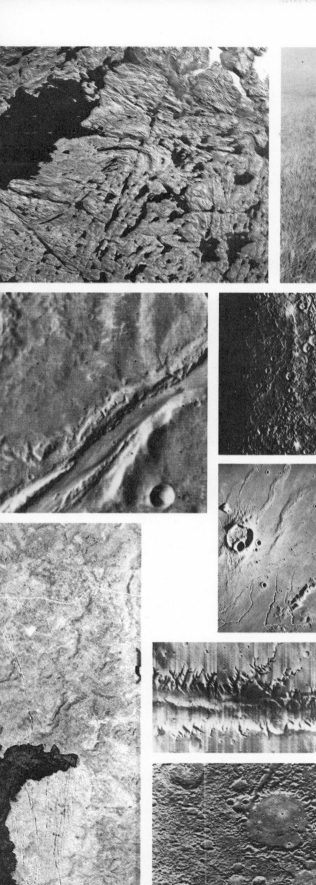

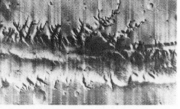
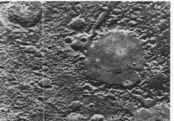

1 The Planet Earth

One of the most significant advances made in the twentieth century concerns our view of the earth and the dynamics of how it changes. Until recently most of our knowledge of the earth has been obtained from "near-sighted" studies of minute details which were plotted on maps and charts which served as models of the "real world." Rivers, mountains, shorelines, and weather patterns were surveyed and studied from hundreds of observation points, but we were never able to see the earth in a regional panoramic view and to observe how it functions as a planet in space. Now, for the first time in human history, space photography permits man to see the earth in one synoptic view and observe large-scale structures of the atmosphere, oceans, and continents, and the regional relationships of river systems, mountain belts, and coastal features. In addition, the Earth Resources Technology Satellite (ERTS) is able to photograph most of the earth on an eighteen-day repetitive cycle. Never before have we had such vivid evidence that the earth is not static or unchanging.

To many, the breathtaking photographs of earth from space are simply items of great beauty, but to the student of geology they provide an important additional perspective of our planet. From these photographs we see much more than outlines of continents and oceans; we see the earth as a dynamic system undergoing constant change. The surface fluids (air and water) are seen in spectacular motion, moving from the oceans to the atmosphere, to the continents, and back to the oceans through the great river systems of the world.

Another expression of mobility in our planet is seen in the folded mountain belts which have been deformed by compression and the great rifts which split the continents. Even though the earth's surface appears to be firm and stationary, there is convincing evidence that the materials which form the "solid" rock on the surface of our planet are in motion. The continents split and drift apart, sea floors are continually being created and destroyed, and the entire earth changes from its own internal energy.

Advances in the study of oceanography have provided another view of part of the earth never before seen by man. Precision depth recorders plot a continuous profile of the topography of the ocean floor and permit us to "see" the landforms beneath the seas as if all the water had been removed.

Our new ability to see the earth as a planet in space has greatly increased our knowledge of how it functions, the processes which operate on it, and the changes which have occurred throughout its history. In this chapter, we will attempt to point out the major geologic characteristics of the earth, utilizing space photography and **bathymetric** data of the oceans. From this study, you should develop a new awareness of the major features seen on the surface of our planet and an appreciation of the earth as a dynamic system.

Major Concepts

1. The **ocean basins** and the **continents** are the principal surface features of the earth.
2. The surface fluids of the earth (atmosphere and water) are in constant motion. The tremendous volume of water circulating at the surface greatly modifies the landscape.
3. The continents consist of three major components: the **shields,** covered shields, and belts of folded mountains, all of which show mobility of the crust. In addition some continental blocks are being split apart by great **rift systems.**
4. River systems and their valleys are the most characteristic landforms.
5. The ocean floor contains several major topographic and structural divisions: (a) the **mid-oceanic ridges,** (b) the **abyssal** floors, (c) the **trenches,** (d) **islands** and **seamounts,** and (e) the **continental margins.**

The Principal Surface Features of the Earth

Statement

Surface features provide the most readily accessible and direct information about a planet. Consider, for example, the moon and what the cratered surface tells us about its origin and history. Impact of meteorites has been the major process on the moon since it became an entity in the solar system. Old craters have been modified by impact of younger meteorites, but there is no modification of the surface by running water or atmospheric gases. Volcanic activity on the moon is evident from the lava flows which cover the dark maria and record a significant thermal event in lunar history.

On this same scale, what are the principal surface features of the earth? Mountains, valleys, plains, rivers, and the sea are all important features, but they are not the major surface features of the earth. If we were to map the earth from an observation point on the moon using a special telescope which could see through clouds and water, we would recognize two principal regions. The largest would be the ocean basins occupying over half of the area of the earth. The ocean basins are not flat featureless plains but contain a variety of spectacular topographic forms, many of which are due to volcanic activity and earth movement. The second major region comprises the continental platforms, most of which are above sea level; however, a little over a fifth of the continental surface is covered with water. The continents also have a variety of topographic forms which are quite unlike those of the ocean basins.

Discussion

We are accustomed to thinking that the ocean basins begin at the shoreline and that the floor of the sea is similar to the continents, differing only in that it is covered with water. This is certainly not true. The ocean basins are completely different from the continents in rock type, structure, age, origin, and history. One way in which this difference is reflected is in their surface elevations. If you examine the **physiographic map,** you will see that there are two distinct levels. The continents have a mean elevation of 270 m (886 ft) above sea level, and the ocean floor has a mean depth of about 4400 m (14,430 ft) below sea level. Only a relatively small percent of the earth's surface rises above the average elevation of the continents or below the average depth of the ocean basins. These levels represent a fundamental difference in **density** between the rocks that form the continental and oceanic **crust.** If continents and ocean basins did not exist and the earth's surface was smooth, water would completely surround the planet as a uniform layer approximately 2400 m (7900 ft) deep.

Why does the earth have continents and ocean basins with distinctly different composition, structure, and history, instead of a uniform crustal layer? We will study this question in sub-

3

sequent chapters, but we must first learn about the characteristics of the earth's surface fluids, continents, and ocean basins.

The Atmosphere and Hydrosphere

Statement

A view of the earth from a distance of approximately 102,000 km (64,000 mi), such as that shown in *figure 1.1*, emphasizes the fact that the earth is a water planet. The most conspicuous features seen from this distance are the brilliant white swirling clouds and patches of blue ocean. We are all aware that air and water are fundamental elements of our environment, but few of us give much thought to the continuously moving surface fluids which blanket the globe and are responsible for so many landforms and surface features which we see on the continents. Without these fluids the surface of the land would be nothing like that with which we are familiar. There would be no rivers, valleys, glaciers, or even continents as we know them, because the movement of surface waters has played a major role in the development of all of these and many more. Geologically, it is natural for water to dominate a distant view of the earth because moving water is the most important agent in developing the surface features of our planet. From this view, we can more fully comprehend the meaning of the terms **atmosphere** (the gaseous envelope that surrounds the earth) and **hydrosphere** (the discontinuous envelope of water that covers most of the earth).

Discussion

The circulation patterns of the atmosphere are shown in *figure 1.1* by the shape and orientation of the clouds. At first glance, the patterns may appear confusing, but they are well organized. If we smooth out the details of local weather systems, the global atmospheric circulation becomes apparent. Solar heat is greatest in the equatorial regions and causes water in the oceans to evaporate and the air to rise. The warm humid air forms an equatorial cloud belt bordered by relatively cloud-free zones to the north and south where the air descends in the middle latitudes. At higher latitudes, low-pressure systems develop where the warm air from the low latitudes meets the polar fronts. The pattern of circulation around the resulting low pressure produces winds moving in a counterclockwise direction in the the northern hemisphere and a clockwise direction in the southern hemisphere. Swirls of clouds showing this circulation are readily apparent in *figure 1.1*.

Less apparent from a photograph, the circulation of water is of the greatest significance in developing the landforms on the earth's surface. In succeeding chapters, we will discuss the geologic processes of moving water as it evaporates from the oceans and other reservoirs, moves through the atmosphere, falls as rain or snow, and returns to the oceans.

The Earth's Dynamic System

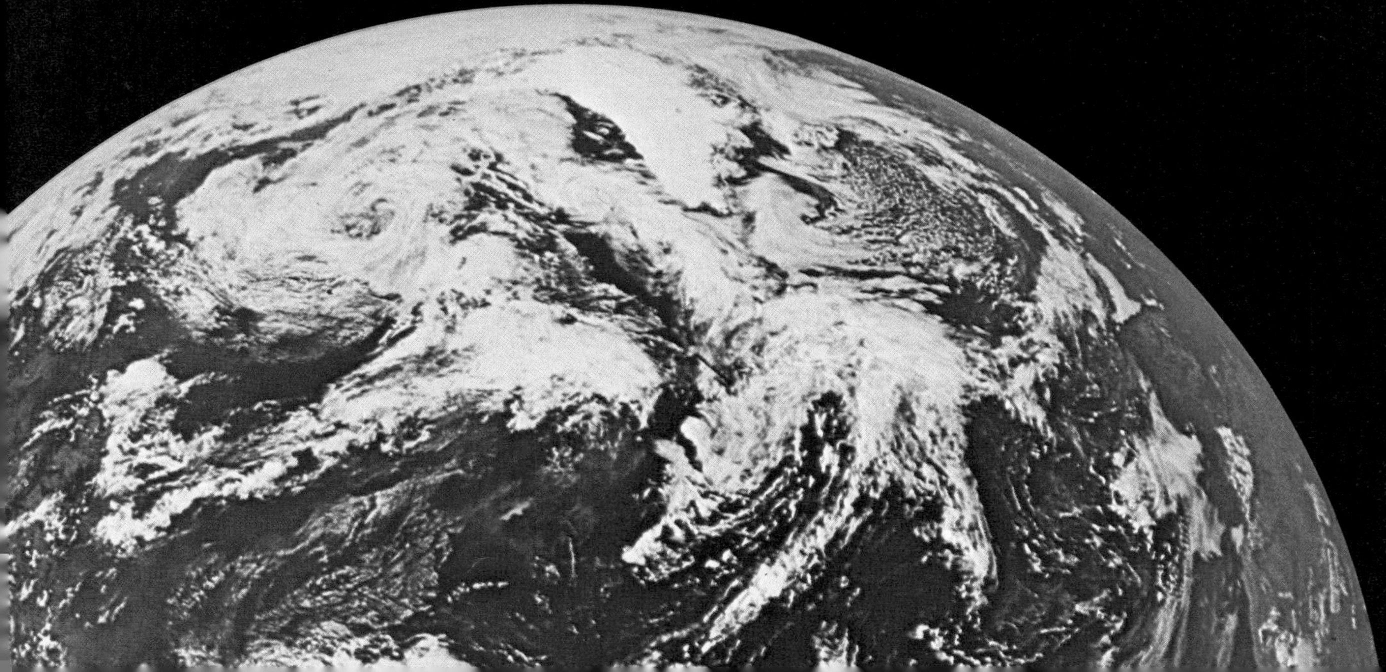

Figure 1.1   A view of the planet earth from space emphasizes the circulating atmosphere and the fact that the surface is predominantly water. Indeed, the earth could be called the blue planet. The upper view is of the Pacific Ocean with part of the western United States visible near the right. In the lower view the northern hemisphere from the Pacific on the left to the Mediterranean on the right can be seen. Greenland is clearly delineated by its ice cap, a principal remnant of the last Ice Age, when similar glaciers covered much of the land shown in this photograph.

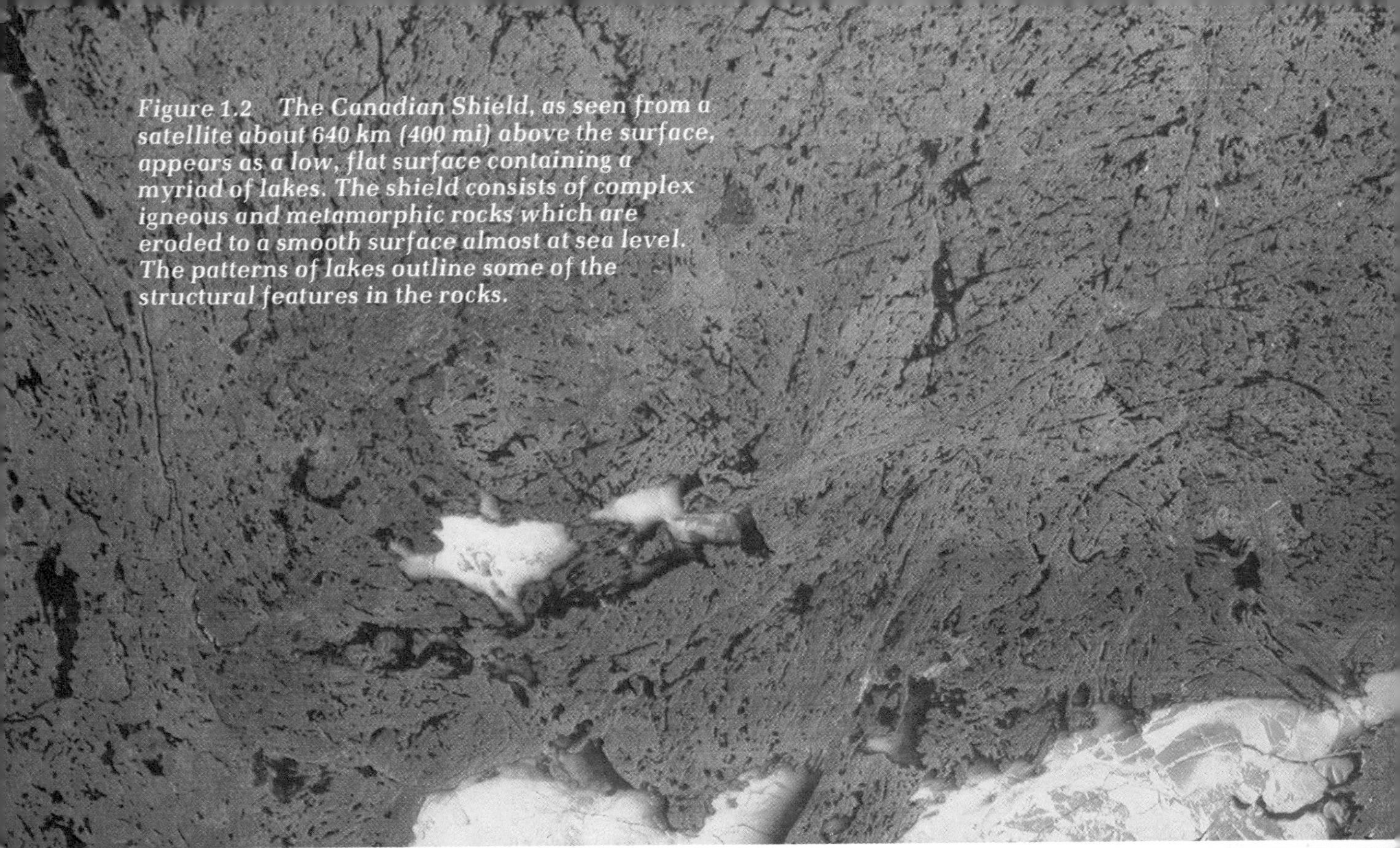

Figure 1.2   The Canadian Shield, as seen from a satellite about 640 km (400 mi) above the surface, appears as a low, flat surface containing a myriad of lakes. The shield consists of complex igneous and metamorphic rocks which are eroded to a smooth surface almost at sea level. The patterns of lakes outline some of the structural features in the rocks.

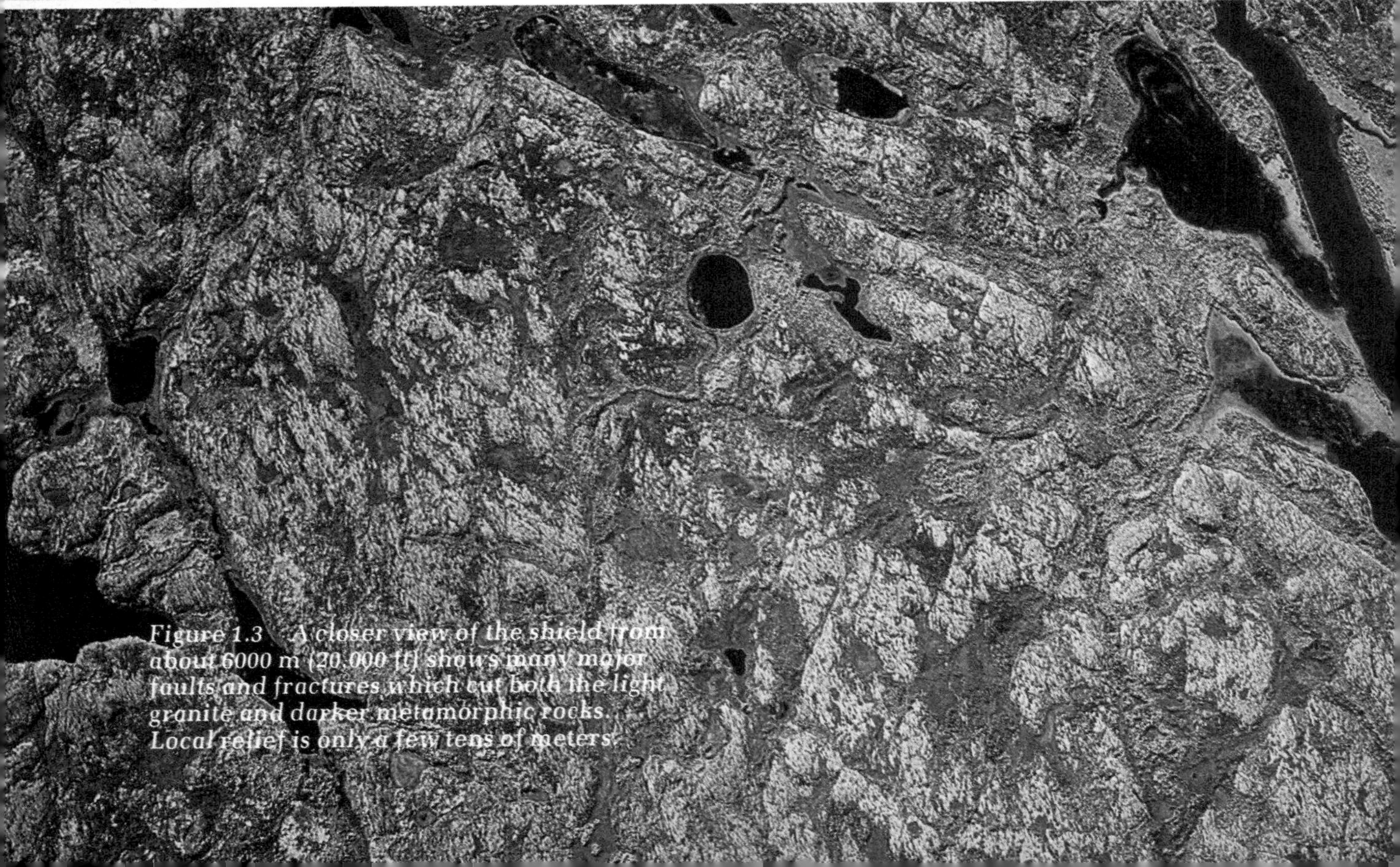

Figure 1.3   A closer view of the shield from about 6000 m (20,000 ft) shows many major faults and fractures which cut both the light granite and darker metamorphic rocks. Local relief is only a few tens of meters.

Figure 1.4 The Black Hills in South Dakota is part of the stable platform where the shield is covered with sedimentary rocks which have been warped into a broad dome. Erosion has removed the sedimentary cover in the central part of the dome (lower left) and has exposed the underlying rocks of the shield. The ridges surrounding the Black Hills are resistant layers of tilted and eroded sedimentary rock.

Figure 1.5 In the Colorado Plateau the rocks of the covered shield are also warped into broad domes and basins. In this view many individual formations are clearly seen by striking color contrasts. The large elliptical structure in the lower central part of the photograph is the San Rafael Swell in central Utah.

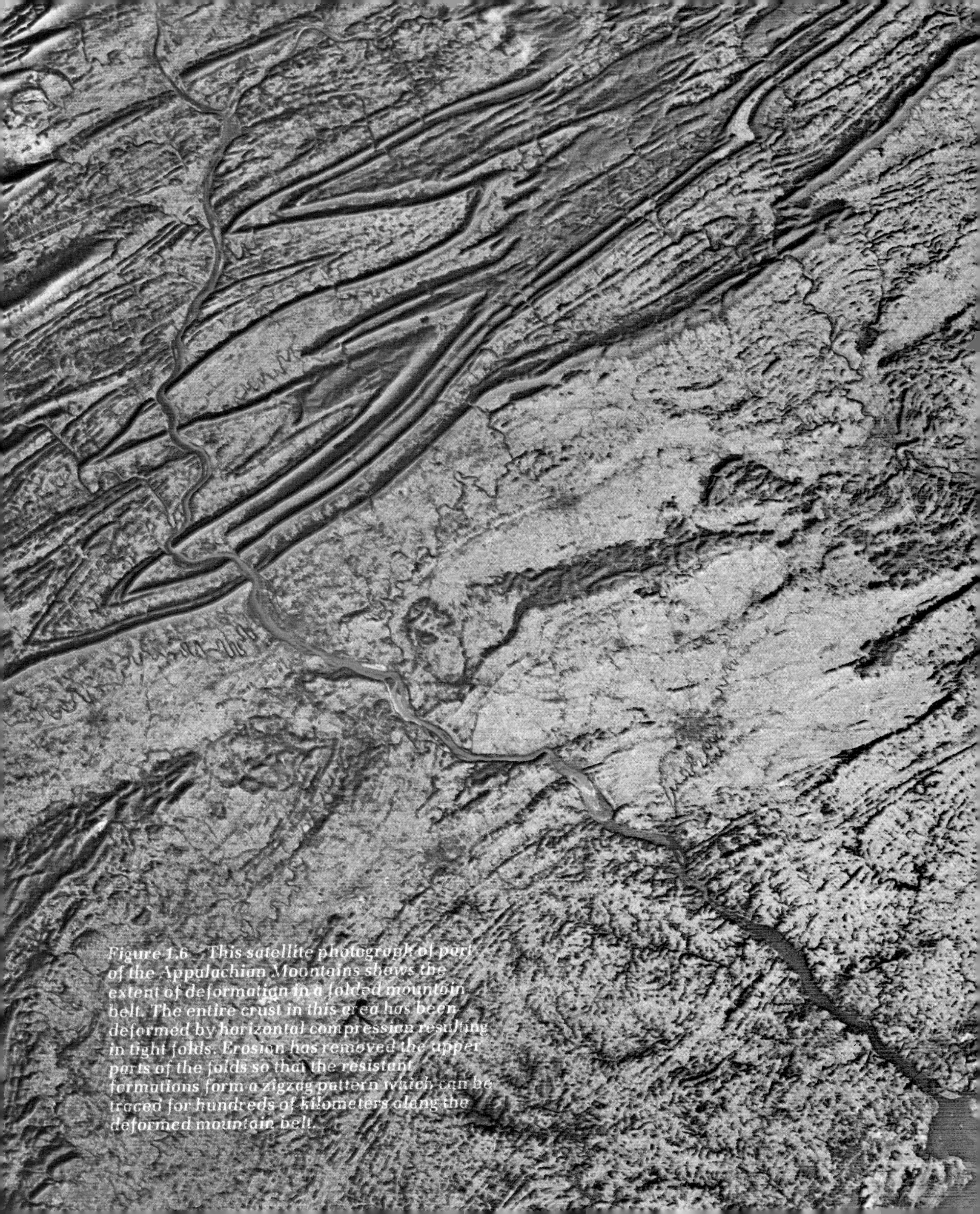

Figure 1.6 This satellite photograph of part of the Appalachian Mountains shows the extent of deformation in a folded mountain belt. The entire crust in this area has been deformed by horizontal compression resulting in tight folds. Erosion has removed the upper parts of the folds so that the resistant formations form a zigzag pattern which can be traced for hundreds of kilometers along the deformed mountain belt.

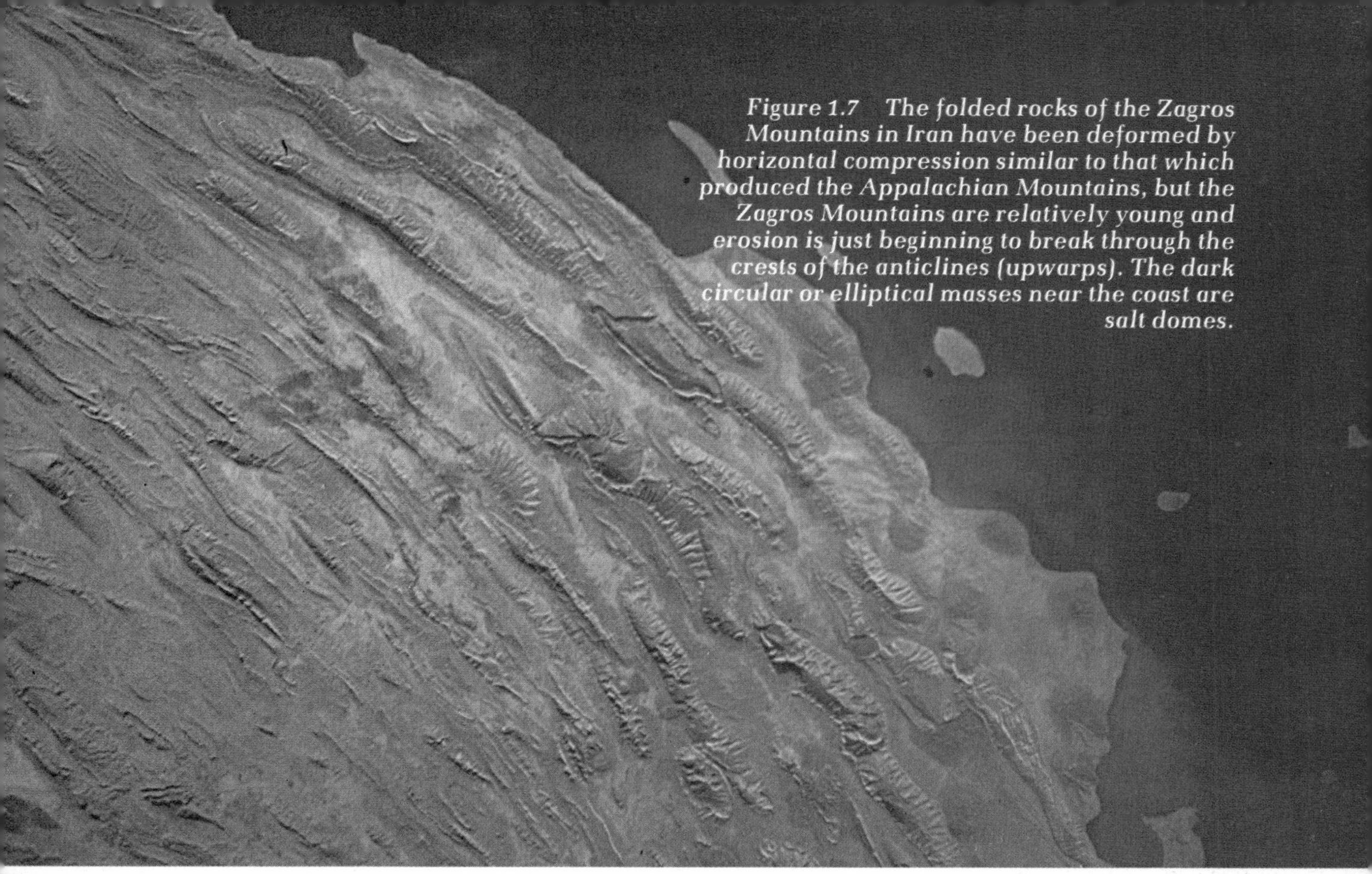

Figure 1.7 The folded rocks of the Zagros Mountains in Iran have been deformed by horizontal compression similar to that which produced the Appalachian Mountains, but the Zagros Mountains are relatively young and erosion is just beginning to break through the crests of the anticlines (upwarps). The dark circular or elliptical masses near the coast are salt domes.

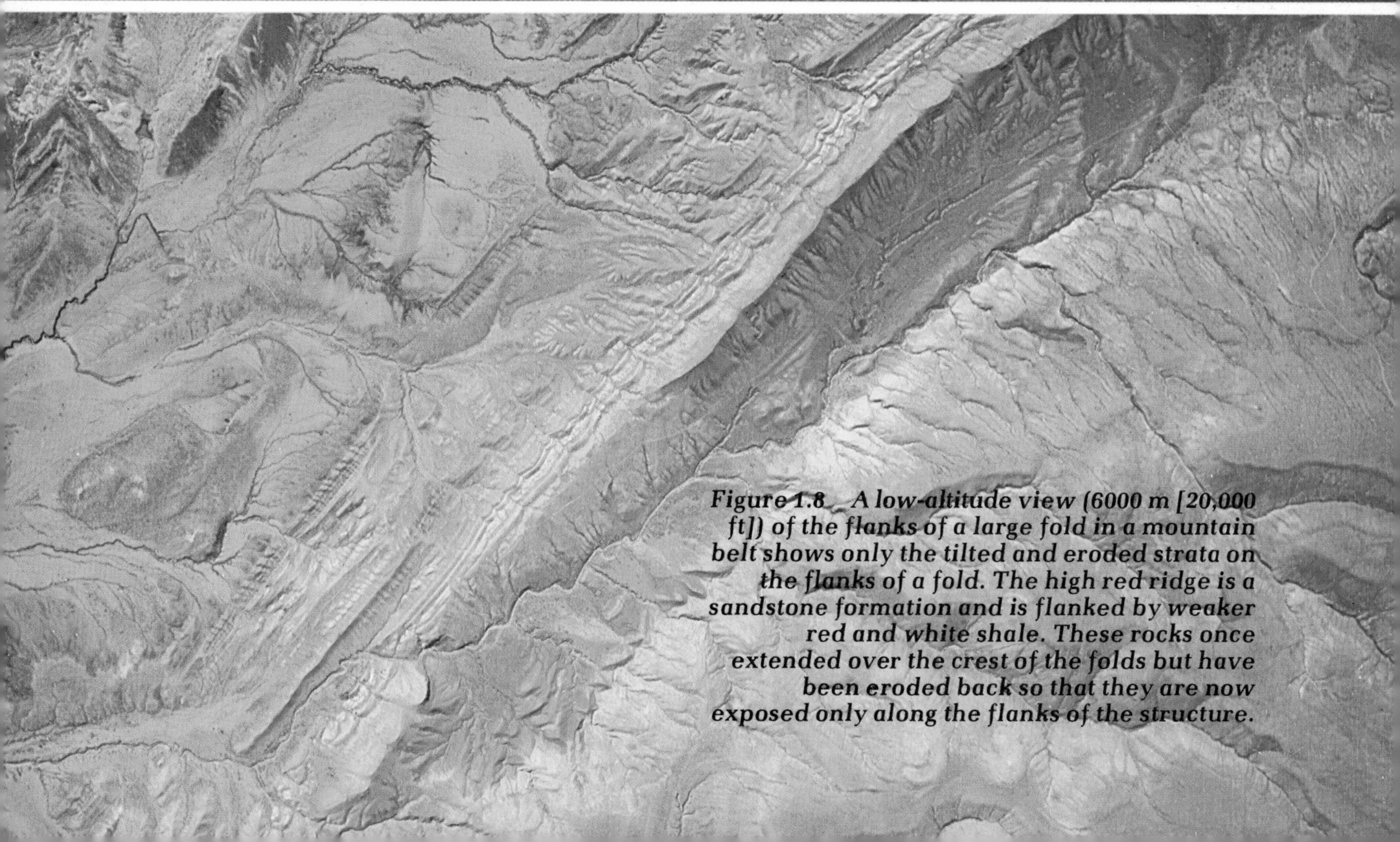

Figure 1.8 A low-altitude view (6000 m [20,000 ft]) of the flanks of a large fold in a mountain belt shows only the tilted and eroded strata on the flanks of a fold. The high red ridge is a sandstone formation and is flanked by weaker red and white shale. These rocks once extended over the crest of the folds but have been eroded back so that they are now exposed only along the flanks of the structure.

Figure 1.9   The great rift of the Red Sea
extends up the Gulf of 'Aqaba and into the
Dead Sea trough. Here the continental crust is
being pulled apart and a new ocean is being
created.

Figure 1.10   The crust in the western United
States and Mexico is also under tensional stress
and is being pulled apart. Large blocks of the
crust have subsided to form fault basins which
are being filled with erosional debris derived
from the adjacent higher blocks.

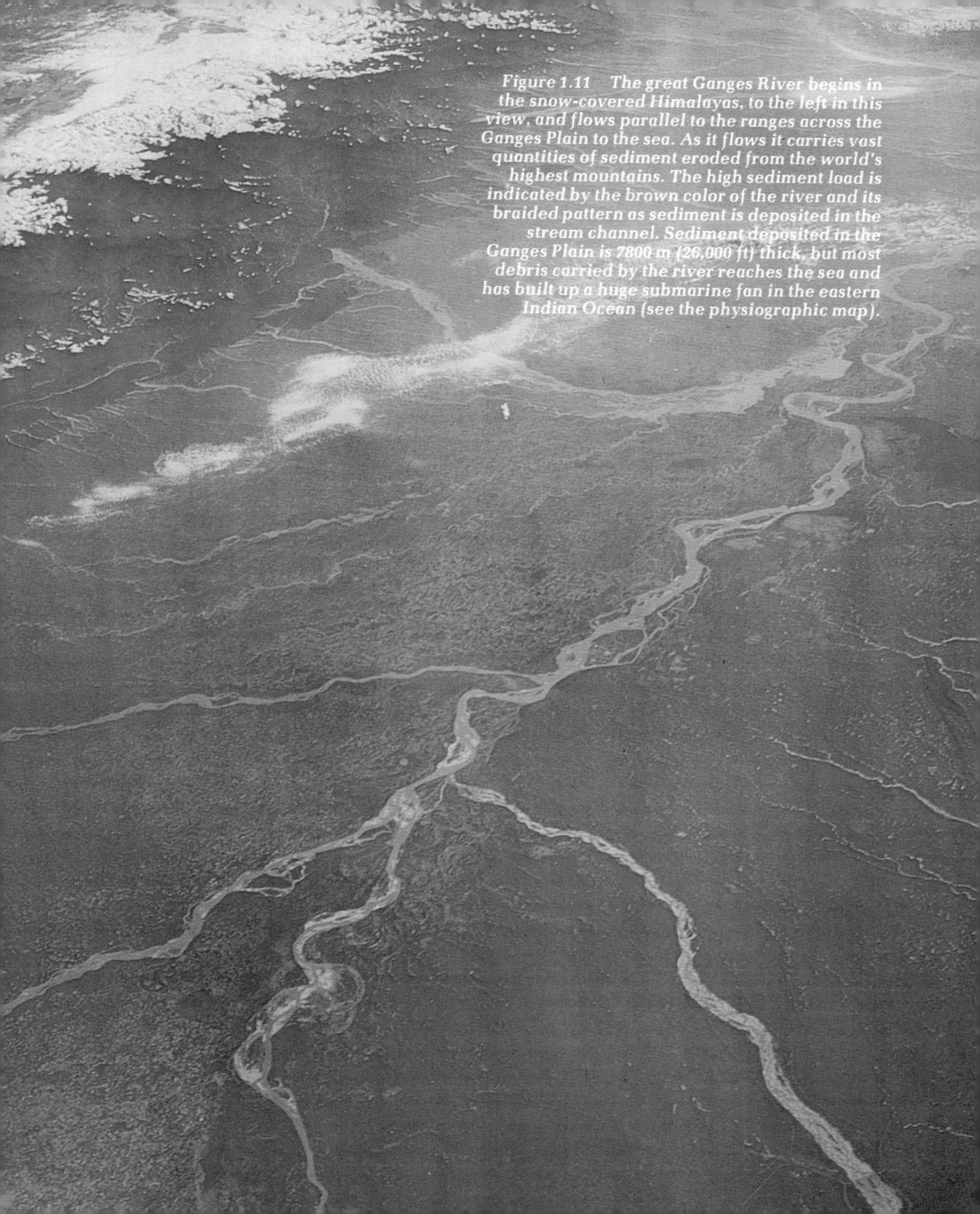

Figure 1.11   The great Ganges River begins in the snow-covered Himalayas, to the left in this view, and flows parallel to the ranges across the Ganges Plain to the sea. As it flows it carries vast quantities of sediment eroded from the world's highest mountains. The high sediment load is indicated by the brown color of the river and its braided pattern as sediment is deposited in the stream channel. Sediment deposited in the Ganges Plain is 7800 m (26,000 ft) thick, but most debris carried by the river reaches the sea and has built up a huge submarine fan in the eastern Indian Ocean (see the physiographic map).

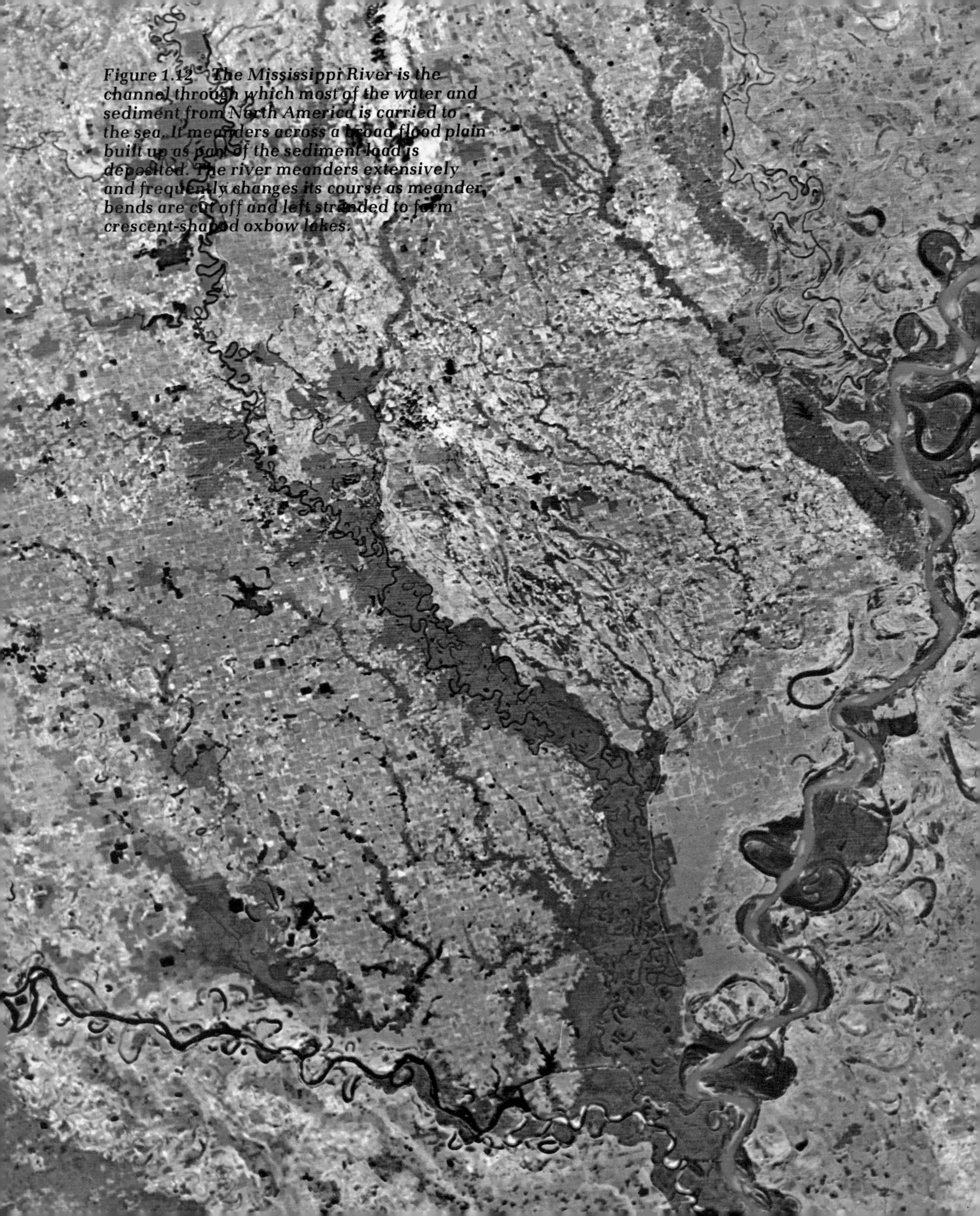

Figure 1.12   The Mississippi River is the
channel through which most of the water and
sediment from North America is carried to
the sea. It meanders across a broad flood plain
built up as part of the sediment load is
deposited. The river meanders extensively
and frequently changes its course as meander
bends are cut off and left stranded to form
crescent-shaped oxbow lakes.

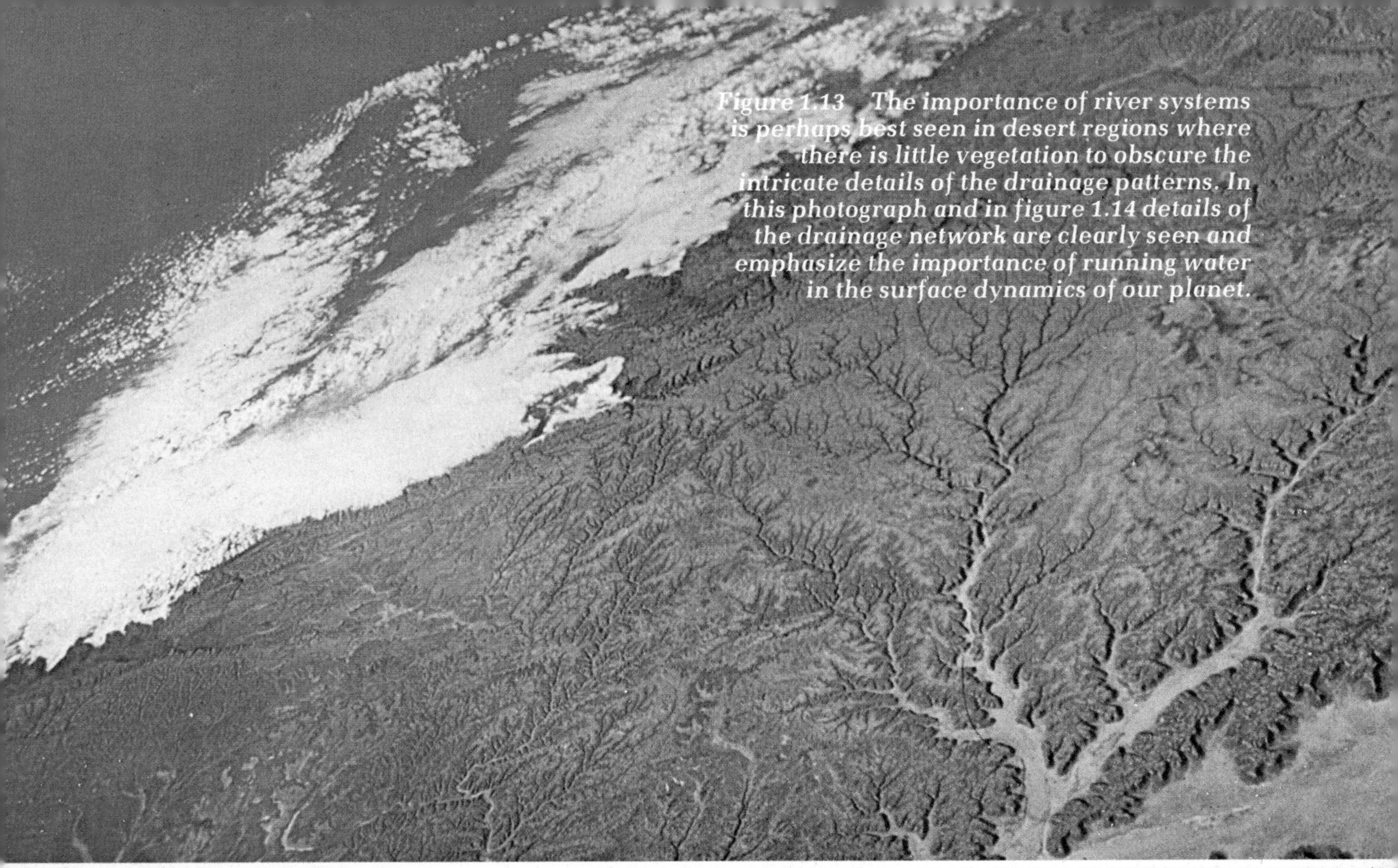

Figure 1.13 The importance of river systems is perhaps best seen in desert regions where there is little vegetation to obscure the intricate details of the drainage patterns. In this photograph and in figure 1.14 details of the drainage network are clearly seen and emphasize the importance of running water in the surface dynamics of our planet.

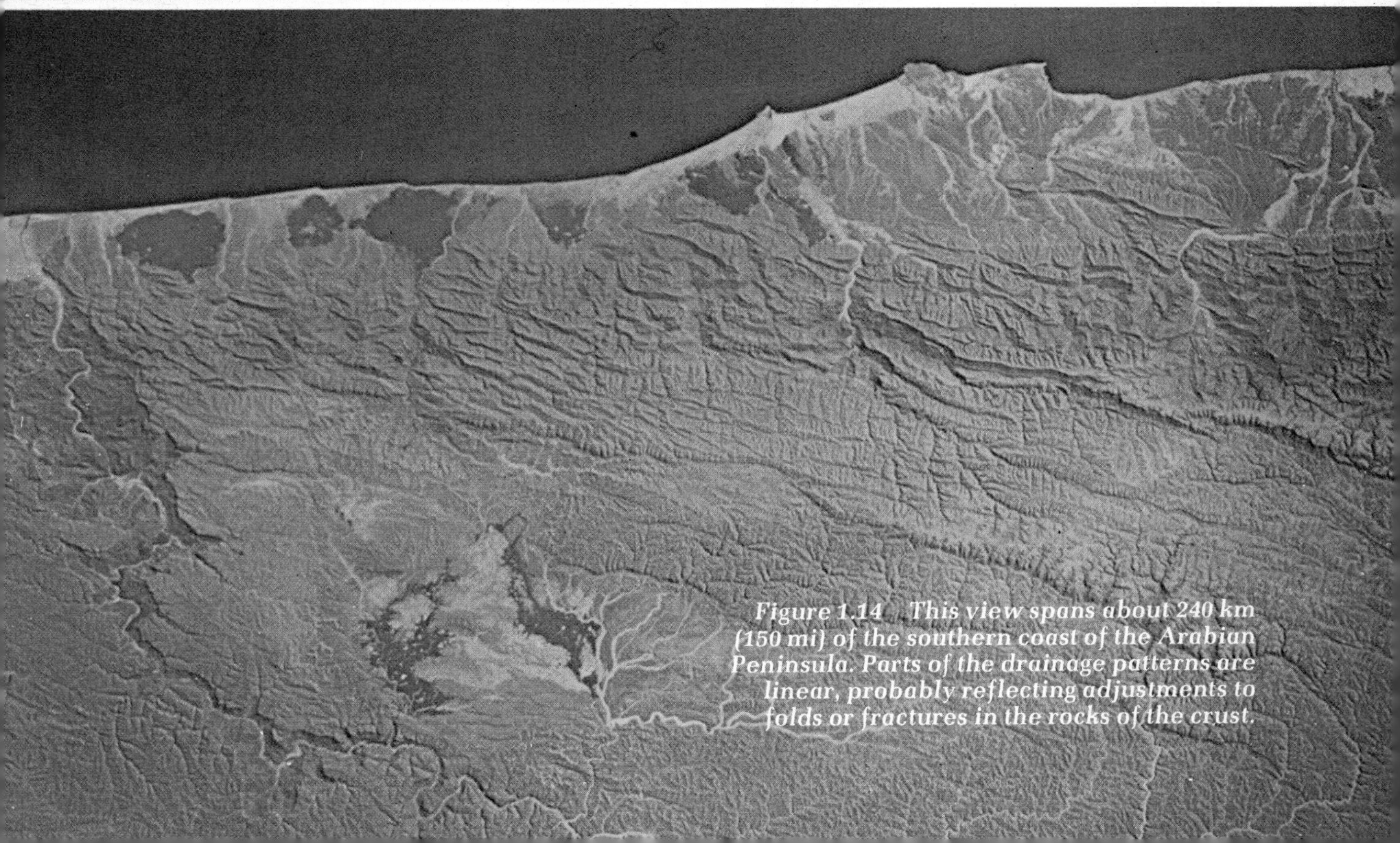

Figure 1.14 This view spans about 240 km (150 mi) of the southern coast of the Arabian Peninsula. Parts of the drainage patterns are linear, probably reflecting adjustments to folds or fractures in the rocks of the crust.

Figure 1.15   The Nile Delta is a huge deposit of sediment built over the centuries by the Nile River. It is the large vegetation-covered triangle extending from Cairo to the coast.

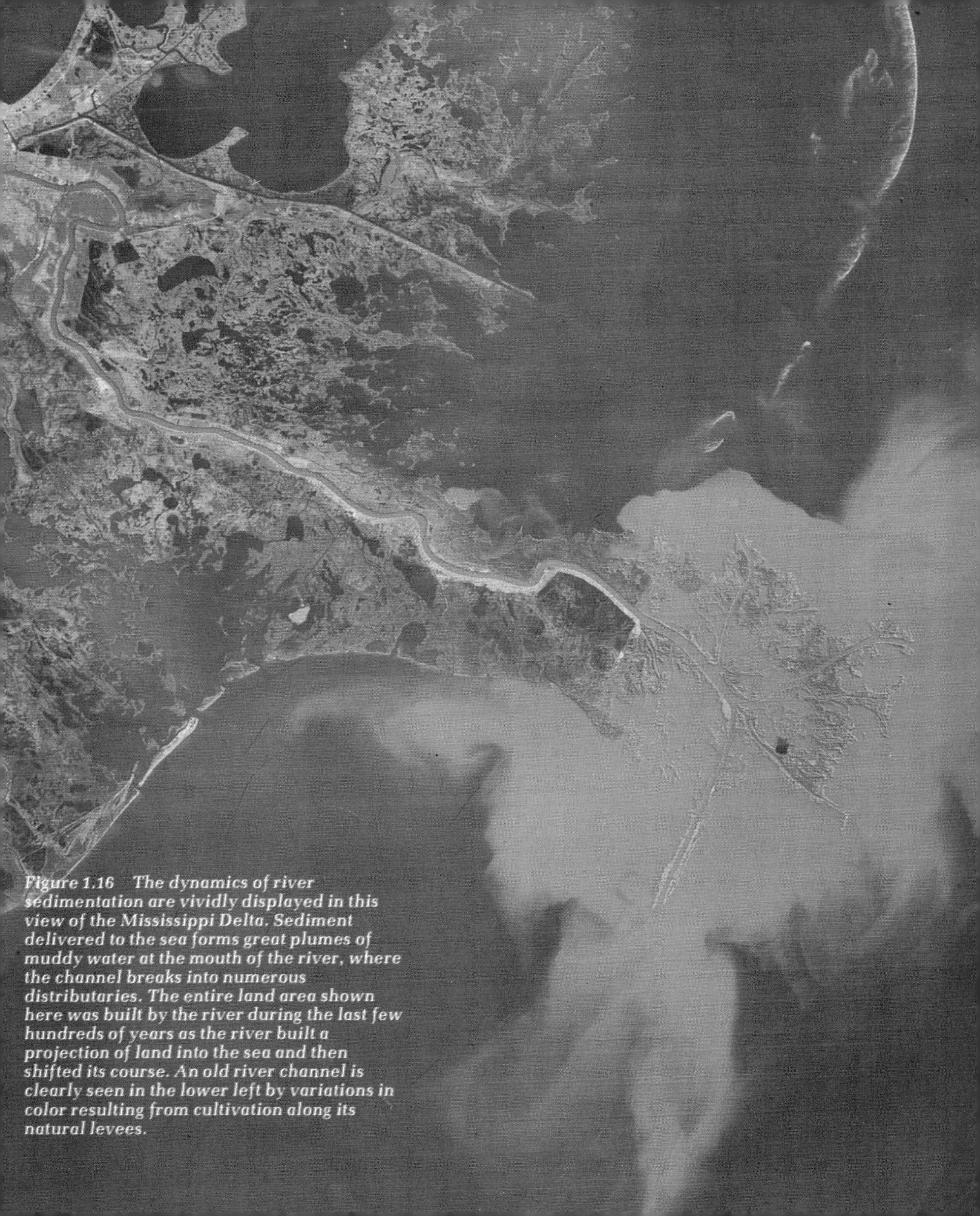

Figure 1.16 The dynamics of river sedimentation are vividly displayed in this view of the Mississippi Delta. Sediment delivered to the sea forms great plumes of muddy water at the mouth of the river, where the channel breaks into numerous distributaries. The entire land area shown here was built by the river during the last few hundreds of years as the river built a projection of land into the sea and then shifted its course. An old river channel is clearly seen in the lower left by variations in color resulting from cultivation along its natural levees.

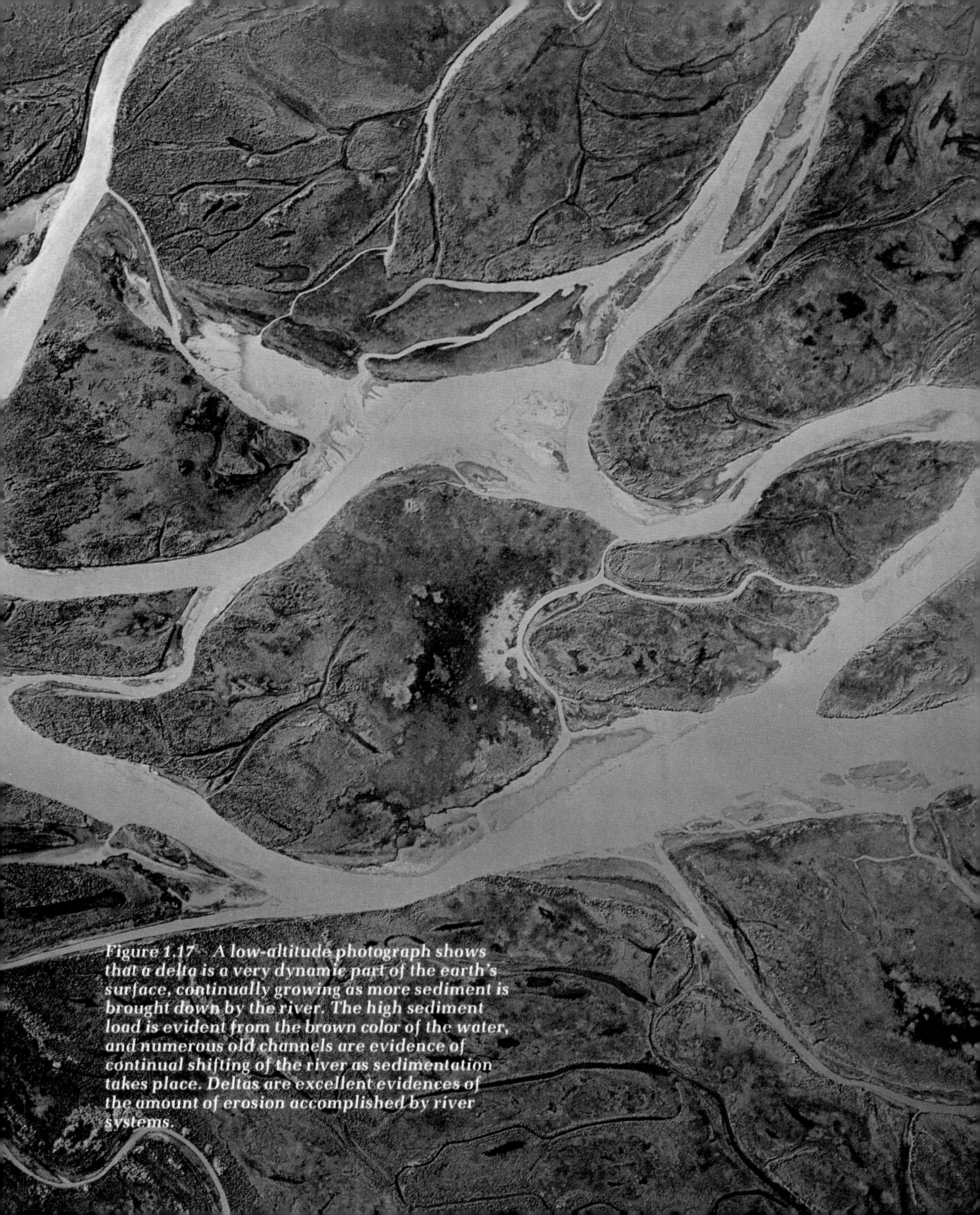

Figure 1.17 A low-altitude photograph shows that a delta is a very dynamic part of the earth's surface, continually growing as more sediment is brought down by the river. The high sediment load is evident from the brown color of the water, and numerous old channels are evidence of continual shifting of the river as sedimentation takes place. Deltas are excellent evidences of the amount of erosion accomplished by river systems.

Figure 1.18 A considerable amount of sediment eroded from the continent is reworked by wave action to form coastal features and offshore deposits. Cape Hatteras is in the center of this view and Pamlico Sound is west (left) of the barrier islands. The most prominent oceanographic feature in this view is the immense sediment plume extending from Pamlico Sound out into the Atlantic. The sediment moving through breaks in the barrier islands and out across the continental shelf is delineated by the line of puffy clouds which show where cold surface water of the eastern edge of the shelf meets warmer surface water of the shallow water near shore. The edge of the shelf is only about 120 m (300 ft) deep.

Figure 1.19   Sand dunes in the Empty Quarter on the Arabian Peninsula show the degree to which the surface of our planet may be modified by wind action. Some of the dunes are 150 m (500 ft) high and extend as unbroken ridges for as much as 640 km (400 mi).

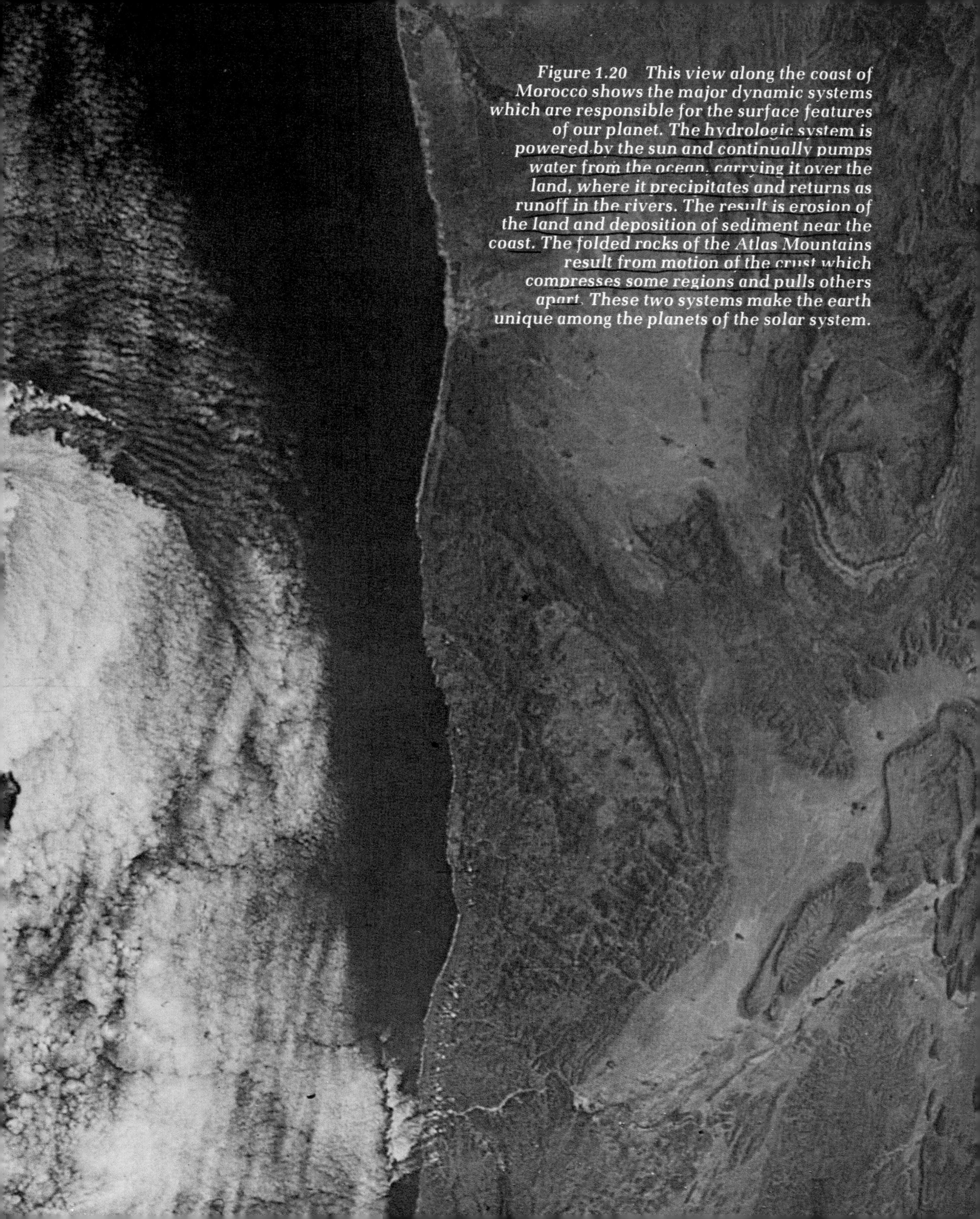

Figure 1.20 This view along the coast of Morocco shows the major dynamic systems which are responsible for the surface features of our planet. The hydrologic system is powered by the sun and continually pumps water from the ocean, carrying it over the land, where it precipitates and returns as runoff in the rivers. The result is erosion of the land and deposition of sediment near the coast. The folded rocks of the Atlas Mountains result from motion of the crust which compresses some regions and pulls others apart. These two systems make the earth unique among the planets of the solar system.

Figure 1.21   The ocean floor is not flat and featureless but contains many fascinating landforms, some of which are discernible on satellite photos. In this view a gorge 1100 m (3700 ft) deep is indicated by the dark blue water as it cuts the shallow Bahama banks.

Features of the Continents

If you examine the physiographic map you will note that the continental platforms rise abruptly (5400 to 6000 m [18,000 to 20,000 ft]) from the deep ocean floor. Although there are a variety of surface features on the continents and some high mountain peaks are nearly 9000 m (30,000 ft) above sea level, large areas of the continents are remarkably flat and very close to sea level. Indeed, much of every continental mass is within 1000 m (3200 ft) of sea level.

Although the various continents may appear unique in size, shape, and surface features, they all have three basic components: (1) a large area of ancient highly deformed crystalline rock (**igneous** and **metamorphic**) known as a shield, (2) broad platforms where the shield is covered with a veneer of **sedimentary** rock, and (3) folded mountain belts located along the continental margins. Geologic differences between continents are mostly in the size, shape, and proportions of these components. A great variety of surface features have been formed on these basic structural features of the continents as a result of moving water and wind, but the most characteristic landform is the slope of river valleys.

Shields. Without some firsthand knowledge of the shield, it is difficult to visualize the nature and significance of this very important part of the continental crust. *Figures 1.2* and *1.3*, showing part of the Canadian Shield, will help you comprehend the extent, complexity, and some of the typical features of shields. The most obvious characteristic seen in these photographs is the vast expanse of the low surface of the shield. Throughout an area of hundreds of thousands of square kilometers, the surface of the shield is within a few hundred meters of sea level. The only features which stand out in relief are the resistant rock formations which rise 30 to 120 m (100 to 400 ft) above the adjacent surface.

The highly deformed rocks clearly indicate that the shields have been subjected to intense horizontal stresses. In addition, the exposure of metamorphic and igneous rocks, which form only under high temperatures and pressure, indicates that a vast amount of erosion must have occurred since the rocks were formed in order for them to now be exposed. Estimates are that erosion has removed at least 8 km (5 mi) of rock from shields in order for the deep-seated metamorphic complex to be exposed.

The shields probably provide the best indication of the true nature of the continental platforms as they constitute the fundamental structural units of continents.

Covered Shields. The interior of the United States from the Appalachian Mountains to the Rockies is a low area where the igneous and metamorphic rocks of the shield are covered with a nearly horizontal layer of sedimentary rocks. Rocks of the shield are rarely exposed in this area, but the geologic relationships are known from thousands of wells which penetrate through the sedimentary cover into the underlying metamorphic and igneous rocks. The sedimentary rocks of the covered shield are commonly warped into broad domes and basins such as the dome which forms the Black Hills in South Dakota (figure 1.4). In the arid Colorado Plateau there is little vegetation to obscure the rock formations, and the broad warps of the sedimentary strata are exceptionally well exposed (figure 1.5). We see from the relationships between the major rock bodies of the covered shield that the continents, being eroded to near sea level, have been flooded during various periods in the geologic past and partly covered with the sea. At the present only the margins of the continents are flooded; in the past shallow seas were much more extensive. In addition the broad warps on the covered shield show that the continental platforms have been deformed by earth movements.

Folded Mountains. One of the most significant aspects of the continents is the belts of young folded mountains which typically occur along their margins. Most people generally think of a mountain as simply a high, more or less rugged landform in contrast to flat plains and lowlands. Mountains, however, are much more than "high country." To a geologist, the term "mountain" refers to a long linear zone in the earth's crust where the rocks have been intensely deformed by great horizontal stresses. They are the result of tremendous forces within the earth which are capable of folding and breaking large segments of an entire continent. The great folds and fractures in mountain belts provide evidence that the earth's crust is in constant motion.

Figures 1.6 to 1.8 vividly illustrate some of the characteristics of folded mountains and the extent to which margins of continents have been deformed. The rocks in these photographs are mostly sedimentary rocks originally deposited in a shallow sea much like that which covers the eastern part of the United States. They are now deformed by compression and have been folded like wrinkles in a rug. Erosion has removed the upper part of the folds so that the resistant layers form a zigzag pattern similar to that which would be produced if the crests of folds in a rug were cut off parallel to the floor.

The mobility of the crust of our planet as indicated by folded mountain belts such as those illustrated in figures 1.6 to 1.8 appears to have started early in earth's history and has continued to the present. We now know that the moon, Mars, and Mercury lack this type of deformation because all impact craters, regardless of their age, are circular. The crust of these planets appears to have been fixed and immoveable, whereas the earth's crust has been in motion.

Rift Systems. Several continents show evidence that certain zones are under tensional stresses and are being pulled apart. Movement of this type is shown in the high oblique view of the

Red Sea and Sinai Peninsula. As can be seen in *figure 1.9*, as well as on the physiographic map, the Arabian Peninsula conforms in geographic outline and in geologic structure to the facing coast of Africa. Various lines of evidence, which will be discussed in subsequent chapters, indicate that the peninsula has been pulled away from Africa and the rift has created the Red Sea. As the Arabian Peninsula moves northward it slides past the Sinai Peninsula; the shear zone has created the long, straight, narrow rift valley occupied by the Jordan River, the Dead Sea, and the Gulf of 'Aqaba.

Another rift system is found in the Basin and Range Province in the western United States (*figure 1.10*). Here a wide zone of tension occurs and has caused large blocks of the crust to drop down, forming a series of structural troughs which are presently being filled with sediment eroded from the adjacent upthrown blocks which form the ranges. Great rift valleys also occur in East Africa, New Zealand, and Iceland. If you will study the physiographic map, you will see that the mid-oceanic ridge extends into all of these areas. This would suggest that there is a worldwide rift system where the earth's crust is being pulled apart.

River Systems. Rivers have been mapped and studied for generations, and their importance as a geologic agent has long been known. However, until space photography became available, we were able to see only small segments of a river from any given viewpoint and had to rely completely on maps and charts for regional studies. With satellite photography taken at elevations ranging from 160 to 640 km (100 to 400 mi) we can now observe many characteristics of an entire river system from a single photograph. The point we wish to emphasize is the overwhelming influence rivers have had in shaping the earth's surface and that the slopes of stream valleys are the most characteristic landforms on the surface of the land. To appreciate this fact, you need only to examine the space photographs shown in *figures 1.11 to 1.17*. From viewpoints on the ground, however, the importance of stream valleys may not be readily apparent, just as the abundance of craters is more obvious from space than on the lunar surface itself.

Figures 1.11 and *1.12* illustrate segments of two great river systems of the world, the Ganges and the Mississippi. The oblique photo in *figure 1.11* is a view to the east. It shows the Ganges Plains beginning at the base of the snow-covered Himalayas and extending across the entire picture. (Locate this area on the physiographic map.) The plains are crossed by numerous streams which contribute to the waters of the Ganges, considered sacred by the Hindus. Most channels, even small tributaries, can be readily traced in detail, and the meanders and braided patterns show that the river system is moving large quantities of both sediment and water. The entire landscape is related to the Ganges River and its system of tributaries which are actively eroding the Himalayas.

In *figure 1.12* the Mississippi River is shown in an ERTS vertical photograph. The Mississippi changes its course frequently as it continually cuts across broad bends, leaving crescent-shaped loops.

Rivers provide other evidence of the dynamics of our planet in the sediment which is deposited in great deltas where rivers enter the sea. The Nile Delta is a classic example and is shown in a single panoramic photograph in *figure 1.15*. For centuries men have considered the Nile Valley as one of the cradles of civilization and have explored, excavated, and interpreted the significance of the river and its delta, but not until 1965 when this photograph was made did anyone see the entire sprawling deposit in one spectacular view. The delta covers approximately 1.3 million square kilometers (500,000 mi²) and is covered with vegetation which makes it stand out in striking contrast to the surrounding desert. This huge wedge of sediment deposited by the river represents a large part of the material eroded from the Ethiopian highlands and carried here over the centuries. Wave action along the coast has constantly redistributed much of the sediment and has eroded arcuate bays along the delta front. The two major distributaries of the Nile have built significant projections out into the sea.

The Mississippi Delta is another excellent example of the great amount of sediment which is carried from the continent by river systems and deposited in the sea (*figure 1.16*). The Mississippi River does not form a fan-shaped deposit like the Nile but flows in a main channel until it almost reaches the Gulf. Only then does it split into a number of distributaries which deliver the sediment load to the sea. The Mississippi River has thus built a long extension of land out into the sea. Ultimately the river shifts its course to seek a more direct route to the ocean, and the projection of land (subdelta) is eroded back by the sea. Previous courses of the Mississippi River can be seen on both sides of the present river.

A low-altitude view of part of a delta (*figure 1.17*) will help to show the vast amount of sediment carried by the earth's rivers. Much of the sediment carried to the sea by rivers is reworked by wave action and transported and deposited as barrier islands such as those shown in *figure 1.18*. The fine mud accumulates in the lagoon behind the barrier whereas currents moving parallel to the shore transport and deposit the coarser sand.

In many desert regions, river valleys still constitute the dominant landform. There is no completely dry place on earth. Even in the most arid regions some rain falls, and climatic patterns change over the years. River valleys are obliterated in large parts of the great deserts by "seas of sand" which cover the landscape. In *figure 1.19* we see such a region in the Empty Quarter on the Arabian Peninsula. Long parallel sand **dunes,** some of which extend unbroken for up to 640 km (400 mi) cover most of the area. The dunes range up from 152 m (500 ft) high and migrate with the prevailing wind. In the background of the

The Earth's Dynamic Systems

photo, parts of Arabia's Hadramawt Plateau can be seen dissected by river valleys, the mouths of which are covered with sand dunes.

Dry river beds in desert regions, through which no water is flowing at present, show particularly well on space photography. Because of the high degree of reflection from the sand-filled beds, these photographs show many details of river systems which may not be apparent in areas covered by vegetation. In figures 1.13 and 1.14 details of the drainage network are most striking. The ubiquitous nature of stream valleys clearly shows the effectiveness of erosion by running water and the importance of the water in the dynamics of our planet. Stream valleys occur in all sizes, shapes, lengths, and inclinations, but all are a response to the erosional process of running water.

In presenting this regional view of the features of the continents the point we wish to emphasize is that one can see quite vividly in space photography the constant change in the surface of our planet, changes on a scale so great that it is difficult to comprehend from limited viewpoints on the ground. Although the surface of the earth may appear to be stable and unchanging from year to year, or even throughout a lifetime, it is continually being modified by geologic processes. The major processes can be seen in action in figure 1.20.

The Ocean Floor

Statement

The ocean floor, not the continents, is the typical surface of the solid earth, and it is the ocean floor that holds the keys to the evolution of the earth's crust. Yet, it wasn't until the 1960s that enough data about the ocean floor were obtained to present a clear picture of regional characteristics. This new knowledge caused a revolution in our ideas about the nature and evolution of the crust. Prior to 1947 most geologists believed that the ocean floor was simply a submerged version of the continental landscape, with huge areas of flat **abyssal plains** covered with sediment derived from the continents. Although echo sounders were used as early as 1922, they were limited to shallow water and required an observer to listen with headphones and time the interval between transmission and return of the signal. A major breakthrough occurred in 1953 with the development of a precision depth recorder which could plot automatically a continuous profile of the ocean floor in any depth of water. Since then, millions of kilometers of profiles have been made, and the information has been used to make maps of the topography of the ocean floor. Some of the features of the ocean floor have been discerned in satellite photos such as that shown in figure 1.21. Here one of the deep channels north of central Cuba can be seen through the clear water.

If you study the physiographic map, it is soon apparent that ocean-floor topography is as varied as that on the continents and, in some respects, more spectacular. Yet, the topographic forms of the ocean floor are in many ways unique. There are no highly folded mountain belts and, in contrast to the continents, erosion is not the underlying process which forms the topography of the ocean floor. Probably the most significant facts are that the composition of the oceanic crust is entirely different from that of the continents and the major topographic features are related in some way to volcanic activity. Moreover, the rocks of the oceanic crust are geologically very young, having been created in the last 165 million years.

Although the topographic forms on the ocean floor may appear to be extremely complex, we can recognize several major divisions: (1) the mid-oceanic ridges, (2) the abyssal floor, (3) trenches, (4) seamounts, and (5) the continental margins. These divisions are especially clear in the Atlantic where the mid-Atlantic ridge extends down the central third of the ocean, flanked by the abyssal floors. Seamounts are conspicuous volcanic cones, some of which project above sea level as islands. The continental margins are cut by deep **submarine canyons** which have stimulated much debate for over two decades. Another fascinating feature on the ocean floor is the deep trenches which are found in every ocean but are most numerous along the margins of the Pacific basin.

Discussion

The Mid-Oceanic Ridges. Probably the most significant result of oceanic research has been the discovery of the mid-oceanic ridge. This feature is one of the largest topographic and structural features of our planet. This huge submarine ridge extends as a continuous feature from the Arctic basin down the center of the Atlantic, into the Indian Ocean, then across the South Pacific, and northward into the Gulf of California. Its total length is nearly 64,000 km (38,400 mi). The mid-oceanic ridge is essentially a broad fractured swell, generally more than 1400 km (840 mi) wide, with higher peaks rising as much as 3 km (1.8 mi) above the ocean floor. A typical profile of the ridge is shown in *figure 1.22* and is compared to the physiographic map showing this feature *(figure 1.23)*. It is apparent from these profiles that the ridge is much larger than the Appalachians and

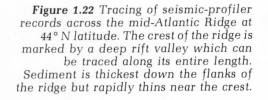

Figure 1.22 *Tracing of seismic-profiler records across the mid-Atlantic Ridge at 44° N latitude. The crest of the ridge is marked by a deep rift valley which can be traced along its entire length. Sediment is thickest down the flanks of the ridge but rapidly thins near the crest.*

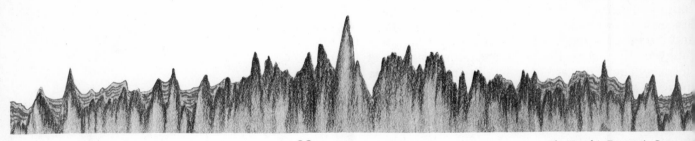

The Earth's Dynamic Systems

Figure 1.23 Physiographic map showing the nature of the mid-oceanic ridge in the Atlantic Ocean.

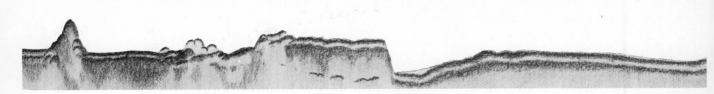

Figure 1.24 *Tracing of seismic-profiler records across the Murray fracture zone in the eastern Pacific Ocean. The fracture is expressed by a pronounced vertical cliff which separates areas of contrasting topography. On the left side of the fault, seamounts are abundant, whereas to the right the sea floor is relatively smooth and featureless.*

more rugged than the Rockies and Alps. The mountains of the mid-oceanic ridge, however, are nothing like the Rockies, Alps, or any of the other continental mountain ranges which are built largely from folded sedimentary rocks. By contrast, the ridge is composed entirely of basalt, a volcanic rock, although in some places there is a thin veneer of oceanic sediment. It is not deformed by extensive folding, such as is common in mountain belts, but has been displaced by **faulting** and great fracture systems.

Close examination of the mid-oceanic ridge shows that the most rugged topography is located along its axis. Along the entire mountain range, a cracklike valley, called the **rift valley,** is located at the crest of the ridge and extends throughout most of its length. In many details, the crest province of the mid-oceanic ridge is identical to the great rift valleys of eastern Africa. The rift valley is 30 to 75 km (18 to 46 mi) wide and is flanked by steep slopes of the adjacent peaks which appear to be fault blocks similar to those of the Basin and Range of the western United States.

Throughout most of its length, the mid-oceanic ridge is offset by great cross-faults resulting from extensive horizontal movement of one crustal block past the other *(figure 1.24).* The fractures are expressed by prominent cliffs and are traceable for hundreds of kilometers. In a series of fractures, the cumulative displacement may be over 4000 km (2400 mi). The topographic expression alone makes it clear that there is some vertical movement, but most of the displacement is horizontal.

According to modern theories of the earth, the mid-oceanic ridge is perhaps the single most important structural feature in the crust. We will have more to say about this in Chapters 15 and 16 where we will see how the mid-oceanic ridge and the deep oceanic trenches (to be described in the following sections) form the boundaries of a system of great crustal plates which move out from the ridges toward the trenches.

The Abyssal Floor. Vast areas of the deep ocean basin consist of broad, relatively smooth surfaces known as the abyssal floor. This surface extends from the flanks of the mid-oceanic ridge to the continental margins and lies generally at depths of

Figure 1.25 *Profiles across abyssal hills in the Pacific Ocean show the local relief is about 300 m (980 ft). The hills are typically elongate or elliptical and cover about 80% of the Pacific floor.*

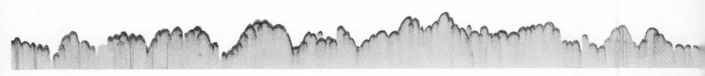

about 3600 m (15,000 to 20,000 ft). The abyssal floor can be subdivided into two sections: (1) the **abyssal hills** and (2) the abyssal plains.

The abyssal hills are relatively small hills rising above the ocean floor 75 to 900 m (150 to 3000 ft) (figure 1.25). They are circular or elliptical and range from 1 to 8 km (5 mi) wide at their base. The hills are found along the seaward margins of most abyssal plains and occur in profusion in parts of the ocean floor isolated from land by ridges, rises, or trenches. In the Pacific, they cover 80% to 85% of the ocean floor. Thus, abyssal hills are the most widespread landform on earth.

Near many continental margins the abyssal hills are completely buried with sediment eroded from the continents. Slopes on this surface are less than 1:1000 so the abyssal plains are much flatter than the Great Plains of North America (figure 1.26).

Trenches. Trenches are the lowest areas on the earth's surface and have attracted the attention of geologists and oceanographers for many years because they must represent a major structural feature in the earth's crust. They are long, deep, narrow depressions of the ocean floor with maximum depths ranging from 8000 to 10,000 m (25,000 to 32,000 ft). As illustrated on the physiographic map, they are almost invariably located adjacent to and on the ocean side of long narrow submarine ridges, or **island arcs,** or coastal mountain ranges of the continents. The most striking trenches occur in the western Pacific where a trench system extends from the vicinity of New Zealand, to Indonesia and Japan, and northward along the south flank of the Aleutian Islands (figure 1.27). Long trenches also occur along the border of Central and South America, in the Indian Ocean west of Australia, in the Atlantic off the tip of South America, and in the Carribean Sea.

Trenches are characterized by relatively steep slopes adjacent to the continental landmass or island arc, and the most gentle slope toward the ocean basin (figure 1.28). As is apparent on the physiographic map, continuity of the deep oceanic trenches is truly remarkable. Individual trenches less than 100 km wide may extend as continuous features across the deep ocean floor for several thousand kilometers.

Islands and Seamounts. Literally thousands of submarine volcanoes occur on the ocean floor, the greatest concentration of which are found in the eastern Pacific (figure 1.29). Some rise above sea level and form islands. They often occur in groups or chains with individual islands being as much as 100 km in diameter and 1000 m high, and are found in all major oceans (note their distribution on the physiographic map). The term **guyot** refers to a special type of seamount, the top of which is a flat mesalike surface rather than a cone.

The Continental Margins. The margins of the continents are covered with water but are not geologically part of the oceanic crust. The submerged part of the continent is referred to

Figure 1.26 *Tracing of seismic-profiler records across the abyssal plains in the eastern Atlantic Ocean. The abyssal plains are areas of the sea floor where the original topography is buried with sediment. They typically occur near the margins of continents where there is a significant supply of sediment.*

as the **continental shelf.** It is a gently sloping platform extending from the shoreline to the start of the steep descent to the deep ocean floor. The width of the shelf ranges from 0 to 1500 km (900 mi) and, at its outer edge, the depth of the shelf ranges from 20 to 550 m (1800 ft). Presently, the continental shelves are equal to 18% of the earth's total land area, but at times in the geologic past they were much larger. The continental shelf in *figure 1.18* extends out to approximately the line of clouds.

The continental shelf is characteristically smooth and flat, but the topography has been greatly influenced by changes in sea level. Large areas once exposed as dry land were subjected to continental erosion. Later, when submerged, they became subjected to submarine processes. Probably the most dramatic period in the history of the shelves has been the last million years during the glacial epoch when sea level fluctuated with the advance and retreat of continental glaciers.

The sea floor descends from the outer edge of the continental shelf as a long continuous slope to the deep ocean basin (*figure 1.30*). This slope is appropriately called the **continental slope** because it marks the edge of the continental mass of granitic rock. Continental slopes are found around the margin

Figure 1.27 *Tracing of seismic-profiler records across the central part of the Aleutian trench. Note the steep flank adjacent to the Aleutian island arc (right) and the gentle slopes toward the ocean basin.*

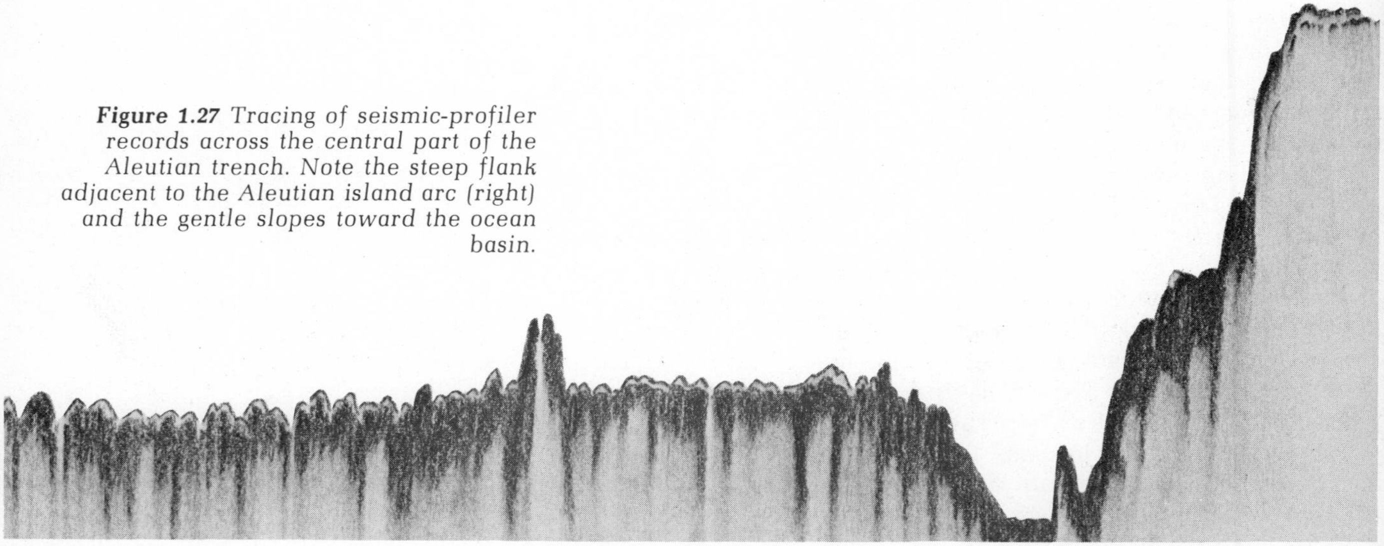

The Earth's Dynamic Systems

Figure 1.28 Physiographic map of the western Pacific Ocean showing the system of trenches and associated island arcs.

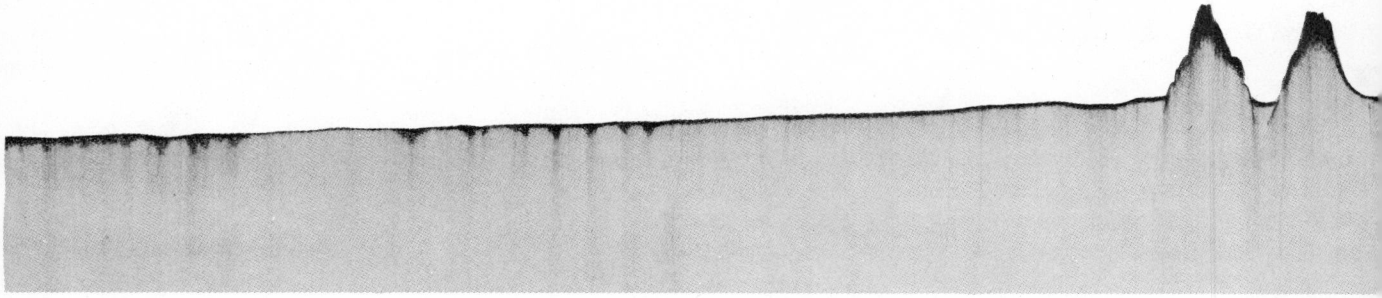

Figure 1.29 *Tracing of seismic-profiler records across a series of seamounts in the central Pacific Ocean. Seamounts are submarine volcanoes which usually occur in groups or chains. Some rise above sea level to form islands (see figure 1.28).*

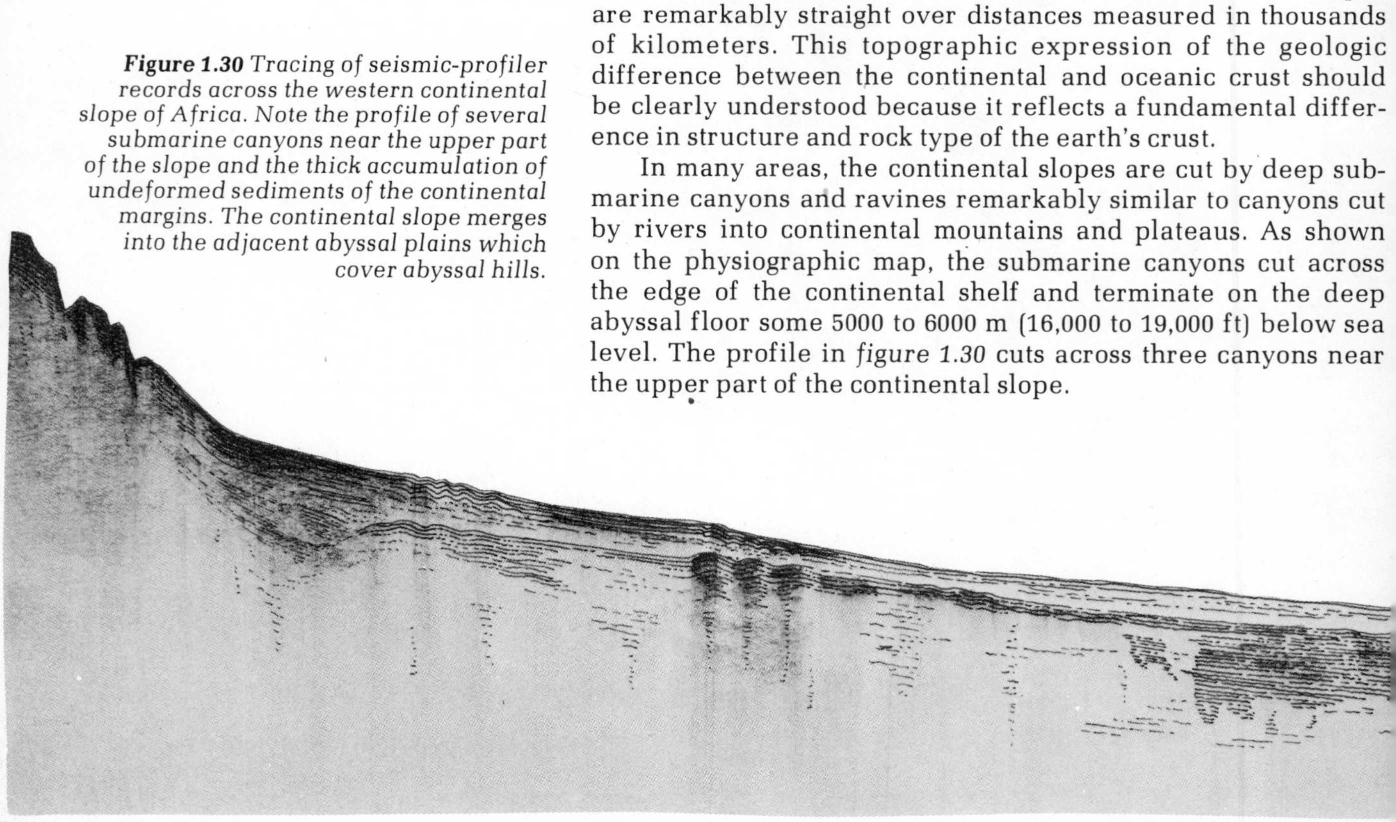

of every continent and around smaller pieces of continental rock such as Madagascar and New Zealand. If you study closely the continental slopes shown on the physiographic map, especially those surrounding North America, South America, and Africa, you will note that the slopes form one of the major topographic features on the earth's surface. On a regional scale, they are by far the steepest, longest, and highest slopes on earth. Within this zone, 20 to 40 km (12 to 25 mi) wide, the average relief above the sea floor is 4000 m (13,000 ft); adjacent to the marginal trenches, relief is as great as 10,000 m (32,000 ft). In contrast to the shorelines of continents, the continental slopes are remarkably straight over distances measured in thousands of kilometers. This topographic expression of the geologic difference between the continental and oceanic crust should be clearly understood because it reflects a fundamental difference in structure and rock type of the earth's crust.

In many areas, the continental slopes are cut by deep submarine canyons and ravines remarkably similar to canyons cut by rivers into continental mountains and plateaus. As shown on the physiographic map, the submarine canyons cut across the edge of the continental shelf and terminate on the deep abyssal floor some 5000 to 6000 m (16,000 to 19,000 ft) below sea level. The profile in *figure 1.30* cuts across three canyons near the upper part of the continental slope.

Figure 1.30 *Tracing of seismic-profiler records across the western continental slope of Africa. Note the profile of several submarine canyons near the upper part of the slope and the thick accumulation of undeformed sediments of the continental margins. The continental slope merges into the adjacent abyssal plains which cover abyssal hills.*

The Earth's Dynamic Systems

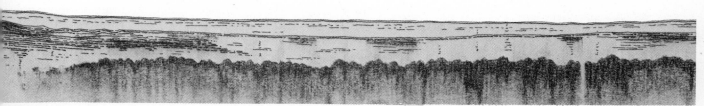

You will note from *figure 1.30* that the lower continental slope is covered with a considerable thickness of sediment, the layering of which is expressed by the numerous horizontal lines. This sediment is largely derived from erosion of the continent. Note that it is relatively uniform and is essentially horizontal in contrast to the folded sedimentary rocks which form the mountain belts on the continents.

A Description of the Continents and Ocean Basins

Statement

With this brief introduction to the major surface features of our planet let us consider some of the characteristics of each of the continents and ocean basins. Although each continent and ocean may at first appear to be unique and different they have fundamental similarities. All continents have three basic structural components: (1) a shield, (2) a covered shield or stable platform, and (3) folded mountain belts. All ocean basins have (1) a mid-oceanic ridge, (2) abyssal floors, and (3) seamounts. Differences between the continents are largely due to differences between the proportions of shield, covered shield, and mountain belts and how these features are modified by weather and erosion. Differences between ocean basins are largely due to differences in proportion and location of the ridge, abyssal floors, and seamounts and the presence or absence of trenches. In studying these features it is important to continually refer to the physiographic map for a visual model of the feature considered.

Discussion

The Continents. The three basic structural elements of the continents can be recognized on the physiographic map by the landforms which are unique to them. Shields are shown by dark complex patterns and appear as lowlands. Covered shields are flat, relatively featureless lowlands with the eroded edges of upturned strata shown where they form ridges. Mountain belts are easily recognized by their relief.

North America has a large shield in Canada, most of which is less than 300 m (1000 ft) above sea level. It extends from the Arctic islands southward to the Great Lakes area and westward to the plains of western Canada. The covered shield extends throughout the central United States and western Canada and is covered with sedimentary rocks which have been slightly warped and eroded. As can be seen on the physiographic map, the style of landforms developed on the stable platform to the south and west, which is covered with sedimentary rocks, is quite different from the eroded igneous and metamorphic terrain of the shield. The eroded edges of the tilted sedimentary rocks throughout most of the central United States and western Canada form a series of ridges at the margins of flat plains. The North American continent is somewhat symmetrical, with the Appalachian Mountains on the east and the Rocky Mountains on the west.

South America consists of a broad shield in Brazil, parts of which are covered in the south and west. The Andes Mountains are an extension of the Rockies on the west coast, but there are no mountain belts on the eastern margin of South America.

Australia is much like South America. Most of the continent is a stable platform, with a thin veneer of sedimentary rocks partly covering the shield. A single mountain range occurs along the eastern margin.

By contrast, Africa consists of an extensive shield, covered only locally with a thin veneer of sediments preserved largely in circular downwarped basins. The only folded mountain belts occur in narrow bands along the southern and northern margins of the continent.

Asia has a much larger area of folded mountains—the Himalayas—which form a wide east-west range. Most of the Asian shield is covered with sedimentary rocks, which form the vast low areas of central Asia, but exposed igneous and metamorphic rock of the shield occur in parts of Siberia. The Ural Mountains form the margin between the Asian and European continents,

The Earth's Dynamic Systems

and the Himalaya Mountains form the margin between Asia and India.

The shield of Europe is exposed in the Baltic area but is covered throughout most of the central and southern parts of the continent. In central Europe the shield is covered with sedimentary rocks which have been warped into broad domes and basins. The eroded edges of these structures, such as the Paris Basin, can be seen on the physiographic map. The Alps form the folded mountains along the southern margin of the European continent.

Most of India is a shield covered locally by horizontal basalt flows and sedimentary rock. The Himalaya Mountains form its northern border. Note that erosional debris from the Himalaya Mountains has been deposited in the broad flood plain of the Ganges River.

The Ocean Basins. The three major ocean basins of the world are quite different in size, shape, and topographic features. These differences are very significant because they indicate a great deal about the age, origin, and evolution of the oceanic basins and provide new insight into the theory of **continental drift** and the internal dynamics of the earth.

The regional topography of the Atlantic floor is basically very simple and shows remarkable symmetry in the distribution of the major features. As you will note from the physiographic map, the dominant feature is the mid-Atlantic ridge, which forms an S-shaped pattern down the exact center of the ocean basin. It separates the ocean floor into two huge north-south-trending parallel troughs characterized by abyssal plains. The abyssal hills occur along the margins of the ridge, and the plains occur along the margins of the continental platforms. The symmetry of the Atlantic basin extends to the continental margins where the outlines of Africa and Europe fit those of South America and North America. This symmetrical distribution of the major features of the Atlantic has attracted scientific attention for many years and is one of the cornerstones for the theories involving continental drift and **sea-floor spreading.**

The Indian Ocean is the smallest of the three "great" oceans and connects with both the Atlantic and the Pacific through broad, open seas south of Africa and Australia. Here, as in the Atlantic, the most conspicuous feature is the mid-oceanic ridge which continues from the Atlantic around South Africa and splits near the center of the Indian Ocean to form a pattern similar to a large inverted Y. The northern segment of the ridge extends into the Gulf of Aden, where it apparently connects with the African rift valley and the rift of the Red Sea. The ridge thus divides the ocean basin into three major parts. Unlike the other oceans, the topography of the Indian Ocean floor is dominated by scattered blocky, and some remarkably linear, plateaus called **microcontinents.** Most are oriented in a north-south direction. Prominent parallel fracture zones are numerous; some of these displace the ridge, whereas others displace the adjacent abyssal floor. The striking northern trend of the parallel fracture zones, together with the trend of the linear microcontinents, imparts a remarkable linear structural fabric to the floor of the Indian Ocean.

The Pacific Ocean is somewhat different from the other ocean basins in that the oceanic ridge occurs near the eastern margin. The basin covers approximately half of the globe and is the largest single unit of oceanic crust. In addition, it is probably the oldest ocean basin and lacks the symmetry of the Atlantic and Indian basins.

The mid-oceanic ridge continues in a broad sweep from the Indian Ocean between New Zealand and Antarctica and then turns northward along the American side of the ocean. The crest of the ridge disappears at the head of the Gulf of California but reappears off the coast of Oregon.

The floor of the western Pacific is studded with more seamounts, guyots, and atolls than all other oceans combined. As is apparent on the physiographic map, many of the seamounts occur in linear chains which extend for a considerable distance. The margins of the Pacific are also different in that they are generally marked by a line of deep arcuate trenches. On the eastern side of the Pacific the trenches lie adjacent to the margins of Central and South America and are parallel to the great mountain systems of the Andes and Rockies. The local relief from the top of the Andes to the bottom of the trench is 14,500 m (48,500 ft), nearly nine times the depth of the Grand Canyon. On the western side of the Pacific a nearly continuous line of trenches extends from the margins of the Gulf of Alaska along the margins of the Aleutians, the islands of Japan and the Philippines, and down to New Zealand.

Summary

In this chapter we have attempted to introduce you to the regional features on the surface of our planet by utilizing the opportunity we now have to look at our planet from space and to "see" the ocean floors with remote-sensing techniques. If you have simply appreciated the beauty of space photography and admired the physiographic map for its artistic detail, you have failed to grasp the significance of what we have tried to introduce.

The facts about the earth which we have discussed and shown geographically may be summarized as follows:

1. The atmosphere and hydrosphere constitute a system of moving surface fluids in which tremendous volumes of water are evaporated from the oceans, move through the atmosphere, and return by precipitation and surface runoff from the land.

2. The continents consist of three major components: (a) the shields composed of complexly deformed and recrystallized

metamorphic and igneous rock eroded down to near sea level; (b) areas of the shield covered with a veneer of horizontal sedimentary rocks; and (c) mountain belts in which sedimentary rocks have been compressed into folds.

3. The oceanic crust is unlike that of the continents in rock type, structure, age, and origin. The major surface features are (a) the mid-oceanic ridge; (b) abyssal hills and plains; (c) seamounts; and (d) deep trenches.

These facts are remarkable because we know that the land is being eroded away by water, wind, and ice. If this process has been operating throughout the history of the earth, why does the land stand above the sea at all? Why is there not a universal ocean circling the globe? Why are there mountains in which rocks are intensely folded?

If we consider the sea floor, there are similar questions. Why are there deep trenches adjacent to mountainous land? Why haven't they been filled with sediment? Why is there a mid-oceanic ridge cut by fracture systems?

The answer to these questions is that the major features of the earth such as continents, ocean basins, mountain belts, and deep trenches are not fixed and permanent but are continually changing and developing. The major processes which cause these changes will be discussed in later chapters but, in order to understand them, we must first consider the interior of the earth and what we know about its structure and composition.

Additional Reading

Lowman, Paul D., Jr. 1972. The Third Planet, Weltflugbild. Zurich, Switzerland: Reinhold A. Muler.

2   The Layered Planet

Speculations regarding the interior of the earth have stimulated man's imagination for centuries, but not until recently have scientists been able to probe the depths of the earth and formulate models of its structure and composition. The earth is commonly described as a **differentiated planet** in which materials are segregated into layers according to **density.** In addition, it has its own source of energy, energy which produces volcanic activity, earthquakes, and mountain-building. Because of this energy, we commonly refer to the earth as a live planet actively changing and evolving.

In this chapter, we summarize our present knowledge of the earth's interior and describe the characteristics of the major structural units.

Major Concepts

1. Evidence of the structure and composition of the earth's interior comes largely from (a) density measurements, (b) studies of **seismic** waves, (c) studies of the earth's magnetic field, and (d) studies of **meteorites.**
2. The earth is a differentiated planet in which the materials are separated into layers according to density.
3. The major structural units of the earth are (a) the central **core** composed predominantly of iron and nickel, (b) the thick surrounding **mantle** composed of silicate minerals rich in iron and magnesium, (c) the **lithosphere** which includes the crust and part of the upper mantle, and (d) the surface fluids of air and water.

Methods of Studying the Earth's Interior

Statement

The nature of the atmosphere, oceans, and the surface of the land is known in considerable detail because they can be studied by direct observation, but the internal structure of the earth presents some of the most difficult problems faced by geologists and geophysicists. The deepest bore holes penetrate no more than 8 km (5 mi) and erosion exposes rocks that were created no more than 20 to 25 km (12 to 15 mi) below the surface. Volcanic eruptions provide samples of material which comes from greater depths, possibly as much as 200 km (120 mi), but aside from these limited data we have no direct knowledge about the nature of the earth's interior. How then are we able to determine the structure and composition of the earth's interior? The evidence comes largely from studies of the physical characteristics of the earth—its density, the way in which it transmits seismic (earthquake) waves, the nature of its magnetic field—and from comparative studies of the composition of meteorites. Although these methods of study do not always provide absolute answers, they do indicate the limits of possibilities of what the interior of the earth may be.

Discussion

Evidence from Density Measurements. A comparison of the earth's average density with the density of the crustal materials provides the first important clue concerning the internal structure of the earth. In the late eighteenth century, Lord Henry Cavendish, using a device known as a torsion balance, found that the overall bulk density of the earth is 5.5 g/cm³ (the density of water is 1.0 g/cm³). Rocks at the surface of the earth, however, are much less dense, averaging 2.0 to 3.5 g/cm³. If the rocks on the earth's surface are only half as dense as the earth as a whole, there is obviously a mass of greater density in the earth's interior and significant changes in the density must occur with depth.

Evidence from Seismic Waves. One of the most important indicators of the nature of the earth's interior is the study of seismic waves, the vibration or shock waves resulting from earthquakes. A natural earthquake is caused by the sudden movement of rocks in the outer part of the earth (down to 700 km [435 mi]). Man-made earthquakes result from explosions capable of generating vibrations in the rocks. The point where the movement occurs, or where an explosion is detonated, is called the **focus.** From the focus, shock waves travel through the earth in all directions. With a network of **seismographs** (an instrument which detects and measures earthquake vibrations) scientists are able, in a sense, to x-ray the structure of the earth's interior.

Several different types of seismic waves are generated by an earthquake shock (*figure 2.1*), and each type travels at

The Layered Planet

41

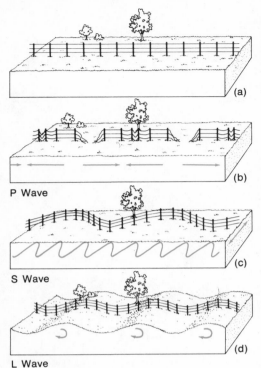

P Wave

S Wave

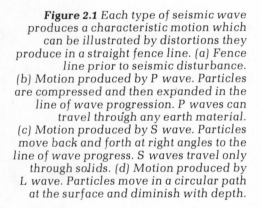

L Wave

Figure 2.1 *Each type of seismic wave produces a characteristic motion which can be illustrated by distortions they produce in a straight fence line. (a) Fence line prior to seismic disturbance. (b) Motion produced by P wave. Particles are compressed and then expanded in the line of wave progression. P waves can travel through any earth material. (c) Motion produced by S wave. Particles move back and forth at right angles to the line of wave progress. S waves travel only through solids. (d) Motion produced by L wave. Particles move in a circular path at the surface and diminish with depth.*

different speeds through the earth so that they arrive at a seismograph hundreds of kilometers away from the focus at different times. The first wave to arrive is called the **primary** or **P wave.** P waves are identical in character to sound waves passing through a liquid or gas. The particles involved in these waves move forward and backward in the direction the waves travel. The waves cause relatively small movement to be recorded on the seismograph.

The second waves to arrive are called **secondary** or **S waves.** In these waves particles move back and forth at right angles to the direction the wave travels. These cause strong movements to be recorded on the seismograph. The last waves to arrive from the focus are **L waves** which travel relatively slowly over the earth's surface.

The P and S waves are important in studies of the internal structure of the earth. Both travel faster through rigid material than through soft or plastic material. Therefore, the velocity at which the P and S waves travel through a specific part of the earth gives some indication of the types of rock involved. One difference between P and S waves is particularly significant. P waves will pass through any substance, solid, liquid, or gas, but S waves are transmitted only through solids having enough strength to return to their former shape after being distorted. They will not pass through a liquid.

Let us consider briefly some of the results of seismic studies of the earth's interior. In *figure 2.2* seismic velocities are plotted against depth. In general, both P and S waves travel more rapidly at greater depths, confirming the conclusion from density studies that the density of the earth increases with depth. The increase in velocity is not uniform, however, as there are several zones where the velocity increases rapidly in very short distances. These are referred to as **seismic discontinuities.** The first discontinuity is at a depth of 5 to 70 km (3 to 40 mi) below the surface. This is called the **Mohorovičić discontinuity** (or simply **Moho**) after the seismologist who first recognized it. The most striking discontinuity occurs at a depth of about 3000 km (1800 mi) where the velocity of the P wave decreases abruptly from 14 km/sec to 8 km/sec. The S wave at this point simply terminates and does not pass into deeper parts of the earth. It is concluded that the deep internal part of the earth, called the **core**, is liquid because S waves won't pass through a liquid. Other discontinuities occur at depths of approximately 100 km (62 mi), 400 km (250 mi), and 5000 km (3100 mi).

These data provide the basis for concluding that the earth has an internal structure consisting of concentric layers or shells: a thin crust, a thick mantle, and a core. These are the large internal divisions of the earth. Within these major units, there are smaller divisions which will be discussed in subsequent sections of this chapter.

Evidence from the Earth's Magnetic Field. The earth is unique among the planets and satellites with which it is closely

The Earth's Dynamic Systems

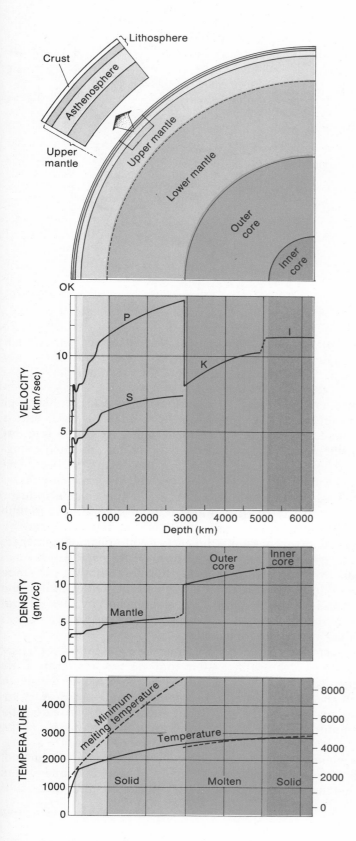

(a)

(b)

(c)

Figure 2.2 The internal structure of the earth
as deduced from variations in seismic waves'
velocities at depth. (a) The velocities of both the
P and S waves increase to a depth of about 3000
km (1800 miles), where both change abruptly.
The S wave disappears and does not travel
through the central part of the earth, and the
velocity of the P wave decreases drastically. This
is the most striking discontinuity in the earth
and is considered to be the boundary between
the core and the mantle. Another discontinuity
occurs at a depth of 5000 km, indicating an
inner core. A low-velocity layer at depths from
100 to 400 km is called the asthenosphere.
(b) Density variations at depth based on seismic
velocities and other geophysical measurements.
(c) Variations in temperature at depth. The
dashed line shows the melting point of the
earth's materials and the solid line, the
temperature. Note that the rocks are near the
melting point at the asthenosphere, but due to
increase in pressure, they are below the melting
point throughout the mantle.

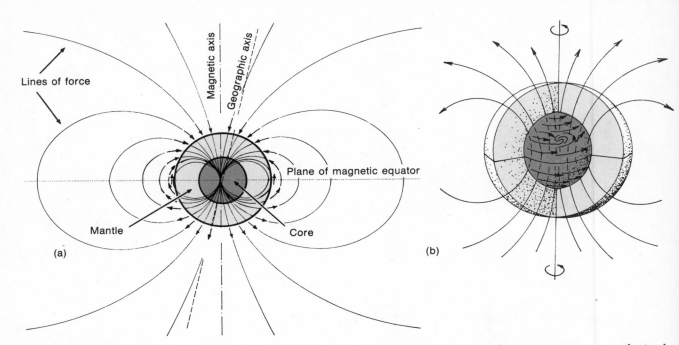

Figure 2.3 *The earth's magnetic field. (a) The lines of force of the earth's magnetic field. A magnetic needle free to move in space would be deflected by the field as shown by the arrows. Near the equator the needle would be horizontal and point toward the poles. At the poles the needle would be vertical. (b) Schematic diagram showing how electric currents in the earth's core could produce the magnetic field. Theoretically, convection in the liquid core generates an electric current, similar to a dynamo, which produces the magnetic field.*

associated in the solar system in that it possesses a relatively strong magnetic field. The moon, Mars, and Venus have only weak magnetic fields, although that of the moon may have been stronger in the distant past.

The magnetic field of the earth (*figure 2.3a*) is like that which would be produced by a simple bar magnet whose axis is inclined at 11° from the earth's geographic axis. The temperature of the mantle and core is far too high for the earth's magnetic field to be produced by iron at the earth's surface. The magnetic field must, therefore, be generated electromagnetically by large-scale motion of the material in the earth's interior (*figure 2.3b*). If the earth had a liquid outer core, as evidence from seismicity suggests, movement of the liquid caused by temperature differences and the earth's rotation could generate strong electric currents which in turn would establish a magnetic field. Many scientists believe, therefore, that the presence of a magnetic field in a planet is very strong evidence that the planet contains a liquid core. This is supported by the fact that S waves, which will not pass through a liquid, are not transmitted through the core of the earth.

Evidence from Meteorites. Meteorites are samples of rock types existing in the solar system that are very different from those that occur in the earth's crust. They have always attracted attention as "visitors from interplanetary space," but to geologists they are especially significant because they provide samples of material which may closely resemble the interior of the earth.

Two major types of meteorites have long been recognized: metallic meteorites composed largely of iron and nickel and stony meteorites composed mostly of silicate minerals. It is

44

believed that meteorites may be fragments of a planet or asteroid which broke up, the metallic meteorites being fragments of the core and the stony meteorites fragments of the mantle. If this is the case, the earth's interior could be considered, by analogy, to consist of a mantle of silicate minerals and a core rich in iron and nickel; this would support the evidence obtained from density, seismic, and magnetic studies.

The space program has stimulated a great resurgence of interest in meteorites and an accumulation of a vast amount of data on these extraterrestrial bodies. It is generally accepted that sometime prior to 4.6 billion years ago masses of material formed by rapid accumulation of the elements. As the material aggregated by gravitational attraction, some grew large enough to generate sufficient heat to melt the material at their interiors, permitting the iron and nickel to migrate to their cores. Subsequently, some of these bodies were disrupted into fragments to form meteorites.

If the assumption that the inner planets of our solar system are made of the same substance is correct, and if meteorites are fragments of a planet, they can provide us with important data about the earth's interior composition.

Structure of the Earth

Statement

The results of geophysical studies described in the previous section indicate that the earth is a differentiated planet in which the minerals are segregated or separated according to density into a series of concentric spherical shells or layers distinguished by their composition and physical properties. *Figure 2.4* summarizes our present understanding of the earth's interior. The major structural units consist of (1) a central core composed predominantly of iron and nickel, (2) a thick surrounding mantle composed of silicate minerals rich in iron and magnesium, (3) a crust composed of relatively light silicate minerals, and (4) surface fluids of water and air.

Discussion

The Core. The core of the earth is a central mass consisting of two distinct parts: a solid inner core and a liquid outer core. It is believed to be nearly twice as dense as the mantle and, although it has only 16.2% of the earth's volume, it has 31.5% of the earth's mass. The conclusion that the outer core is liquid is based on the fact that S waves, which will not pass through a liquid, are not transmitted through the core. A solid inner core is suggested by the high pressure and density which exist near the center of the earth. Presumably, the increase in pressure at depth causes the core to change from liquid to solid. The magnetic field further supports the assumption that at least part of the core is liquid.

Figure 2.4 *Diagram summarizing our present understanding of the structure of the earth. The major components are: (a) A dense solid inner core; (b) A liquid outer core; (c) A thick solid mantle. The upper mantle contains two significant layers: (1) The asthenosphere, a soft low-velocity layer; (2) The lithosphere, a rigid layer which includes the crust. Surrounding the entire planet is a blanket of surface fluids—water and air.*

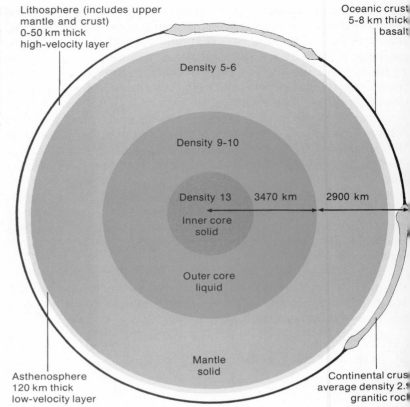

Lithosphere (includes upper mantle and crust) 0-50 km thick high-velocity layer

Oceanic crust 5-8 km thick basalt

Density 5-6

Density 9-10

Density 13   3470 km   2900 km

Inner core solid

Outer core liquid

Mantle solid

Asthenosphere 120 km thick low-velocity layer

Continental crust average density 2.9 granitic rock

The composition of the core is assumed to be iron with minor amounts of nickel, sulfur, silicon, and magnesium. The composition of meteorites supports this conclusion as the metal phase in stony meteorites consists of 94% iron and 6% nickel.

The Mantle. The earth's mantle extends from the base of the crust to the outer margin of the core, a distance of 2900 km (1800 mi). This layer constitutes the great bulk of the earth (82.3% of the volume and 67.8% of the mass) and is of special interest because it is in this area that the earth's internal energy is released. On the basis of the velocity at which it transmits seismic waves, the mantle can be divided into three major zones: (1) the lower mantle extending from the core up to about 650 km (400 mi) from the surface, (2) a transition zone extending from 650 km to 400 km (250 mi) below the surface, and (3) the upper mantle which extends from 400 km to the base of the crust.

The lower mantle and transition zone are so inaccessible and the pressures and temperatures are so high that it is difficult to determine precisely their composition and physical characteristics. The materials are most likely dense oxide and silicate minerals, high in iron and magnesium, which grade upward through the transition zone to the upper mantle composed of iron and magnesium minerals of lower density. The velocity of seismic waves decreases upward through the lower mantle and transition zone.

The upper mantle is believed to be composed mostly of the rock **peridotite**, consisting of silicate minerals high in iron and magnesium. This conclusion is based upon density measurements and velocities at which seismic waves are transmitted through the mantle. Volcanic material, presumably derived from the upper mantle, and diamond pipes formed at great depths are composed largely of the rock peridotite, further substantiating this conclusion. In addition, peridotite is similar in texture and composition to many stony meteorites believed to be fragments of the mantle of another planet.

Two zones in the upper mantle have attracted special interest because of their importance in earth dynamics and the evolution of the crust (*figure 2.5*). A zone ranging in depth from 50 to 200 km (30 to 125 mi) is called the **asthenosphere,** meaning plastic or weak. Seismic velocities within the asthenosphere are distinctly lower, suggesting that the material is soft and will yield to plastic flow. The temperature in the asthenosphere is probably near the point at which rocks of the mantle will begin to melt at the pressure existing at that depth. If partial melting of the asthenosphere occurs, there is probably significant pore fluid which could account for its soft plastic nature.

Above the asthenosphere is a rigid solid layer called the lithosphere. This zone includes the crust and behaves as a single mechanical unit. The boundary between the rigid lithosphere and the soft asthenosphere is gradational and, although the depth is well established, details of the structure and composition are not completely understood. The boundary between the two layers may simply represent a change in physical properties as the rock approaches the melting point. This change might be similar to the difference in properties of a steel rod, one end of which is held in a furnace. As the end of the bar

Figure 2.5 *The relationship of the asthenosphere, lithosphere, and crust. Above the soft asthenosphere is the lithosphere, which includes both the rigid upper mantle and the crust. Two types of crust are recognized: a thin dense oceanic crust and a thicker light continental crust. The base of the crust marks the first worldwide seismic discontinuity.*

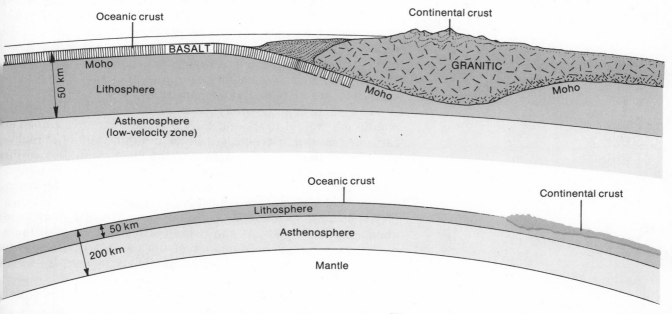

approaches the melting point, it will become soft and pliable but, toward the cool end, the bar will remain hard and brittle. In the same manner, below the asthenosphere, the rock is somewhat plastic but will fracture as stresses are suddenly applied.

There is increasing evidence that the layers of the uppermost mantle are not homogeneous throughout but vary laterally. This is probably due to partial melting and the removal of the liquid by volcanic activity.

The Crust. Geologists use the term "crust" in reference to the outermost layer of the earth. Once the earth was thought to be completely molten in its early stages: as it cooled, a hard crust formed, enveloping the still liquid interior. Though this concept has been obsolete for nearly a century, the term "crust" is still popular. However, it has acquired another generally accepted meaning. Today, the term crust designates the outer layer of the earth extending from the solid surface down to the first major discontinuity in seismic-wave velocity, the Moho (*figure 2.2*). The crust of the continents, however, is distinctly different from the crust beneath the ocean basins (*figure 2.5*). The continental crust is much thicker (as much as 48 km [30 mi] thick) and is composed of relatively light granitic rock which includes the oldest rock of the crust. By contrast, the oceanic crust is only about 8 km (5 mi) thick and is composed of basalt, a volcanic rock having a density much greater than granite. The oceanic crust is young and relatively undeformed. The differences between the continental and oceanic crust, as we shall see, are of fundamental importance in understanding the earth.

It should be emphasized at this point that the lithosphere and the crust behave as a single rigid unit in earth dynamics, although the crust is not considered part of the mantle.

The Surface Fluids. The gas and liquid of the atmosphere and oceans are as much a part of the earth's system as the rocks of the crust. They not only constitute part of the material of our planet but, as we have seen in the previous chapter, they are of prime importance in the origin and development of the landscape. The atmosphere and water, the surface fluids of the earth's system, are concentrated above the solid part of the earth because of their lighter density.

The earth's atmosphere consists of a mixture of highly mobile gases held to the earth by gravitational attraction. The gases are most dense near sea level and thin rapidly at higher altitudes so that 97% of the mass of the atmosphere lies within 29 km (18 mi) of the surface. It is difficult to draw a sharp boundary for the outer limit of the atmosphere because the density of gas molecules decreases almost imperceptibly into interplanetary space. Many scientists consider the outer boundary to be about 9500 km (6000 mi) above the earth's surface, a distance nearly as great as the diameter of the earth. Only the lower part of the atmosphere is visible as clouds formed by water vapor.

Water is present as oceans, lakes, rivers, ice, ground water and vapor in the atmosphere. It is constantly moving from reservoir to reservoir through the hydrologic cycle and passing from one physical state (gas, liquid, and solid) to another as it moves. The movement of water is responsible to a large degree for the erosional and depositional features of the landscape.

Summary

Our knowledge of the earth's interior comes from studies of the physical characteristics of the earth, such as its density, the manner in which seismic waves are transmitted through the earth, the nature of the earth's magnetic field, and from comparative studies of meteorites. These studies establish definite limits as to the possibilities of what the earth's interior may or may not be. According to our present knowledge, the earth has a central core composed predominantly of iron and nickel. The central part of the core is believed to be solid, but the outer core is liquid. Surrounding the core is a thick mantle composed of silicate minerals rich in iron and magnesium. The upper mantle is of special interest because it is the zone where heat is generated, producing energy for earth dynamics. The upper mantle and the crust behave as a rigid mechanical unit and are called the lithosphere. Below the lithosphere is a zone of less strength called the asthenosphere where convection is believed to occur. The crust is the outer layer extending from the solid surface down to the first major seismic discontinuity. The continental crust, however, is distinctly different from the oceanic crust. The surface fluids are as much a part of the planet as the solid rock and, although the atmosphere may extend out into space some 9500 km (6000 mi), most of it is concentrated within 29 km (18 mi) of the surface.

Movement of the surface fluids and the material in the asthenosphere and lithosphere result in constant change in the earth's surface. We will consider these dynamic systems in the next chapter.

Additional Readings

Bolt, D. A. 1973. "The Fine Structure of the Earth's Interior," Sci. Amer. 228(3): 24-33.

Bott, M. H. P. 1971. The Interior of the Earth. New York: St. Martin's Press.

Clark, S. P., Jr. 1971. Structure of the Earth. Englewood Cliffs, N.J.: Prentice-Hall.

Gaskell, T. F. 1970. Physics of the Earth. New York: Funk and Wagnalls.

Phillips, O. M. 1968. The Heart of the Earth. San Francisco: Freeman, Cooper and Company.

Sumner, John S. 1969. Geophysics, Geologic Structures and Tectonics. Dubuque, Iowa: William C. Brown Company, Publishers.

Takeuchi, H., S. Uyeda, and H. Kanamori, 1970. Debate About the Earth. San Francisco: Freeman, Cooper and Company.

3 The Earth's Dynamic Systems

We have seen in the previous chapters that the earth is a differentiated planet; that is, the materials are segregated and separated according to density into a series of concentric layers or shells. The material within each of these layers is in motion. The most obvious motion is in the surface fluids (air and water), which is the hydrologic system of our planet. Drawing energy from the sun, water circulates from the oceans, to the atmosphere, and back to the earth and the oceans. The precipitation of water on the earth furnishes a fluid medium for many processes which shape the earth's surface features. The rivers systematically carve away the land as they flow to the seas. **Ground water** percolating through the pores of rocks carries with it dissolved **minerals** from the earth. Where temperature drops low enough, glaciers form and move over large parts of the continents, modifying the surface by erosion and deposition as they move outward from centers of accumulation. Even though the earth's crust appears to be stationary and stable, there is convincing evidence that it, too, is in constant motion. Thermal energy is transferred from the asthenosphere to the surface, probably by **convection.** At places where **convection currents** reach the lithosphere and move laterally, the crust splits apart. New crust is created in the rift zone, so the earth's crust is not only in motion but is continually being created and destroyed.

The **tectonic** system of the earth involves those changes resulting from energy coming from within the earth and includes mountain-building, **vulcanism**, earthquakes, **continental drift**, and **sea-floor spreading**. These changes are independent of the hydrologic system.

This chapter summarizes the basic elements of these two fundamental geologic systems of moving material which continually modify the earth. Throughout the remainder of this book, the material in each chapter is concerned largely with the physical processes in these systems.

Major Concepts

1. The earth is a dynamic planet as evidenced by the fact that the materials are differentiated and segregated into distinct layers or zones (the core, mantle, lithosphere, and surface fluids). The most dynamic layers are (a) the surface fluids (water and atmosphere) and (b) the soft asthenosphere in the upper mantle.
2. The system of moving water on the earth's surface (the hydrologic system) involves movement of water in rivers, as ground water, in glaciers, and in oceans. The volume of water in motion is almost incomprehensibly large and, as it moves, it erodes, transports, and deposits sediment which continually modifies the earth's surface. The source of energy for the hydrologic system is heat from the sun.
3. The tectonic system involves movement of material in the earth's interior which results in sea-floor spreading, creation of new crust, continental drift, vulcanism, earthquakes, and mountain-building. **Radiogenic heat** in the upper mantle is probably the source of energy for the tectonic system.
4. The earth's lithosphere is buoyed up and "floats" on the denser plastic mantle beneath and will rise or sink in an attempt to establish isostatic equilibrium.
5. The earth's surface and the hydrologic system constitute the basis of our **ecology,** and our natural resources are ultimately the result of the hydrologic and tectonic systems.

The Hydrologic System

Statement

Surface processes such as erosion and deposition by running water, ground water, glaciers, and waves are the result of the system of moving fluids at or near the earth's surface. The basic elements of this system are very simple and are shown diagrammatically in *figure 3.1*. Heat from the sun evaporates water from the oceans, the principal reservoir of the earth's water. Most of the water returns directly to the oceans as rain; the rest drifts over the continents and is precipitated as rain or snow. The water which falls on the land may take a variety of paths back to the ocean. The greatest quantity returns to the atmosphere by evaporation, but the most obvious return is by surface runoff in river systems back to the seas. Some water also seeps into the ground and moves slowly through the pore spaces of the rock. Plants use part of the ground water, expelling it back into the atmosphere through a process called evapotranspiration, but much ground water slowly seeps into streams or lakes or migrates in the subsurface back to the ocean. Water may be temporarily trapped as glacial ice, but glaciers move from cold centers of accumulation into warmer areas and ultimately melt, returning their water to the system. Water in the hydrologic system, moving as surface runoff, ground water, glaciers, or waves and currents, erodes and transports rock material and ultimately deposits it as deltas, beaches, or other types of sedimentary deposits.

Discussion

The idea of a complete cycle in the movement of water was recorded in Biblical times (Ecclesiastes 1:7), but it was not demonstrated as a fact until the mid-seventeenth century when two

The Earth's Dynamic Systems

Figure 3.1 *Schematic diagram showing circulation of water in the hydrologic system. Water evaporates from the oceans, circulates around the globe with the atmosphere, and ultimately returns to the surface by precipitation in the form of rain or snow. The water which falls upon the land returns to the ocean by surface runoff and ground-water seepage. Variations in the major flow patterns of the system include temporary storage of water in lakes and glaciers.*

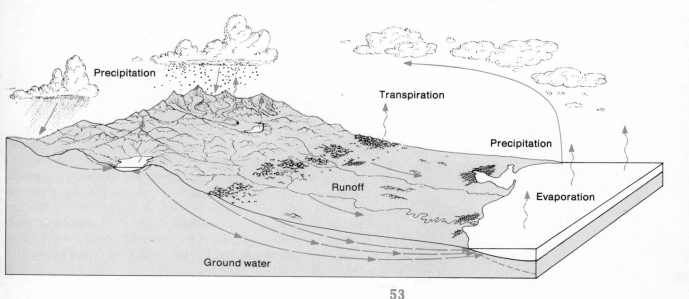

French scientists, Pierre Perrault and Edmé Mariotte, independently measured precipitation in the drainage basin of the Seine River and then measured the discharge into the ocean during a given interval of time. Their measurements proved that precipitation alone could produce not only enough water for river flow but also enough for springs.

The importance of this system is so great that it might well be considered the most fundamental and significant geologic system operating on the surface of the earth. Your problem as a student new to geology will be not in understanding the process, but in conceiving the worldwide scope of this system which influences in both subtle and dramatic fashion the development of the earth's surface features. The space photos of the earth shown in *figures 1.1* to *1.21* provide one of the best ways to gain an accurate concept of the magnitude of the hydrologic system because they permit us to see the system in action.

Another way of grasping the magnitude of the hydrologic system is to consider the volume of water involved. From measurements of rainfall and stream discharge, together with measurements of heat and energy transfer into bodies of water, scientists have calculated that 400,000 km³ (95,000 mi³) of water evaporate each year, or 1100 km³ (264 mi³) per day. Of this, about 336,000 km³ (80,000 mi³) are taken from the oceans, and 63,000 km³ (15,000 mi³) evaporate from water on the land. Of the total 400,000 km³ of water evaporated each year, about 101,000 km³ (24,000 mi³) fall as rain or snow upon the continents. Most of the precipitation (about 60 to 80%) returns directly to the atmosphere by evaporation and transpiration, with about 38,000 km³ (9000 mi³) of water flowing back to the oceans over the surface or beneath the ground. These figures may or may not be impressive in themselves, but when related to more familiar volumes and rates, they give some idea of the magnitude of this system.

From these rates we know that if the hydrologic system were interrupted and water did not return to the oceans, sea level would drop 1 m (3.3 ft) per year, and all the ocean basins would be completely dry within 4000 years. The recent glacial epoch demonstrates this point very clearly as it partly interrupted the hydrologic cycle by freezing much of the water that fell upon the northern hemisphere and preventing it from returning immediately to the sea. Consequently, sea level dropped over 100 m (328 ft).

We can also get some idea of the magnitude of this system by considering the volume of water running off the earth's surface. At any given instant there is an average of 1260 km³ (300 mi³) of water flowing in the world's rivers. Gauging stations on the major rivers of the world indicate that, if the ocean basins were empty, enough runoff occurs to completely fill them in 40,000 years. On the other hand, if all the evaporated water returned by runoff, with no return to the atmosphere by immediate evaporation and evapotranspiration, the ocean basins would be filled in slightly more than 3000 years.

The Earth's Dynamic Systems

The source of energy for the hydrologic system is heat from the sun which evaporates the water from the ocean. The water then moves with the circulating atmosphere and ultimately condenses under proper temperature and pressure conditions and falls as rain or snow. Under the force of gravity it then flows back to the sea through the river and ground-water systems. Without the heat there would be no hydrologic system.

The Tectonic System

Statement

Geologists have long recognized that the earth has its own source of internal energy which is manifested repeatedly by earthquakes, volcanic activity, and mountain-building, but it wasn't until the late 1960s that a unifying theory of earth dynamics was developed. This theory, known as **plate tectonics**, has radically transformed our thinking about the crust and its movements. Abandoned is the old idea that the crust is a fixed and rigid sphere, with movements being largely vertical. We now have evidence that the crust is in continual motion, with individual fragments or plates moving thousands of kilometers. As the plates move, split apart, collide, and descend back into the mantle, they create earthquakes, vulcanism, mountain-building, and other features in the "solid" part of the surface.

The basic elements of the plate tectonic theory are quite simple and can easily be understood by studying the diagram in *figure 3.2*. As we have seen in the previous chapter, the lithosphere which includes the crust and upper mantle behaves as a single mechanical unit, whereas the underlying soft asthenosphere yields to plastic flow. The fundamental idea of plate tectonics is that the lithosphere moves in response to flow in the asthenosphere. Although there are many unsolved questions, radiogenic heat in the upper mantle is considered to be the basic source of energy. The heat causes the material in the asthenosphere to move slowly in a **convection cell**, with the hot material rising to the base of the lithosphere where it then moves laterally, cools, and descends to become reheated, beginning the cycle again. Where the convecting mantle rises it arches the lithosphere to form the mid-oceanic ridge and, as it moves laterally, it pulls the rigid lithosphere apart. The lithosphere is thus broken into a series of fragments or **plates** which are several thousand kilometers in diameter. As the plates move apart, molten rock from the hot asthenosphere rises into the rift zone and cools to form new crust. The continental blocks, composed of relatively light granitic rock, float passively on the denser lower part of the lithosphere, sometimes splitting (when a convection cell rises beneath a continent) and sometimes colliding. The continents do not drift through the lower lithosphere; they are carried by it. Since the earth is a sphere, the shifting plates are in collision with each other. Plates con-

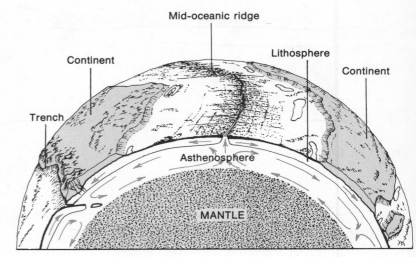

Figure 3.2 *Schematic diagram showing the major elements of the tectonic system. The material in the asthenosphere is thought to be moving in a series of convection cells resulting from heat generated by radioactivity. Where the convecting mantle rises and moves laterally, it pulls the overlying rigid lithosphere apart and creates new crust in the rift zone. The lithosphere, which may contain a block of continental crust, is carried by the convection cell, moves as a single mechanical unit, and descends back into the mantle at the deep sea trenches. The continents carried by the plates may split and drift apart but, being less dense than the mantle, cannot sink back into the asthenosphere. As a result, the continental margins are deformed into mountain ranges where two plates collide. The plate margins are the most active areas and are the sites of most intense vulcanism, earthquake activity, and crustal deformation.*

taining dense oceanic crust move via the descending convection currents down into the asthenosphere at the deep oceanic trenches and are consumed. By contrast, plates containing light continental crust cannot sink back into the mantle. Instead, continental margins adjacent to the descending plates are deformed into linear folded mountain belts. The boundaries of plates thus coincide with ridges and trenches and mark zones of earthquakes and volcanic activity. According to the plate tectonic theory, the lithosphere is moving at a rate of 2 to 16 cm per year.

Discussion

The evidence for such a revolutionary theory of crustal movement comes from many sources and includes data on the structure, topography, and magnetic patterns of the ocean floor, the location of earthquakes, patterns of heat flow in the crust, location of volcanic activity, the structure and geographic fit of continents, and the nature and history of mountain belts. These will be discussed and evaluated in Chapters 15 through 17.

Let us consider now the present structural features of the earth and how they fit into the plate tectonic theory. The boundaries of the plates are delineated with dramatic clarity by the belts of active earthquakes and volcanoes (*figure 3.3*). Six major lithospheric plates are recognized, together with several smaller ones. The spreading center, where the lithosphere is pulled apart, is marked by the mid-oceanic ridge which extends from the Arctic down through the central Atlantic into the Indian and Pacific oceans. Movement of the plates is away from the crest of the mid-oceanic ridge. The American plate includes both North and South America plus the western Atlantic Ocean. It is moving in a westward direction and encounters the eastern Pacific plate along the west coast of the Americas. The Pacific plate consists only of oceanic crust and is moving from the ocean ridge northwestward to the system of deep trenches in

The Earth's Dynamic Systems

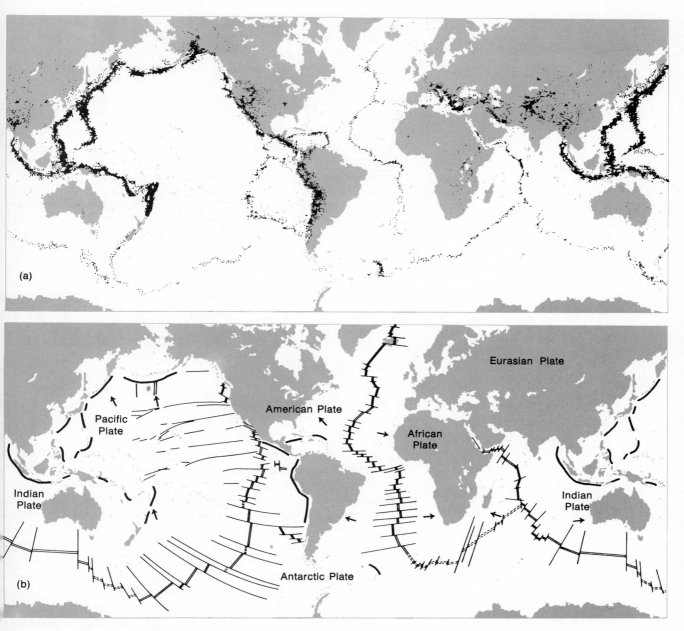

(a)

(b)

Pacific Plate

Indian Plate

American Plate

African Plate

Eurasian Plate

Indian Plate

Antarctic Plate

the western Pacific basin. The Indian plate includes Australia, the northeastern Indian Ocean, and India. It is moving northward, causing India to collide with Asia to produce the Himalaya Mountains. The long linear ridges in the Indian Ocean very dramatically show the direction of plate movement. The African plate includes the continent of Africa plus the southeastern Atlantic and western Indian oceans. It is moving eastward and northward. The Eurasian plate is the largest and is moving eastward.

Many major tectonic features of the earth can be explained nicely by this theory. The Rockies and Andes mountain chains

Figure 3.3 Maps showing the six major plates of the lithosphere and the geologic activity associated with their margins: (a) The plate margins are outlined with remarkable fidelity by zones of earthquake activity. Earthquakes (black dots) occur along the crests of the oceanic ridges where plates are being pulled apart and along the deep-sea trenches and mountain belts where plates are colliding. (b) The major lithospheric plates and patterns of relative movement away from the oceanic ridges and toward the trenches.

result from the encounter of the American and eastern Pacific plates. Earthquakes that consistently rock Chile, Peru, and Central America result from the encounter of the Pacific and American plates. The Alps and Himalayan mountain systems result from the collision of the African and Indian plates with Eurasia, which also produces the vulcanism and earthquakes that torment the Mediterranean and Near East. The great system of deep-sea trenches in the Pacific marks the zone where the Pacific plate descends down into the mantle. Earthquakes and volcanic activity also mark this plate boundary where the lithosphere is being destroyed.

A closer look at the ridges, trenches, and mountain ranges shown on the physiographic map reveals many other features which fit into the plate tectonic theory. The Red Sea and the Gulf of Aden appear to be an embryo ocean formed by the separation of Saudi Arabia from Africa. The rift system from the Indian Ocean ridge extends into this area and divides into two branches: one extending up the Red Sea into the Dead Sea-Jordan River rift valley and the other extending southward into East Africa to form the African rift valleys. The rift of the eastern Pacific plate can be traced in a similar manner into the Gulf of California and up into the western United States where the rift system of the Basin and Range occurs.

The Concept of Isostasy

Statement

In addition to changes brought about by the hydrologic and tectonic systems, the earth's crust is continually responding to the force of gravity in an effort to reach a gravitational balance or **isostasy** (Greek: *isos*, "equal"; *stasis*, "standing"). According to this concept, the earth's crust is considered to be buoyed up and "floating" on a denser plastic mantle beneath, with each portion of the crust displacing the mantle according to its bulk and density; denser crustal material sinks deeper into the mantle than crustal material of low density. Similarly, high mountains and plateaus having a large vertical thickness sink deeper into the mantle than areas of low elevation. Any change in an area of the crust, such as removal of material by erosion or addition of material by sedimentation, volcanic extrusion, or accumulations of large continental glaciers, will cause an isostatic adjustment.

Discussion

The concept of isostasy is fundamental to studies of major features of the crust, such as continents, ocean basins, mountain ranges, and the response of the crust to erosion, sedimentation, and glaciation. We will consider some of the evidence upon which the concept is based by briefly reviewing how the concept developed.

In the 1850s, the British conducted a highly precise triangulation survey to establish reference points for the mapping of India. By checking the triangulation against astronomical observations, they discovered a discrepancy of 152 m (500 ft) between stations in central India and at the base of the Himalayas. Much too large to result from inaccuracies of geodetic work, the error was found to be consistent by subsequent surveys.

Seeking to explain the discrepancy, J. H. Pratt proposed that the error resulted from deflection of the plumb bob of the surveying instruments by gravitational attraction of the mass of the Himalayas. Such deflection would lead to errors in the astronomic survey, since the plumb line would not point to the earth's center as is assumed in astronomic computations of latitude. Calculating the expected deflection of the plumb bob, Pratt found it to be three times greater than that which actually occurred. He attributed the distortion in deflection to a difference in density between the rocks in and beneath the mountains and those in the adjacent plains and proposed that both were "floating" on denser material beneath. A simple illustration of Pratt's concept is shown in *figure 3.4a.*

Less than two months after Pratt presented his ideas, G. B. Airy submitted an alternate proposal which accounted equally well for the observed measurements. Airy concluded that no rocks were strong enough to support the weight of high mountains and plateaus and that they would yield to plastic flow until a balance was reached. He suggested, however, that the floating crustal blocks had the same density but were of unequal thickness. High mountain ranges with the same density as the adjacent lowlands must "sink" deeper into the denser material

Figure 3.4 *The concept of isostasy. Isostasy is neither a force nor a process but the universal tendency to establish a condition of gravitational balance between segments of the earth's crust. (a) J. H. Pratt proposed that mountains were high because they were composed of lighter materials than the surrounding lowland. (b) G. B. Airy concluded that mountains are of the same density as adjacent lowland but are high because they are thicker.*

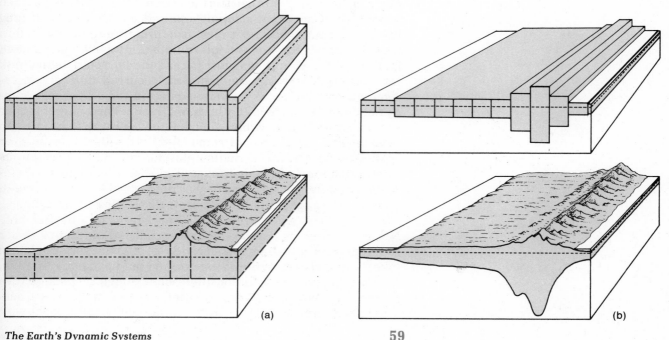

(a)

(b)

below the crust. This concept is illustrated in *figure 3.4b*. Both views are valid concepts of the tendency of the crust to reach isostatic equilibrium, and neither mechanism operates to the exclusion of the other.

Isostatic adjustment in the earth's crust may be thought of as though the surface layers were floating on denser material like ice on a lake. If you were to skate on the ice, your weight would cause the ice layer to bend down beneath you, displacing water equal to your weight. As you leave that point the ice will rebound and the displaced water will flow back.

Another example might be a block of floating ice which is melted by heat from the sun only at the exposed surface. As the upper part of the ice melts, the submerged part rises to maintain a floating balance. The same processes may be implied in the earth's crust. When weight is added in some area, such as the delta at the mouth of a river, the area subsides, displacing the subcrustal material. Similarly, as a mountain is eroded and the weight of the upper rocks removed, the area rises to compensate for the removed material.

Man-Made Reservoirs and Isostasy. The construction of Hoover Dam on the Colorado River provides an excellent illustration of isostatic adjustment because of the added weight of the water and sediment in the reservoir. From the time of its construction in 1935, 24 billion metric tons (26 billion short tons) of water plus an unknown amount of sediment accumulated in Lake Mead. In a matter of years, this weight added to the earth's surface caused the crust to subside in a roughly circular area centered around the lake. Maximum subsidence was 1.7 m (5.6 ft).

Ice Sheets and Isostasy. Continental glaciers provide another clear example of isostatic adjustment of the crust, as the weight of an ice sheet several thousand meters thick disrupts the crustal balance and causes the crust beneath to be depressed. The weight of the ice in both Antarctica and Greenland has depressed the central part of the landmass below sea level. A similar isostatic adjustment occurred during the Ice Age, when continental glaciers existed in the Baltic region of Europe and the Hudson Bay area of Canada. Parts of both areas are still below sea level but, now that the ice is gone, the crust is rebounding at a rate of 5 to 10 m (16 to 33 ft) per 1000 years. Geologists measure the extent and rates of rebound by mapping tilted shorelines of ancient lakes (see *figure 11.20*).

Ancient Lakes and Isostasy. Tilted shorelines of ancient lakes provide a means of documenting isostatic rebound (*figure 3.5*). Lake Bonneville, for example, was a large lake in Utah and Nevada during the ice age but has since dried up to such small remnants as Utah Lake and Great Salt Lake. Shorelines of this lake were level when they were formed but have been tilted in response to unloading as the water was removed. The lake was only 305 m (1000 ft) deep and covered a much smaller area than a continental glacier, but still this relatively small weight was

The Earth's Dynamic Systems

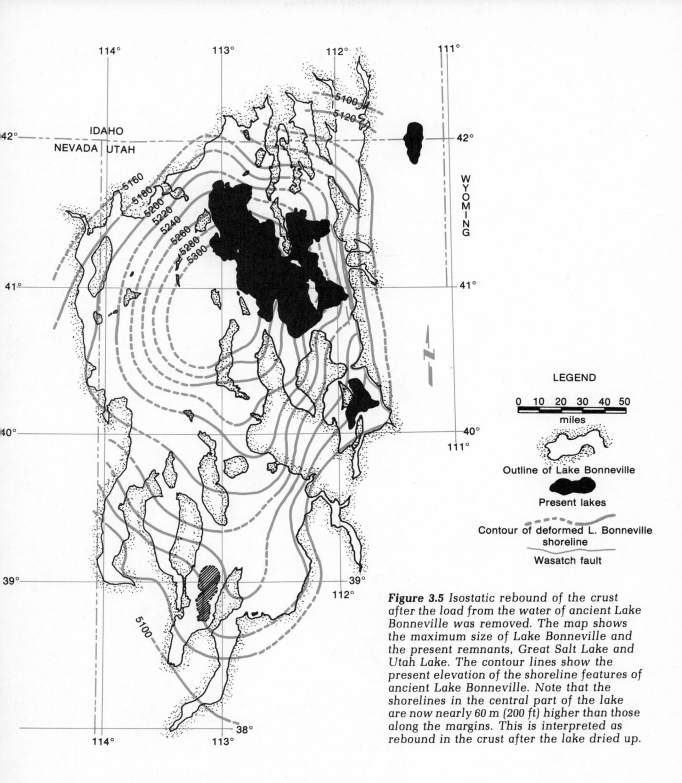

Figure 3.5 *Isostatic rebound of the crust after the load from the water of ancient Lake Bonneville was removed. The map shows the maximum size of Lake Bonneville and the present remnants, Great Salt Lake and Utah Lake. The contour lines show the present elevation of the shoreline features of ancient Lake Bonneville. Note that the shorelines in the central part of the lake are now nearly 60 m (200 ft) higher than those along the margins. This is interpreted as rebound in the crust after the lake dried up.*

LEGEND

0 10 20 30 40 50
miles

Outline of Lake Bonneville

Present lakes

Contour of deformed L. Bonneville shoreline

Wasatch fault

sufficient to depress the crust. Since it has been removed, the shorelines near the deepest part of the lake have rebounded nearly 60 m (200 ft).

The Earth's Systems as the Basis of Our Ecology

Statement

Studies of the earth's systems are not limited to academic interest but are basic to understanding our environment and utilizing our resources. The surface of the earth and the hydrologic system constitute the basis of our ecology. We are all aware of how weather affects our lives but, surprisingly enough, few people understand the interrelationships of the various parts of the hydrologic system and how it is impossible to alter one part of it without affecting the others. Similarly, the new theory of plate tectonics is more than an interesting "mystery unveiled." It is of daily concern to millions of people who live along the plate boundaries and are confronted with the earthquakes and volcanic activity which result from plate movement. We know a great deal about how and why the major dynamic systems operate and must use this knowledge to live within our environment.

Our natural resources are ultimately the result of the hydrologic and tectonic systems for there is essentially nothing on the surface of the earth that is not intimately involved with the dynamics of the surface fluids or the lithospheric plates.

Discussion

Water, of course, is essential to life; therefore, the hydrologic system must be considered as fundamental to any problems of the environment. Modification of it may invoke unsuspected consequences. Man began to harness natural forces only after a long period of time in which the diverse natural processes had achieved a delicate balance. Can dams be built across rivers without affecting the movement of ground water or the amounts of local evaporation? Can we seed the clouds and order rain without disrupting the flow of rivers, the level of lakes, and the level of the water table? Can we drill wells to supply cities with water without affecting the flow of rivers?

Increasingly, man's knowledge of the hydrologic system shows that it cannot be modified indiscriminately. The hydrologic system is a system in balance. If it is disturbed or upset, the geologic processes involved may be reversed, stopped, or otherwise altered in a number of unanticipated ways.

The tectonic system is somewhat different. We can do little to modify it and, therefore, must live with it. Studies of plate movements are leading us to an understanding of earthquakes, volcanic activity, and the way in which our mineral resources are created and distributed. With this new understanding, it may soon be possible to predict the occurrence and magnitude of earthquakes and ultimately exert some control over them. Plates moving 3 to 6 cm per year are relatively harmless but, when they are held firm by friction and then move several meters in an instant, disaster follows. It may be possible in the

The Earth's Dynamic Systems

near future to lubricate the plate margins, thus permitting them to slowly and continually move without major earthquakes.

The movement of plates has other economic values, for plate movement controls the distribution of the earth's surface material. Finding ore deposits on one side of the ocean has led to the discovery of related deposits on the facing continent. More fundamental is the way plate movement directly affects the process of mineral concentration. For example, many important ores are produced by heat generated at plate margins. Large deposits of copper, iron, manganese, gold, zinc, and silver have recently been discovered in "hot spots" in the Red Sea—an area heated by convection currents now separating Africa from Saudi Arabia. Similar deposits may be found in other spreading centers such as the Labrador Sea between Greenland and Canada.

In the following chapters, as we consider the basic processes of the hydrologic and tectonic systems, we will attempt to cite a few examples of man's involvement with these systems, a subject commonly called environmental geology. In Chapter 18 we will cover this subject in greater depth.

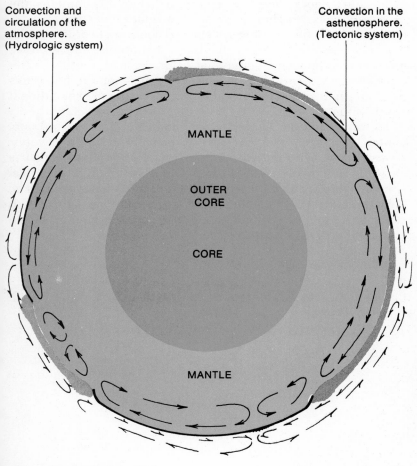

Convection and circulation of the atmosphere. (Hydrologic system)

Convection in the asthenosphere. (Tectonic system)

MANTLE

OUTER CORE

CORE

MANTLE

Figure 3.6 *Schematic diagram showing the two major dynamic systems of the earth which create and modify the surface features. The circulation of water in the hydrologic system is powered by energy from the sun and involves the processes of weathering, running water, ground water, glaciation, wind action, and shoreline processes. The energy for the tectonic system comes from within the earth, presumably from radioactive heat in the asthenosphere. Convection in the asthenosphere causes the lithosphere to break up into plates and move great distances. Earthquakes, vulcanism, mountain-building, sea-floor spreading, and continental drift are among the major results of the tectonic system.*

The Earth's Dynamic Systems

63

Summary

In this chapter, we have outlined the mechanics of the two major dynamic systems which continually modify the earth's surface (*figure 3.6*). The hydrologic system involves all possible paths of water movement, through the atmosphere and oceans, over the lands, and below the ground. The system operates as a gravity flow system with solar energy constantly lifting water from ocean basins where it is carried in the atmosphere to elevated land surfaces down which it flows back to the sea.

The tectonic system is an internal energy system; the upper mantle is in motion as a result of radiogenic heat inherited from the time of the earth's formation. Heat within the upper mantle produces convection. As it moves, it splits the rigid lithosphere and carries it laterally. New crust is created as molten mantle material moves into the fracture zone and cools. As the convection current descends, it carries with it the lithosphere which is ultimately consumed in the hotter mantle.

This is an irreversible or one-way system, in which energy inherited from the beginning of the earth is continually being dissipated into space. Ultimately, this energy will run out, just as the sun's energy must be consumed. When this happens, the earth will cease to change by its own internal system of heat, but, until then, the tectonic system will continue to create ocean basins, cause continental drift, build mountains, and produce new crust by volcanic activity.

As these systems operate, isostatic adjustment in the earth's crust continues to occur in an effort to establish gravitational equilibrium.

In order for the dynamic systems of the earth to operate, matter is changed back and forth from solid to liquid or vapor states. Rocks are created and destroyed, a process which is basically one involving the growth and destruction of minerals. A knowledge of the nature and origin of minerals and rocks is, therefore, fundamental, not only to understand the composition of the earth but also to understand the processes which operate upon and within the earth.

Additional Readings

Bates, D. R. 1964. The Planet Earth. New York: Pergamon Press.

Cailleux, A. 1968. Anatomy of the Earth. New York: McGraw-Hill Book Company.

4 Rocks

In all processes involved with the earth's dynamic systems, rocks are created and destroyed, matter changes from one state to another, and minerals grow, melt, dissolve, or are broken and modified by abrasion. As the earth's surface is weathered and eroded, minerals are created and destroyed. As sediments are deposited in the sea, minerals grow from solution, or mineral fragments are united into a coherent mass to form a rock. In volcanic activity, minerals grow from a liquid **melt.** Under great stress, elements are removed from the **crystal structure** of one mineral and recombine in the structure of another. As continents drift, rocks are created and destroyed through a variety of physical and chemical processes. Some knowledge of the major rock types and their constituent minerals is, therefore, essential in understanding earth dynamics.

A rock may be defined quite accurately as an aggregate of minerals which form the solid part of the earth but, if we go beyond this formal definition and consider how rocks are formed, we soon realize that rocks are products of geologic processes and, as such, constitute records of past events, documents of earth's history. Indeed, it is in careful observation and study of rocks that the facts, principles, and theories of geology are based.

Our major object in this chapter is to introduce to you the three major groups of rocks—**igneous, sedimentary,** and **metamorphic**—and to show some examples of how a rock is a record of some past event in the history of the earth. A secondary objective is to define and describe the most common rocks in the earth's crust as a necessary background to the study of the earth's processes. In order to accomplish this, we will first consider some of the more important characteristics of minerals, because minerals are the particles from which rocks are made.

Major Concepts

1. Minerals are natural inorganic solids in which atoms occur in specific proportions and are arranged in a definite structural framework.
2. Minerals grow by adding atoms to the crystal structure and dissolve or melt by removing atoms from the structure.
3. Minerals have definite physical properties as a result of their chemical composition and internal structure.
4. Minerals grow under specific conditions and can be used as indicators of past physical and chemical environments.
5. The most important rock-forming minerals are the feldspars, olivines, pyroxenes, amphiboles, quartz, clay minerals, and calcite.
6. The events recorded by rocks are preserved in (a) the mineral composition, (b) the texture, and (c) field relations of the rock body.
7. The most important rocks in the earth's crust and the history they record are as follows: (a) Basalt—volcanic eruptions and crustal spreading along the mid-oceanic ridge; (b) Granite—intrusions of **magma** associated with mountain-building; (c) Sandstone—sedimentary environments of relatively high energy such as beaches, river bars, dunes, etc.; (d) Shale—sedimentary environments of relatively low energy such as lagoons, flood plains, shallow offshore marine areas; (e) Limestone—mostly shallow-water environments; (f) Slates—low temperatures and pressure with minor recrystallization and mountain-building; (g) Gneiss and Schists—high temperature and pressure with considerable recrystallization and mountain-building.

The Nature of Minerals

Statement

To define the term "mineral" precisely is not easy but, with few exceptions (mercury and water), minerals are solid at surface temperatures and pressures and occur in the crystalline state. Therefore, they have a definite internal structure and definite chemical composition which varies only within certain limits. The difference between minerals arises from the kinds of atoms they contain and the way in which the atoms are arranged. The internal structure or pattern of atoms in a mineral repeats itself in three dimensions indefinitely. A crystal of a mineral may be so small that it cannot be seen even with a high-powered microscope, or it may be as much as several meters long.

If a mineral is allowed to grow in an unrestricted environment, crystals will have a perfect geometric shape: their faces will be parallel to the planes of the atomic framework of the crystalline structure. Each mineral has a characteristic form and, although the size and shape of the crystal may vary, similar pairs of crystal faces always meet at the same angle. This fact is known as the *Law of Constancy of Interfacial Angles*. With modern methods of x-ray diffraction, it is now possible to determine the exact geometric arrangement of atoms (or ions) within a crystal.

Discussion

Groups of atoms may exist in any of three states of aggregation: as a gas, a liquid, or a solid. Any given sample of matter may change from one of these states to another. In doing so it will assume completely different properties, but its chemical composition will remain the same. At the earth's surface temperature and pressure, water can change through all three states in a temperature range of only 100° C. The minerals which form the rocks of the earth are capable of the same change, but their transition from solid to liquid to gas usually occurs at higher temperatures and pressures than those normally found at the earth's surface.

The states of matter may be characterized by: (1) the degree of order in the arrangement of atoms or molecules, (2) the space between atoms, and (3) the degree of particle motion. Temperature and pressure are the two factors which determine the physical state of matter. Increasing temperature causes expansion; that is, the space between atoms enlarges so the same number of atoms occupy a greater volume of space. At "low" temperatures (less than 300° C) most substances tend to be solid, with the atoms held in more or less rigid positions. At "moderate" temperatures (300° to 1000° C) most substances melt to a liquid state in which the thermal vibrations of the atoms overcome bonding strength and the atoms can move past one another but are still in random contact rather than linked to a crystalline structure. At "high" temperatures (over 1000° C) much material vaporizes to a gaseous state in which the atoms or molecules separate and drift apart and only for brief instants make contact with each other. (With still higher temperatures, far beyond any on earth, electrons may be freed, even to the extent of atomic nucleic dissociation.)

Crystal Form. When a mineral crystallizes, the atoms arrange themselves in a definite three-dimensional framework. The resulting solid is a crystal which continues to grow as long as the proper environmental conditions are maintained and there is an adequate supply of ions.

Crystals show varying degrees of symmetry. For example, an imaginary plane passing through the crystal divides it into two identical halves. You can visualize a line through a crystal, such as that shown in the drawings, about which the crystal can be rotated 90°, bringing it into a position with identical appearance. Further rotation of 90° produces another position that appears the same. Other axes can be found passing through garnet, in which a rotation of 360° produces identical appearances two or three times. All perfect crystals of the same mineral, regardless of size or place of origin, have the same elements of symmetry—an additional expression of the uniformity of internal structure for each mineral.

X-Ray Studies of Crystals. Precise study of the internal structure of minerals was made possible by the discovery of x-rays. When a very thin beam of x-rays is passed through a mineral, it is diffracted or dispersed by the framework of atoms. The dispersed rays produce an orderly arrangement of dots on photographic film placed behind the crystal (*figure 4.1*). By measuring the relationship among the dots, the systematic orientation of planes of atoms within the crystal can be reduced to a mathematical formula. From these patterns detailed models of crystal structures can be constructed and analyzed. The x-ray instrument is now the most basic device for determining the internal structure of minerals and is used extensively for precise mineral identification and analysis.

The importance of internal structure in the definition of a

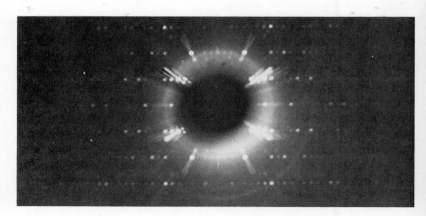

Figure 4.1 *X-ray diffraction patterns of a mineral.*

The Earth's Dynamic Systems

mineral may be illustrated by the minerals diamond and graphite. Both are composed of the element carbon so their chemical composition is identical. However, the internal structure is very different in the two substances. In diamond the structure is compact and the atoms are tightly bound together, but in graphite the structure consists of open sheets.

To a geologist, the crystal structure provides clues to how the rock formed and the environment in which it originated. However, to identify a mineral the geologist also makes use of other physical properties, and we shall discuss these in some detail.

Physical Properties of Minerals

Statement

Because a mineral has a definite chemical composition and an internal crystalline structure that varies only within definite limits, all specimens of a given mineral, regardless of when or where it was created, will have the same physical and chemical properties. This means that one piece of quartz, for example, will be as hard as any other piece of the same quartz mineral, it will have the same specific gravity, and it will break in the same manner.

The most significant and readily observable physical properties of minerals are **cleavage, hardness, specific gravity, color,** and **streak.**

Discussion

Cleavage. Cleavage is the tendency of a crystalline substance to split or break repeatedly along one plane or a set of parallel planes. This phenomenon results from variations in the strength of bonds between particles in the crystal lattice. Cleavage tends to be parallel to the planes of weak bonding and is commonly expressed by a series of incipient cracks in the mineral. On large crystals, cleavage is commonly expressed as a series of small parallel surfaces rather than a single large planar surface. The number and direction of cleavage planes is restricted by the structure of the mineral and varies from one to six. If bonds are especially weak, as in mica and halite, cleavage occurs with ease, and it is difficult to break the mineral in any direction other than along the cleavage plane. Cleavage may be poor or imperfect where the difference in bond strength is not great.

Some minerals, having no weak planes in their crystalline structures, lack cleavage and break along various types of fracture surfaces. Thus, the type of fracture can be useful in identifying some minerals. Quartz, for example, lacks cleavage and characteristically breaks along curved surfaces (**conchoidal fracture**) similar to the curved surface in chipped glass.

Hardness. Hardness is a property of a mineral which is

easily recognized and widely used in field identification of minerals. It is a measure of a mineral's resistance to abrasion. Over a century ago, Friedrich Mohs, a German mineralogist, assigned an arbitrary relative number to ten common minerals according to their hardness. Diamond, the hardest mineral known, was assigned the number 10. Softer minerals were ranked in descending order, with talc, the softest mineral, assigned the number 1. Mohs' hardness scale (*table 4.1*) provides a standard for testing minerals for preliminary identification in the field.

Specific Gravity. Specific gravity is the ratio between the weight of a given substance and the weight of an equal volume of water. For example, if a quart of molten lead were poured into a mold and allowed to cool, the solid lead when cooled would weigh a little over 11 times more than a quart of water. Thus, the specific gravity of lead is 11.

Specific gravity is one of the more precisely defined properties. It depends upon the kinds of atoms making up the mineral and the closeness of atoms packed into the crystal structure. Clearly, the more numerous and compact the atoms, the higher the specific gravity. Most common rock-forming minerals have a specific gravity from 2.65 (for quartz) to about 2.37 (for olivine).

Color. Color is one of the most obvious properties of a mineral but is *not* diagnostic because most minerals are found in various hues, depending on such things as variations in composition, inclusions, and impurities. Quartz, for example, ranges through a spectrum of colorless clear crystals to purple, red, white, and jet black.

Streak. When a mineral is powdered, it usually exhibits a much more diagnostic color than when it is in large pieces. The color of the powdered mineral is referred to as streak. In the laboratory, streak is obtained by vigorously rubbing the mineral

Table 4.1 Mohs' Hardness Scale

Hardness	Mineral	Test
1	Talc	Fingernail
2	Gypsum	
3	Calcite	Copper coin
4	Fluorite	Knife blade or
5	Apatite	glass plate
6	Orthoclase feldspar	
7	Quartz	Steel file
8	Topaz	
9	Corundum	
10	Diamond	

The Earth's Dynamic Systems

on an unglazed porcelain plate. The streak of a mineral may be different from the color seen in the hand specimen so one should never anticipate the color of streak by visual examination of a mineral fragment. Furthermore, the hardness of porcelain is about that of glass, and minerals of greater hardness will scratch the plate. Most minerals with a nonmetallic luster have a white or pastel streak, and for this reason streak is not very useful in distinguishing nonmetallic minerals.

The Growth and Destruction of Minerals ρ 58

Statement

One of the most important characteristics of a mineral is its ability to grow. In order for growth to occur, atoms must have a chemical affinity for each other and must be able to migrate so they can arrange themselves in the structural framework. Crystal growth is a time-dependent process; that is, time is required for the atoms to move to the crystal face and become "locked in" on the structure. Growth is essentially a process of accretion, the addition of new atoms on the crystal face.

Mineral crystals grow from molten solids (melts), from solutions in which ions are dissolved, and from sublimation (vapors carrying atoms to the crystal faces). A crystal structure is never completed and can be extended indefinitely. Therefore, a single crystal, containing all physical and chemical properties of the mineral, may be so small it cannot be identified with a high-power microscope, or it may be more than several meters in length. Most crystals in rocks of the crust are small, less than a centimeter in length, and a large percent are microscopic. Many people are, therefore, unaware that rocks are in fact aggregates of minerals.

Crystals may dissolve or melt in a manner which is the reverse of the growth process, that is, by removal of the atoms from the crystal structure. The most significant factors controlling the growth and destruction of minerals are heat, pressure, and concentration of ions.

Discussion

Crystal Growth. Growth occurs by addition of atoms to the crystal face because the outer layers of atoms on a crystal are capable of forming additional bonds. The structural framework of a crystal is incomplete on the margins and can be extended indefinitely. An environment suitable for crystal growth includes: (1) proper concentration of the kinds of atoms required for a particular mineral, such as elements in a liquid or gaseous state, and (2) the proper temperature and pressure. The atoms in solution are attracted to and interlock with the bond-forming atoms on the surface of the crystal.

The time-lapse pictures in *figure 4.2* show the manner in which crystals grow in an unrestricted environment. Although

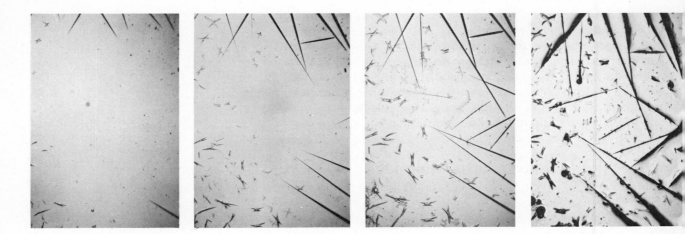

Figure 4.2 *Time-lapse photographs showing the growth of crystals.*

the size increases, the form of the initial crystal remains the same. New atoms are added to the outer edge of a crystal parallel to the planes of atoms in the basic structure, thus producing a perfectly symmetrical crystal form.

However, some crystal faces grow faster than others and, in an environment where spaces are restricted, a crystal face may not grow symmetrically. Where a crystal face encounters a barrier, it ceases to grow. This process is illustrated in *figure 4.3.* The resulting crystal will thus assume the shape of the confining space, and well-developed crystal faces will not form. In such a case the external form of the crystal may be practically any shape, but the internal structure of the crystal is in no way modified because growth of the crystal face is restricted. The internal structure remains the same, the composition of the mineral is unaffected, and there is no change in the physical and chemical properties. The only modification is in the relative size of the crystal faces.

The process of crystal growth in restricted space is especially important in rock-forming minerals, for in a melt or a solution many crystals grow at the same time and must soon compete for space. As a result, in the later stages of growth, crystals in rocks commonly lack unique, well-defined crystal

Figure 4.3 *Diagrams showing growth of crystals in a restricted space. (a) Where growth is unrestricted each crystal face grows with equal facility and the perfect crystal is enlarged by growth on each crystal face. (b) Restricted space may eliminate growth on certain crystal faces (a,b,c,d) whereas growth on faces w,x,y, and z is eliminated as available space is used. The final shape of the mineral is determined by the geometry of the space available for growth. In the example, a hexagonal crystal developed the form of a square. The internal structure of the crystal, however, remains the same regardless of the growth space.*

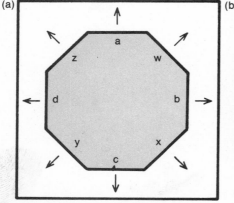
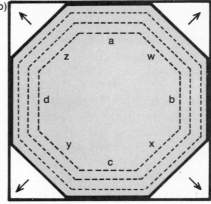
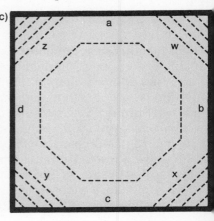

The Earth's Dynamic Systems

faces and typically interlock with adjacent crystals to form a strong, coherent mass (*figure 4.4*).

Most crystals are rather small, measuring from a few tenths of a millimeter to several centimeters in diameter, some so small that they can be seen only with an electron microscope capable of enlargements of one million times actual size. On the other hand, in an unrestricted environment, crystals may grow to enormous sizes (*figure 4.5*).

Destruction of Crystals. Crystals may dissolve or melt by removal of the atoms from the crystal structure, permitting a return to a fluid state (liquid or gas). The heat which causes a crystal to melt increases atomic vibrations enough to break the bonds holding the atom to the crystal structure. Under such conditions, the atoms are more stable in the fluid state than in a solid. The breakdown or dissolution of a crystal begins at the surface and moves inward.

Growth and destruction of crystals are of paramount importance in geologic processes because rocks originate and change as minerals are created and destroyed. Crystal growth occurs in volcanic processes when lava cools and solidifies. It occurs in the sea where minerals grow from solution to form limestone, salts, and other mineral deposits. Destruction and growth of crystals also occur deep within the crust, where heat

Figure 4.4 *Sketch showing how crystals may grow to form an interlocking texture.*

Figure 4.5 *Large crystals of gypsum.*

and pressure cause crystal structures to break down and new minerals to form in their place, and at the surface where minerals react chemically with elements in the atmosphere.

Ionic Substitution

Statement

An important characteristic of some minerals is that different kinds of **ions** may be substituted for others in the three-dimensional framework of the mineral structure. The suitability of substitute ions in the crystal structure is determined by the size and electrical charge of the ion in question. The substitution of one type of ion for another results in a chemical change in the mineral without a change in the crystal structure. There are, however, definite limits within which the substitution can occur.

Discussion

The substitution of ions is in a general way analogous to the substitution of a plastic brick for a clay brick of equal size in a wall. Because the size of the substitute brick is the same as the original, the structure of the wall is not affected, but there is a change in composition.

Ionic substitution occurs widely in the common rock-forming minerals and is responsible for mineral groups which have the same structure but variable composition. For example, in the olivine group $(Mg, Fe)_2SiO_4$, a common green, glassy mineral found in black volcanic rocks, ions of Fe and Mg may substitute freely one for another. The total number of Fe and Mg atoms is constant relative to the number of Si and O atoms in the olivine, but the ratio of Fe to Mg may be different in different samples. The common minerals such as feldspar, pyroxene, amphibole, and mica each constitute a family (group of related minerals) in which atomic substitution produces a range of chemical composition.

Rock-forming Minerals

Statement

Rock-forming minerals rarely have well-developed crystal faces because: (1) they grow or crystallize from a melt (magma) or from solution (sea water) and vigorously compete for space, (2) they are abraded as they are transported as sediment, or (3) they are deformed under intense pressure. In addition, most rock-forming mineral groups have variable composition because of ionic substitution in the crystal structure. Their color, hardness, and other physical properties thus may be variable.

It is important for you to become familiar with each of the major rock-forming mineral groups (feldspars, micas, quartz, olivine, pyroxene, amphibole, clays, calcite, dolomite, halite, and gypsum) and to know something about their mode of origin and the environment in which they form, and their genetic significance. You will find the following summary of each mineral group to be much more significant if you will examine a specimen of the rock containing the mineral while studying the written description. Constant reference to the illustrations is encouraged.

Discussion

Let us begin by carefully examining the minerals which make up a granite. *Figure 4.6a* shows a polished surface of a typical granite rock. You can readily see that the rock is composed of a myriad of mineral grains having different size, shape, and color (*figures 4.6b* and *c*). Although the minerals interlock to form a tight, coherent mass, each one has definite distinguishing properties or characteristics.

The Feldspars. A large part of the granite specimen is composed of a pink porcelainlike mineral, having a rectangular shape, and a milky white porcelainlike mineral, which is somewhat smaller but similarly shaped. These are feldspars (German, "field" crystal), the most abundant mineral group in the earth's crust. The feldspars have good cleavage in two directions, a porcelain luster, and a hardness of about 6 on Mohs' hardness scale.

The crystal structure of the feldspars permits considerable ionic substitution, giving rise to two major types: **potassium feldspar** and **plagioclase feldspar**.

Potassium feldspar ($KAlSi_3O_8$), most commonly pink in rocks, is shaded light brown in *figure 4.6*. Plagioclase (stippled gray) permits complete substitution of Na for Ca in the crystal structure, giving rise to a compositional range from $NaAlSi_3O_8$ to $CaAl_2Si_2O_8$. White plagioclase in granite is rich in sodium. Feldspars are common in most igneous rocks, in many metamorphic rocks, and in some sandstones.

The Micas. The tiny, black, shiny grains with a tabular shape in *figure 4.6* are mica. This is the name of a group of minerals readily recognized by their perfect one-directional cleavage which permits the mineral to be easily broken into thin elastic flakes. The mineral is a complex silicate with a sheet structure that is responsible for its perfect cleavage. Two common varieties occur in rocks: biotite, which is black mica high in magnesium and iron, and muscovite, which is white and lacks iron and magnesium. Mica is abundant in granites and many metamorphic rocks and is also a significant constituent in many sandstones.

Quartz. The glassy grains with irregular shapes in *figure 4.6* are quartz. Because quartz is the last mineral to form in a granite, it lacks well-developed crystal faces. The mineral simply fills the space between early-formed feldspars and mica. Quartz is abundant in all three major rock types. It is the second

Figure 4.6 Mineral grains in granite
(a) Polished surface of granite. The minerals
have grown into a tight, interlocking texture.
(b) Sketch emphasizing grains of individual
minerals. Potassium feldspar is shaded light
brown, plagioclase is stippled gray, quartz
is white, and biotite is black. (c) "Exploded"
diagram of figure b, showing size and shape
of each mineral grain.

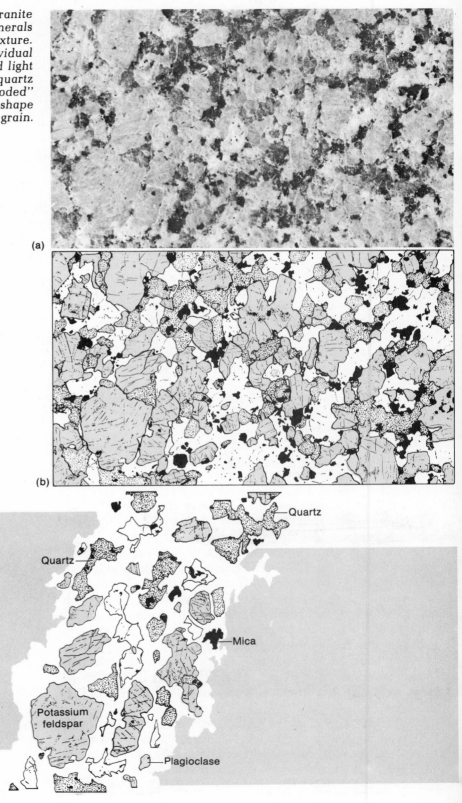

(a)

(b)

Quartz

Quartz

Mica

Potassium
feldspar

Plagioclase

(c)

most abundant mineral in the continental crust. It has the simple composition SiO_2 and is distinguished by its hardness (7), conchoidal fracture, and glassy luster. Pure quartz crystals are colorless, but slight impurities may produce a variety of colors. Where crystals are able to grow freely, they crystallize in a six-sided pyramid, but well-formed crystals are rarely found in rock aggregates. In sandstone, the mineral is abraded into rounded sand grains. Quartz is stable both mechanically (it is very hard and lacks cleavage) and chemically (it will not readily react with elements at or near the earth's surface). Therefore, it is a difficult mineral to alter or destroy.

Ferromagnesian Minerals. In addition to the minerals in a

(a)

Figure 4.7 *Mineral grains in basalt. (a) Hand specimen. (b) Sketch showing microscopic view of texture and mineral grains. (c) "Exploded" diagram showing size and shape of individual grains.*

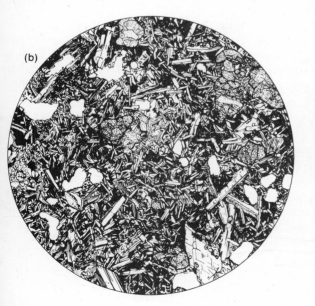

(b)

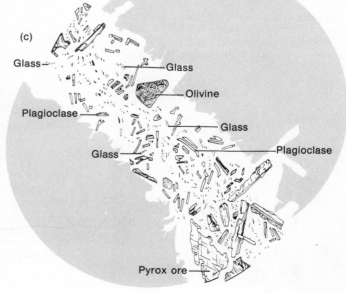

(c)

Glass

Glass

Olivine

Plagioclase

Glass

Glass

Plagioclase

Pyrox ore

Figure 4.8 *Photograph showing crystals of dark-colored amphibole in a granitic rock.*

Figure 4.9 *A clay mineral as seen under the electron microscope magnified 5000 times.*

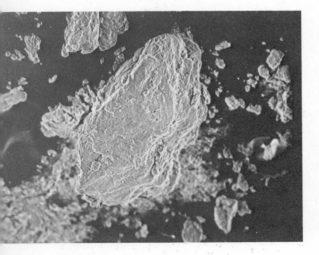

granite described above, there are a number of silicate minerals referred to as **ferromagnesian minerals** because they contain appreciable amounts of iron and magnesium. These minerals are generally dark green to black in color and have a high specific gravity. Biotite is classified in this general group, together with the olivines, pyroxenes, and amphiboles. Biotite is common in granite, but the others are rare or absent. The ferromagnesian minerals, however, are common in basalt (*figure 4.7*). The plagioclase and most other minerals are microscopic in size and cannot be seen in the hand specimen, but under the microscope small crystals of plagioclase and several ferromagnesian minerals are apparent.

Olivine. The only mineral clearly visible in the hand specimen in *figure 4.7* is the green glassy mineral called olivine. The olivine family is a group of silicates in which iron and magnesium substitute in the crystal structure. The composition is expressed as $(Mg,Fe)_2SiO_4$. This hard mineral is characterized by an olive-green color and glassy luster and, in rock aggregates, rarely forms crystals larger than a millimeter in diameter. As a high-temperature mineral, olivine is common in basalt (volcanic rock) and is probably a major constituent of the material beneath the crust. It is also a common mineral in meteorites—one reason for believing meteorites are fragments of other planets.

Pyroxene. Pyroxene and olivine commonly occur in the same types of rocks. In *figure 4.7* pyroxene occurs as microscopic crystals, but some samples contain larger grains of this mineral which are typically dark green to black. Pyroxenes are relatively high-temperature minerals found in igneous and metamorphic rocks.

Amphiboles. Figure 4.8 shows a rock similar to a granite, but it has less quartz and contains appreciable amounts of the mineral amphibole. The amphibole minerals have much in common with the pyroxenes. Their compositions would be identical except that amphibole contains hydroxyl ions (OH^-), whereas pyroxene does not. The structure of the amphiboles, however, differs from pyroxene and produces elongate crystals which cleave perfectly in two directions, not at right angles. The color of amphibole ranges from green to black. This mineral is common in rocks closely related to granite and may be more abundant in that rock than biotite. It is especially common in some metamorphic rocks known as amphibolite. Hornblende is the most common variety of amphibole.

Clay Minerals. We may encounter the clay minerals more frequently than other minerals in everyday experience because they form at the earth's surface through the interaction of air and water with various silicate minerals mentioned above. Like the micas, the clay minerals are sheet silicates, but the crystals are microscopic and can usually be detected only with an electron microscope (*figure 4.9*). More than a dozen clay minerals can be distinguished on the basis of their crystal structure and variations in composition.

The Earth's Dynamic Systems

Calcite. Calcite is calcium carbonate, the principal mineral in limestone. It may be precipitated directly from sea water, removed from sea water by organisms and used to make their shells, or it may be dissolved by ground water and re-precipitated as a new crystal in caves and fractures in rock. The color is usually white or transparent, but aggregates of calcite crystals which form limestone are mixed with various impurities, giving them gray or brown hues. Calcite, a common mineral at the earth's surface, is easy to identify. It is soft enough to scratch with a knife (hardness of 3) and effervesces in hydrochloric acid. It has perfect cleavage in three directions but not at right angles, so cleaved fragments are in the form of a rhombohedron (*figure 4.10*). In addition to being the major constituent of limestone, calcite is the major mineral in the metamorphic rock marble.

Dolomite. Dolomite is a carbonate of calcium and magnesium. Large crystals form rhombohedrons, but most dolomite forms granular masses of small crystals. Dolomite is widespread in sedimentary rocks, having been altered from the original calcite through the activity of solutions of magnesium carbonate in sea water or ground water. Dolomite can be distinguished from calcite in that it will effervesce in dilute hydrochloric acid only in a powdered form.

Halite and Gypsum. Halite and gypsum are two of the most common minerals formed by evaporation of sea water or saline lakes. Halite, common salt (NaCl), is easily identified by its taste, and has one of the simplest of all crystal structures: the

[handwritten margin notes: clean shallow. Precipitates warm tropical water florida keys Calcite = Limestone P.65]

[handwritten margin notes: mica is found in Igneous Rocks]

Figure 4.10 *Cleavage in the mineral calcite.*

sodium and chlorine ions are united in a cubic structure. All physical properties are related to this structure. Halite crystals cleave in three directions at right angles to form a cubic or rectangular fragment. Salt, of course, is very soluble and readily dissolves in water.

Gypsum is composed of calcium sulfate and water and forms crystals that are generally clear white, with a glassy or silky luster. It is a very soft mineral and can easily be scratched with a fingernail. It cleaves perfectly in one direction to form thin nonelastic plates. The mineral occurs as single crystals, as aggregates of crystals in compact masses, or in a fibrous form known as satinspar.

Mineral Resources

Statement

Many minerals other than the common rock-forming minerals described in the previous section are of prime importance to man and have played a prominent role in human affairs. They constitute some of the most important resources upon which our civilization depends; yet they occur in such minute quantities compared with the rock-forming minerals that one might wonder how and where they are concentrated in mineable quantities. Let us consider briefly the percentages of some of the important metals and energy-producing minerals in the earth's crust. As *table 4.2* notes, oxygen and silicon account for 74% by weight of the crust, whereas six additional elements, aluminum, iron, calcium, sodium, potassium, and magnesium account for an additional 24%. All other elements combined constitute less than 2% of the earth's crust. Copper, nickel, lead, zinc, and uranium all occur in such small quantities that their concentrations are measured in parts per million and, in most cases, a very few parts per million. Understanding how and where these important elements are concentrated are problems which geologists have to solve. Concentration of mineral resources is not a haphazard or random process but the result of geologic processes operating in the earth's system. Essentially, every geologic process, including **weathering**, stream action, wind and waves, as well as internal processes such as volcanic activity, metamorphism, and earth movements, plays a part in the concentration of minerals.

Discussion

We will discuss mineral resources and their relation to human affairs in some detail in subsequent chapters, but in order to help you appreciate how rare minerals are concentrated into economic deposits, we will simply present several examples and consider details later.

Weathering processes concentrate certain minerals by decomposing the more soluble minerals in rocks, leaving the less

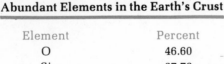

Table 4.2. Concentration of the Most Abundant Elements in the Earth's Crust

Element	Percent
O	46.60
Si	27.72
Al	8.13
Fe	5.00
Ca	3.63
Na	2.83
K	2.59
Mg	2.09
Ti	0.44
H	0.14
P	0.12
Mn	0.10
S	0.05
C	0.03

After Mason, p. 48

The Earth's Dynamic Systems

soluble material as residue. By this process, minerals such as aluminum, iron, copper, silver, and mercury, which occur in only a few parts per million, are concentrated as residues when the more soluble material is decomposed and removed by weathering and erosion.

Geologic processes in rivers and along beaches are effective in concentrating heavy or dense minerals because the more abundant, lighter fragments of the same size are transported with greater ease. The heavy minerals tend to accumulate where the water velocity decreases (at the inside of meander bends in a river or along sheltered segments of a beach). The minerals in these concentrations, referred to as **placer** deposits, must be resistant to weathering and have a higher specific gravity than quartz. Gold, diamonds, and platinum are examples of placer deposits which are mined not only from modern rivers and beaches but also from ancient stream channels.

Evaporation concentrates minerals such as salt, potassium, magnesium, and gypsum which are mined from the beds of salt lakes and the floors of ancient **evaporite** basins. Important salt deposits are found in Michigan, New York, and Kansas, and the evaporation process is used to concentrate salts from modern salt lakes such as the Dead Sea and the Great Salt Lake.

In molten rock material, rare elements such as gold, silver, copper, lead, and zinc occur as only a very few parts per million. These metals are concentrated during later phases of crystallization of the melt because the size and electrical charge of their atoms do not permit them to fit well into the structures of the silicate rock-forming minerals. They thus remain in the melt while O, Si, Ca, Na, K, Fe, and Al unite to form the common silicate minerals. The fluids containing these rare elements are then injected into fractures of the surrounding rock and crystallize to form veins of ore in which gold, silver, copper, lead, etc., constitute a relatively high percentage of the rock. Careful geologic mapping and study of igneous rock association can be used to locate these rich veins.

Metamorphic processes also tend to concentrate mineral deposits, especially in the zone of metamorphism surrounding rocks which cooled from a melt underground. The fluids of the molten rock react chemically with the surrounding rock and replace it with concentrations of important minerals. Limestone replacement is especially important because it results in many deposits of lead, silver, zinc, and some iron.

An understanding of how geologic processes concentrate rare minerals is important not only in exploration for new deposits but in appreciating the limits to our resources and the time required to replenish them. Proper exploration and utilization of our natural resources is one of the most pressing problems facing man and may well be the limiting factor to the growth of human society as we know it. We shall have more to say about this in Chapter 18.

Igneous Rocks

are Crystalline = interlocking frame work.

Statement

Igneous rocks (Latin, *ignis*, "fire") are formed by the cooling and crystallization of liquid rock material called magma. This results in a distinctive texture in which the mineral grains generally form a tight interlocking network of crystals which show no signs of abrasion, such as is common in many sedimentary rocks, nor evidence of stress, which characterizes metamorphic rocks. The composition of igneous rocks is largely the silicate minerals: olivine, pyroxene, amphibole, feldspar, mica, and quartz. These, of course, are minerals which crystallize at relatively high temperatures ranging from 700° to 1200° C. Low-temperature minerals are absent.

The only igneous processes that man can observe directly are volcanic extrusions where magma works its way to the surface and erupts from vents or fissures. On the surface, the lava cools rapidly and forms a rock characterized by small crystals. This is called **extrusive rock**. Basalt is the most abundant extrusive rock and is typically black, dense, and hard and is composed of microscopic interlocking crystals of plagioclase feldspar, pyroxene, and amphibole. It forms from extrusions of magma along the rift zones, where crustal plates split and drift apart, and constitutes the bedrock for all of the oceanic basins.

Less spectacular, but just as important, is the great volume of magma which never surfaces. Magma originates by partial melting in the lower crust, commonly in deformed mountain belts and rift zones, and may remain trapped deep in the crust where it slowly cools and solidifies. The solidified magma is referred to as **intrusive** or **plutonic rock**. Granite, composed of quartz, potassium feldspar, and sodium plagioclase, is the most common variety of plutonic rock. It is exposed as large masses in the cores of many mountain ranges only after erosion has removed the overlying rock beneath which the magma cooled. (See Appendix for classification of igneous rocks.)

Silica in Magma (ranges 37 – 75%)

Discussion

Extrusive rocks. Volcanic eruptions are among the most spectacular of all geologic processes observed by man, and the extrusion of lava to form new rock has been witnessed many times. As lava cools, the resulting rock retains many features indicative of its igneous origin, permitting the geologist to distinguish volcanic rocks from other major rock types. To emphasize some of the unique characteristics of volcanic rocks, let us consider some of the features of recent flows in Hawaii.

Upon extrusion, the lava flows downslope and may fill a valley or other topographic depression. Gas plays an important part in the behavior of the flow and the nature of the cooled rock. If the gas content is high, the lava will flow in a manner similar to molten steel. A thin glassy crust commonly forms on the surface and develops flow marks resembling coiled rope

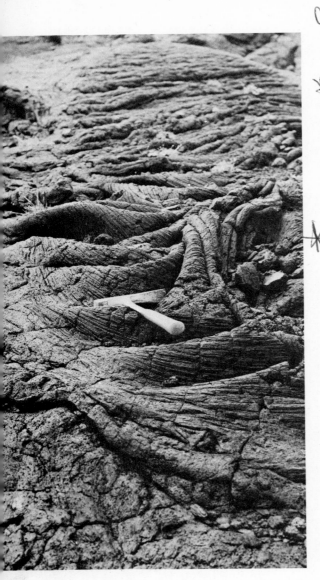

Figure 4.11 *The surface of a recent pahoehoe lava flow on the flanks of Kilauea, Hawaii, showing the ropy flow structures. Pahoehoe flows are produced from fluid lava and are characteristically very thin.*

P no

Igneous rocks record thermal history of Earth.
• Origin associated with Tectonic Plates.
Sea floor Spreading,
Mt. Building - Continental Crust formation.

(figure 4.11). These flows are called **pahoehoe**, a term derived from Hawaii where this type of flow is common. If the gas content is low, the flow moves much slower and may attain a thickness of several tens of meters. As the flow moves (commonly less than a meter per hour), the hardened crust is broken into a jumbled mass of angular blocks and cinders (figure 4.12). The Hawaiian term for this type of flow is **aa**. Gas within the flow migrates to the top and may be trapped beneath the hardened crust. As the lava cools, the gas bubbles form small voids called **vesicles**, causing the top of the flow to be very porous (figure 4.13). With the gas bubbles migrating to the top, the interior of the flow is commonly massive and nonvesicular. As the lava cools, it contracts and forms a system of polygonal cracks called **columnar joints** which are similar to mud cracks (figure 4.14). Columnar joints develop perpendicular to the cooling surface and are generally vertical. In some flows, the fluid interior may break through the crust and flow out, leaving a long lava tube or tunnel. More commonly, pressure from the fluid interior causes

[handwritten margin notes:]
× magma = parent rock material.
Under neath surface.
magma = silicate slush.
mixture of solid, liquid
& Gas.
Chemical Elements = Silicon, Oxygen
Iron, Magn, Potassium, Alum.
Amount of Silica in magma
determines Explosive behavior
Combined with water.

Figure 4.12 The surface of an aa flow at Craters of the Moon, Idaho, showing the jumbled mass of angular blocks formed when the congealed crust is broken as the flow slowly moves. Aa flows are viscous and much thicker than pahoehoe flows.

[handwritten note:] Water in form of Steam Cause Explosiveness / High Silica Magma + High Water Mag. Explosive.

Figure 4.13 *Vesicles formed by gas bubbles trapped near the top of a basalt flow.*

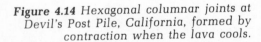

Figure 4.14 *Hexagonal columnar joints at Devil's Post Pile, California, formed by contraction when the lava cools.*

the crust of the flow to arch up in a pressure ridge which is characterized by a central crack through which gas and lava escape (*figure 4.15*).

If we look closely at the texture of a lava flow, we see that the rock is typically hard and dense, with most crystals too small to be seen with the naked eye. Under the microscope, the rock is seen to consist of small interlocking crystals of high-temperature minerals such as feldspar, pyroxene, and olivine.

In summary, the distinguishing features of rock formed by cooling of molten lava are (1) aa or pahoehoe surface features, (2) vesicles near the top of the rock unit, (3) columnar joints, (4) microscopic interlocking crystals, and (5) high-temperature minerals.

Intrusive or Plutonic Rocks. Rocks which are formed by magma cooling beneath the surface are also characterized by a texture of interlocking crystals of high-temperature minerals but, because cooling is very slow, the crystals grow much larger (several millimeters to several centimeters in length). Evidence that the rock was originally in a fluid state is found in the relationships of the rocks near the margins of the intrusive body. The magma cools rapidly near the contact with the surrounding rock so the grains along the contact are smaller than those in the interior of the intrusive. In addition, tongues and stringers of magma may be injected into fractures in the surrounding rock and cool to form veins of igneous rock. In places, fragments of the surrounding rock may break off and be included in the plutonic mass, a feature difficult to explain if the magma were not liquid at one time. As final evidence, the adjacent rock is typically altered and baked by heat. The minerals are partly or completely recrystallized and, in places, new minerals are added to the surrounding rock. Indeed, many concentrations of ore are located along contacts with intrusive bodies.

Field Relations of Igneous Rocks. The size and shape of igneous rock bodies are usually quite different from those of sedimentary or metamorphic rocks. Some of the most familiar signs of ancient igneous activity are the relatively small bodies of igneous rock which collect at the shallow depths within the crust. As magma migrates upward, it exerts considerable pressure upon the surrounding rock and may squeeze into fractures where it cools rather rapidly and forms a tabular body of igneous rock called a **dike** (*figure 4.16*). Dikes cut across sedimentary layers and are therefore said to be discordant. Some dikes radiate from ancient volcanic necks or fill fissures associated with fissure eruptions. Others simply fill fractures at depths where magma never reached the surface.

If selected bedding planes offer zones of least resistance, the magma may be injected between beds, forming a **sill**, an intrusive body parallel to the stratification. Sills may form as local offshoots from dikes, or they may be connected directly to larger intrusive bodies. Characteristically, they are formed from basaltic magma which is highly fluid and capable of being

The Earth's Dynamic System

squeezed between layers of older rock. The more viscous granitic magma rarely forms sills.

When viscous magma is injected between layers of strata, it tends to arch up the overlying strata to form a dome. The resulting intrusive body is lens-shaped, with a flat floor and an arched roof. These intrusions, called **laccoliths** (figure 4.17), are not exceedingly common but may occur in blisterlike groups in areas of weak flat-lying strata such as shale. They may be several kilometers in diameter and thousands of meters thick. Rocks in laccoliths are typically **porphyritic**.

The most significant intrusives are large bodies of magma which slowly cool thousands of meters below the surface; these are called **batholiths**. They are composed mostly of granitic rock and constitute the largest rock bodies in the earth's crust. For example, the Idaho batholith is a huge body of granite exposed over an area of nearly 41,000 km² (16,000 mi²); the British Columbia batholith to the north is over 2000 km (1200 mi) long and 290 km (180 mi) wide and at least 3000 m (9800 ft) thick. The true three-dimensional form of batholiths, however, is difficult to determine because of the uncertainty regarding its shape and extension at depth. Evidence showing the layered nature of the crust upon the mantle indicates that batholiths are limited in depth to the thickness of the crust and do not extend down into the mantle. Thus, they are less than 60 km (37 mi) thick. The nature of the base or floor of a batholith, however, remains a

Figure 4.15 *Pressure ridge arched up by trapped gas in a recent flow in southern Idaho.*

Figure 4.16 *Basalt dike cutting across horizontal sediments at Hance Rapids in the Grand Canyon.*

(a)

Figure 4.17 Laccoliths. (a) The Henry
Mountain laccolith in the Colorado
Plateau. Note the light-colored, tilted, and
eroded sedimentary strata along the flanks of
the intrusion which forms the central mass
of the mountain. (b) Schematic diagram
showing the shape of a typical laccolith and
its relation to the sedimentary rocks which it
intrudes.

(b)

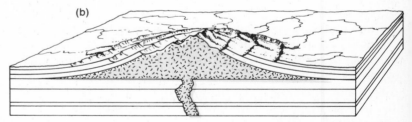

matter of conjecture. Indirect evidence such as gravity measurements suggests that batholiths may increase in size at depth for
some distance and then possibly taper off at greater depth,
somewhat like the root of a tooth. The diagrams in figure 4.18
give a rough idea of the geometric form of some of the better-
known batholiths in North America.

Igneous Rocks and Plate Tectonics. The generation of
magma and the origin of igneous rocks are integral parts of the
tectonic system. Two types of magma are believed to be generated along the plate margins: (1) basaltic magma along the
spreading centers where the plates move apart and (2) granitic
magma along the subduction zone where the crust descends into
the mantle. Basaltic magma is formed where convection in the

The Earth's Dynamic Systems

upper mantle brings hot mantle material close to the surface beneath the mid-oceanic ridge. As the plates move apart, basaltic magma is intruded as dikes or extruded as lava which may build up large cones. Iceland is the prime example of an area where extrusives along the mid-oceanic ridge have built a large accumulation of basalt which rises above sea level. Where continents are split apart by moving plates, floods of basalt typically accumulate along the rift. The basalt of the Columbia River Plateau, the Ethiopian Plateau, and India's Deccan Plateau are good examples.

Granitic magma, which is richer in silica than basaltic magma, is generated where two plates collide and the oceanic crust moves back down into the mantle. As it moves, the friction generates heat which, when added to the heat from the asthenosphere, is sufficient to melt some of the minerals in the basaltic crust. The first minerals to melt are the lower-temperature silica-rich minerals so that the magma generated by partial melting of basalt is granitic in composition (rich in silica). The granitic magma, being more viscous, moves slowly upward and intrudes into the deformed rocks of the folded mountain ranges to form batholiths.

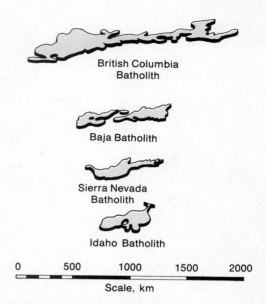

Figure 4.18 *Diagrams showing the relative size and shape of several well-known batholiths in North America.*

Sedimentary Rocks

Statement

Sedimentary rocks form at the earth's surface through the interaction of the atmosphere and water with preexisting rocks. This process can be observed and studied in great detail in streams, deltas, beaches, and shallow seas. Therefore, studies of a rock type forming at the present time can substantiate the interpretation of ancient sedimentary rock (see Chapter 14).

Solid rock is constantly decomposed and broken down by weathering, and the resulting material is transported continuously by running water, ground water, wind, and glaciers. Most of the sediment is ultimately deposited in the sea, but some accumulates on land—in river channels, flood plains, lakes, and deserts.

During transportation, the dissolved and disintegrated rock material is differentiated and sorted on the basis of size and composition. Large rock particles are deposited as gravels; fine-grained material (mostly quartz) accumulates as sand; and clay minerals settle out as mud. These sediments are ultimately consolidated into solid rock to form conglomerate, sandstone, and shale. Dissolved mineral matter may remain in solution for long periods of time and ultimately precipitate as limestone, salt, or other chemical deposits. (See Appendix for classification of sedimentary rocks.)

The features of a sedimentary rock which indicate the most about its history are texture, composition, internal structures, and fossils. Fossils commonly reveal the most detail about the

Figure 4.19 *The sequence of sedimentary rock formations exposed in the Grand Canyon. Each major rock unit erodes into a distinctive landform. The resistant formations such as sandstone and limestone erode into vertical cliffs. Nonresistant rocks such as shale erode into slopes.*

environment in which the sediment formed and furnish a remarkable record of life on earth. Sedimentary rocks record the physical events in the history of the earth's surface, including such things as erosion of the continents, uplift and mountain-building, changes in climate, and the constantly changing patterns of land and sea.

Discussion

Sedimentary rocks are probably more familiar to most people than the other major rock types because they cover approximately three-fourths of the surface of the continents and, therefore, form most of the landscape. However, few people are aware of the nature and extent of sedimentary rock bodies. Let us briefly consider the Grand Canyon in which so many features of sedimentary rocks are well exposed (*figure 4.19*).

The most obvious characteristic of the sedimentary rocks in the Grand Canyon is that they occur in distinct layers, many of

The Earth's Dynamic Systems

Figure 4.20 Diagram showing how the rock units in figure 4.19 would appear in cross section.

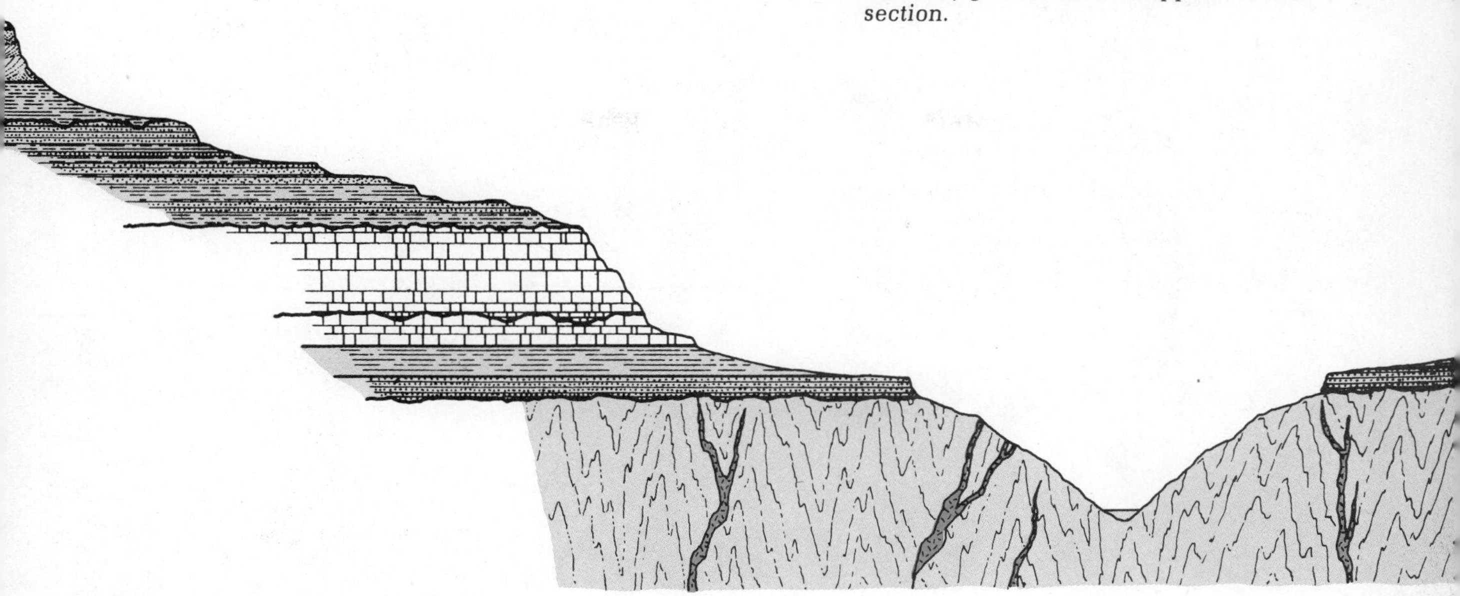

which are more than 100 m thick. The resistant rock types form cliffs, whereas the nonresistant rocks erode into gentle slopes. From *figure 4.19* you should be able to recognize the cross section of the rock formations shown in the diagram in *figure 4.20*.

A high-altitude aerial photograph of the canyon (*figure 4.21*) shows that the major layers can be traced across much of northern Arizona and cover an area of more than 256,000 km² (100,000 mi²).

A closer view of sedimentary rocks shows that their texture, composition, and internal structure is quite different from the other major rock types. In *figure 4.22* we see that the major layers actually consist of smaller units separated by bedding planes; these bedding planes are marked by some change in composition, grain size, color, or other physical features. Fossil animals or plants are common in most of the units and may be preserved in great detail (*figure 4.23*). The texture of most sedimentary rocks consists of mineral grains or rock fragments, which show evidence of abrasion (*figure 4.24*), or interlocking grains of the mineral calcite. In addition, many layers show **ripple marks** (*figure 4.25*), **mud cracks** (*figure 4.26*), and other evidence of water deposition preserved on the bedding planes. All of these features support the conclusion that sedimentary rocks form at the earth's surface in environments similar to present-day deltas, streams, beaches, tidal flats, lagoons, and shallow seas.

Field Relations of Sedimentary Rocks. The primary structure of sedimentary rocks, those features which formed at the time of deposition, reflects to a considerable degree the nature

Figure 4.21 High-altitude photograph of the Grand Canyon. The uppermost cliff-forming unit, the Kaibab limestone, forms the bedrock for the entire plateau shown in this view.

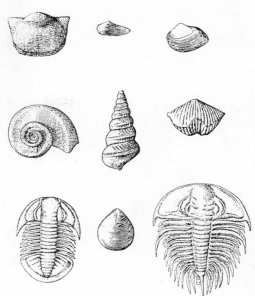

of the sedimentary environment and the conditions under which the sediment was deposited. **Stratification** is probably the most significant sedimentary structure; it forms in response to changes occurring during sedimentation and provides important clues as to changes which have occurred in the past. Close examination of rocks shows that stratification, or layering, occurs on many different scales. In large exposure, the larger scale of stratification is expressed by major changes in rock type (e.g., major cliffs of limestone or sandstone alternating with slopes of weaker shale in *figure 4.19*). Within each of these major rock units, bedding occurs on several smaller scales and is expressed by differences in texture, color, or composition of the rock.

A very important aspect of stratification is that the rock layers do not occur in a random fashion but overlie one another in definite sequences and patterns. One of the simple and very common patterns in a vertical sequence of sedimentary rock is the alternation of two related rock types such as sandstone and shale or limestone and shale which commonly originate in closely related environments. Conglomerate commonly alternates with sandstone but is rarely found associated with limestone. This illustrates a definite order in the building up of sedimentary deposits, not a haphazard or random accumulation of sedimentary debris. In many areas more complex patterns or cycles of stratification are apparent, such as the repeated sequence of shale, sand, and coal in the coal regions of Pennsylvania or the cycles of limestone types in the midcontinent.

According to studies of modern sedimentary environments, stratification results from changes or fluctuations in the processes of erosion, transportation, and deposition. Major changes from one formation to another are generally a result of lateral

Figure 4.22 (left) Stratification in the Tapeats Sandstone (the lowest horizontal formation shown in figures 4.19 and 4.20) consists of numerous layers 0.5 to 1 m thick, each of which contains cross-bedding or other types of layering.

Figure 4.23 (right) Typical invertebrate fossils found in the sedimentary rocks of the Grand Canyon.

Figure 4.24 (below) Close-up view of sand grains in a typical sandstone. The grains are rounded and sorted to approximately the same size.

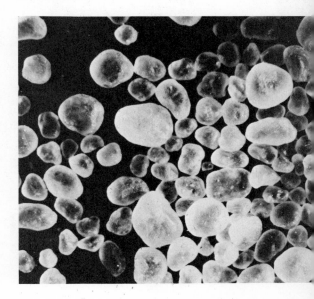

Figure 4.25 Ripple marks in a sandstone, indicating that the rock formed in water agitated by currents.

Figure 4.26 Mud cracks formed where sediment may be temporarily exposed to dry, such as in a tidal flat or shallow lake bed.

shifting or migration of a major sedimentary environment. This is commonly accomplished by the expansion or contraction of the sea. Smaller units of stratification generally represent changes within the environment and may represent fluctuation in the amount and type of sediment delivered to the environment—a result of changes in climate or seasonal changes and variations in the amount of runoff. Other forms of stratification could result from normal variation in the direction and intensity of depositing currents.

In brief, stratification is a record of changes ranging from normal fluctuation which may occur in a day or season to major reorientation of geographic units. It thus provides important bench marks of events in earth history.

Cross-bedding is a general term for layers within a bed which are inclined to the bedding planes of the major stratum (figure 4.27). It is one of the most common and significant sedimentary structures because it occurs in practically all **clastic** rocks and forms in response to the moving currents active during the time of deposition. As clastic particles are moved by turbulent flow of water or wind, the particles tend to collect in ridges or mounds in the form of ripples or dunes called sand waves. Sand waves range in scale from small ripples less than a centimeter high to giant sand dunes several hundred meters high. They are typically asymmetrical with the steep slope facing in the direction of the moving current. As the particles migrate up and over the sand wave, they accumulate on the steep downcurrent face and form inclined layers. The direction of flow of ancient currents that formed sets of cross-strata can be determined by measuring the direction in which the strata are inclined. It is, therefore, possible to determine the patterns of ancient current systems by mapping the direction of cross-bedding in a sedimentary rock.

A very different type of stratification is produced in lakes and in the deep sea where sediment is carried in suspension by **turbidity currents**. Water containing a large amount of suspended sediment is much denser than clear water and will flow under the layer of less dense water through the force of gravity. This forms a turbidity current (muddy current).

As the turbidity currents spread out over the bottom of the basin, the coarsest material is deposited first, followed by successively finer sediment. Each turbidity current thus forms a layer characterized by a gradational decrease in grain size from the base to the top. This type of layering is called **graded bedding.**

In addition to stratification, many sedimentary rocks contain various structures which provide important information concerning the environment of deposition. Ripple marks may be preserved on the bedding surfaces and indicate shallow water and current directions. Mud cracks show that certain sedimentary rocks were formed in an environment which was occasionally exposed to the air during the process of deposition. The

The Earth's Dynamic Systems

presence of this feature in rocks suggests deposition in shallow lakes, tidal flats, or exposed stream banks.

Tracks, trails, and borings of animals are commonly associated with ripple marks and mud cracks and may provide important clues about the environment in which the sediment accumulated. Imprints of raindrops may even be preserved on bedding planes if the succeeding sediment settles out without destroying the bedding surface.

Figure 4.27 *Cross-bedding in sedimentary rocks.*

Metamorphic Rocks

Statement

Metamorphic rocks are those which have been altered by heat and pressure and the chemical actions of **pore fluids** to such an extent that the diagnostic features of the original sedimentary or igneous rocks are obliterated or modified. Most original structural and textural features such as stratification, orientation of grains, fossils, vesicles, and porphyritic textures, are generally changed to such an extent that they cannot be recognized.

Rocks

Metamorphic rocks form in response to high temperature and pressure applied to the continental crust. As a result, atoms of the original minerals recombine and form new minerals, with crystal structures which are stable in the new environment. A suite of minerals, indicative of the intensity or grade of metamorphism, thus forms. An important feature in metamorphic rocks is that the new minerals grow in the direction of least stress, a phenomenon which produces a preferred orientation in the mineral grains. This imparts to the rock a distinct planar element called **foliation**. Most metamorphic rocks may be classified on the basis of type of foliation and mineral composition. (See Appendix for classification of metamorphic rocks.)

We cannot observe metamorphic processes in action in the crust, but indirect evidence of stress and movement in the crust is quite clear from the texture and orientation of metamorphic minerals. We can, moreover, duplicate the extreme pressure and temperature necessary for metamorphism in the laboratory and study the conditions under which metamorphic minerals grow. Information from these studies, together with field observations, provides a firm basis for interpreting metamorphic rocks.

Metamorphic rocks constitute a large part of the continental crust below the veneer of sedimentary rocks and indicate that the continents have been mobile and dynamic throughout geologic time, being repeatedly subjected to great horizontal stresses.

Discussion

Many people have some knowledge of various igneous and sedimentary rocks but understand only vaguely the nature of metamorphic rocks. Perhaps the best way to become acquainted with this group of rocks and to appreciate their significance is to carefully study *figures 4.28* through *4.31*. The vertical aerial photo of a portion of the Canadian Shield (*figure 4.28*) shows that the rocks are extensively deformed, folded, twisted, squeezed, and smashed. Originally, these were sedimentary layers deposited in a horizontal position, but they have been so intensely deformed that it is difficult to determine the original base or top of the rock sequence. Intrusions of granitic rocks have been injected into the metamorphic series, and the entire mass has been broken, fractured, and displaced by numerous faults.

Figure 4.29 shows a more detailed view of metamorphic rocks. The alteration and deformation of the rock is clearly evident in the contorted bands of ferromagnesian minerals, and the pattern of distortions shows that the rock has been subjected to compressive forces while in a plastic or semiplastic state.

The degree of deformation resulting from metamorphism is perhaps easier to comprehend by comparing the shape of pebbles in conglomerate with those in a **metaconglomerate.** As is evident from the photograph (*figure 4.30*), the original spherical pebbles in the conglomerate have been stretched into long,

The Earth's Dynamic Systems

Figure 4.28 *Aerial photograph showing the nature of metamorphic rocks. The rocks shown here have been compressed and deformed to such an extent that many original features have been obliterated. This occurred at great depths. The area was then eroded so the complex rock sequence is now exposed. Note that in addition to the tight folding, the rocks are broken by fractures and are intruded by granitic rocks (light tones).*

Figure 4.29 Outcrop of highly deformed metamorphic rocks in British Columbia, Canada.

Figure 4.30 Outcrop of metamorphosed conglomerate showing how pressure has stretched and deformed the original spherical boulders.

elliptical blades (as much as thirty times their original diameter), all oriented in the same direction.

Even on a microscopic scale, distortion and deformation of the individual grains can be seen. A definite preferred orientation of the grains in *figure 4.31* shows that they either recrystallized under stress or responded to pressure and "flowed" as a plastic. It is important to note that the typical texture of metamorphic rocks does not show a sequence of formation of the individual minerals such as is shown in igneous rocks; all grains in metamorphic rocks apparently recrystallize at roughly the same time.

Metamorphic Processes. Metamorphism is essentially a series of changes in the minerals of a rock which tend to reestablish equilibrium with the high temperature and pressure induced by stresses in the crust. The original sequence of rock may, for example, consist of a series of marine sandstones, shales, and limestones, formed at surface temperatures and pressures, and interbedded basalt flows, which crystallized at very high temperatures. The original components (texture and composition) of these rocks may be unstable under the temperatures and pressures of mountain-building forces. Therefore, the minerals have a tendency to react in such a way as to reestablish equilibrium. The atoms recombine to form new minerals whose structure is stable under higher temperature and pressure. Most metamorphic changes take place in chemically closed systems; that is, elements are neither added nor removed by the process.

Changes in mineral assemblages resulting from alterations of relatively low temperatures and pressures are called **low-grade metamorphism**. **High-grade metamorphism** refers to mineral alterations and development of foliation in response to more intense heat and pressure.

The fluid phase plays an important role in metamorphism. Whenever a single atom breaks the bonds which hold it to the crystal structure of a mineral and moves from that mineral to some other place, part of the rock is essentially in a fluid state. With a small amount of pore fluid, the bulk of the rock may remain solid during the time the original mineral slowly breaks down and forms new crystal structures. The metamorphic rock may not have much pore fluid, but the small amount which does exist provides a medium for material in solution that can be diffused through the rock and rearranged into new minerals.

Temperature is one of the most important factors in metamorphism. At temperatures below 200° C only a small amount of fluid phase exists; therefore, *minerals may remain somewhat out of equilibrium for millions of years.* As the temperature increases, the amount of pore fluid in the rocks increases, reactions become more vigorous, and new mineral assemblages begin to appear. At temperatures greater than 700° C, more fusible components of the rock become fluid, and there is considerable evidence that partial melting during intense meta-

The Earth's Dynamic Systems

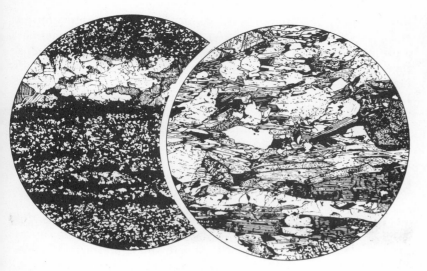

Figure 4.31 *Microscopic view of a metamorphic rock showing the degree to which originally spherical or equidimensional grains have been stretched and deformed.*

morphism approaches the magma stage. If a considerable part of the rock is melted, layers of solid material probably exist mixed with layers of fluid, giving rise to a rock which is transitional between igneous and metamorphic.

Since different minerals are in equilibrium at successively higher temperatures, the mineral composition of the rock provides a key to the temperature at which the rock formed. Field observations have disclosed that a zonal arrangement of mineral assemblages exists around an igneous intrusion and around areas of greatest stress, expressing variations in temperature on a regional scale.

High pressure within the earth's crust causes various changes in the physical properties of many rocks. Pressure tends to reduce the space occupied by the mineral components and may produce new minerals with closer atomic packing. An increase in pressure may be brought about by deep burial, but this is seldom sufficient to produce a complete change in the nature of the rock. The great bulk of metamorphic rocks results from stress of directed pressure due to structural movement within the crust.

Metamorphic Rocks and Plate Tectonics. Although heat and pressure may result from igneous intrusions and deep burial, there are many good reasons to believe that most metamorphic rocks of regional extent are the result of horizontal stress associated with mountain-building. The foliation in regional metamorphic rocks is generally vertical or nearly vertical so that horizontal stresses are strongly implied. Also, regional metamorphic rocks occur in linear belts and are closely associated with granitic igneous intrusions.

According to the plate tectonic theory, metamorphism occurs in the deep roots of folded mountain belts and results from the collision of lithospheric plates. A new belt of metamorphic rock is generated along the continental margin with each mountain-building event. As the mountains are eroded down to near sea level, the metamorphic roots are elevated by isostatic adjustment and eroded to a flat surface to become an additional segment of the continental shield. Continents, therefore, appear to grow by accretion of new material along their margins. The area of the Canadian Shield shown in *figure 4.28* is believed to be the roots of ancient mountain systems formed more than 2 billion years ago by metamorphism resulting from the collision of tectonic plates. We will consider the theory of continental growth in greater detail in Chapter 17.

Summary

Minerals are natural inorganic solids which constitute the fundamental particles or building blocks of rocks. They grow and are destroyed as matter changes to and from the solid state. Minerals are therefore products of specific physical and chemical environments and contain many important clues concerning the origin and history of rocks.

Rocks are aggregates of minerals which form the solid part of a planet. They originate in three main ways: (1) by cooling of a melt (igneous rocks), (2) by erosion, transportation, and deposition of sediment (sedimentary rocks), and (3) by changes resulting from heat, pressure, and reaction of fluids (metamorphic rocks). In addition to mineral composition, the texture of a rock (size, shape, and arrangement of the constituent grains) is of fundamental importance in rock classification.

All of the features of rocks which we have briefly considered in this chapter imply that rock-forming processes are time dependent. That is, time is required for minerals to grow from a melt. Crystallization is not instantaneous but takes place over a period of time. Likewise, time is required for lava to erupt, flow downslope, and solidify. Time is required for weathering of bedrock and for rivers to transport sediment to the sea. Time is necessary for the resulting sediments to solidify and become rock. In a similar manner, time is necessary for metamorphism to occur, for new minerals to grow, and for rock deformation to occur. The element of time is of great importance in considering geological processes and is discussed in some detail in the next chapter.

Additional Readings

Ahrens, L. H. 1965. Distribution of the Elements in Our Planet. New York: McGraw-Hill Book Company.

Deer, W. A., R. A. Howie, and J. Zussman. 1966. An Introduction to the Rock-Forming Minerals. New York: John Wiley and Sons.

Ernst, W. F. 1969. Earth Materials. Englewood Cliffs, N.J.: Prentice-Hall.

Simpson, B. 1966. Rocks and Minerals. New York: Pergamon Press.

U. S. Geological Survey. Atlas of Volcanic Phenomena. Washington, D.C.: U.S. Geological Survey.

Zim, H. S., and P. R. Shaffer. 1957. Rocks and Minerals. New York: Golden Press.

5 Geologic Time

S ome sciences deal with incredibly large numbers, others with great distances, still others are concerned with infinitely small particles. In every field of science, the student must expand his concept of reality, a sometimes difficult but very rewarding adjustment to make.

In geology, the student is introduced to new concepts regarding the duration of time. Because our life is so short, we conceive a "long time" to be 20 or 50 years. A hundred years in most frames of reference is a "very long time." Yet, in the study of the earth and the processes that operate on it, we must attempt to comprehend time spans of a million years, 100 million years, and even several billion years.

How do scientists measure such long periods of time? Nature contains many types of time-measuring devices or "clocks." The earth itself acts like a clock as it rotates on its axis once every 24 hours. Rocks are records of time and, from their interrelationships, the events of the earth's history can be arranged in their proper chronological order. Rocks also contain radioactive clocks which permit us to measure with remarkable accuracy the finite number of years that have passed since the minerals that formed the rocks crystallized. Fossils embedded in the rock constitute a separate "organic clock" from which we can identify synchronous events in earth history.

Geologists use all of these methods to date geologic events and with them have devised a "geologic calendar" which enables them to organize and coordinate long periods of geologic time into workable units, thus permitting a systematic study of the earth's history.

Major Concepts

1. The interpretation of past events in the earth's history is based upon the principle that the laws of nature do not change with time.
2. Relative dating (determining the chronologic order of a sequence of events) can be made by applying the principles of:
   a. **superposition;**
   b. **faunal succession;**
   c. **cross-cutting relations;**
   d. **inclusions;**
   e. **succession** in landscape development.
3. The standard **geologic column** was established from studies of the sequence of rocks in Europe and can be used worldwide by correlating rocks on the basis of their contained fossils.
4. Absolute or finite time designates a specific duration in units of hours, days, or years. In geology, long periods of finite time can be measured by radioactive decay.

Geologic Time

Constancy of Natural Laws

Statement

The interpretation of rocks as products and records of past events in the earth's history is based upon one of the few assumptions which make the scientific enterprise possible: the principle of **uniformitarianism**—*that the laws of nature do not change with time*. That is, the chemical and physical laws which operate today are assumed to have operated throughout all time. The physical attraction (gravity) between two bodies (which Sir Isaac Newton measured and formulated mathematically in his laws of motion) operated in the past as it does now. Oxygen and hydrogen which today combine under given conditions to form water did so in the past. In brief, although man's ability to understand and explain how or why physical nature operates has improved and changed over the centuries, the physical laws or processes which he studies are constant and do not change. All chemical and physical actions and reactions occurring presently have the same causes that operated to produce those actions and reactions 100 years or 5 million years ago.

Discussion

The principle of uniformitarianism may best be understood in the historical context in which it originated. In the late sixteenth century, before modern geologic science was fully established, the prevailing concept of the origin and history of the earth throughout much of the world was based upon literal interpretations of the account of the creation given in the Bible. The earth was considered to have been created in six days and to be approximately 6000 years old. The creation involving such a short period of time was thought to have involved tremendous violent events surpassing anything experienced in modern times. This type of creation has been referred to as **catastrophism**, and the theory was supported by many learned men of the time, foremost among whom was Baron Georges Cuvier, a noted French naturalist. Cuvier, an able student of fossils, recognized many extinct forms of life and concluded that each group of fossils was unique to a given sequence of rocks and was the result of a special creation which was subsequently destroyed by a catastrophic event.

This theory was generally supported by theologians until 1785, when it was challenged by James Hutton who proposed the principle of uniformitarianism, a concept which maintains that the earth has evolved by uniform gradual processes over an immense span of time. It attempts to explain past geologic events in light of present-day natural processes such as erosion by running water, vulcanism, and gradual uplift of the earth's crust. Hutton assumed that these processes occurred in the distant past in much the same way as they presently occur. The first expression

101

of the concept of uniformitarianism, based on Hutton's observations of the rocks of Great Britain, visualized "no vestige of a beginning—no prospect of an end." Sir Charles Lyell (1797-1875), who used Hutton's uniformitarianism as a basis for his book *Principles of Geology*, accepted this conclusion. Lyell in his writings established uniformitarianism as the accepted philosophy for interpreting the geologic and natural history of the earth. Charles Darwin accepted Lyell's uniformitarianism in formulating his theory of the origin of the species and the descent of man. Lyell, however, was perhaps a bit too adamant in not admitting the possibility that the rates at which processes operate could change with time, as more recent studies have indicated.

Modern Views of Uniformitarianism (Naturalism). With the help of modern scientific instruments geologists have studied much more of the geologic record than did Hutton and Lyell and have observed many subtle details of the rocks which the earlier scientists could not see. Modern science is making significant advances in understanding the earth, its long history, and how it was formed. By applying discoveries in kinetics, thermodynamics, electromagnetism, chemistry, and related scientific disciplines, we are discovering more specific clues to the nature of the earth's genesis and evolution.

What then are the modern concepts of uniformitarianism? The assumption on which the idea of uniformitarianism rests is the concept of continuity: natural phenomena (chemical and physical reactions) do not change with time (our ability to express the relations of natural phenomena may change, but the phenomena do not). As explained earlier, this is not a principle unique to geology; it is basic to the scientific enterprise. Without constancy in natural law, verification of conclusions by repetition of experiments would be impossible. We could not assume that water will freeze at the same temperature tomorrow as it did today, nor could we be sure that oxygen and hydrogen would combine to form water; they might form alcohol or even hydrochloric acid.

However, we must modify the assumptions made by Hutton that the geologic past involved only those activities or processes observed today, operating at essentially the same rates. The earth has evolved throughout its history and has undergone a continuous dissipation of energy; with this dissipation, rates may necessarily change. Some processes may have been much more important at times in the past (for example glaciation and tides), and the rates at which the processes function may have varied with time. Ultimately, the earth will exhaust its energy, and we may anticipate cessation of volcanic activity and earthquakes at some time in the far distant future.

In summary, the basic assumption endorsed by essentially all geologists today in deciphering the history of the earth is that natural laws do not change with time. In other words, the method of the geologist is the inductive (scientific) method (observe, collect data, experiment, make a hypothesis, test the hypothesis, etc.).

The assumptions of constancy in natural law are not unique to the interpretations of geologic history but constitute the logical essentials in deciphering recorded history as well. We observe only the present and interpret past events on inferences based on present observations. Thus, we conclude that books or other records of history such as fragments of pottery, cuneiform tablets, flint tools, temples, and pyramids, which were in existence prior to our arrival, have all been the works of man despite the fact that postulated past activities have been outside the domain of any possible present-day observations. Having excluded supernaturalism, we draw these conclusions because man is the only known agent capable of producing the effects observed. Similarly, in geology we conclude that ripple marks in a sandstone formation in the folded Appalachian Mountains were in fact formed by currents or wave action, or that coral shells found in limestones exposed in the high Rocky Mountains are indeed the skeletons of corals which lived in a now nonexistent sea.

Many features of rocks serve as records or documents of past events. The mineral composition and the texture, as well as the internal structures, of the rock body preserve clues to how the rock formed.

Concepts of Time

Statement

We are all aware of growth and change in the physical and biological worlds. Were things unchanging and motionless, we would not be aware of time. Time is measured by change. There are many clocks or mechanisms for measuring time, the most fundamental of which utilize some periodic physical phenomenon: the swing of a pendulum, the flow of sand through an hourglass, the revolution of the moon around the earth and the earth around the sun. For most practical and scientific purposes, the earth is our ultimate timepiece, for, as it revolves, distinct changes in the day, the seasons, and the year may be observed and experienced. It is by these changes that we are aware of time. *Efforts to determine the age of the earth are basically efforts to determine how long the earth has been revolving around the sun.*

Discussion

A clock is simply a mechanism for telling time. We are generally familiar only with mechanical clocks constructed to measure hours, minutes, and seconds, but there are a great variety of natural clocks which provide as good or better measures of time than the standard "clock on the wall." Natural clocks, however, measure other intervals of time. Vibrations of atoms, for example, provide a means of measuring extremely small intervals of time with great precision. Electric waves provide another "clock" to measure intervals of time useful to man and his activities. There are also biologic clocks which measure intervals of time between various biologic activities (breathing, heartbeat, hunger, menstruation, and life span or generation). Natural clocks include layers of sediment deposited during specific seasons (layer of sand in the spring and summer, layer of mud in the winter when water freezes), tree rings, and growth marks on corals. For spans of time of a much greater interval, the thickness of sediment in the rock record is a measure of time much like the sand in an hourglass. The most effective clocks for measuring long periods of time, however, are the radioactive clocks which will be discussed later.

Early Estimates of Geologic Time

Statement

The first serious efforts to estimate the magnitude of geologic time were made in the late nineteenth century. Prior to Hutton and Lyell, few people even recognized geologic processes or thought about the age of the earth. After Hutton presented his arguments for uniformitarianism and Lyell further developed the concept, much interest was generated in the magnitude of geologic time. Early attempts to estimate the age of the earth were based on: (1) salinity of the oceans, (2) thickness of the total sequence of sedimentary rocks, and (3) heat loss from the earth. Each attempt showed evidence of a considerable period of time, but none proved to be accurate.

Discussion

Estimates Based on Salinity. In 1899, John Joly concluded from calculations based on the salinity of the oceans that the earth was 90 to 100 million years old. It was assumed that the original ocean water was fresh and that the present salinity is a result of rivers bringing salt to the sea where it is concentrated by evaporation. From chemical analysis of river water, Joly simply estimated the amount of sodium delivered to the sea annually. This divided by the amount of salt now in the ocean gave the age of the ocean.

Joly's estimates were far too low because they failed to take into account the interchange of salt in the sea and the rocks of the crust. Salt is removed from the sea and deposited as part of marine sediment. Later, the salt may be exposed by uplift, eroded, and recycled back to the sea.

Estimates Based on Thickness of Sediment. Soon after sedimentary rocks were recognized as a record of erosion, transportation, and sedimentation, geologists reasoned that the time represented by the sequence of sedimentary rocks in the earth's crust could be estimated if the average rate of sedimentation in modern seas could be determined and the total thickness of ancient sedimentary rocks measured. Early measurements of maximum thickness of sediment preserved in the geologic record range from 25,000 to 112,000 m (82,000 to 367,000 ft). With later mapping, new rock units have been discovered, and the thickness of fossiliferous rocks is now considered to be at least 150,000 m (490,000 ft). Rates of sedimentation vary from place to place, but most estimates place the average rate of sedimentation at about 0.3 m (1 ft) per 1000 years. This rate would set the age of the first abundantly fossiliferous rocks at 500 million years.

Two problems make this method inaccurate but not completely useless. First, accurate estimates of the average rate of sedimentation are difficult to obtain because different kinds of sediments accumulate at vastly different rates. Secondly, there are many interruptions in the sequence of sedimentary rocks. In a given area, sedimentation will be followed by uplift and erosion, subsidence, and then renewed sedimentation. Thus, an unknown amount of time is not recorded by uplift and erosion, and during this process, part of a previously formed record is removed.

Estimates Based on Heat Loss. Lord William Kelvin reasoned that the earth cooled from a molten state and that the entire rate of cooling could be determined by measuring the present rate of heat flow. Estimates based on this method range from 20 to 40 million years.

This method is also inaccurate: (1) Evidence indicates that the earth formed by gravitational attraction and was never in a completely molten state and (2) the earth generates heat by radioactive decay. It thus does not necessarily lose heat at a constant rate.

Relative Dating (To Determine the Order of Events)

Statement

Relative dating is simply determining the chronologic order of a sequence of events. We employ relative dating when we determine that one child is older than another or that some event occurred before another, such as a war or birth of a famous person. In relative dating, no quantitative or absolute number of days is deduced, only that one event occurred earlier or later than another.

In studying the earth, relative dating is important because many physical events such as vulcanism, canyon cutting, deposition of sediment, or upwarping of the crust can be identified. To establish the relative age of these events is to determine their proper chronologic order. This can be done by applying several principles of remarkable simplicity and universality. The most significant are:

1. *The principle of superposition* which states that in a se-

quence of undeformed sedimentary rocks, the oldest beds are on the bottom and the youngest are on the top.

2. *The principle of faunal succession* which states that groups of fossil plants and animals occur in the geologic record in a definite and determinable order and that a period of geologic time can be recognized by its respective fossils.

Discussion

Superposition and Original Horizontality. The principle of superposition is the most basic guide in determining the relative age of rock bodies. This concept is easily understood, even though a casual observer may have difficulty recognizing rocks as records of events and grasping the scale and implications of the rock bodies involved.

In applying the principle of superposition, two assumptions are made. The first is that the rock layers were orginally deposited in an essentially horizontal position; the second, that the rocks have not been deformed to a degree that the beds are overturned.

The Principle of Faunal Succession. In addition to superposition, the sequence of sedimentary rocks in the earth's crust contains another independent element which can be used to establish the chronologic order of events: the upward change in the assemblage of fossils contained in the rocks. Fossils are the remains of ancient organisms, such as bones and shells, or evidence of organisms, such as trails and tracks. Their abundance and diversity are truly amazing. Some rocks such as coal, chalk, and certain limestones are composed almost entirely of the remains of former life. Others contain literally millions of specimens. In some areas, fossil shells are so abundant that they have been used as road gravel. Invertebrate marine forms are found most often, but even large vertebrate fossils such as mammals and reptiles are plentiful in many formations. For example, it is estimated that to date over 50,000 fossil mammoths have been discovered in Siberia, and many more remain covered. A truly remarkable record of ancient life is preserved in the rocks.

The extensively studied fossil record, showing that plants and animals have evolved with time, provides an independent timepiece or document of earth's chronology. With it, the relative age of a rock body formed during the last 600 million years can be established independent of superposition (fossils are very rare in rocks older than 600 million years). This was recognized some 150 years ago by William Smith, a British surveyor, even before Darwin developed the theory of organic evolution. Working throughout much of southern England surveying the courses of roads and canals, Smith carefully studied the fresh exposures of rocks in quarries, road cuts, and excavations and collected fossils from the strata. By correlating types of fossils with certain kinds of rock, he developed a practical tool which he could use to predict the location and properties of rocks beneath the surface. In a succession of interbedded sand-

The Earth's Dynamic Systems

stone and shale formations, the several shales were very much alike, but the fossils they contained were not. Each shale had its own particular group of fossils.

Soon after Smith announced that the fossil assemblages of England change systematically from the older beds to the younger, others discovered the same thing to be true in locations throughout the world. Parallel succession of fossils was found in many countries, even those separated by oceans.

Today, the principle of faunal succession has been confirmed beyond doubt. It has been used extensively to locate valuable natural resources such as petroleum and mineral deposits and is the foundation for the standard geologic column (see figure 5.2). The geologic record shows that very few species have existed longer than 20 million years, the average being about 5 million.

Fossils provide a means of establishing relative dates in much the same way as artifacts do. Both show evolution and change with time. For example, in a city dump where refuse is buried in succession, one could recognize a period of time prior to the automobile by remains of wagon wheels, saddles, etc. A layer containing abundant scraps and pieces of the Model T Ford would be recognized as being older than one containing remains of a Model A. A layer containing newer models such as the Mustang would be recognized as being one of the youngest layers even though it might not rest upon layers containing any of the older materials.

Cross-Cutting Relations. The relative age of certain events is shown by cross-cutting relationships. **Faults** and igneous intrusions are younger than the rocks they cut, a fact so obvious that it hardly needs mentioning. However, cross-cutting relations can be complex, and careful observation may be required to establish the correct sequence of events in complex areas.

Figure 5.1 *The sequence of events in a schematic diagram may be determined by the use of the principles of superposition and cross-cutting relationships.*

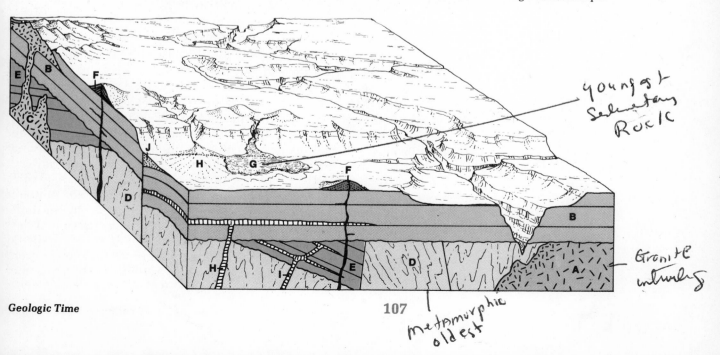

The scale of cross-cutting features is highly variable, ranging from large faults with displacements of hundreds of kilometers to small fractures less than a millimeter long. The sequence of events in *figure 5.1* can be worked out by applying the principles of superposition and cross-cutting relations.

Inclusions. The relative age of intrusive igneous rocks (with respect to the surrounding country rock) is commonly apparent from inclusions, fragments of older rocks in the younger. As a magma moves upward through the crust, it dislodges and engulfs large fragments of the surrounding material which remain as unmelted foreign inclusions.

The principle of inclusion is also clear in many conglomerates in which relatively large pebbles and boulders derived from some preexisting rocks have suffered erosion and transportation and have been deposited in a new formation. The conglomerate is obviously younger than formations from which the pebbles and cobbles were derived. In areas where superposition or other methods do not indicate relative ages, a limit to the age of a conglomerate may be determined from the rock formation represented in the pebbles and cobbles.

Succession in Landscape Development. Surface features of the earth's crust are continually being modified by erosion and commonly show the effects of successive events through time. Many landforms evolve through a definite series of stages so that the relative age of the feature can be determined from the degree of erosion. This is especially obvious in volcanic features such as cinder cones and lava flows created during a period of volcanic activity and then subjected to the forces of erosion until they are completely destroyed or buried by erosional debris.

The Standard Geologic Column

Statement

Using the principles of superposition and faunal succession, geologists have determined the chronologic sequence of rocks on a regional basis and constructed a standard geologic time scale. Most of the original scale was pieced together from sequences of strata studied in Europe during the mid-nineteenth century. Each major unit of rock was named after the area where it is well exposed. The rock units are distinguished one from the other by major changes in rock type, unconformities, or sudden vertical changes in the groups of fossils they contain. In effect, the original subdivision of the geologic column was simply based on the sequence of rock formations in their superposed order as they were found in Europe. Rocks in other parts of the world, containing the same assemblage of fossils as those in a given part of the succession in Europe, are considered to be the same age and are commonly referred to by the same name.

The Earth's Dynamic Systems

Discussion

The standard geologic column, shown in *figure 5.2*, is based upon the principles of superposition and faunal succession and indicates only the relative age of the rocks. The column by itself tells us nothing about the specific duration of time represented by each period nor the age (time before the present) of each period.

For the purpose of studying geologic processes, we will merely name and describe briefly the major divisions so the student will have some idea of the record and magnitude of geologic time. The nomenclature for the geologic column may at first seem nothing more than a collection of meaningless

RELATIVE GEOLOGIC TIME			*ATOMIC TIME
ERA	**PERIOD**	**EPOCH**	
CENOZOIC	Quaternary	Holocene	
		Pleistocene	2-3
	Tertiary	Pliocene	12
		Miocene	26
		Oligocene	37-38
		Eocene	53-54
		Paleocene	65
MESOZOIC	Cretaceous	Late Early	136
	Jurassic	Late Middle Early	190-195
	Triassic	Late Middle Early	225
PALEOZOIC	Permian	Late Early	280
	Carboniferous Systems: Pennsylvanian	Late Middle Early	
	Carboniferous Systems: Mississippian	Late Early	345
	Devonian	Late Middle Early	395
	Silurian	Late Middle Early	430-440
	Ordovician	Late Middle Early	500
	Cambrian	Late Middle Early	600
PRECAMBRIAN Longest (Fewest Fossils)			3600

*Estimated ages of time boundaries (millions of years)

Based on fossil

Figure 5.2 *The standard geologic column.*

names, but the standard geologic column serves as a type of calendar for the earth's history and is the basic language used to designate large time intervals. An understanding of the origin and meaning of these names will be helpful.

The Precambrian. Precambrian time is represented by a group of highly complex metamorphic and igneous rocks which form a large volume of the earth's crust. The rocks represent great thicknesses of sedimentary and volcanic rocks which were intensely folded and faulted and intruded with granitic rock. Because they contain only a very few fossils of the more primitive forms of life, arrangement of individual rock layers within this general group into their proper stratigraphic sequence is difficult, if not impossible.

The Paleozoic Era. Rocks younger than Precambrian are much less complex and contain great numbers of fossils, a fact which permits their recognition on a worldwide basis. The term Paleozoic means "ancient life," and these rocks contain fossils of primitive invertebrate marine organisms, primitive fish, and amphibians. The era is subdivided into periods based largely on the rock formations of Great Britain.

Cambrian comes from *Cambria,* a Latin word for "Wales," where these rocks were first studied. In most areas of the world, Cambrian rocks rest upon the highly deformed Precambrian metamorphic complex.

Ordovician is derived from "Ordovices," an ancient tribe of Wales, and designates the strata overlying the Cambrian but differing in the types of fossils contained in the rocks.

Silurian is a term designating rocks exposed on the border of Wales, a territory originally inhabited by a British tribe, "Silures."

Devonian is named for rocks exposed in Devonshire, England.

Carboniferous. Above the Devonian rocks lies a sequence of coal-bearing formations first studied in England and named the Carboniferous. In the United States, these rocks are subdivided into two major units: the *Pennsylvanian* (from the state of Pennsylvania) and *Mississippian* (from the upper Mississippi Valley.)

Permian is a term referring to rocks exposed over much of the province of Perm, Russia, just west of the Ural Mountains. Corresponding rocks in England lie above the Carboniferous.

The Mesozoic Era. Mesozoic means "middle life" and is used for this period of geologic time because fossil reptiles and a significant number of more modern invertebrates dominate these rocks. This era includes only three periods: the Triassic, Jurassic, and Cretaceous.

Triassic is a term which does not refer to a geographic location but to the striking three-fold division of the rocks overlying the Paleozoic in Germany.

Jurassic is the term first introduced for strata outcropping in the Jura Mountains.

Cretaceous refers to the chalk formations in France and England and is derived from the Latin *creta*, meaning "chalk."

The Cenozoic Era. Cenozoic refers to "recent life," and fossil forms found in these rocks include many types with close relationship to the modern forms, including mammals, modern plants, and invertebrates.

Tertiary is a term held over from the first attempts to subdivide the geologic record into three divisions referred to as Primary, Secondary, and Tertiary. The companion terms Primary and Secondary have been replaced by terms referring to types of fossilized life forms found in the rocks.

Quaternary is the name proposed for the very recent deposits which contain fossils with living representatives.

Radiometric Measurements of Absolute Time

Statement

Unlike relative time, which signifies only the chronologic relationship among events, absolute or finite time designates specific duration measured in units of hours, days, or years. Duration of time is measured by any regularly recurring event. One very useful natural clock measures time by the processes of radioactive decay. Radioactive decay is a process in which the atoms of an element lose particles from their nucleus and break down to form atoms of other elements. Because the rate at which they decay is unaffected by conditions such as pressures, temperature, and chemical binding forces, radioactive decay is a very precise and accurate geological measuring device. The time elapsed since a radioactive element was locked into the crystal of a mineral may be determined if the rate of decay is known and the proportions of the original element and the decayed product can be measured. The most important radioactive clocks for geologic studies are uranium, thorium, rubidium, and potassium.

Discussion

The rate of radioactive decay is measured in terms of **half-life**. In one half-life, half of the original atoms decay; in a second half-life, one-half of the remainder, or one-quarter of the original atoms decay; in a third half-life, half of the remaining quarter, and so on (see *figure 5.3*). Knowing the rate at which a particular radioactive element decays, the time elapsed since the formation of the crystal containing the element can be calculated by comparing the amount of the radioactive element remaining in the crystal (the parent) to the amount of disintegration products (daughter).

There are numerous radioactive isotopes, but most have rapid rates of decay (that is, short half-lives) and lose their radioactivity within a few days or years. However, some decay very slowly, with a half-life of hundreds of millions of years,

Figure 5.3 *The contrast between linear and exponential rates of depletion. Most processes have a uniform straight line depletion such as sand moving through an hour glass. If half of the sand is gone in one hour, all will be gone in two hours. Radioactive decay, in contrast, is exponential. If half is depleted in one hour, half of that remaining, or one-fourth, will be depleted in two hours.*

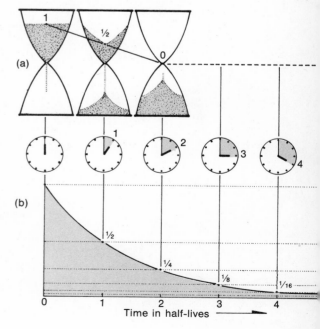

Table 5.1. Radioactive Isotopes Useful in Determining Geologic Time

Parent Isotope	Daughter Product	Half-Life
Uranium - 238	Lead - 206	4.5 billion years
Uranium - 235	Lead - 207	713.0 million years
Thorium - 232	Lead - 208	13.9 billion years
Rubidium - 87	Strontium - 87	50.0 billion years
Potassium - 40	Argon - 40	1.5 billion years

and can be used as atomic clocks for very long periods of time. The parent isotopes and their daughter products most useful for geologic dating are listed in the table below.

The theory of **radioactive dating** is simple enough, but the laboratory procedures are complex, the principal difficulty being precise measurement of minute amounts of isotopes. The accuracy of the method depends upon the accuracy with which the half-life is determined. (U^{235} to Pb^{207} measurements are considered accurate within 2%.)

At present, the potassium-argon method is of great importance since it can be used with the micas and amphiboles which are widely distributed in igneous rocks. It can also be used on rocks as young as a few thousand years or on the older known rocks. Whenever possible, two or more radioactive elements from the same specimen are analyzed to confirm or verify the results.

Another important radioactive clock is based on the decay of carbon-14 (C^{14}), which has a half-life of 5692 years. C^{14} is produced continually in the earth's atmosphere as a result of the bombardment of nitrogen-14 (N^{14}) by cosmic rays. This newly formed radioactive carbon becomes mixed with ordinary carbon atoms in carbon dioxide gas. Plants absorb the carbon dioxide, and animals eat the plants so that both have a fixed proportion of C^{14} while they are alive. After death, no additional C^{14} can replenish the supply in their tissues, and radioactive decay begins, causing C^{14} to revert to N^{14}. The time elapsed since the organism died may be determined by measuring the amount of C^{14} remaining. The longer the time elapsed since death, the less C^{14} will remain. Inasmuch as the half-life of C^{14} is 5692 years the amount of C^{14} remaining in organic matter older than 50,000 years is so small that it cannot be measured accurately. Therefore, this method is useful for dating very young geologic events plus most archeologic material.

Radiometric Time Scale

Statement

Absolute dates of numerous geologic events have been determined and are used in combination with the standard geologic column to provide a radiometric time scale from which the absolute age of a geologic event may be estimated. Unfortu

The Earth's Dynamic Systems

nately, radiometric dates cannot be determined for any given layer of sedimentary rock because sediments are composed of eroded debris of preexisting rocks from various sources. We may determine the radiometric dates of minerals in sedimentary rocks, but the date indicates when the mineral formed, not when the bed of sediment was deposited. Radioactive isotopes can be used to date the time igneous rocks crystallized and the time heat and pressure developed new minerals in metamorphic rocks because in each case the mineral and rock formed together. The problem in developing a reliable radiometric time scale is to accurately place radiometric dates of igneous and metamorphic rocks in their proper position in the relative time scale established by sedimentary rocks. Layers of volcanic rocks and **bracketed intrusions** are the most suitable rocks for radiometric age determinations which can be used as finite time markers in the standard geologic column. These time "bench marks" accurately placed in the geologic column constitute the basis for the radiometric time scale.

Discussion

Layered Volcanics. The best reference points for the radiometric time scale are probably volcanic ashfalls and lava flows. They are deposited instantaneously as far as geologic time is concerned and are commonly interbedded with fossiliferous sediments so that their exact position in the geologic column can be determined. These rocks provide the main basis for establishing an absolute time scale within the geologic column.

Bracketed Intrusions. As shown in figure 5.4, a molten rock may cool within the earth's crust without ever breaking out onto the surface. Subsequent erosion may expose this rock; later, younger sediments may be deposited on top. In some cases, the entire sequence of events may take only a few million years; in others, the events may require a much longer time. The stratigraphic age of the igneous rock falls in the bracket between the oldest sediments (A) and the younger sediments (D). Unfortunately, the span of time between A and D is commonly too long to permit the date of the intrusion to be useful in detailed geochronology, but radiometric dates on such rocks do establish the time of major igneous events.

The presently accepted geologic time scale is based on both the standard geologic column, established by faunal succession and superposition, plus the finite radioactive dates of rocks which can be precisely placed in that column. Each system of dating provides a cross-check on the other inasmuch as one is based on relative time and the other on absolute time. In most instances, there is remarkable agreement between the two systems, and discrepancies are few. In a sense, the radioactive dates act as the scale on a ruler and provide accurate reference markers between which some interpolation can be made. The geologic column in figure 5.5 shows the standard geologic column together with radiometric dates of layered volcanics

Figure 5.4 *Diagrams showing how radioactive dates of intrusive rocks may be used in developing a radiometric time scale. The sequence of sedimentary rocks (A) are deposited and subsequently intruded by the igneous body (B). Erosion removes part of the sequence (A) and (B). Subsequent deposition of the sediment sequence (D) occurs, followed by lava flow (E) and younger sedimentary rocks (F). A radiometric date of lava flow E would provide an excellent age for rocks in that position of the geologic sequence because the flow occurred as part of the normal sequence of rocks. A date for the intrusive (B) is more difficult to place in the column. We only know that it occurred sometime after A and before C and D.*

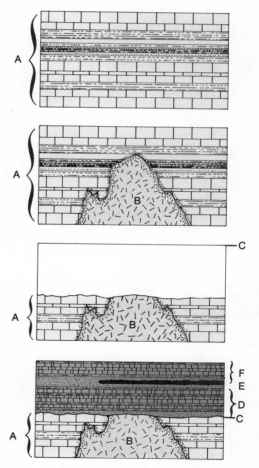

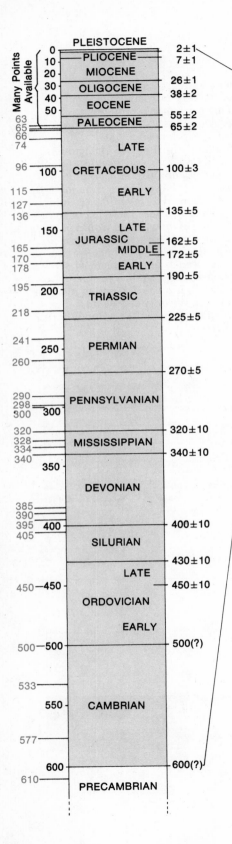

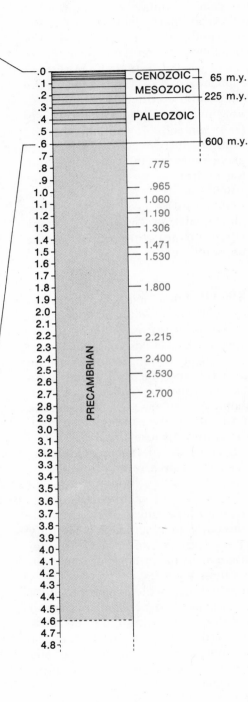

Figure 5.5 Radiometric geologic time scale. Radiometric dates of rocks whose precise stratigraphic position is known are used to establish a scale of absolute time for the standard geologic periods. To obtain the age of a rock using this scale, one needs to know its exact position in the geologic column; from this an approximate absolute or finite date can be interpolated.

and bracketed intrusions in the sedimentary sequence. Enough dates have been established so that the time span of each geologic period can be estimated with considerable confidence. To determine the age of a rock, one need only to determine its location in the geologic column and interpolate between the nearest radiometric time marks.

From this radiometric time scale, several general conclusions can be made about the history of the earth and geologic time: (1) Present evidence indicates that the age of the earth is around 4.5 to 4.8 billion years. (2) Precambrian time constitutes more than 80% of geologic time. (3) Phanerozoic time (Paleozoic and later) began about 570 million years ago. Rocks deposited since Precambrian time can be correlated on a worldwide basis by means of fossils, and many important events during this time can be established on the basis of radioactive dates. (4) Some major events in the earth's history may be difficult to place in their relative positions in the geologic column but can be dated by radioactive methods.

The Magnitude of Geologic Time

Statement

Concepts involving great time spans are difficult for most of us because norms established through sensory experience are the familiar short-time measurements such as the day, the week, and the changing seasons. References to extremely long periods of time are thus difficult to comprehend. Therefore, as a student of geology, you must continually attempt to enlarge your time norms to encompass the magnitude of geologic time. Otherwise, the extremely slow geologic processes considered in terms of human experience will have little meaning.

Discussion

Perhaps the best way to begin is to consider geologic time in reference to something tangible and familiar rather than in terms of large numbers. In *figure 5.6*, the length of the yardstick represents the length of time from the beginning of the earth to the present. A scale of absolute time is plotted on the right, and the standard geologic periods are shown on the left. This diagram reveals several facts. Subdivisions of geologic time are not equal. Precambrian time constitutes the great bulk of the earth's history, whereas the Paleozoic and younger periods are equivalent to only the last five inches on the scale. In order to show events with which most people are familiar, the last few inches on the scale must be enlarged. The first abundant fossils occur five inches from the top. The great coal swamps are about two inches from the top. The dinosaurs became extinct about one-half inch from the top, and the Ice Age would be shown in the uppermost 1/100 of an inch. Recorded history would be shown only in the upper 1/1000 of an inch on this scale.

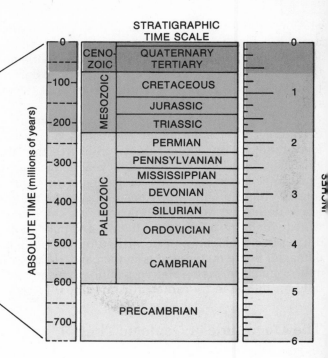

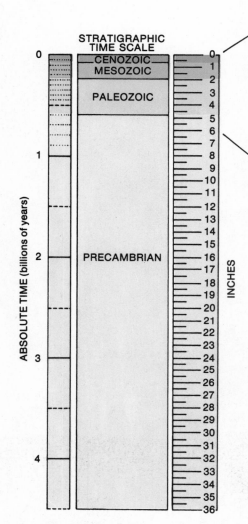

Figure 5.6 *The length of geologic time compared to a yardstick. On this scale, Precambrian time represents the first 31 inches and all events since the beginning of the Paleozoic are compressed into the last 5 inches. Dinosaurs first appeared 2 inches from the top. The glacial epoch occurred in the last fraction of an inch, and historic time is so small that it cannot be represented, even on the enlarged part of the figure.*

Summary

We have seen in previous chapters that mineral types form under very specific environmental conditions and that the shape of the mineral grain reveals many things about its history. Rocks, being aggregates of minerals, thus constitute records of events in our changing planet.

In this chapter, we have seen how the events recorded in rocks may be arranged in their proper sequence by using the principles of superposition and faunal succession and how the date of some events can be determined in terms of a finite number of years. In the following section of this book, we will study the details of the hydrologic system and how the surface features of the earth are formed, modified, and changed with time.

Additional Readings

Berry, W. B. N. 1968. Growth of a Prehistoric Time Scale. San Francisco: W. H. Freeman and Company.

Deevey, E. S., Jr. 1952. "Radiocarbon Dating." Sci. Amer. 186(2): 24-28

Eicher, D. L. 1968. Geologic Time. Englewood Cliffs, N.J.: Prentice-Hall.

Faul, H. 1966. Ages of Rocks, Planets and Stars. New York: McGraw-Hill Book Company.

Harbaugh, J. W. 1968. Stratigraphy and Geologic Time. Dubuque, Iowa: William C. Brown Company, Publishers.

Hurley, P. M. 1959. How Old Is the Earth? Garden City, N.Y.: Anchor Books (Doubleday and Company, Inc.).

Toulmin, S., and J. Goodfield. 1965. The Discovery of Time. New York: Harper and Row, Publishers.

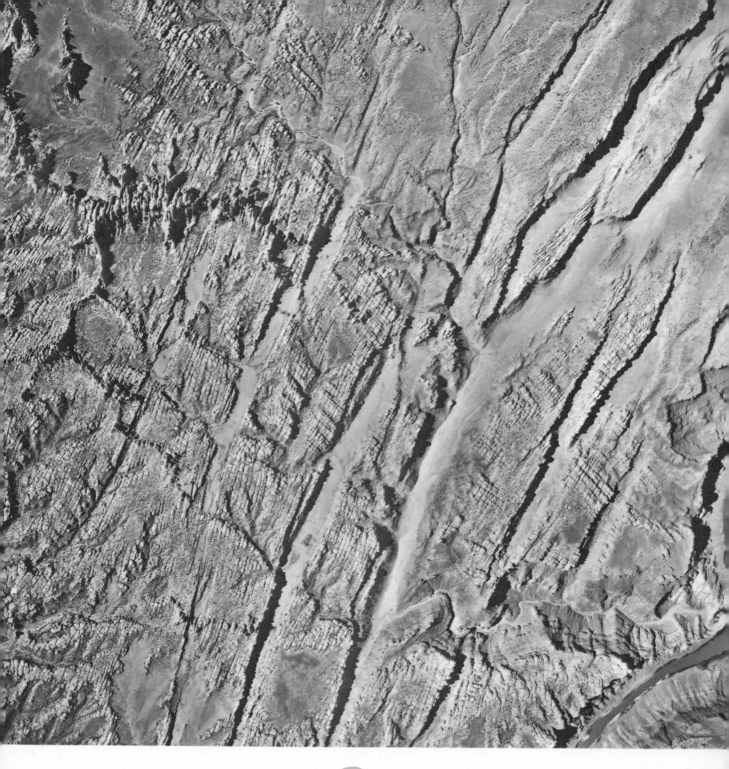

6 Weathering

Weathering is the interaction of the atmosphere with the rocks exposed at the surface of the earth. Because most rock-forming minerals were formed under high temperatures and high pressure conditions, they are not stable at the low temperatures and pressures which exist at the surface, especially in the presence of water, oxygen, and carbon dioxide. Oxygen and water tend to combine with carbon dioxide, and this mixture reacts with rock-forming minerals to form new minerals which are stable under atmospheric conditions.

The atmosphere tends to weather or break down the rocks in two ways: (1) by mechanical disintegration in which rocks are broken by physical forces and (2) by chemical decomposition in which there is a chemical reaction between elements in the atmosphere and those in the minerals of the rocks. These two processes usually occur together and the effects of each may not be clearly separated.

Weathering is an important process in the transformation and modification of the landscape, for it is the weathered rock material that is removed by processes of erosion. In addition, the products of weathering form a blanket of soil, the basis of most terrestrial life, over the **bedrock** surface. We should therefore consider weathering as part of the geologic system which has tremendous ecological significance.

Major Concepts

1. The products of weathering form a blanket or cover of decomposed and disintegrated rock debris known as regolith.
2. **Mechanical weathering** includes various mechanical stresses, foremost among which is **frost wedging**.
3. The major types of chemical weathering are **oxidation**, **solution**, and **hydrolysis**.
4. **Joints** are important in weathering in that they permit the atmosphere to attack a rock body at considerable depth. They also greatly increase the surface area of a rock where chemical reactions can occur.
5. There is a universal tendency for weathering processes to produce rounded or spherical surfaces on decomposed rocks.

Bedrock and Regolith

Weathering processes create a blanket or cover of loose, decayed rock debris known as **regolith** (Gr. *rego*, "blanket," and *lithos*, "rock") which forms a cover over the solid bedrock below. The regolith ranges from a few centimeters to many meters thick depending upon the climate, type of rock, and length of time uninterrupted weathering has proceeded. There are two types of regolith: (1) residual regolith which forms in place and (2) transported regolith in which the weathered debris is moved from its source by streams, glaciers, or wind. Soil is the upper part of the residual or transported regolith which is decomposed and altered sufficiently to support the growth of plants.

The bedrock beneath the regolith consists of largely unaltered rock bodies: layers of sedimentary strata or large units of igneous and metamorphic rock. In most areas much of the bedrock is concealed beneath the regolith, but where weathering processes are limited, as in deserts or cold climates, large areas of bedrock are exposed.

Discussion

To understand the nature of bedrock and regolith, carefully study the photograph and diagrams of part of the Wasatch Mountains. It is clear from the photograph (*figure 6.1a*) and the diagram (*diagram 6.1b*) that exposures of bedrock are limited to certain areas of resistant limestone and sandstone strata which form discontinuous cliffs along the upper part of the mountain front. In the canyons very little soil is retained and bedrock is exposed from the base to the top of the canyon walls.

The sketch in *figure 6.1c* was made from the photograph and outlines the surface covered with regolith. In this diagram, the outcropping bedrock is not shown so that the regolith appears as a thin, discontinuous blanket over the surface. The holes in the regolith are where bedrock is exposed. The regolith on the mountain front is largely residual, decomposed rock debris. Transported regolith fills the valley in the foreground but is not shown on the diagram.

In *diagram 6.1d* only the outcropping bedrock is shown. This is, of course, the area of "holes" in the regolith of *figure 6.1c*. If you carefully study the areas of exposed bedrock in *figures 6.1a* and *6.1d*, you will see that the strata are warped into broad folds (*figure 6.1e*). These folds form the structure of the mountains. Erosion has cut its canyons into the bedrock, and weathering has produced a partial cover of regolith.

Mechanical Weathering

Statement

Various mechanical stresses act upon rocks and break them into small particles. This is a physical process, called mechanical

119

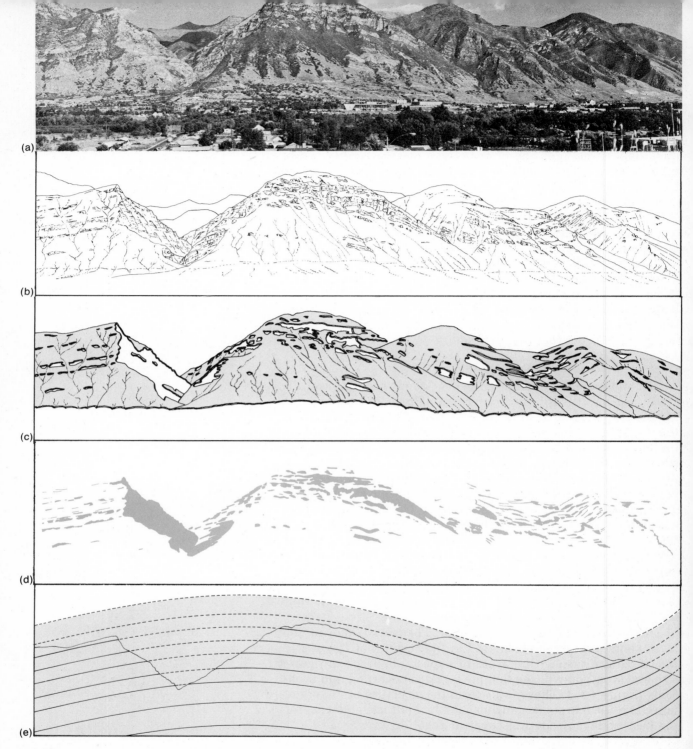

Figure 6.1 The relationship between bedrock and regolith. (a) Photograph of the Wasatch Mountains in Central Utah. (b) A field sketch showing outcrops of bedrock in cliffs and canyons and the slopes covered with regolith. (c) The discontinuous blanket of regolith. Outcrops of bedrock form "holes" in the regolith cover. Some formations are almost completely covered with regolith, whereas others are exposed as discontinuous cliffs. (d) Areas where bedrock is exposed outline the structure of the rocks. (e) The structure of the bedrock. The rock layers are warped into broad folds, some of which are cut by canyons. Compare with photo in figure 6.1a.

weathering or **disintegration.** The rock does not change in chemical composition but is simply broken into smaller fragments. The most important type of mechanical weathering is frost action or ice wedging in which expansion of freezing water in cracks or bedding planes wedges the rock apart.

Expansion caused by release of pressure as overburden is removed is called **sheeting** and develops large fractures parallel to the earth's surface. Many other types of mechanical weathering may occur but are only locally significant.

Discussion

Frost Action (Ice Wedging). Figure 6.2 illustrates the process of **frost wedging.** Water from precipitation or from melting snow easily penetrates open joints, bedding planes, planes of foliation, or other openings in the rock and expands approximately 9% when it freezes. If the water is even partly confined, a great pressure is exerted on the enclosing rock. Each time water within a crack freezes, stresses are exerted so that over a period of time the rock is literally hammered apart. The force generated by each freeze is roughly equivalent to one produced by dropping a ball of steel weighing 7 kg (16 lb) from a height of 3 m (10 ft). At $-22°$ a pressure of 108 kg/cm^2 (30,000 lb/in.^{2}) is exerted; this pressure is far greater than the tensile strength of the rock.

The principle of ice wedging was successfully utilized in some early quarrying operations before more refined stone-cutting processes were developed. The quarry operators simply drilled a series of holes in the rock along the line of desired cut, filled the holes with water, and allowed it to freeze. The stresses set up within the rock were sufficient to break the rock into a large block.

Conditions necessary for effective frost action include: (1) an adequate supply of moisture, (2) fractures, cracks, or other voids within the rock into which the water can enter, and (3) temperatures which rise and fall across the freezing point. Temperature is especially important because pressure is applied with each freeze. Frost action is much more effective in areas where freezing and thawing occur many times a year. It is less effective in exceptionally cold areas where water is more or less permanently frozen. Therefore, frost action occurs most frequently above the timber line and is especially active in the steep slopes above alpine glaciers where melt water produced during the warm summer days seeps into cracks and joints and freezes during the night.

Sheeting or Unloading. Rocks originally formed deep within the earth's crust are under high confining pressure from the weight of thousands of meters of overlying rocks. When the overburden is removed by erosion, the pressure is released and the rock body *tends* to expand. The internal stresses set up by expansion may develop large fractures or expansion joints parallel to the earth's surface. This process is commonly observed

Figure 6.2 *The process of ice or frost wedging. Water seeps into fractures in a rock and expands with great force when it freezes. This wedges the rock apart and produces shattered angular fragments.*

in granite quarrying when the removal of large blocks is followed by rapid, almost explosive, expansion of the quarry floor. A sheet of rock several centimeters thick bursts up, and at the same time numerous new parallel fractures appear deeper in the rock body. The same process occasionally causes rock bursts in mines and tunnels when the confining pressure is released during the tunneling operation.

Other Types of Mechanical Weathering. Animals and plants play a variety of minor roles in mechanical weathering. Worms and burrowing animals such as rodents and termites mechanically mix the soil and loose rock particles, a process which facilitates further breakdown by chemical means. For example, it is estimated that there are 150,000 worms per acre in the average soil, capable of reworking 9 to 14 metric tons (10 to 15 short tons) of rock material in each acre per year.

Pressure from growing roots widens cracks and contributes to the rock breakdown. Lichens can live on the surface of bare rock and extract nutrients from the rock minerals by ion exchange. The effect is the alteration of minerals both mechanically and chemically. These processes may seem trivial, but the work of innumerable plants and organisms for long periods of time adds significantly to the disintegration of the rock.

The products of mechanical weathering commonly accumulate at the base of a cliff as a pile of angular rock fragments known as **talus** (*figure 6.3*). Where rainfall is abundant the weathered rock material is washed into streams and transported as sediment to the sea.

Chemical Weathering

Statement

Chemical weathering consists of several important chemical reactions between the elements in the atmosphere and

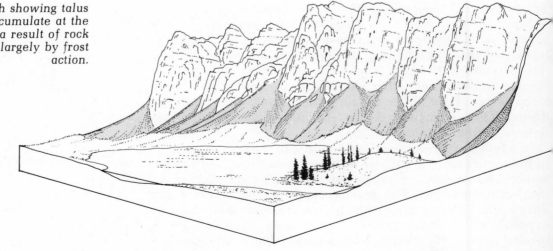

Figure 6.3 *Sketch showing talus cones which accumulate at the base of a cliff as a result of rock falls produced largely by frost action.*

The Earth's Dynamic Systems

those in rocks and minerals of the earth's crust. During chemical weathering, rocks are decomposed, internal mineral structure is destroyed, and new minerals are created. Thus, there is a significant change in the chemical composition and physical appearance of the rock. Water is of prime importance because it contains ions of atmospheric gases in solution. It readily seeps into the cracks and pore spaces of the rocks, bringing the elements of the atmosphere into direct contact with the minerals of the rocks. Oxygen ions are readily available and combine with metallic ions in the minerals to form new minerals (oxides) which are stable under atmospheric conditions. In addition, carbon dioxide from the air and from plant activity in the soil is soluble in water and forms carbonic acid (H_2CO_3), a weak acid capable of dissolving calcium carbonate in limestones. In these and other chemical reactions between elements in the atmosphere and those in the rocks, water plays a dominant role. The rates of chemical weathering, therefore, are greatly influenced by the amount of precipitation. No area of the earth's surface is continually dry, for even in the most arid deserts some rain does fall and causes chemical weathering. Therefore, chemical weathering is essentially a worldwide process. However, it is least effective in cold climates where water is frozen the entire year.

The major types of chemical reactions in the processes of weathering are hydrolysis, solution, and oxidation.

Discussion

Hydrolysis. The chemical union of water and a mineral is known as hydrolysis, a process involving not merely absorption of water, as in a sponge, but a specific chemical change in which a new mineral is produced from the original material. The reaction is between the H^+ or OH^- ions of the water and the ions of the mineral.

A good example of hydrolysis is found in the chemical weathering of potassium feldspar and is described chemically as follows:

$$2KAlSi_3O_8 \ + \ H_2CO_3 \ + \ nH_2O \ \rightarrow \ K_2CO_3 \ + \ Al_2(OH)_2Si_4O_{10} \cdot nH_2O \ + \ 2SiO_2$$

(potassium feldspar)   (carbonic acid)   (water)   (potassium carbonate— readily soluble)   (clay mineral)   (soluble hydrated silica or finely divided quartz)

The hydrogen ion displaces the potassium ion and combines with the aluminum silicate radical of the feldspar to form a new clay mineral. The potassium is released as a free positive ion in the soil water. Silica is also released but may remain in solution. The new clay mineral does not contain the potassium present in the original feldspar and has a new crystal structure consisting of sheets of silica tetrahedra which form submicroscopic crystals.

The importance of hydrolysis in weathering cannot be overstressed since it acts on the very abundant feldspars and ferromagnesian minerals, the dominant minerals in most rocks.

Solution. Solution is commonly one of the first stages of chemical weathering, occurring wherever water makes contact with soluble minerals. Salt is extremely soluble, surviving at the earth's surface only in the most arid regions. Gypsum is less soluble but is easily dissolved by surface water. Limestone is also soluble in water, especially if the water contains carbon dioxide.

An idea of the effectiveness of solution activity in the weathering of rocks can be gained from analysis of the composition of water in rivers. Fresh rainwater has relatively little dissolved mineral matter, but running water soon dissolves the more soluble minerals in the rock and transports them in solution. The amount of dissolved minerals carried by the rivers of the world to sea each year is approximately 3.9 billion metric tons (4.3 billion short tons).

Oxidation. Oxidation is the combining of atmospheric oxygen with a mineral to produce an oxide. It is especially important in minerals having a high iron content. The iron in silicate minerals unites with oxygen to form hematite (Fe_2O_3) or limonite [$FeO(OH)$]. Hematite is deep red and when dispersed in sandstone or shale imparts a red color to the entire rock.

The Importance of Joints in Weathering

Statement

Almost all rocks are broken by systems of fractures called joints resulting from strains established by earth movements, expansion due to the release of overburden, or contraction related to cooling of lava. In fact, finding a large mass of rock which is not fractured is unusual. Joints influence the weathering of rock bodies by: (1) effectively cutting large blocks of rocks into smaller ones, thereby greatly increasing the surface area available for chemical reactions, and (2) acting as channelways whereby water can penetrate and break down the rock by frost action.

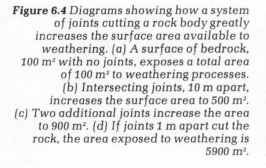

Figure 6.4 *Diagrams showing how a system of joints cutting a rock body greatly increases the surface area available to weathering. (a) A surface of bedrock, 100 m² with no joints, exposes a total area of 100 m² to weathering processes. (b) Intersecting joints, 10 m apart, increases the surface area to 500 m². (c) Two additional joints increase the area to 900 m². (d) If joints 1 m apart cut the rock, the area exposed to weathering is 5900 m².*

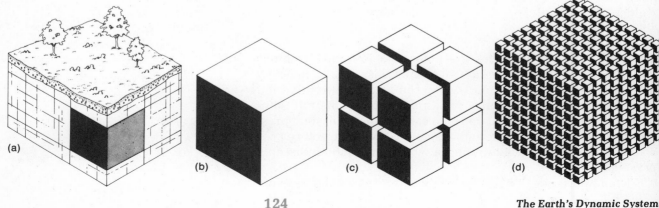

(a)   (b)   (c)   (d)

Discussion

The importance of jointing in weathering processes may be better appreciated by considering the amount of new surface they produce for chemical weathering. For example, consider a cube of rock which measures 10 m (33 ft) on each side (figure 6.4). If the upper surface of the cube were exposed and the rock were not jointed, weathering could attack only the exposed surface of 100 m² (1076 ft²). If the block were bounded by intersecting joints 10 m apart, the surface area would be 500 m² (5380 ft²)—the base of the cube is not exposed. If two additional joints cut the cube into eight smaller cubes, the surface area available for weathering would be 900 m² (9688 ft²). If joints 1 m apart cut the rock, 5900 m² (63,500 ft²) of rock surface would be exposed to weathering.

Figure 6.5 Joint systems cutting a sandstone formation near Arches National Park, Utah. Weathering along the joints has produced deep, narrow crevasses so the joint planes are greatly emphasized.

Figure 6.6 *Ground view of an intersecting system of joints, dividing the rock into rough columns.*

In addition to providing a much larger surface area for chemical decomposition, joints act as a system of channels through which water can penetrate below the surface, permitting mechanical and chemical weathering processes to attack the rock from several sides hundreds of feet below the surface.

Some of the most striking examples showing the influence of jointing on weathering are found in the Colorado Plateau where thick sandstone formations have developed prominent joint systems because the rock is so brittle. Weathering proceeds along each joint surface and cuts the rock into large slabs (*figure 6.5*). In other formations intersecting joints divide the rock into large columns and play an important role in the sculpture of columns and pillars (*figure 6.6*). Jointing likewise plays a dominant role in the weathering of large masses of granite and provides the initial shape for further rock breakup.

Spheroidal Weathering

Statement

In the weathering process, there is an almost universal tendency to produce rounded or spherical surfaces on decaying rock. The rounded shape results because weathering attacks an exposed rock from all sides at once. Therefore, the depth of decomposition is much greater along the corners and edges of the rock, and soon the weathered corners crumble, producing a rounded edge. As the process continues, the block is eventually reduced to an ellipse or a sphere.

Discussion

Figure 6.7 illustrates the mechanics of **spheroidal weathering.** Joints separate the rock into blocks with sharp corners. As water penetrates the fractures, it attacks the rock from all sides, and the block soon becomes completely decomposed on the edge

Figure 6.7 *Diagram showing spheroidal weathering. (a) Joint systems cut a rock body into angular blocks. (b) Weathering proceeds inward on each block from the joint face. (c) The corners of the block are soon completely decomposed so the unweathered rock assumes a spherical or elliptical shape.*

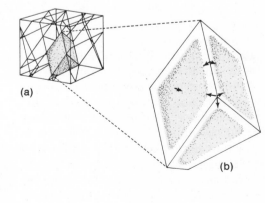

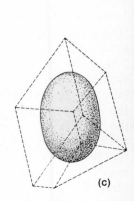

The Earth's Dynamic Systems

Figure 6.8 *Spheroidal weathering in the building blocks of the Parthenon, Athens, Greece. The original blocks were rectangular and closely fit together like the restored section shown in the left part of the photo. Weathering proceeding inward from the sides of each block has produced a spheroidal form.*

Figure 6.9 *Sketches showing evolution of spheroidal boulders. (a) Rock cut by a joint system. (b) Preliminary stages of spheroidal weathering. (c) Late stages of spheroidal weathering.*

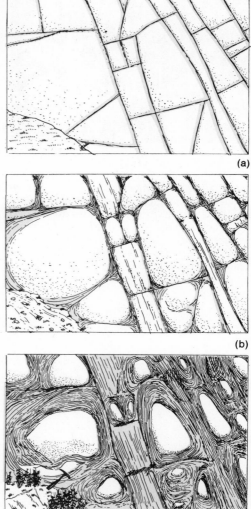

(a)

(b)

(c)

and corners. As this material falls off, the corners become rounded. The sphere is the geometric form that has the least amount of surface area per volume and, once the block attains this shape, it simply becomes smaller. Examples of spheroidal weathering can be seen in rounded blocks in ancient man-made structures such as the Parthenon shown in *figure 6.8*. The original blocks had sharp corners and were fitted together with precision, as can be seen in the unweathered block used in part of the restoration. The edges of the original blocks, in contrast, have completely decomposed and have assumed an elliptical or spherical shape.

In nature, spherical weathering is produced both at the surface and at some depth. A series of sketches (*figure 6.9*) shows how an intersecting joint system produces angular blocks which ultimately decompose into elliptical or spheroidal boulders.

Exfoliation. **Exfoliation** is a special type of spheroidal weathering in which the rock breaks apart by separation along a series of concentric shells or layers like cabbage leaves (*figure 6.10*). The layers, essentially parallel to each other and to the surface, develop by both chemical and mechanical means. Sheeting or unloading may play an important part, as deeply buried rocks like granite have a tendency to expand upward and outward as the overlying rock is removed. Frost wedging along the sheeting joints plays an important part in the gradual removal of successive layers. The volume increase associated with decomposition of feldspar is also thought to function in development of exfoliation.

(a)

(b)

Figure 6.10 Exfoliation Domes. (a) Exfoliation domes in the Sierra Nevada Mountains, California. The Sierras are composed of massive granite cut by joints which separate the rock into large blocks. The exfoliation domes developed as huge slabs of rock spalled off. The joints separating the rock slabs probably resulted from expansion as erosion removed the overlying rock. (b) Sketch of Independence Rock, Wyoming.

Climate and Weathering

Statement

The type of weathering and the rate at which it operates is profoundly influenced by the climate of a region, the major controls being related to water and temperature. Chemical weathering is extreme in humid tropical climates and develops thick soils by decomposing the rock to depths sometimes exceeding 70 m (230 ft). Under such conditions, the feldspar in granites and related rocks in completely altered to clay minerals. In arid regions, chemical weathering is less intense and takes place at a much slower rate. By contrast, frost action is insignificant in tropical areas but, in cold climates where the temperature fluctuates about the freezing point, frost action may be the dominant agent in rock destruction.

Discussion

Water. Water is the most important agent in almost all forms of weathering, involving not only the total amount of precipitation, but also the intensity of rain, percentage of runoff, and evaporation-precipitation rates. Total rainfall is probably the most important factor, but infiltration, runoff, and rate of evaporation are also significant. For example, 8 cm (3 in.) of rain a year in Arizona results in a dry landscape, but in northern Alaska the same amount of rain is sufficient to form swamps because of the impermeable permafrost below the surface. Also, weathering in savanna climates with alternating wet and dry seasons is much different from that in the tropics where rainfall is more evenly distributed throughout the year. The distribution of the rainfall, whether it occurs evenly or in drenching showers, is also significant, as it determines how much and what kind of weathering occurs.

Temperature. Temperature greatly influences the rate of chemical reactions as 10° C increase in temperature commonly doubles the reaction rates. Temperature factors in weathering include the average temperature, temperature range, and fluctuations of temperature about the freezing point. Fluctuations of temperature about the freezing point have a profound influence on the effectiveness of frost action. Minimal frost action occurs where temperature is too warm for freezing or too cold

The Earth's Dynamic Systems

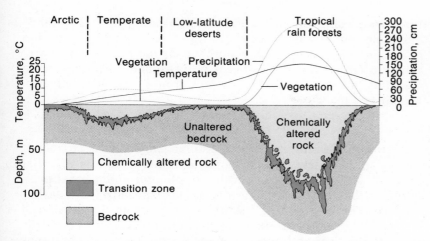

Figure 6.11 *Schematic diagram showing variations in weathering from the Arctic to the tropics. The graphs at the top of the diagram show variations in precipitation, temperature, and vegetation. The generalized cross section in the lower part of the diagram shows the relative depths of weathering resulting from fluctuations of these factors. Weathering is most pronounced in the tropics where precipitation, temperature, and vegetation reach a maximum; conversely, a minimum of weathering is found in deserts and polar regions where these factors are also minimal.*

for thawing. For example, in cold areas, south-facing slopes are exposed to the sun and experience more melting and a greater number of cycles of freezing and thawing than north-facing slopes which are in the shade. The south-facing slopes thus undergo more mechanical weathering and may be noticeably more rugged.

Figure 6.11 summarizes the interaction of some of the major factors affecting weathering. The cross section is a generalized diagram from the north pole to the equator showing fluctuations in precipitation, evaporation, temperature, and amount of vegetation. In polar climates, weathering is largely mechanical, producing a thin, coarse veneer of angular fragments altered very little chemically. In temperate climates (cold to warm and subhumid to subarid) weathering is both mechanical and chemical, and soil develops on decomposed rock material. Chemical weathering decreases to a minimum in deserts and semidesert regions; however, mechanical weathering is evident in fresh angular rock debris which covers most slopes. Exposures of bare rock are abundant, and soil cover is very thin. High evaporation commonly produces alkaline deposits on the surface where ground water is drawn upward by capillary action and evaporation.

Weathering Characteristics of Major Rock Types

Statement

The weathering of rocks is influenced by a number of variables and is a complicated process. Mineral composition is of prime importance; some minerals such as quartz are very stable and remain essentially unaltered for long periods of time; others such as olivine and the feldspars are very unstable and begin to decompose almost immediately. The texture of the rock is also very significant because of its influence on **porosity** and **permeability** which govern the ease with which water can enter the rock and attack the mineral grains. Precipitation and tempera-

Weathering   129

ture are the chief climatic controls but weathering will be influenced not only by the total rainfall but also by the distribution of precipitation through time, percent of runoff, rate of evaporation, etc. Therefore, a given rock will respond to weathering in a variety of ways, but the major rock groups weather in characteristic fashion.

Discussion

Granite. Granite is a massive homogeneous rock composed of feldspar, quartz, and mica, with minor amounts of other minerals. It forms at considerable depth and under great pressure, so it is out of equilibrium when exposed at the earth's surface. The release of pressure resulting from the erosion of the overlying rocks produces expansion joints which aid in the development of exfoliation. Although granite is known as a rock of great strength, chemical weathering is readily apparent on most outcrops.

Feldspars weather rapidly by chemical reaction with water and are altered to various clay minerals. Calcium plagioclase is least resistant, followed by sodium plagioclase. Although potassium feldspars are most resistant, all feldspars readily break down into clay. Mica weathers somewhat more slowly than most feldspars but is easily attacked by water along its cleavage planes, and ion exchange is common. Micas alter with little change in structure to chlorite and clay minerals. Quartz, in contrast, is very resistant to both chemical and mechanical weathering and remains essentially unaltered as the other minerals are decomposed. Therefore, it constitutes the most significant particle or fragment produced by the weathering of a granite.

Basalt. Basalt is a fine-grained rock composed mostly of feldspar, olivine, and pyroxene. The surface of a basalt flow is generally vesicular and very porous, and the interior of the rock body is commonly broken by a system of columnar joints. Therefore, flows are highly permeable and susceptible to decomposition. Quartz is not present in basalt, so all minerals in this rock are eventually converted to clay and iron oxides. The ultimate weathering product is a red or brown soil.

Sandstones. Sandstones are composed mostly of quartz grains with varying amounts of small rock fragments, feldspar, and clay minerals. The quartz in sandstone is highly resistant to chemical weathering so that chemical decomposition of the rock consists largely of an attack on the cement. The major cementing materials in sandstones are calcite, iron oxide, and quartz.

Limestone. Limestone is composed mostly of the mineral calcite although it generally contains some clay and other impurities. It is the most soluble of the common rock types and, except in extremely dry climates, solution is the dominant weathering process. In pure water calcite is not very soluble; if carbon dioxide is present, it combines with water to form carbonic acid which is capable of dissolving much more calcite than pure water. The formation of carbonic acid in water is expressed by the following equation:

$$H_2O + CO_2 \rightarrow H_2CO_3$$

(water) (carbon (carbonic acid)
dioxide)

In the solution of limestone, the acid reacts with calcite to form calcium bicarbonate, which remains in solution and is removed by ground water. This reaction may be expressed as follows:

$$H_2CO_3 + CaCO_3 \rightarrow Ca(HCO_3)_2$$

(carbonic (calcite) (calcium bicarbonate)
acid)

In most limestone regions in humid climates, solution activity enlarges joints and bedding planes and forms a network of caverns and caves; the limestone formations in such regions typically form valleys. In arid regions, where solution activity is at a minimum, limestones form cliffs.

Rates of Weathering

Statement

Rates of weathering can be calculated by measuring the amount of decay on rock surfaces of known age. Tombstones, ancient buildings, and monuments, for example, provide datable rock surfaces for estimating rates of weathering. In addition, weathering rates can be determined from recent lava flows that can be dated by radioactive means. Analysis of drainage water reveals how much mineral matter is being removed in solution and provides another approach for estimating rate of weathering.

Discussion

Weathering of the Great Pyramids. Interesting weathering rates for several rock types are provided by studies of the Egyptian pyramids. The Great Pyramid of Cheops was originally faced with polished, well-fitted blocks of limestone which protected the inner layers of core rock from weathering until they were removed about 1000 years ago to build mosques in Cairo. Since then, weathering has attacked all four main rock types used in construction of the pyramid. The most durable rock (least weathered) in the pyramid is a hard gray limestone that still retains marks of the quarry tools. The shaly limestone used in other blocks weathered rapidly, and many have a zone of decay as deep as 20 cm (8 in.). Most of the weathered debris has remained as talus on individual tiers and around the base of the pyramid (*figure 6.12*). The volume of weathered debris produced from the pyramid during the last 1000 years has been calculated to be 50,000 m³ (1.8 million cubic feet), or an average of 50 m³ (1800 ft³) per year. This is a loss of approximately 3 mm per year over the entire surface of the pyramid. Some of the older stepped pyramids, built 3000 years ago, show much greater weathering, and many large blocks are nearly completely decayed or reduced to small spheroidal boulders. In addition, high piles of debris have accumulated on each major terrace (*figure-6.13*).

Recent Volcanic Flows. On a longer time scale, rates of decomposition have been determined from volcanic ash and basaltic flows which have been dated by radiometric means. In 4000 years, a layer of clay 1.8 m (6 ft) thick formed on a deposit of volcanic ash in the West Indies. Similarly, weathering of basalt flows in the arid Colorado Plateau shows a complete breakdown of the original rough flow surface in less than a million years. In the more humid regions of Hawaii, weathering of new basaltic flows has produced enough soil within one year for cultivation.

The Products of Weathering

Statement

It may appear from previous sections of this chapter that weathering simply produces decomposed and fragmented particles of rock. This is basically true, but these products of weathered rock are of prime importance to man. **Soil** is probably the most important weathering product, for we depend

Figure 6.12 *Weathering of the great Pyramids of Egypt. The weathering process is expressed in two ways: (a) The rectangular building blocks have been modified into an elliptical form, and (b) the weathered debris from the blocks has accumulated as talus on each step.*

The Earth's Dynamic Systems

Figure 6.13 Weathering of the ancient stepped pyramids. Each step is completely covered with talus, and many individual blocks are weathered to a spherical shape.

Discussion

Soil is widely distributed and so economically important that it has acquired a variety of definitions. Basically, it is the upper part of the regolith. It is composed chiefly of small particles of rocks and minerals, plus varying amounts of organic matter. Gradations from fresh rock, to partly decomposed fragments, to completely decayed rock material mixed with organic matter at the surface can be seen in road cuts or other excavations; these are referred to as the **soil profile** (figure 6.14).

A mature soil profile shows a rather consistent sequence of zones, distinguished by composition, color, or texture, between the surface and unbroken bedrock.

Horizon A is the topsoil layer often visibly divisible as three layers: A_0 is a thin surface layer of leaf mold, especially obvious on forest floors; A_1 is a humus-rich dark layer; and A_2 is a light, bleached layer.

Horizon B is the subsoil containing fine clays and colloids washed down from the topsoil.

Horizon C is a zone of broken and partially weathered bedrock.

The nature of the soil depends largely on climate, especially the amount and kind of rainfall and the range of temperature. Where mechanical weathering dominates, as in deserts, arctic

Figure 6.14 Sketch showing the major layers in a soil profile.

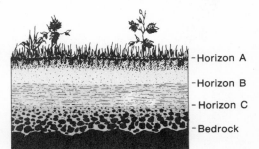

Weathering

133

tundras, and high mountain regions, the soil is thin and consists largely of broken fragments of bedrock. Organic matter is minimal. In equatorial regions where rainfall is heavy and temperature is high, chemical processes are much more rapid. In Brazil, the zone of weathered rock reaches a thickness of 130 m (427 ft). Elsewhere, in tropical to subtropical climates, layers of decomposed rock 60 m (197 ft) thick are not uncommon. Sodium, potassium, calcium, and magnesium form relatively soluble compounds and are soon removed by surface and subsurface water. This concentrates the insoluble hydrous iron and aluminum oxides which form a reddish soil called **laterite**. In the humid tropics or subtropics where intense and prolonged weathering occurs, concentrations of iron or aluminum may be so great that they can be mined. The aluminum-rich laterites are called **bauxites** after the little town of Les Baux, France, where they were first recognized. The formation of bauxite occurs in low-lying, relatively flat areas in the tropic or semitropical regions. Under such conditions, large volumes of water are available for solution activity and the warm temperatures speed up the chemical reactions.

Summary

Weathering processes act upon a rock to break it into small pieces and decompose the minerals which are unstable. The fact that most rocks are fractured by stresses in the crust greatly influences weathering by increasing the surface area of exposed rock and by providing channels through which weathering agents such as air and water have access to rock below the surface. As weathering proceeds, there is a universal tendency to produce spheroidal fragments. Climate exerts a profound influence upon weathering and controls the type and rate of weathering processes. One of the main products of weathering is soil.

Weathering processes do not erode and transport the fragmented and decomposed rock materials; they produce them. Other agents such as running water pick up the weathered rock fragments and transport them as sediment toward the sea. It is the process of erosion by running water to which we will direct our attention in the following three chapters.

Additional Readings

Carroll, D. 1970. Rock Weathering. New York: Plenum Press.

Goldich, S. S. 1938. "A Study in Rock Weathering." J. of Geology 46:17-58.

Keller, W. D. 1957. Principles of Chemical Weathering. Columbia, Mo.: Lucas Bros. Publishing Company.

Loughnan, F. C. 1969. Chemical Weathering of the Silicate Minerals. New York: American Elsevier Publishing Company.

Ollier, C. D. 1969. Weathering. New York: American Elsevier Publishing Company.

7 River Systems

A river system consists of a network of connecting channels through which surface water is collected and funneled back to the ocean. Estimates are that at any given time there are about 1300 km³ (300 mi³) of water flowing in the world's rivers. As the water flows, it carries with it a tremendous volume of sediment.

By measuring the volume of sediment in motion within a river system, we can measure the rates of erosion. Yet, how can we prove that streams have formed the valleys and canyons through which they flow? How could the Colorado River have cut the Grand Canyon over 2 km deep in solid bedrock? Since the erosion of a canyon cannot be observed from start to finish, geologists must approach the problem of erosion and the development of landforms indirectly. Two methods have been used: (1) They have analyzed the drainage network and found it to have system and order susceptible to mathematical analysis, and (2) they have studied the processes operating in drainage basins today, processes infinitely slow, but over a period of several million years capable of eroding the earth's surface to the form in which we see it today.

In this chapter, we will consider the order in river systems and the mechanics by which they erode the surface.

Major Concepts

1. Running water is part of the hydrologic system of our planet and is the most important eroding agent.
2. A river system consists of a main channel and all of the **tributaries** which flow into it. It may be subdivided into three subsystems: (a) tributaries which collect and funnel water and sediment into the main stream, (b) a main trunk which is largely a transporting system, and (c) a dispersal system at the river's mouth.
3. There is a high degree of order between the various elements of a river system.
4. The capacity of running water to erode and transport sediment is largely dependent upon stream velocity, although characteristics of the channel are also significant.
5. There is a universal tendency for a river system to establish equilibrium between the various factors which influence stream flow (velocity, volume of water, **gradient** of stream, and volume of sediment).

The Geologic Importance of Running Water

Running water is by far the most important agent of erosion and is responsible (together with **mass movement** of debris downslope) for the configuration of the landscape of most continental surfaces. Other agents such as ground water, glaciers, and wind are locally dominant but sculpture only limited parts of the land surface. Even where other agents are especially active, the effect of running water will still most likely be seen.

To a visiting astronaut, the surface of the moon appears as an irregular broken landscape cluttered with rock debris. The crater systems and patterns of terrain, so striking when viewed from space, are not at all apparent from vantage points on the surface. Indeed, crater rims may appear as rounded hills. Without the aid of maps or aerial photographs it is difficult to recognize some of the larger craters as being circular when seen only from the moon's surface.

The problem of appreciating the significance of stream valleys as systems of landforms covering the entire continent is much the same. When viewed from the ground, stream valleys may appear only as relatively insignificant irregular depressions between rolling hills, mountain peaks, or broad plains but, when viewed from space, they dominate the landscape on earth in much the same way as craters dominate the landscape of the moon. The ubiquitous nature of stream valleys and the importance of running water as the major agent of erosion may be best appreciated by considering a broad regional view of the continent and their major river systems obtained by high altitude aerial photography. The photographs in *figure 7.1* illustrate that throughout broad regions of the continents the surface is little more than a complex of valleys created by stream erosion. Even in the desert where it may not rain for tens of years, the network of valleys fashioned by streams is commonly the dominant landform.

Major Features of a River System

A river system consists of a main channel together with all of the tributaries which flow into it. It is bounded by a **divide** (ridge) beyond which water is drained by another system. The surface of the ground slopes toward the network of tributaries so that the drainage system acts as a funneling mechanism for removing precipitation and weathered rock debris.

A typical river system may be separated into three sub-

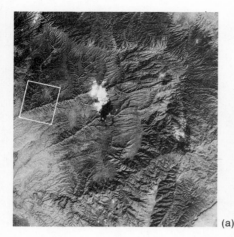

(a)

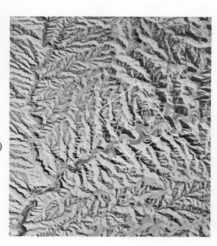

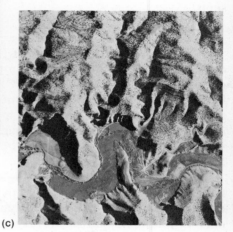

(c)

Figure 7.1 *(a) A Skylab photograph of an area in the arid southwestern United States shows the regional patterns of a river system and its valleys. (b) A high-altitude aerial photograph of a portion of the area shown in picture a shows that an intricate network of streams and valleys exists within the tributary regions of the larger streams. (c) A low-altitude photograph of part of the area shown in b shows many smaller streams and valleys in the drainage system.*

Figure 7.2 *Details traced from aerial photographs show the intricacy of a drainage system.*

systems: (1) a network of tributaries which collect and funnel water and sediment into the main stream (natural gathering system), (2) the main river channel (a transporting system), and (3) a network of **distributaries** at the mouth of the river where the sediment and water leave the river system (a dispersal system).

Discussion

The major features of a drainage system are clearly shown on a regional aerial photograph or a topographic map. The map pattern of most river systems is **dendritic** (treelike), with the trunk (main river) and numerous branches (tributaries). In theory, the drainage system includes all the minor rills and channels which are definite water courses, but these are too small to show on most maps. To appreciate the intricate network of channels integrated into a drainage system consider the details shown in *figure 7.2*.

The subsystems of an idealized river system are shown in *figure 7.3*. The material-gathering subsystem is the primary source of both water and sediment and is the area of most vigorous erosion. The transporting subsystem consists of the main river and larger tributaries. The major process of this segment is transporting water and sediment out of the drainage area, although additional water and sediment may be collected in this area. Also, some deposition may occur in the transporting sys-

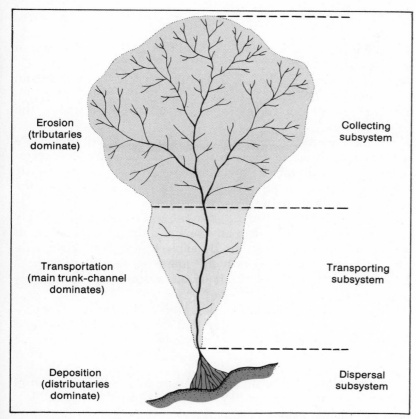

Erosion
(tributaries
dominate)

Collecting
subsystem

Transportation
(main trunk-channel
dominates)

Transporting
subsystem

Deposition
(distributaries
dominate)

Dispersal
subsystem

Figure 7.3 *Idealized diagram showing the major parts of a river system. The tributaries in the headwaters constitute a collecting subsystem where water and sediment are collected and funneled into a main trunk stream. Erosion is dominant in this area. The main trunk river functions as a transporting subsystem. Both erosion and deposition occur in this area. The lower end of the river is a dispersal subsystem where most sediment is deposited in a delta or an alluvial fan and water is dispersed into the sea. Deposition is the dominant process in this part of a river.*

Figure 7.4 *Sketches showing various dispersal systems of rivers. (a) Where a river enters a dry basin it generally splits into distributaries and deposits most of the sediment it carries to form an alluvial fan. (b) Where a tributary enters a major stream some sediment may be deposited temporarily, and some is carried away by the major stream. (c) Where the main river enters a lake or the sea, it splits into distributary channels which disperse the sediment across the delta and out into the shallow water.*

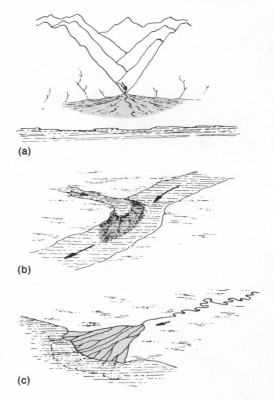

(a)

(b)

(c)

tem when the stream overflows or the channel migrates. The dispersing subsystem of a river is near the point where a river enters the sea, a lake, another river, or a dry basin. Here the river tends to break up into a series of small channels called distributaries (*figure 7.4*).

The characteristics of a river system, such as the length of stream segments, gradient (slope), and amount of discharge, can be measured and thus provide the quantitative basis for a study of the origin and processes of landscape evolution. These measurements are described in the following section and show that river systems have a high degree of organization.

Order in Stream Systems

Statement

Studies of drainage systems show that where a stream system is able to develop freely on a homogeneous surface there is a definite mathematical relationship (ratio) between the tributaries and the size and gradient of the stream and the valley through which it flows. Some of the more important relationships are as follows:

1. The number of stream segments (tributaries) increases upstream in a definite geometric progression.

2. The length of tributaries becomes progressively longer downstream.
3. The gradient of a stream channel decreases downstream in a constant ratio.
4. The stream channels become progressively deeper and wider downstream.
5. The size of the valley is proportional to the size of the stream.

Thus, if we were to survey the headwater of a drainage system and then proceed downstream, we would discover a series of very small tributaries flowing into larger ones having slightly lower gradients, longer lengths, and greater widths than the preceding ones. Farthest downstream we would reach the main river having the lowest gradient and largest valley of all. This relationship applies to any part of a drainage network, as well as the entire river system. At each stream junction the tributaries merge smoothly with the major stream to produce a hydrologic system of progressively increasing size and capacity downstream.

These relationships of an integrated drainage system constitute the basis for the assumption that streams erode the valleys through which they flow.

Discussion

The first meaningful observations about the relation of stream systems to their valleys was made in 1802 by John Playfair, an English geologist who stated:

> Every river appears to consist of a main trunk, fed from a variety of branches, each running in a valley proportional to its size, and all of them together forming a system of valleys connecting with one another, and having such a nice adjustment of their declivities that none of them join the principal valley at either too high or too low a level; a circumstance which would be infinitely improbable if each of these valleys were not the work of the stream which flows in it.

This statement has become known as *Playfair's Law.*

The relationship between various segments of a drainage system is pictured in *figure 7.5*, a map of a typical drainage system. We can subdivide the drainage network shown into a hierarchy of tributaries and analyze the relationship between various parts of the river system and its valleys. In the headwaters of the river system, the fingertip (unbranched) tributaries at the uppermost reaches of the stream are designated as *first-order streams.* Two first-order tributaries join to form a second-order and so on, with the major river or the trunk stream of any river system being the *highest-order number* of the entire system. (It should be noted that a first-order segment may join the second-, third-, or any higher-order channel without an increase in order at that point.)

The division of a river system into segments according to

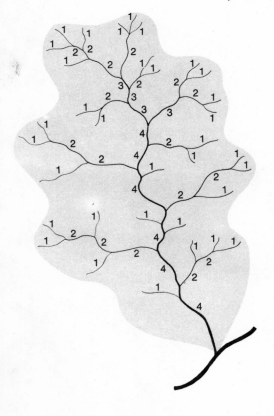

Figure 7.5 *Map showing the order of stream tributaries. In assigning orders to stream tributaries within a drainage system, the smallest tributaries are designated as first-order segments. The junction of two first-order tributaries produces a second-order segment. Two second-order segments join to produce a third-order segment and so on down to the largest stream channel of the system. Note that the junction of a first-order stream with a higher-order channel does not increase the order of the channel segment it joins.*

The Earth's Dynamic Systems

the method described above makes possible the demonstration of some important mathematical relationships concerning the form, dimensions, and other characteristics of a river system. For example, when traced upstream the channel branches into tributaries of a lower order at a constant ratio so that the number of tributaries in each succeeding order of stream forms a geometric progression. Thus, if a stream has a branching ratio of 3 and a main trunk of the sixth order, the number of tributaries will be one sixth-order, three fifth-order, nine fourth-order, 27 third-order, 81 second-order, and 243 first-order. This is, of course, geometric progression, showing a high degree of system and order in tributaries.

The length of tributaries also varies in a definite geometric progression beginning with the first-order segments and increasing downstream at a constant ratio. Studies of many stream networks confirm this principle and show that in areas of uniform climate, rock type, and state of development, the length of each tributary increases nearly three times in each succeeding higher order. Employing this principle, geologists have calculated that there are approximately 5.2 million kilometers (3.2 million miles) of river channels more than 1.5 km long in the United States. If the minor streams, creeks, and channels .5 to 1 km long are added to this, the network of rivers and valleys in the United States is over 28 million kilometers (17 million miles) long. This system of tributary development holds true regardless of scale. For example, a small area of the Badlands shown in *figure 7.6* is almost identical to a larger area shown in *figure 7.1.* Numerous observations support this deduction and show that the laws of stream numbers and lengths hold true regardless of whether the first-order drainage is measured in meters or kilometers.

Other relationships of stream segments and their valleys which can be demonstrated by statistical studies include the following: (1) the gradient of a stream or channel slope decreases in constant ratio downstream in each successively higher order, (2) the channels become deeper and wider downstream, and (3) the area drained by successive stream orders increases in a constant ratio (*figure 7.7*).

The mathematical relationship between orders of stream tributaries shows that there is a systematic development of drainage basins. It is a modern extension of observations made by Playfair who pointed out two important relationships between streams and their valleys: (1) The size of a valley is proportional to the size of the stream that flows in it and (2) the tributaries join a larger stream at an even level (the stream's own water level). Playfair reasoned that if valleys were "ready-made" for them by some other process such as faulting or other earth movements, these two relationhips would be "infinitely improbable." You may easily confirm Playfair's observations by carefully studying the aerial photographs in *figure 7.1.* Does each tributary have a steeper gradient than the stream into

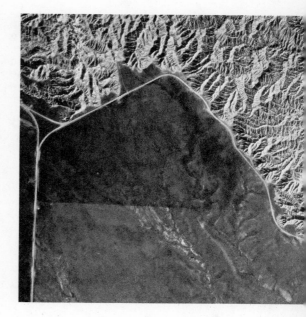

Figure 7.6 The system of tributaries developed by a small drainage system develops an intricate pattern similar to that of a major river draining a large area.

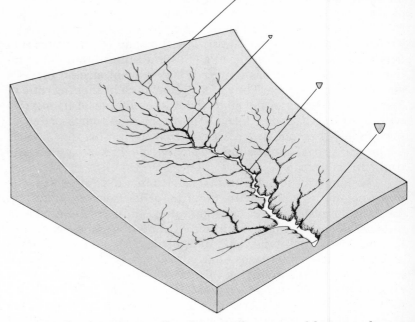

Figure 7.7 *Schematic diagram showing the systematic change in characteristics of a river downstream. The gradient decreases, as indicated by the curve of the diagram. In addition, the channel becomes larger, the volume of water increases, and the size of valley increases.*

which it flows? Does each tributary flow smoothly into a larger stream without an abrupt change in gradient? Are the tributary valleys smaller than the valleys into which they drain?

Stream erosion has been studied in great detail since Playfair's time, and over the last 100 years we have been able to observe and measure many aspects of stream development and erosion by running water. The origin of valleys by erosion is well established, and it is clear that running water constitutes the most significant process shaping the earth's surface.

The Flow of Water in Natural Streams

Statement

The flow of water through rivers is turbulent, moving with numerous secondary eddies and swirls in addition to the main downslope flow. **Laminar flow** (water moving in straight lines), when it occurs, is ~~restricted~~ *Not very* to a thin film a few millimeters thick on the bottom and walls of the stream channel. The greatest velocity of flow in a straight channel is usually found near the middle and above the deepest part of the channel. In curved or meandering channels maximum velocity is on the outside curve.

The discharge of water in rivers is influenced by a number of variables such as: (1) gradient (the slope), (2) the volume of flowing water, (3) the shape of the channel, and (4) the roughness of the channel.

Discussion

The nature of flow in rivers may be studied by injecting dye into a stream and observing its path of movement. In the

Figure 7.8 *Diagram showing types of flow: (a) Laminar flow—water particles move in parallel lines. (b) Turbulent flow—many secondary eddies are superposed on the main stream flow.*

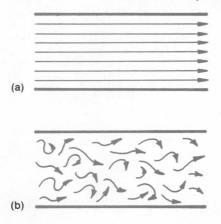

(a)

(b)

142

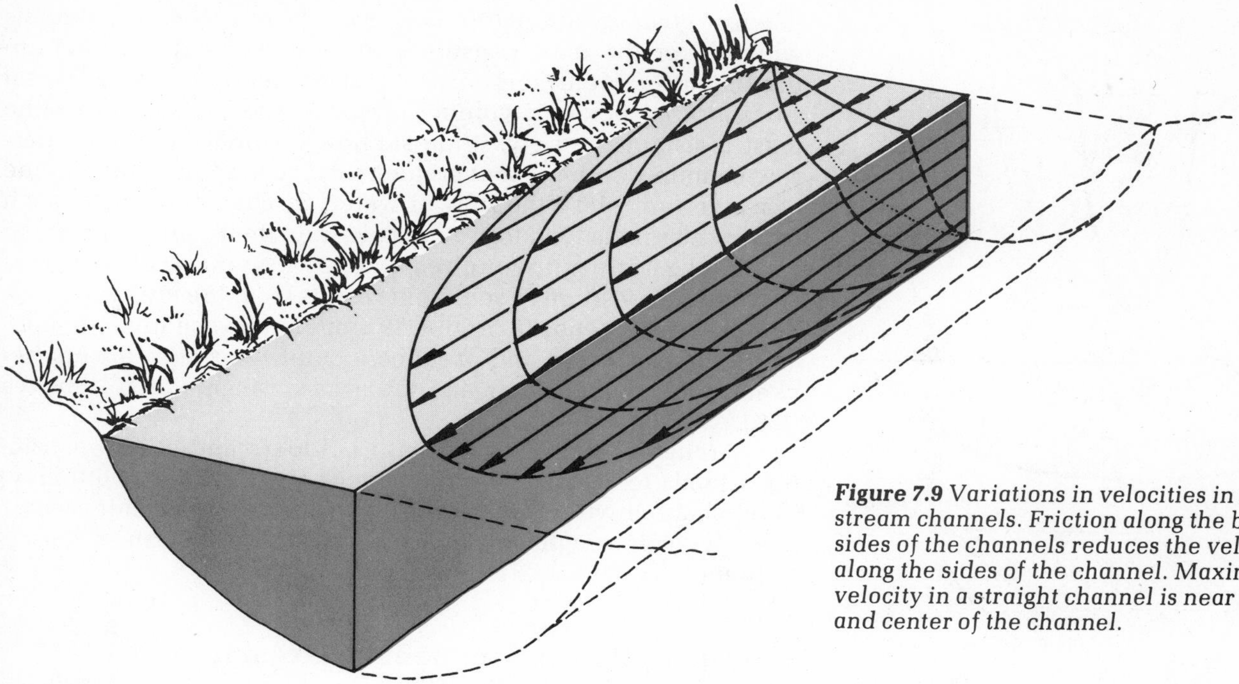

Figure 7.9 *Variations in velocities in natural stream channels. Friction along the base and sides of the channels reduces the velocity along the sides of the channel. Maximum velocity in a straight channel is near the top and center of the channel.*

laboratory, water moving in a glass tube at very low velocities moves in laminar flow, with parallel layers of water shearing over one another. Theoretically, there is no mixing. Laminar flow is not normally found in natural streams except for a very thin layer along the bottom and banks of the channel. However, it is common in ground-water flow.

Turbulent flow is a type of irregular movement, with eddies superposed on the main downstream flow (*figure 7.8*). Dye injected in turbulent flow moves down, sideways, and upward and shows that the water is capable of lifting loose sediment from the river bed and transporting it downstream.

The velocity of flowing water in a river is greatest near the center of the channel above the deepest part and away from the frictional drag of the sides and bottom of the channel (*figure 7.9*). As the channel curves the zone of maximum velocity shifts to the outside of the bend, as shown in *figure 7.10*.

The velocity of flowing streams is obviously related to the gradient of the stream channel. Steep gradients produce very rapid flow, seen commonly in high mountain streams. Where slopes are very steep, waterfalls and rapids develop and velocity approaches that of free fall. Low gradients produce slow sluggish flow and, where a stream enters a lake or the sea, the velocity is soon reduced to zero.

The velocity of flowing water also depends upon the volume of water within the channel. The greater the volume in a given channel, the faster the flow.

The cross-sectional shape of the river channel also influences the velocity at which a given volume of water will flow

Figure 7.10 *Patterns of flow in a curved channel. Water on the outside of the bend is forced to flow faster than that on the inside of the curve. This, together with normal frictional drag on the channel walls, produces a corkscrewlike pattern of flow. As a result, erosion occurs on the outer bank, and deposition occurs on the inside of the bend. This produces an asymmetrical channel which slowly migrates laterally.*

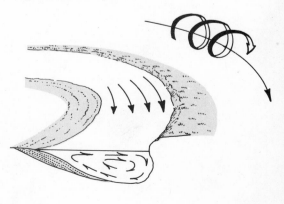

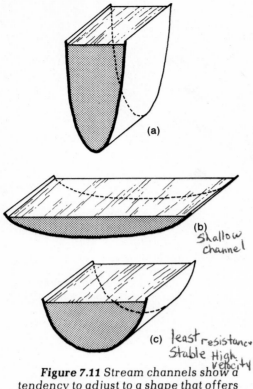

Figure 7.11 *Stream channels show a tendency to adjust to a shape that offers minimum resistance to flow. (a) Deep channels present a large surface area per unit volume of water and will be eroded wider. (b) Flat channels also present a large surface area and will be eroded deeper. (c) A semicircular channel presents a minimum surface per unit volume of water and is the optimum channel shape.*

down a given gradient. Velocity will be greatest in channels which offer the least resistance to flow. A channel which approaches an ideal semicircular cross section has the least surface area per unit volume of water and, hence, offers the least resistance to flow. Wide shallow channels and deep narrow channels both present greater surface area per unit volume of water and retard the flow of water. Streams flowing through narrow channels tend to crowd and erode their banks so that the channel shape approaches that of a semicircle. In flat channels, deposition of sediment results from the low velocities, and the stream tends to produce a more narrow channel. In both cases changes occur and produce a balance or equilibrium between the flowing water and the area through which it moves (figure 7.11).

A rough channel strewn with boulders and other obstructions tends to retard flow and reduce velocity. A smooth clay-lined channel will offer the least resistance and permit greater velocity. This roughness factor is difficult to measure, but its influence is obvious.

Transportation of Sediment by Streams

Statement

Running water is the major agent of erosion not only because of its ability to erode but because of its enormous power to transport loose sediment produced by weathering. The volume of erosional debris transported by a river is almost incomprehensibly large. The Mississippi River alone carries 454 million metric tons (500 million short tons) of sediment a year, a quantity roughly equivalent to a layer of sediment 0.3 m (1 ft) deep spread over an area of 870 km^2 (336 mi^2). In some rivers, the sediment carried at a given time may even greatly exceed the volume of water. Therefore, it is probably more accurate to consider a river as a system of moving water and sediment rather than one of flowing water alone. Within this system, sediment is transported in three ways: (1) solution, (2) suspension, and (3) traction (rolling, sliding, or bouncing).

Discussion

Suspended Load. The suspended **load** is the most obvious and generally the largest fraction of the material moved by a river. It consists of mud, silt, and sand and can be seen in most parts of the transporting segment of a river system. In arid regions the volume of sediment may exceed the volume of water; therefore, there are many gradations between a clear stream and a mudflow.

The mechanics by which sediment is transported in suspension can be understood by taking a handful of dirt and dropping it into a container of water. The large, coarse particles will sink rapidly, whereas the smaller particles will

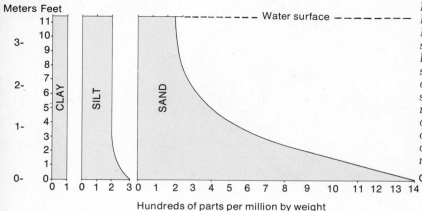

Figure 7.12 *Distribution in the sediment load from the top to the bottom of the Missouri River at Kansas City. The vertical scale shows depth of water, and the horizontal scale shows the amount of sediment being transported in hundreds of parts per million (by weight). Clay and silt are transported in suspension and are nearly equally distributed throughout the channel. Sand-size particles increase conspicuously towards the bottom of the channel because they are transported mostly by saltation and traction.*

Hundreds of parts per million by weight

stay in suspension for hours. Each size of particle has its own settling rate, the constant velocity at which the particle falls through a still fluid. The settling velocity is determined not only by the size of the particle but also by the specific gravity and shape of the particle, as well as by the density of the fluid. More time is required for a given particle to settle in muddy water or salt water because it is denser than clear water. Particles of gold, iron, or lead will settle faster than quartz because they are heavier. This principle is obvious to miners, who have "panned" for minerals in streams for ages.

In the turbulent flow of streams, a particle will remain in suspension if the upward component of turbulence exceeds the settling velocity. Therefore, fine silt and mud will remain in suspension in the transporting segment of a stream and will settle out only where velocity is reduced, e.g., where the stream enters a quiet body of water. Samples of sediment moved by the Missouri River at Kansas City show that clay-size particles are equally distributed from top to bottom, but sand-size particles occur in larger amounts near the bottom of the channel (*figure 7.12*).

The suspended load carried by large rivers represents a natural adjustment to existing conditions and is of considerable ecological importance. A dam across a river effectively traps all sediment behind it: the sediment will eventually fill the entire water storage basin and, unless removed by dredging or other means, will end the useful life of the reservoir. At the same time, depriving a river of its sediment load may seriously upset the balance of river activity downstream. In the design of canal systems, for example, the form of the channels must be adjusted for the quantity of sediment carried; otherwise deposition (*figure 7.13*) or abnormal erosion along the banks may occur.

The Traction Load. Particles of sediment too large to remain in suspension collect on the bottom of the stream and are moved by sliding, rolling, or **saltation** (short leaps). The movement of this **bed load** is one of the major tools of stream erosion, for as

Figure 7.13 Sedimentation in a canal, resulting in part from the universal tendency for a stream to develop a meandering pattern. The flow pattern illustrated in figure 7.10 causes variations in velocity to occur and be accentuated on the inside of a bend. Deposition occurs and the stream attempts to develop a meandering pattern.

the sand and gravel move they abrade (wear away) the sides and bottom of the stream channel. As the velocity and turbulence increase, these smaller particles are picked up from the stream bed and moved for a short distance in suspension. There is, thus, a transition from the movement of sediment by saltation to movement by traction.

Bed load may constitute 50% of total load in some rivers but in most rivers it ranges from 7 to 10% of the total **detrital** load. Although the movement of this material is not easily seen in most streams and rivers, one can feel and sometimes hear the dull thundering impact of boulders rolling over the channel bottom in large, rapidly flowing streams.

Dissolved Load. Dissolved matter is essentially invisible and is transported in the form of chemical ions. All streams carry some dissolved material derived principally from ground water which emerges as seeps and springs along the river banks. The most abundant compound in solution is calcium bicarbonate [$Ca(HCO_3)_2$], but sodium, magnesium, chlorine, ferric

The Earth's Dynamic Systems

and sulfate ions are also common. Various amounts of organic matter are also present, and some streams are brown with organic acids derived from decay of plant material. The velocity of flow, which is so important to the transportation of the suspended and traction load, has little effect on the capacity of the river to move material in solution. Apparently once mineral matter is dissolved it remains in solution, regardless of velocity, and is precipitated and deposited only upon changes in the chemistry of the water.

Chemical analysis shows that most rivers carry less than a thousand parts per million dissolved load, but some streams in arid regions carry several thousand parts per million and are distinctly salty to the taste. Although these amounts of dissolved material seem small, they are far from trivial. Sampling indicates that in some rivers, when rainfall is abundant, more than half of all material carried by rivers to the ocean is in solution. In areas of low relief such as the Atlantic and Gulf Coast states, runoff is slow, and solution activity lowers the surface at a rate of 52 metric tons (57 short tons) per square kilometer, or about 1 m in 25,000 years. The dissolved load is somewhat less in mountainous terrain (1 m in 100,000 to 3 million years), but it is by no means negligible.

Relation of Velocity to Erosion, Transportation, and Deposition

Statement

The capacity of running water to erode and transport sediment is largely dependent upon the stream velocity. If the velocity is doubled, the size of particles that can be transported increases four times. This is because each particle of water not only hits a particle of sediment twice as hard, but twice as much water strikes the face of the fragment during a given interval of time.

Discussion

The results of experimental studies of water's capacity to erode, transport, and deposit sediments are summarized in figure 7.14. The velocity at which a particle of a given size will be picked up and moved is the zone of stippled brown. On the graph, this is a zone instead of a line because the value or erosional velocity varies according to the characteristics of the water and the shape and density of the grain to be moved (shape has a definite effect on the suspension of flat mica flakes, for example). Moreover, erosional velocities vary with depth and density of the water. The lower curve (dashed line) shows the velocity at which the particle will settle out and come to rest. It is interesting to note that particles such as fine silt and clay require a relatively high velocity to lift them into suspension.

Figure 7.14 *Diagram showing threshold velocities of sediment in stream flow. The upper curve shows the velocity necessary for a stream to pick up and move a particle of a given size. This is a zone on the graph, not a line, because of variations resulting from stream depth, etc. The lower curve indicates the velocity at which a particle of a given size will settle out and be deposited. Note that fine particles will stay in suspension at velocities much lower than those required to lift them from the stream bed surface.*

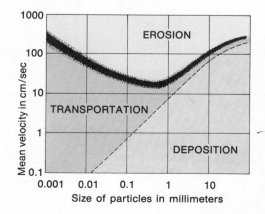

Small particles tend to stick together. However, once in suspension, fine particles will remain suspended with a minimum velocity.

The period of greatest erosion and transportation of sediment is during floods. The increase in velocity results in an increase in both the maximum size of material to be transported and the total load. Exceptional floods may cause an unusually large amount of erosion and transportation during a brief period of time. In arid regions many streams carry little or no water throughout the year, but during a cloudburst great quantities of sediment may be moved.

The graph in *figure 7.14* also shows that sediment will be deposited according to size whenever the velocity decreases below a critical settling velocity. Thus, on gentle slopes where the stream's velocity is reduced, a significant part of the sediment load will be deposited along the channel or on the flood plains of the river. Most of the remaining sediment is deposited when velocity is reduced at the point where the river enters a lake or the ocean.

Equilibrium in River Systems

Statement

In previous sections, we have emphasized that streams do not occur as separate, independent entities, but constitute a drainage system with each segment intimately related to the others. Every stream has tributaries, and every tributary has smaller tributaries extending down to the smallest gully. The entire river system functions as a unified whole, and any change brought about in one part of the system will affect other parts.

Adjustments within a river system are therefore continually being made so that ultimately the slope or gradient of a stream is altered to accommodate the volume of water available, channel characteristics, and velocity necessary to transport the sediment load. A change in any of these factors will cause compensating adjustments within the entire drainage system to restore equilibrium.

A profile of complete equilibrium in a river is one in which the channel form and gradient are delicately balanced so that neither erosion nor deposition occurs—an ideal condition toward which rivers are continually adjusting. The concept is important in understanding the natural evolution of the landscape. It also has some very practical applications since man is continually attempting to modify the rivers to better suit his needs, and he should know how the river system will respond to man-made modifications.

Discussion

The concept of equilibrium in a river system may be best appreciated by considering the slope or longitudinal profile of

The Earth's Dynamic Systems

a hypothetical stream in which equilibrium has been established (*figure 7.15*). In *diagram a*, the volume of water and the gradient of the stream are just enough to transport the sediment provided by the drainage basin so that neither erosion nor sedimentation is taking place. The energy within the stream system is in balance. The stream profile approximates a smooth curve.

Diagram b shows a disruption in the stream's gradient resulting from faulting. The steep gradient at the fault causes erosion to occur upstream (*diagram c*). The eroded sediment added to the stream on the dropped block is more than that segment of the stream can transport inasmuch as the stream was in equilibrium before faulting occurred. Therefore, the river deposits part of its load and builds up the gradient of the channel (shaded area in *diagrams c, d,* and *e*) until a new profile of equilibrium is established (*e*).

An example of the adjustments described above is found in Cabin Creek, a small tributary to the Madison River north of Hebgen Dam in Montana. In 1959, a 3 m fault scarp formed across the creek during the Hebgen Lake earthquake. By June, 1960, erosion by Cabin Creek had erased the waterfalls at the cliff formed by the fault and developed a small rapid. By 1965, the rapids were completely removed and equilibrium was reestablished.

The tendency for river systems to establish a profile of equilibrium is also illustrated by construction of dams. Dams are built to store water for industrial and irrigation use, to control floods, and to produce electric power. Yet, when a dam is built the balance of the river system established over thousands or millions of years is instantly upset. Many unforeseen long-term effects occur both upstream and downstream from the new structure and its great reservoir of water. In the reservoir behind the dam, the gradient is reduced to zero; hence, where the stream enters the reservoir, the sediment load carried by the stream is deposited as a delta and as layers of silt and mud over the floor. This well-known fact places an effective limit on the life of the reservoir because the lake basin is ultimately destined to be filled with sediment rather than with water (*figure 7.16*).

Less obvious are the changes in the gradient upstream as the delta builds out into the reservoir. To adjust to the new gradient, deposition occurs over the valley floor upstream from the delta. For example, upstream from Elephant Butte Reservoir in New Mexico, the Rio Grande buried the village of San Marcial with 3 m (10 ft) of sediment, while at Albuquerque 160 km (100 mi) further upstream 1.2 m (4 ft) of sediment was deposited.

Disturbance of the balance between gradient, sediment load, and discharge are among the downstream consequences of dam building. Since most sediment is trapped in the reservoir, the water released downstream has practically no sediment load. Therefore, it is capable of much more erosion than the previous river which carried a sediment load adjusted to

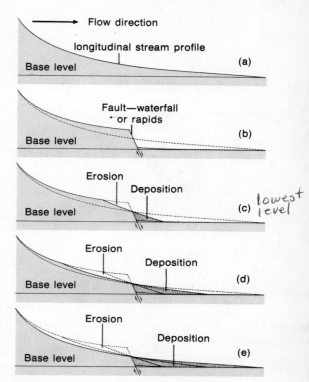

Figure 7.15 *Adjustments of a stream profile to reestablish equilibrium. (a) When the stream profile is at equilibrium, velocity, load, gradient, and volume of water are in balance so that neither erosion nor deposition occurs. (b) Faulting disrupts equilibrium by decreasing the gradient downstream and increasing the gradient at the fault line. (c) Erosion proceeds upstream from the fault, and deposition occurs downstream. (d) Erosion and deposition continue to develop a new stream profile. (e) A new profile of equilibrium is eventually established in which neither erosion nor deposition occurs.*

Figure 7.16 Mono Reservoir, California. This reservoir was built as a debris basin to protect the Gebaltar Reservoir downstream. It was built in 1935 following a severe fire which burned off much of the vegetation in much of the watershed area. Very heavy rains occurred during the following two winter seasons before vegetation could be reestablished, and the accelerated erosion provided enough silt and sand to completely fill the Mono Reservoir and partly fill the lower Gebaltar Reservoir. This serves as a vivid reminder that a river system is a system of moving water and sediment, not just water alone.

the gradient, and extensive scour and erosion commonly results downstream.

Another important downstream effect of a dam is that the sediment supply to the sea is cut. The delta built by the river ceases to grow and is commonly eroded by wave action.

The Aswan Dam of the Nile provides a good example of the many consequences of modifying river systems. For centuries the Nile River has been the sole source of life to the country of Egypt. The principal headwaters of the Nile are located in the high plateaus of Ethiopia. Once a year for approximately one month, the Nile normally rises to flood stage and covers much of the fertile farm land in the Nile Delta area. The Aswan Dam, completed in the summer of 1970, was to provide Egypt with irrigation water for 1 million acres of arid land, generate 10 billion kilowatts of power, double the national income, and industrialize the state. The dam has drastically modified the equilibrium of the Nile, and many adjustments to this major change in the river have resulted.

The Nile is not only the source of water for the delta, it is also the source of sediment. When the dam was finished and began to trap the sediment in Lake Nasser, it destroyed the physical and biological balance in the delta area. Without the annual "gift of the Nile," erosion is eating the delta coastline which now is exposed to the full force of marine currents. Some parts of the delta are receding several meters a year.

150

The sediment previously carried by the Nile was an important link in the aquatic food chain nourishing marine life in front of the delta. The lack of Nile sediment has reduced plankton and organic carbon to a third of what they used to be, thus either killing off sardines, scombroids, and crustaceans or driving them away. The annual harvest of 16,000 metric tons (18,000 short tons) of sardines and a fifth of the previous fish catch have been eliminated.

The sediment of the Nile also acted as a natural fertilizer; without this annual addition of soil nutrients, Egypt's million cultivated acres need artificial fertilizer.

Without its load of sediment, the water discharged from the dam flows downstream much faster and is vigorously eroding the channel bank. This scouring process has already destroyed three old barrier dams and 550 bridges built since 1953. Ten new barrier dams between Aswan and the sea must now be built at a cost equal to one-fourth the cost of the dam itself.

The annual flood of the Nile was also important to the ecology of the area because it washed away the salts which formed in the soil. Already soil salinity has increased not only in the Delta but throughout the middle and upper Nile and, unless corrective measures are taken at a cost of over $1 billion, millions of acres will revert to the desert within a decade.

The change in the river system has permitted double-cropping, but there are no periods of dryness which previously helped limit the population of bilharzia, a blood parasite carried by snails which infects the intestinal and urinary tracts of humans. One out of every two Egyptians now has the infection, and one out of every ten deaths in the country is caused by it.

Problems also occur in the lake behind the dam. The lake was to have reached a minimum level in 1970, but it may actually take 200 years to fill up. More than 15 million cubic meters (530 million cubic feet) of water escape underground into the porous Nubian Sandstone which lines 480 km (300 mi) of its western bank. It is capable of absorbing an almost unlimited quantity of water. Moreover, the lake is located in one of the hottest and dryest places on earth, and evaporation losses are staggering. This was expected, but additional loss from transpiration of plants growing along the lake shore and increased evaporation from high wind velocity have nearly doubled the expected loss of water from the lake. This loss equals half the total amount of water that once was "wasted" while flowing unused to the sea.

Flood Control. Of course, there is another side to the story. In terms of property damage and loss of life, floods are considered to be one of the most disastrous of natural phenomena and are most effectively controlled by dams. Floods are caused primarily by high precipitation over a short period of time in a given drainage basin. Rapid thawing of snow and ice may also furnish flood waters during the spring thaw. In some areas like the lower Nile, it may be at fairly regular intervals; in other

Parasites in Snails causes infections.

Lake behind the Dam was supposed to be resort. = Nile River Crocodile

rivers, years may pass between floods. However destructive they may seem to be, floods are natural stream processes which have operated throughout geologic time and are important in maintaining the balance in a river system.

A long-range problem with flood control measures such as dams and levees is that they upset the balance of the river system. The gradient is modified, and the sediment is impounded behind dams and levees. Ultimately, the reservoirs will be filled with sediment, and meandering rivers will change courses.

Summary

A river system is a network of connecting channels through which surface water is collected and funneled back to the ocean. It may be divided into three subsystems: (1) tributaries which collect and transport water and sediment into the main stream, (2) a main trunk stream which is largely a transporting system, and (3) a dispersal system at the river's mouth.

Studies of river systems shows that there is a high degree of order between tributaries and the size and gradients of their valleys, indicating that streams erode the valleys through which they flow.

A river system is best thought of as a system of moving water and sediment because an enormous volume of sediment is constantly being transported to the sea by running water. The sediment is moved in solution, in suspension, and by traction and is visible in most rivers, especially where it is being deposited as deltas near the mouth of the river. The capacity of running water to erode and transport sediment is largely dependent upon the stream's velocity.

A river system functions as a unified whole and shows a universal tendency to establish equilibrium between the various factors which influence flow (velocity, volume of water, gradient of stream, and volume of sediment). When one of these factors is changed, adjustments in the river system will be made to reestablish equilibrium.

Running water is the most important eroding agent on our planet, and stream valleys are the most characteristic landforms on continental surfaces. In the next chapter we will consider how streams erode their valleys and canyons and how the resulting sediment is ultimately deposited as deltas, alluvial fans, and other landforms.

Additional Readings

Leopold, L. B., and W. B. Langbein. 1966. "River Meanders." Sci. Amer. 214(6): 60-70 (Offprint No. 869).

Leopold. L. B., M. G. Wolman, and J. P. Miller. 1964. Fluvial Processes in Geomorphology. San Francisco: W. H. Freeman and Company.

Morisawa, M. 1968. Streams—Their Dynamics and Morphology. New York: McGraw-Hill Book Company.

8 Processes of Stream Erosion and Deposition

Erosion of the land, one of the major effects of the hydrologic cycle, is a process which will continue as long as the continents are exposed above the sea. The products of rock weathering, broken and decomposed rock fragments, are sooner or later removed from their place of formation and carried by wind, water, or ice and ultimately deposited as sediment. Erosion and its by-product, sedimentation, are two processes of prime importance in the circulation of matter on the earth's surface.

Because man lives from the soil, erosion and removal of the soil have always been considered a catastrophe, something to be stopped or at least reduced. Various attempts have been made, and will continue to be made, to slow down this inevitable process. Therefore, an understanding of the mechanics of erosion is of fundamental importance in understanding our ecological setting. However, erosion is part of a natural system which if checked in one area may produce undesirable effects in another. Attempts to control stream erosion in the upper part of a drainage basin may result in the destruction of land by marine erosion in the delta. Attempts to control erosion must take into consideration the entire river system, including the potentially disastrous chain of cause-and-effect events which may result from the alteration of a seemingly isolated feature in a delicately balanced system.

To understand erosional processes and their results, we must attempt to analyze them as independent isolated activities, but we must always remember that they are interdependent and intimately related to a single system.

In the development of a drainage system and subsequent erosion of a landscape, there is a universal tendency for the river system to (1) deepen its channel by downcutting if there is sufficient gradient, (2) extend the drainage network upslope by **headward erosion**, (3) widen its valley by mass movement and **slope retreat,** and (4) extend its drainage channel downslope as sea level (base level) recedes.

Where stream velocity is reduced, deposition of the sediment load produces a variety of landforms such as **flood plains, deltas,** and **alluvial fans.**

Major Concepts

1. Downcutting of stream channels is accomplished by **abrasion, solution,** and **hydraulic** action.
2. Drainage systems grow by headward erosion, which commonly results in **stream piracy.**
3. Slope retreat causes valleys to grow wider and is associated with downcutting and headward erosion. It results from various types of mass movement under the pull of gravity.
4. As a river develops a low gradient, it releases part of its load by deposition on point bars of meanders, natural levees, and across the surface of its flood plain.
5. Where rivers empty into lakes or the ocean, most of the sediment is deposited. This commonly builds a delta at the river's mouth. In arid regions many streams deposit their loads as alluvial fans at the base of a steep slope.

Downcutting of the Stream Channel

Statement

The basic process of erosion in all streams, from the small gulley on a hillside to the great canyons of the major rivers, consists of the abrasive action of sand, cobbles, and boulders as they are moved by running water along the channel floor. Other erosional processes in stream channels are solution activity and direct hydraulic action.

Discussion

Abrasion. When we clearly realize that river systems are not only systems of flowing water but systems of moving water and sediment, we can better comprehend the effect of abrasive action on the river channel as gravel and sand are swept along the river bed. Although the sediment may be deposited temporarily along the way during low water, it is picked up and moved again during high water or floods, and as it moves it continually abrades the surface of the stream channel. The process of abrasion in river beds is not unlike that of a wire saw used in quarries to cut and shape large blocks of stone. An abrasive such as garnet, corundum, or quartz, when dragged across the rock by a wire, is able to cut through the block. Steep, near-vertical gorges in many canyons of the southwestern United States are clear examples of the power of streams to cut downward (*figure 8.1*). With the abrasive tools of sand and gravel the hardest bedrock can be excavated.

An effective and interesting type of stream abrasion is the drilling action of pebbles or cobbles trapped in a depression and swirled around by the moving water. As they are worn away, new ones take their place and continue to drill in the stream channel, eventually developing **potholes** which may range in diameter from a few centimeters to several meters (*figure 8.2*).

In addition to the abrasion and downcutting of the stream channel, pebbles and cobbles themselves are worn down as they strike one another and the channel bottom. The corners and edges are chipped off, and the particles become smooth and rounded. Thus, large boulders which fall into the stream, or are transported only during a flood, are broken and worn down to smaller fragments and ultimately washed away as grains of sand.

Stream abrasion also effectively undercuts the stream banks, and large blocks of rock and regolith **slump** off and become part of the stream load. Undercutting is particularly effective at the base of a waterfall where water is particularly turbulent and hydraulic action especially effective (*figure 8.3*).

Solution. Although solution activity of running water is generally not visible, the chemical content of rivers shows that it is a very effective and important means of lowering the land surface. Estimates based on measurements of the chemical

Processes of Stream Erosion and Deposition

Figure 8.1 *Sand and gravel are the tools of stream erosion. Transported by a river, they act as powerful abrasives and cut through the bedrock as they are moved by flowing water. The abrasive action of sand and gravel has cut this vertical gorge through the resistant limestone formation in the Grand Canyon, Arizona.*

Figure 8.2 *Potholes are eroded in the hard rock of a stream bed by sand, pebbles, and cobbles whirled around by eddies of the stream.*

Figure 8.3 *The retreat of Niagara Falls upstream with the resulting formation of Niagara Gorge. The Niagara River originated as the last glacier receded from the area and water flowed from Lake Erie to Lake Ontario over the Niagaran Cliffs. Erosion caused the waterfalls to migrate upstream at an average rate of 1.3 m (4 ft) per year.*

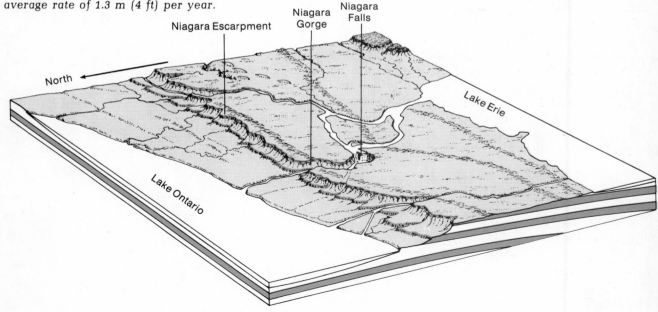

Niagara Escarpment

Niagara Gorge

Niagara Falls

North

Lake Erie

Lake Ontario

content of the world's major rivers indicate that each year a total of 3.5 billion metric tons (3.9 billion short tons) of dissolved material are carried from the continents by running water. The effectiveness of chemical erosion is illustrated by the fact that in humid areas limestone, which is the most soluble of the major rock types, forms valleys and is considered a weak rock, but in arid regions it is resistant to erosion by solution and forms cliffs.

Most of the dissolved matter is probably derived from ground-water seepage into the stream, but chemical erosion, especially in limestone terrains, also occurs in the stream channel and in the small particles of soluble rock carried by the stream.

Hydraulic Action. The force of running water alone, without tools of abrasion, can quickly erode soft unconsolidated material such as soil, clay, sand, and gravel and is especially effective when water is moving at high velocities. Although directed hydraulic action is unable to rapidly erode large quantities of solid bedrock, pressure exerted by moving water into cracks and bedding planes may be sufficient to remove slabs or blocks from the channel floor or sides.

Headward Erosion and Growth of the Drainage System Upslope

Statement

In the process of stream erosion and valley evolution there is a universal tendency for each stream to erode headward, or upslope, and increase the length of its valley until it reaches the divide. Every tributary is involved in this process until the length of the valley is extended upslope as far as possible.

Discussion

The processes of headward erosion can be analyzed by references to figures 8.4 and 8.5. The reason erosion is more vigorous at the head of a valley than on its sides will be apparent when you understand the relationship which exists between the valleys and regional slope. Above the heads of valleys, water flows as a sheet (sheet flow) down the regional slope and converges to a point where a definite stream channel begins (figure 8.4). As soon as the water is concentrated into a channel, its velocity and erosive power increase to a point far beyond that of a slower moving sheet of water on the surrounding and ungullied surface. This additional volume and velocity erode the the head of the valley much faster than the sheet did on ungullied slopes, so the head of the valley is extended upslope. In addition, ground water moves downslope toward the valley, making the head of the valley favorable for the development of **springs** and **seeps.** This, in turn, tends to undercut overlying resistant rock and causes the head of the valley to grow upslope much faster than the retreat of the valley walls.

Figure 8.4 *Headward erosion and extension of the drainage system upslope. Water flows down the regional slope (away from the viewer) as a sheet and converges toward the head of the tributary valleys. The tributary valley is thus eroded headward or up to the regional slope.*

Figure 8.5 *Headward erosion is caused by the concentration of sheet flow into a channel at the head of tributary valley. At this point this greatly increases the velocity of the water, thus increasing its ability to erode. The tributaries therefore erode upslope.*

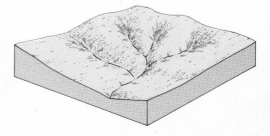

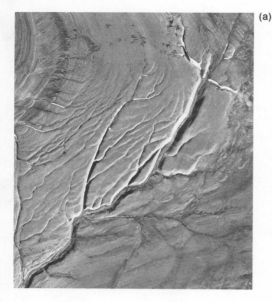

(a)　　　　　　　　(b)　　　　　　　　(c)

Figure 8.6 *Stream piracy occurs where tributaries with high gradients erode headward and capture the tributaries of another stream. Many of the southwest flowing streams shown in the photograph (a) have been captured by the tributaries shown in brown in the sketch (b). A sketch showing the drainage before stream capture is shown in diagram (c). Where will stream piracy likely occur in the near future?*

Careful study of the relationship between the drainage network and regional slope (*figure 8.5*) shows that the position of the tributary is fixed at the point where it joins the next larger segment of the stream. Continued erosion along this stream segment is restricted to downcutting or valley widening. Thus, it is possible for the drainage system to grow only by extending existing gullies up the regional slope or by developing new tributaries along the walls of the larger stream. Inevitably the gullies will extend up the slope of the undissected surface toward the gathering grounds of the waters that feed the stream. Thus, headward erosion is a universal tendency in the evolution of drainage systems.

Stream Piracy. With the universal tendency for headward erosion, the tributaries of one stream may extend upslope and intersect the middle course of another, thus diverting the headwater of one stream to another. The process is known as stream piracy or stream capture and is illustrated in *figure 8.6*.

Stream piracy is most likely to occur in a situation where the headward erosion of one stream is favored by a steeper gradient or by a course in more easily eroded rocks. Some of the most spectacular examples occur in the folded Appalachian Mountains where nonresistant shale and limestone are easily eroded, whereas the hard quartzite formations with which they are interbedded are very resistant. The process of stream capture and the evolution of the drainage system in this region is shown in the series of diagrams in *figure 8.7*. The original streams flowed in a dendritic pattern on horizontal sediments which covered eroded folds. As erosion removed the horizontal sediments, the drainage pattern became superposed or placed upon the folded rocks, cutting across the weak and resistant rocks alike. As the major stream cut a valley across the folded rocks, new tributaries rapidly extended themselves headward along

The Earth's Dynamic Systems

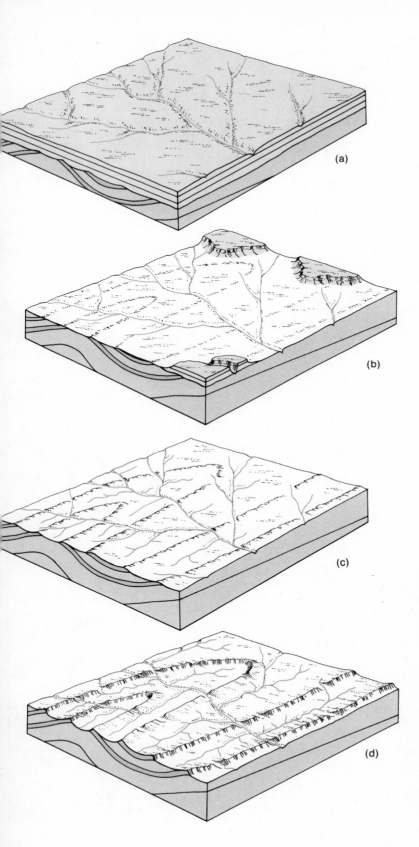

Figure 8.7 Superposition of a dendritic
drainage pattern across a series of folded
rocks with the subsequent development of
a trellis drainage pattern resulting from
headward erosion. (a) The original
dendritic pattern forms on horizontal
sedimentary rocks which cover the
ancient eroded folds. (b) Erosion removes
the horizontal sediments exposing the
older folded rocks at the surface. The
dendritic drainage pattern is then
superposed or placed upon the folded
rocks. (c) Streams cut across resistant
and nonresistant rocks alike because
erosion is restricted to downcutting,
headward erosion, and slope retreat.
(d) Rapid headward erosion along
exposures of weak rocks results in stream
capture and modification of the original
dendritic pattern to a trellis pattern.

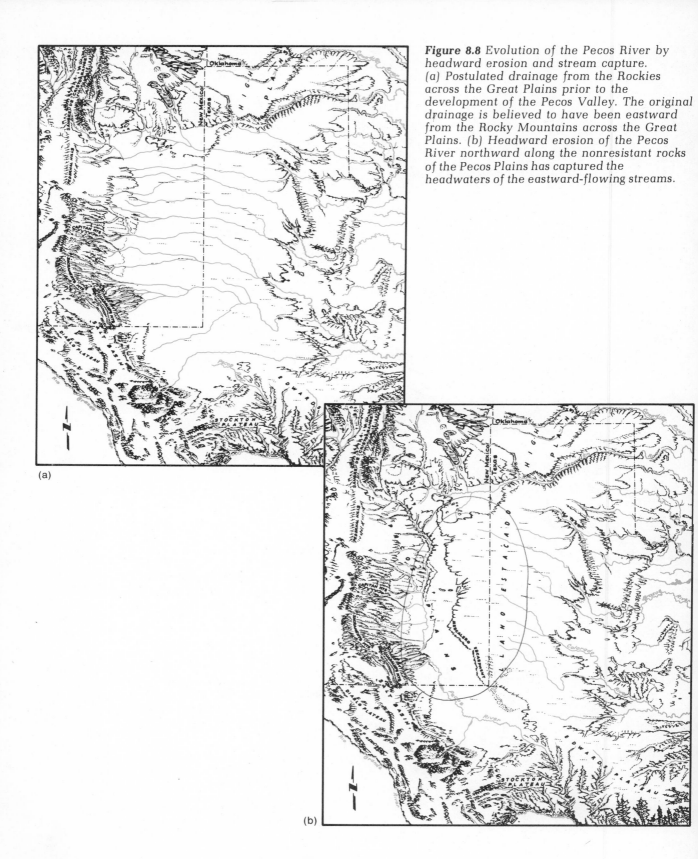

Figure 8.8 Evolution of the Pecos River by headward erosion and stream capture. (a) Postulated drainage from the Rockies across the Great Plains prior to the development of the Pecos Valley. The original drainage is believed to have been eastward from the Rocky Mountains across the Great Plains. (b) Headward erosion of the Pecos River northward along the nonresistant rocks of the Pecos Plains has captured the headwaters of the eastward-flowing streams.

(a)

(b)

the nonresistant formations. In doing so, they progressively captured the superposed tributaries and changed the dendritic drainage to a trellis pattern.

Extensive stream piracy and development of a trellis drainage pattern can be seen almost anywhere folded rocks are exposed at the surface. In the folded Appalachian Mountains the major streams that flow to the Atlantic, such as the Susquehanna and Potomac, are all superposed across the folded strata, but their tributaries flow along the nonresistant rocks parallel to the geologic structure and have captured many superposed streams.

Another example is the Pecos River in New Mexico which, by extending itself headward along the weak shale and limestone, has captured a series of east-flowing streams which once extended from the Rockies across the Great Plains. The original eastward drainage shown in *figure 8.8A* resulted from uplift of the Rocky Mountains. Now as a result of headward erosion and stream piracy the headwaters of the southern rivers flow to the Pecos (*figure 8.8B*).

Headward erosion proceeds until the tributary meets a stream flowing in the opposite direction at the crest of a dividing ridge. Further headward extension is effectively eliminated when no undissected slope remains. Subsequent erosion in the tributary areas is restricted to downcutting and valley widening. Therefore, a drainage system grows in a specific way and, once sheet flow is concentrated into a channel, the future development of the drainage is predictable.

Mass Movement and Slope Retreat

Statement

In addition to downcutting and headward erosion, a valley is subjected to a variety of slope processes which cause the walls to recede and retreat away from the river channel. These processes, commonly referred to as mass movement, may be rapid and devastating, such as a great landslide, or imperceptibly slow, such as creep down the gentle slope of a grass-covered field.

Gravity is the driving force behind all slope processes. In waterless areas, such as local regions of Antarctica and on the moon, gravity alone operates to move unconsolidated material downslope, but in most areas on earth water is an important factor because it lubricates the unconsolidated material and adds weight to the mass. The force of gravity is, of course, continuous, but gravity is able to move material only when it is able to overcome inertia. Any factor that tends to reduce resistance to motion aids mass movement. These include saturation of the material by water, over-steepening of slopes by undercutting by streams or waves, alternating freezing and thawing, and vibrations from earthquakes.

Processes of Stream Erosion and Deposition　　　　　**161**

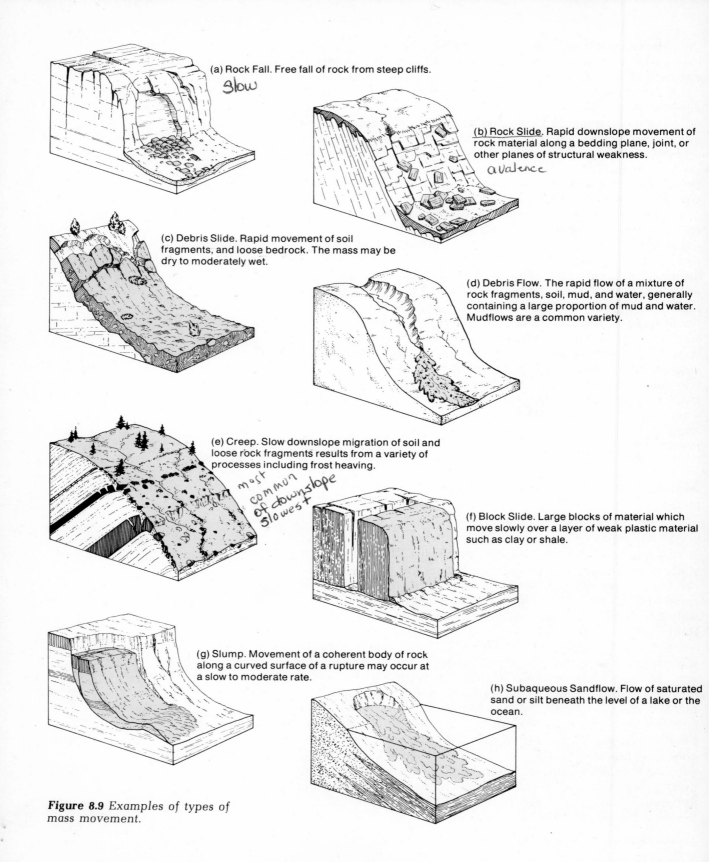

(a) Rock Fall. Free fall of rock from steep cliffs.

Slow

(b) Rock Slide. Rapid downslope movement of rock material along a bedding plane, joint, or other planes of structural weakness.

avalence

(c) Debris Slide. Rapid movement of soil fragments, and loose bedrock. The mass may be dry to moderately wet.

(d) Debris Flow. The rapid flow of a mixture of rock fragments, soil, mud, and water, generally containing a large proportion of mud and water. Mudflows are a common variety.

(e) Creep. Slow downslope migration of soil and loose rock fragments results from a variety of processes including frost heaving.

most common of downslope slowest

(f) Block Slide. Large blocks of material which move slowly over a layer of weak plastic material such as clay or shale.

(g) Slump. Movement of a coherent body of rock along a curved surface of a rupture may occur at a slow to moderate rate.

(h) Subaqueous Sandflow. Flow of saturated sand or silt beneath the level of a lake or the ocean.

Figure 8.9 *Examples of types of mass movement.*

Mass movement of surface material downslope may be considered as the most universal of all erosional processes. It occurs under all geologic conditions—on the steep slopes of high mountains, the gentle rolling plains, the sea cliffs, and on the slopes beneath the oceans and, maybe even more important, as an erosional process on the moon, Mars, and other planets.

Discussion

The mechanics of mass movement may be very complicated; however, for convenience we will consider examples from two groups: (1) predominantly rapid downslope movements and (2) predominantly slow movements. Some examples of each are shown in the diagrams in *figure 8.9*.

Rock falls are the most rapid, ranging from large masses of rock that break loose from the face of a cliff to small fragments loosened by frost action and other weathering processes. In areas where rock falls are common the debris accumulates at the base of a cliff as talus.

In a **rock slide** movement occurs along a plane of structural weakness within the rock body, such as a bedding plane or fracture. The rock may move as a large block for a short distance, but more frequently it breaks up into smaller blocks and rubble. Rock slides are likely to occur on steep mountain fronts but may develop on slopes with gradients as low as 15°. They are commonly among the most catastrophic of all forms of mass movement, sometimes involving millions of tons of rock which may plunge down the side of a mountain in a few seconds (*figure 8.10*).

Debris slides consist of dry to moderately wet loose rock fragments which move rapidly over the surface of underlying bedrock.

Figure 8.10 *A recent rock slide in the Swiss Alps.*

Processes of Stream Erosion and Deposition

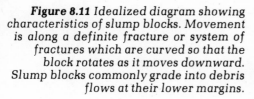

Figure 8.11 *Idealized diagram showing characteristics of slump blocks. Movement is along a definite fracture or system of fractures which are curved so that the block rotates as it moves downward. Slump blocks commonly grade into debris flows at their lower margins.*

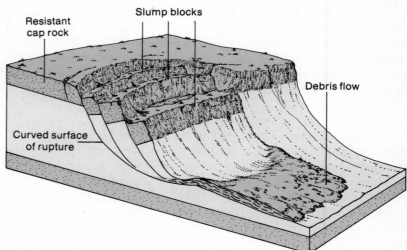

Debris flows consist of a mixture of rock fragments, mud, and water which flow downslope as a thick viscous fluid. Depending on viscosity, movement may range from that similar to the flow of freshly mixed concrete, to a stream of fluid mud in which rates of flow are equal to those of rivers. Many debris flows begin as slumps and continue as a flow near the distal margins of the slump block (*figure 8.11*).

Mudflows, a variety of debris flow consisting of a large percentage of silt and clay-sized particles, almost invariably result from unusually heavy rain or sudden thaw. Water content may amount to as much as 30%. As a result of the predominance of fine-grained particles and high water content, mudflows tend to follow stream valleys. They are common in arid and semiarid regions and originate typically in steep-sided gullies where there is abundant loose weathered debris. As they reach the open country at a mountain front, they tend to spread out in the shape of a large lobe or fan. Because of their great density, mudflows can transport large boulders over slopes as gentle as 5° and have been known to move homes and barns from their foundations. Many of the disastrous landslides in southern California are really mudflows which move rapidly down a valley for considerable distance. Mudflows may vary in size and rate of flow depending on water content, slope angle, and available debris, but many are over 100 m (320 ft) thick, and some may be as much as 80 km (50 mi) in length.

Creep is an extremely slow, almost imperceptible downslope movement of soil and rock debris. It is expressed in weakly consolidated grass-covered slopes as bulges or low wavelike swells. In road cuts or stream banks, evidence of creep may be expressed by the bending of steeply dipping strata in a downslope direction or transported blocks of a distinctive rock type downslope from their outcrop. Additional evidence of creep consists of tilted trees and posts, displaced monuments, deformed roads or fence lines, and tilted retaining walls.

164

Many factors contribute to the cause of creep. Soil moisture is undoubtedly important in that it weakens the soil's resistance to movement. In cold regions creep is produced by frost heaving, a process in which water percolates into the pore space of the rock, freezes, and expands, causing the ground surface to be lifted at right angles to the slope. When thawing occurs, each particle tends to drop vertically, coming to rest slightly down-slope from its original position (*figure 8.12*). Repeated freezing and thawing causes the particles on the slope to move in a discontinuous series of zigzags. Wetting and drying may cause a similar motion of loose particles as moisture causes expansion of clay minerals.

Many other factors contribute to creep, including growing plants which exert a wedgelike pressure between rock particles in the soil, thus causing them to be displaced downslope. Burrowing organisms also displace particles, permitting the force of gravity to move them. Creep may involve the entire slope of a valley, although the rate of movement varies from one place to another, being greatest where water is most abundant.

Landslide is a term used for a wide variety of predominantly rapid earth movements including the subtle slumping of stream banks or sea cliffs as well as the more obvious sliding (slipping) of mountain sides. Movement may include both the unconsolidated overburden of soil and the solid bedrock. A landslide block moves as a unit or series of units along a definite plane, in contrast to debris flows which move as a viscous fluid. The development of a definite fracture or system of fractures in the rock body due to gravitational stresses leads to slumping. The material acts as an elastic solid in which a large block moves downward and outward along a curved plane (*figure 8.11*).

Slumping commonly occurs where resistant rock such as limestone, sandstone, or basalt caps a much weaker shale formation. Rapid erosion of the weak underlying shale undermines the cap rock to produce an unstable condition. Slump blocks in bedrock may be as much as 4.8 km (3 mi) long and 150 m (500 ft) thick and move in a matter of seconds or gradually slip throughout a period of several weeks.

Solifluction, a special type of earth flow, occurs in arctic and subarctic regions where the ground is frozen to considerable depth. During the spring and early summer, the ground begins to thaw from the surface downward. Since the melt water cannot percolate downward into the impermeable permafrost, the upper zone of soil is always saturated and tends to slowly flow down even the most gentle slopes.

Rock Glaciers. In regions of cold temperatures such as Alaska, the high Rockies, and other alpine areas, tonguelike masses of angular rock debris resembling a glacier move as a body downslope at rates ranging from 3 cm (1 in.) a day to 1 m (3.3 ft) a year. These bodies are known as **rock glaciers** and represent a type of mass movement involving slow flowage. Evidence of flow includes concentric ridges within the body, its

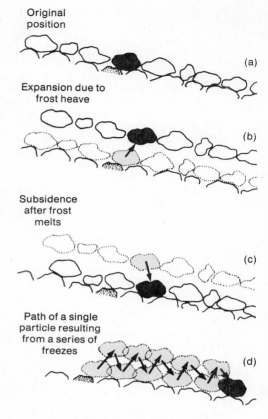

Original position

Expansion due to frost heave

Subsidence after frost melts

Path of a single particle resulting from a series of freezes

Figure 8.12 *Idealized diagram showing how creep results from repeated frost heaving. (a) Water seeps into the pore spaces between loose rock debris. (b) As the water freezes and expands, the soil and rock fragments are lifted perpendicular to the ground surface. (c) As the ice melts, the particles move down vertically through the pull of gravity and are displaced slightly downhill. (d) Repeated freezing and thawing causes a significant net displacement downslope.*

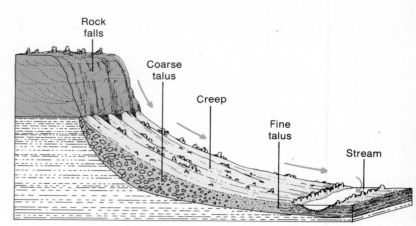

Figure 8.13 *Diagram showing the system by which loose debris is moved downslope into a drainage channel. The talus accumulates at the base of the cliff as coarse fragments from rock falls. It is then transformed into smaller and smaller particles by mechanical and chemical weathering and moves downslope by creep. Some of the debris may be collected by minor tributaries and moved to the main stream by running water. The remaining fine-grained debris continues to move downslope by gravity and is ultimately fed into a stream which carries it out through the drainage system.*

Figure 8.14 *Diagrams illustrating the importance of mass movement in valley development and slope retreat.*
(a) Downcutting by stream abrasion produces a vertical gorge the width of the stream valley but, unless rocks are extremely resistant, the valley walls are widened by mass movement. (b) As the stream cuts deeper, the valley grows wider. (c) In most valleys, downcutting accounts for only a small part of volume of material removed. These diagrams do not take into consideration the material moved by small tributaries along the valley slope.

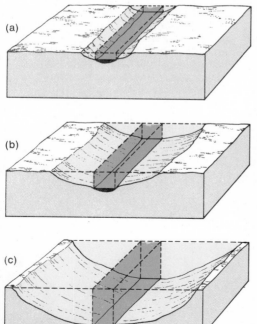

lobate form, and its steep front. Excavations into the rock glaciers reveal a considerable amount of ice in the pore space of the rock. The ice is presumably responsible for the movement. With a continuous supply of rock fragments from above, the increased weight causes the interstitial ice to flow. Favorable conditions for development of rock glaciers include a cold climate which keeps interstitial ice frozen and steep cliffs which supply coarse rock debris that creates large interstitial spaces for the ice. Some rock glaciers, however, may be debris-covered, formerly active glaciers (no longer active because rate of flow is below the minimum accepted as "active").

Talus Slopes. Having described some of the ways in which rock debris is transported downslope through the pull of gravity, we will now consider a typical slope and the combined effects of weathering and mass movement. The diagram in *figure 8.13* shows a resistant sandstone overlying a weaker shale. In areas where weathering is dominantly mechanical, the resistant sandstone will break up into blocks which will accumulate at the base of the cliff in piles of rock fragments referred to as talus. The material composing the talus moves chiefly by rock falls, rock slides, and by rolling. Subsequently, the talus moves slowly downslope by creep. During this process, the rock particles are weathered and broken into smaller fragments which continue to move downslope toward the stream. The fine-grained rock debris gradually enters the river where it is carried away by the running water.

The talus slope may thus be thought of as a local open system with an input of coarse rock debris at the base of the cliff and an output of fine rock fragments into the stream. As the system operates, the valley slope gradually retreats away from the stream channel although the shape of the slope profile may remain essentially constant.

The volume of material transported by mass movement in the process of eroding a valley is shown in *figure 8.14*. Downcutting accounts for the volume of material directly above the stream channel, whereas mass movement is largely responsible for moving the material along the slopes.

The Earth's Dynamic Systems

Extension of Drainage Systems Downslope

In addition to downcutting, headward erosion, and slope retreat, a drainage system may grow in length by simply extending its course downslope as sea level drops. This process is probably fundamental in determining the original course of many major streams, especially in the interior lowlands where the sea covered much of the continent many millions of years ago. As the sea withdrew, drainage lines were extended down the newly exposed slope and later modified by headward erosion and stream piracy.

An example of the beginning of a new segment of a drainage system resulting from sea withdrawal can be seen in tidal channels along coastal plains. The pattern of land and tidal drainage is shown in *figure 8.15.* Because the material upon which this drainage is established consists of recently deposited horizontal sediments, the drainage pattern is characteristically dendritic. However, if the slope is pronounced, the tributaries, as well as the major streams, flow parallel for a long distance. If sea level were to drop, the streams would continue to flow downslope, as shown in *figure 8.16.*

Figure 8.15 *Tidal channels develop a characteristic dendritic drainage pattern between high and low tides.*

Figure 8.16 *Extension of drainage downslope as a result of shoreline retreat. (a) Original position of the shore with tidal channels between high and low tide. (b) As the sea level recedes, tidal channels become part of the permanent drainage system. (c) With each sucessive retreat of the shoreline, new tidal channels develop, and drainage is again extended downslope. The drainage pattern typically produced on the homogeneous tidal-flat material is dendritic.*

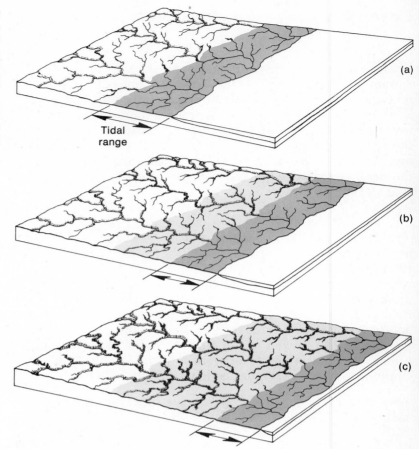

The development of a drainage system by downcutting, headward erosion, slope retreat, and extension of the drainage net downslope can be seen in the photograph in *figure 8.17*. This area was once occupied by a prehistoric lake, the remnants of which can be seen in the background. The well-defined terrace midway up the mountain front marks the highest level of the lake. Lower terraces formed as the làke receded.

The lake levels provide important reference lines with respect to the growth of the drainage network. Above the highest lake level, stream erosion has continued without interruption. The drainage was established before the lake was created, and downcutting, headward erosion, and slope retreat have produced the dissected mountain front. Note that the valleys above the lake level are much wider than those below as a result of the greater amount of slope retreat. Slope retreat, downcutting, and minor headward erosion are all operating in this area at the present time.

When the lake came into existence the drainage system ended at the upper shoreline and sediment carried by the streams was deposited in the lake. These deposits were redistributed by waves to form the prominent terrace in the middle

The Earth's Dynamic Systems

of the photograph. As the lake level receded, the major streams were extended downslope, and new subparallel streams originated on the newly exposed lake bed. Many of these terminated at the lower shoreline because the lake level remained stationary for some time at this elevation. The lowest zones have been exposed only since the lake completely receded to its present level. This surface is marked by extension of the major drainage lines downslope.

Within this small area all major processes of drainage development can be observed: (1) headward erosion, (2) downcutting, (3) slope retreat, and (4) extension of the drainage system downslope as the shoreline retreats.

Figure 8.17 *The shoreline of ancient Lake Bonneville illustrates all the major processes involved in dissection of the land by running water. Downcutting and slope retreat occur in all segments of the streams. Headward erosion is extending the drainage network upslope in the mountains above the shoreline. As the shore of the ancient lake retreated, the drainage was also extended downslope.*

Models of Stream Erosion

Statement

Geologists have recently studied the interaction of downcutting and slope retreat by using a computer model of the Colorado River acting upon the rock sequence in the Grand Canyon area. Variables such as rates of downcutting and slope retreat on various rock formations were analyzed, and a series of profiles showing a series of changes in the canyon with time were calculated and printed by the computer. Erosion of the model canyon was not uniformly fast or slow but occurred in a series of pulses. Downcutting was slow where resistant rocks were encountered, and rapid erosion occurred on nonresistant strata. The rate of slope retreat was shown to be intimately related to the rate of downcutting.

Discussion

This study produced a series of hundreds of computer-calculated profiles from the time that Colorado River erosion began cutting through the Kaibab Plateau in the Grand Canyon region; several of these profiles are shown in *figure 8.18*.

Initial downcutting produced a narrow gorge in the very resistant limestone exposed at the surface. When nonresistant rocks were encountered, rapid downcutting occurred, followed by an interval of slope retreat and canyon widening.

Although this model cannot be verified directly, we can get a glimpse of the stages of canyon development by studying the nature of the canyon longitudinally. Upstream near Lee's Ferry, the river flows on the top of the Kaibab Limestone and the entire sequence of strata exposed farther downstream in the Grand Canyon is below the surface in this area. The canyon cut into the resistant formation is a narrow gorge in the Kaibab Limestone (*figure 8.19*). Farther downstream uplift has permitted the river to cut much deeper into the sequence of rocks, and a profile very similar to that developed by the computer model has

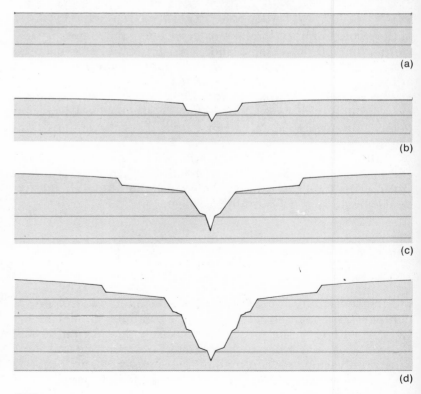

(a)

(b)

(c)

(d)

Figure 8.18 Series of profiles showing the evolution of the Grand Canyon as determined by a computer model. (a) Original undissected surface. Underlying bedrock contains a sequence of alternating resistant and nonresistant rock types. (b) Initial dissection and slope retreat. Downcutting is slow on resistant rock units, and slope retreat causes nonresistant rocks to recede back from the river, leaving a terrace on the resistant rock layers. (c) Downcutting is accelerated as the river cuts through the resistant formation and rapidly erodes the weaker underlying rocks. (d) Downcutting continues as differential slope retreat produces alternating cliffs and terraces.

Figure 8.19 At Lee's Ferry in the eastern Grand Canyon, the river is just beginning to cut through the rock sequence and has produced a profile like that in figure 8.18b. Downstream, uplift has permitted the river to cut deeper, producing profiles similar to 8.18c and d.

been formed. Thus, the canyon itself is evidence of the evolution of slope morphology and corroborates the findings of the computer model.

The result of slope retreat is that all valley walls gradually recede from their respective river channels. The slopes eventually meet, and the general surface is gradually worn down. As erosion occurs, new tributaries form on the new slope and in turn cut a new generation of valleys that retreat laterally away from the stream channel.

Flood Plain Deposits

Statement

Erosion by a river occurs predominantly in the headwater areas of the drainage system where both water and sediment are collected by systems of tributaries. In the transporting segment of a river or where rivers flow across gently inclined surfaces such as are common in the continental interior, the velocity is low so the river is commonly unable to transport all of its sediment load. Deposition becomes a major process in this segment of rivers and typically develops broad flood plains, **meanders** and **point bars**, **natural levees**, and **backswamps**.

Discussion

Figure 8.20 shows a block diagram of a segment of a river flowing across a gently sloping surface and illustrates many features developed by low-gradient streams.

Meanders and Point Bars. There is a universal tendency for a river to flow in a curved or sinuous pattern even on a pro-

Figure 8.20 Schematic diagram showing the major features of a flood plain. As a stream flows around a meander bend, it erodes on the outside curve and deposits sediment on the inside to form a point bar. The meander bend migrates and is ultimately cut off to form an oxbow lake. Natural levees build up the banks of the stream, and backswamps develop in the lower surfaces of the flood plain. Yazoo tributaries find it difficult to enter the main stream and flow parallel to it for considerable distances. Slope retreat continues to widen the low valley which is partly filled with river sediment.

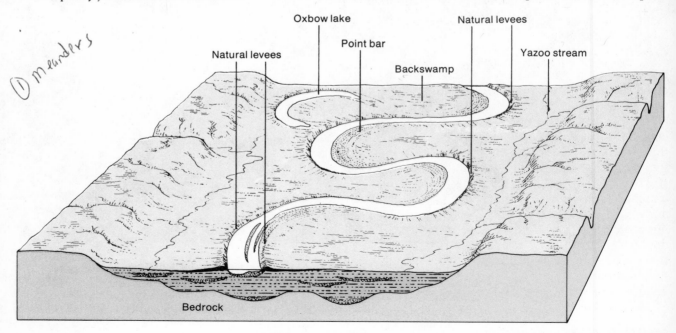

The Earth's Dynamic Systems

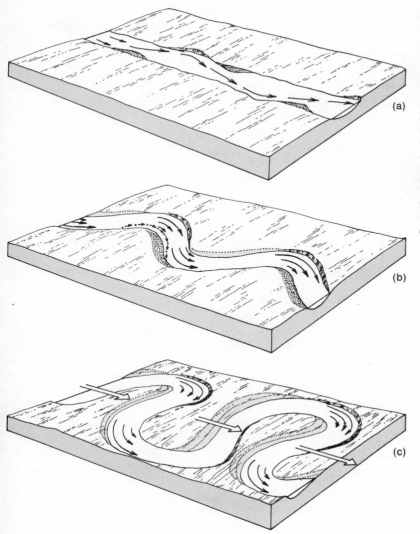

Figure 8.21 Development of meander bends and point bars. (a) Patterns of stream flow are deflected by any irregularity and move to the opposite bank where erosion begins. (b) Once the bend begins, centrifugal force continues to thrust the flow of water toward the outside of the bend. Erosion continues to form a meander loop. At the same time deposition occurs on the inside of the bend as a result of the lower stream velocities in that area. (c) The meander is enlarged and migrates laterally with contemporaneous growth of the point bar. There is a general downslope migration of the meanders as they grow larger and ultimately cut themselves off to form an oxbow lake.

nounced slope. This is because any bend or irregularity in the channel will deflect the flow of water toward the opposite bank where the force of the current causes undercutting. This forms a bend in the river which becomes accentuated with time as the current is deflected and continues to impinge on the outside of the bend (figure 8.21). On the inside of the meander bend, velocity is at a minimum, so deposition occurs to form a point bar (figure 8.22). The major processes involved in water flowing in a meandering channel are (1) erosion on the outside of the meander and (2) deposition on the inside. By these processes the meander loop grows larger and migrates laterally as well as downstream; ultimately, the meander grows large enough for the channel to meet and cut off the meander loop. The abandoned meander bend stands partly filled with water, forming an **oxbow lake**.

Natural Levees. As a river overflows its banks during flood

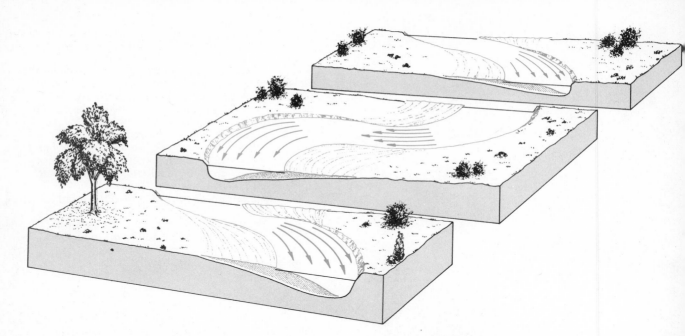

Figure 8.22 *The process of meandering involves erosion on the outside of a curve in the stream channel where velocity is greatest and deposition on the inside of the curve where velocity is at a minimum.*

Figure 8.23 *Natural levees are wedge-shaped deposits of silt which taper away from the stream banks toward the backswamp. They form during flood stages because, as the water overflows its banks, the velocity is reduced, causing the silt to be deposited. As the levees grow higher, the stream channel also rises so the river may be higher than the surrounding area.*

stage, the water is no longer confined to a channel and is permitted to flow as a sheet. This significantly reduces the velocity of flow and the coarsest material is deposited close to the channel, building up an embankment known as a natural levee on either side of the river (*figure 8.23*). Beyond the levee, the surface slopes down and away from the river. Through the process of building natural levees, a river actually may build its channel many feet higher than the surrounding areas.

Backswamps. Because the natural levees are built up higher than the surrounding area, a considerable part of the flood plain may be below the level of the river. This area, known as the backswamp, is poorly drained and is commonly the site of marshes and swamps. Tributary streams are unable to flow up the slope of the natural levee. They are therefore forced to either empty into the backswamp or flow parallel to the main stream for many miles. Strangely enough, then, the highest surface on the flood plain is along the natural levees immediately adjacent to the river.

There are many excellent examples of flood plain deposits, for most streams which flow across the lowland of the continental interior have a low gradient and deposit part of their sediment load. The lower Mississippi River is a well-known example, as much of the sediment it carries is deposited along a broad flood plain between Cairo, Illinois, and the Gulf of Mexico. An aerial photograph of the river (*figure 8.24*) displays most of the features illustrated in the graphic model shown in *figure 8.20*. The dynamics of the river and the change which it can bring about by deposition is illustrated by the fact that during the period from 1765 to 1932, the river abandoned 19 meanders between Cairo, Illinois, and Baton Rouge, Louisiana. Now the river is controlled by dams and artificial levees, which

The Earth's Dynamic Systems

have modified the hydrology in a manner similar to that of the Nile and Colorado Rivers.

Braided Rivers. Where streams are supplied with more sediment than they can carry, the excess material is deposited as sand and gravel bars on the channel floor. This deposition forces the stream to split into two or more channels and to form an interlacing network of channels and islands (figure 8.25). The term **braided river** is used to describe the pattern of a low-gradient stream in which deposition is the dominant process.

Braided rivers have wide, shallow channels and are best developed where a river is heavily laden with erosional debris from a high mountain range. When the river emerges from the mountain front, its velocity is rapidly reduced by abrupt change in gradient, and part of the load, including all of the coarser debris, is dropped.

Melting ice caps or glaciers also produce favorable conditions for braided streams because the low gradients in front of the glaciers are insufficient to transport the exceptionally large load of reworked glacial debris.

Stream Terraces. We have examined in this section how a stream may deposit much of its sediment load across a flood plain and build up a deposit of alluvium. The general process of deposition may be initiated by any change that reduces the capacity of a stream to transport sediment. These include (1) a

Figure 8.24 *Flood plain features of the Mississippi River valley include (a) meander bends, (b) point bars, (c) natural levees, (d) oxbow lakes, and (e) backswamp.*

Figure 8.25 *A braided river. This pattern commonly results when a river is supplied with more sediment than it can carry. Deposition occurs, causing the river to repeatedly develop new channels.*

Figure 8.26 *The evolution of stream terraces. (a) A valley eroded by normal processes. (b) Changes in climate, base level, or other factors which would reduce energy cause the stream to partly fill its valley with sediments, forming a broad flat floor. (c) An increase in energy causes the stream to erode through the previously deposited alluvium, leaving a pair of terraces which are remnants of the former flood plain. (d) The stream shifts laterally and forms lower terraces as it erodes through the valley fill.*

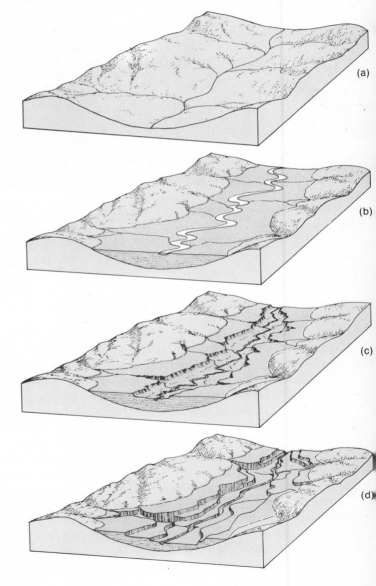

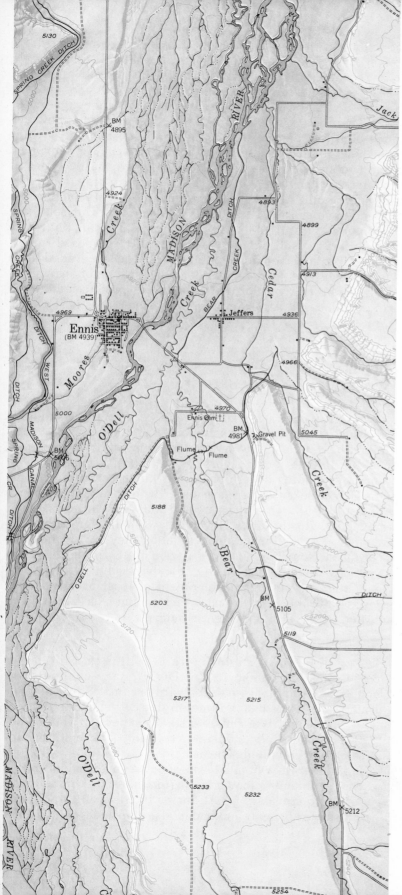

Figure 8.27 Stream terraces near Ennis, Montana. The Madison River and its tributaries have cut through previous valley fill and have developed a series of well-defined stream terraces.

reduction in discharge (climatic changes; stream piracy), (2) a change in gradient (rise of sea level; regional tilting), or (3) an increase in sediment load. During the last ice age, melt water from the glaciers carried large quantities of sediment, and the associated climatic changes produced significant changes in the hydrology of many rivers. As a result, many streams show evidence of filling part of their valley and subsequently cutting through the fill to form stream **terraces**. The basic steps involved in the evolution of stream terraces are shown in the series of diagrams in *figure 8.26*. In *block a* a stream cuts a valley by downcutting and slope retreat. In *block b* changes such as described above cause the stream to deposit part of its sediment load and build up a flood plain which forms a broad, flat valley floor. In *block c* subsequent changes (uplift, increased runoff, etc.) result in renewed downcutting into the easily eroded flood plain deposits and the development of a single set of terraces on either side of the river. Further erosion may produce additional terraces (*block d*). An example of stream terraces is shown in *figure 8.27*.

Deltas

Statement

As a river enters a lake or ocean its velocity is suddenly diminished, and most of its sediment load is deposited to form a delta. The details of how a delta grows may be very complex, especially for the larger rivers which deposit a huge body of sediment extending several hundred kilometers out into the ocean. This is because there are many distinct subenvironments of deposition in large deltas, such as tidal flats, beaches, lakes, swamps, and flood plains. However, two major processes are fundamental in the formation of a delta. They are (1) the splitting of the stream channel into a system of distributaries, channels which extend themselves out into the open water in a semi-circular pattern, and (2) development of local breaks or crevasses in natural levees through which sediment is diverted and deposited in the area between tributaries.

Discussion

Deltas. The diagrams in *figure 8.28* illustrate the major processes in delta construction. Distributary channels are formed and extend themselves seaward because sediment tends to accumulate at the channel mouth due to the sudden loss of velocity where the river enters a standing body of water (*diagram a*). This causes the channel to split into two smaller channels which are extended seaward.

The manner in which the area between distributaries is filled with sediment is shown in *diagram b*. A local break in the levee, called a **crevasse**, forms during high runoffs, and a significant volume of runoff and bed load is diverted from the main

Figure 8.28 Development of distributaries and splays. (a) A bar develops at the mouth of a river channel when the water previously confined to the channel loses velocity as it enters the sea. The bar then diverts the water coming from the main stream into two distributary channels which grow seaward. This process is repeated to form branching tributaries. (b) A break in the natural levee permits part of the stream to be diverted to the backswamp where reduction in velocity causes deposition of sediment in a fan-shaped splay.

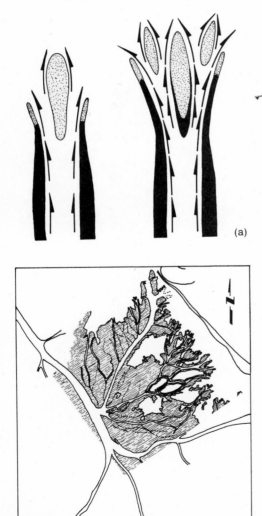

(a)

(b)

The Earth's Dynamic Systems

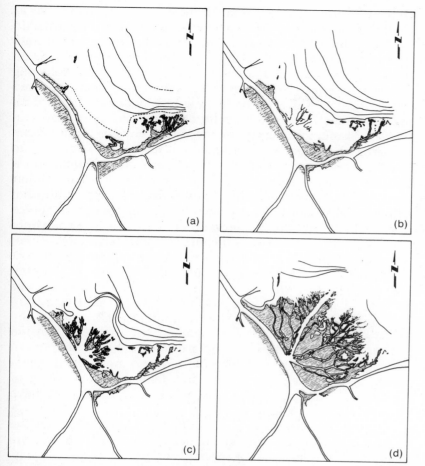

Figure 8.29 The evolution of a splay on the Mississippi Delta. (a) 1838. Configuration of main channel prior to the development of a crevasse. (b) 1870. About 10 years after opening of a crevasse in the natural lines, a splay began to grow as sediment was deposited adjacent to levees. (c) 1877. About 17 years after the crevasse opened. Note the extent of minor splays and distributaries. (d) 1903. About 43 years after the crevasse opened. The extensive development of subsplays and minor distributaries caused the splay to fill in much of the area adjacent to the channel. The splay has continued to grow by subsplays and distributary channels. See figure 8.28b.

stream through the crevasse; here it spreads out to form a **splay** which is in essence a small delta itself, with small distributaries and systems of subsplays. An excellent example of delta construction is provided by a sequence of maps of the Mississippi Delta shown in *figure 8.29*.

The manner in which a delta grows depends upon a number of factors, including rate of sediment supply, rate of subsidence, and removal of sediment by waves and tides. If deposition dominates, the delta is extended seaward. If waves and tides dominate, the sediment delivered by the river is transported along the coast and deposited as beaches and bars. Two examples will illustrate this point.

The Mississippi Delta is an example of a delta in which processes of river deposition dominate. The delta is fed by the extensive Mississippi River system which drains a large part of North America and discharges an annual sediment load of approximately 454 million metric tons (500 million short tons) per year. The river is confined to its channel throughout most of its course, except during high floods. Therefore, most of the sediment reaches the sea through two or three main distributary channels and has rapidly extended the river channels far out into the

Gulf of Mexico (*figure 8.30*). These extensions are known as the **birdfoot subdelta**. The present birdfoot has formed within the last 500 years. The extension of the distributaries out into the sea cannot continue indefinitely because of the low gradient; thus, the river is eventually diverted into a new course having a steeper gradient. The old birdfoot subdelta is abandoned and is reworked by wave action. A new birdfoot extension is built up in another sector of the delta complex and the cycle is repeated. Through this process, the entire river is shifted from one area to another, building a salient (projection) out into the sea. As the river shifts again, the salient is abandoned and reworked by marine processes after the river shifts to another area. It is apparent from river studies that the present birdfoot extension is as long as the balance of natural forces will permit and, without continued intervention by man, the Mississippi River will shift to the present course of the Atchafalaya River. When this happens, New Orleans will cease to be a seaport.

The Nile Delta differs from the Mississippi in several important ways. Instead of being confined to one channel, a series of distributaries begins at Cairo more than 160 km (100 mi) inland and fans out over the entire delta (*figure 8.31*). Two of the larger distributaries have built major lobes extending beyond the general front of the delta. Before construction of the Aswan dam, the annual flood of the Nile covered much of the delta during a brief period each year and deposited a thin layer of silty mud. In the past 3000 years, 3 m (9.8 ft) of sediment have accumulated near Memphis.

Strong wave action in the Mediterranean redistributes the sediment at the delta front and deposits it in a series of arcuate barrier bars which close off segments of the sea to form lagoons. The lagoons form a subenvironment and soon become filled with fine sediment.

The difference between the Nile and the Mississippi deltas is due largely to the balance between influx of sediment, which tends to build birdfoot subdeltas, and the strength of wave action which redistributes the sediment to form barrier bars.

Figure 8.30 *The Mississippi Delta and its most recent subdelta. The active distributary system has built a major "birdfoot" subdelta during the last 500 years. The water is confined to the main distributary, and deposition is concentrated in a small sector of the delta front. After one sector is built far out into the sea the entire flow is shifted to some other sector and the process is repeated. Wave action then erodes back the "birdfoot" subdeltas.*

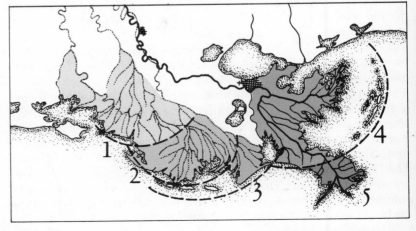

The Earth's Dynamic Systems

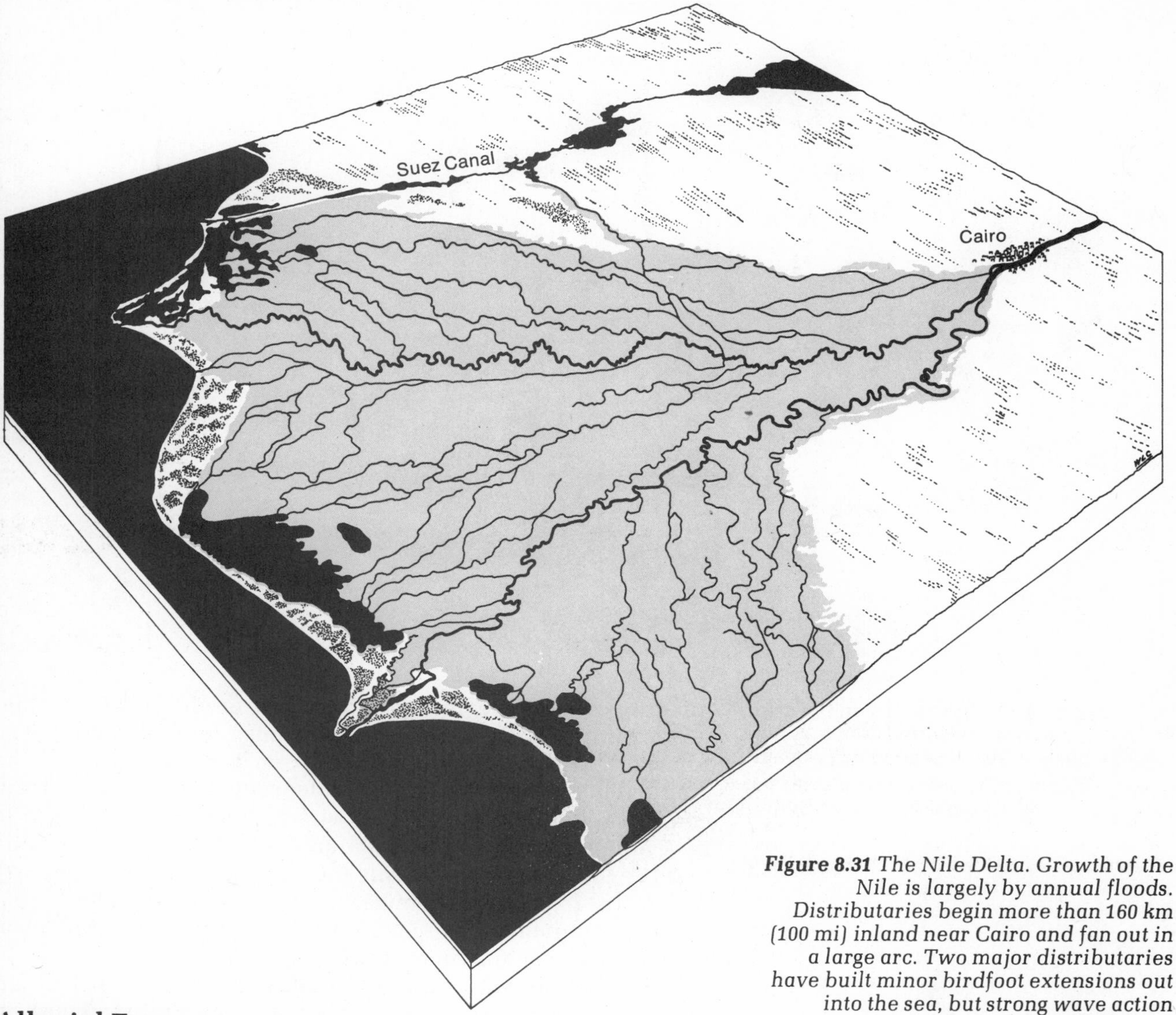

Figure 8.31 *The Nile Delta. Growth of the Nile is largely by annual floods. Distributaries begin more than 160 km (100 mi) inland near Cairo and fan out in a large arc. Two major distributaries have built minor birdfoot extensions out into the sea, but strong wave action reworks the sediment and deposits it as spits and barrier islands.*

Alluvial Fans

Statement

Alluvial fans are, in one sense, land deltas. They are stream deposits consisting of gravel and sand which accumulate in a dry basin at the base of a mountain front. Deposition results from the sudden decrease in velocity as a stream emerges from the steep slopes of an upland onto a gently sloping plain. Alluvial fans are most abundant in arid regions where runoff is insufficient to produce through-flowing streams.

Discussion

Alluvial fans form mostly in arid regions where the streams flow only intermittently. In such areas there is usually a large

Figure 8.32 Alluvial fans in the arid southwestern United States. A fan develops as a stream enters a dry basin and deposits its sedimentary load because of decrease in gradient.

quantity of loose weathered rock debris on the surface so that when rain falls, the streams have a huge load of sediment to transport. Where a river emerges from a mountain front, the gradient is sharply reduced, and a deposition occurs. The stream channel soon becomes clogged with sediment and is forced to seek a new course. In this manner, the stream shifts from side to side and builds up an arcuate fan-shaped deposit (figure 8.32). As several fans build basinward at the mouth of adjacent canyons, they ultimately merge to form a broad slope of alluvium at the base of the mountain range (figure 8.33).

Although alluvial fans and deltas are somewhat similar, they are different in mode of origin and internal structure. In deltas, stream flow is checked by standing water and sediment is deposited largely in an aqueous environment. The level of the sea or lake effectively forms the upper limit to which the delta can be built. In contrast, a fan is deposited in a dry basin, and the upper surface is not limited by water level. The coarse, unweathered sands and gravels of the alluvial fan also contrast with the sand, silt, and mud that predominate in the delta.

Summary

We have seen in this chapter that a river system is continually changing. It erodes primarily in the headwater regions by downcutting, slope retreat, and headward erosion. Some sedi-

The Earth's Dynamic Systems

ment is deposited in the transporting segment of the river system and forms a variety of flood plain features. Most of the sediment load, however, is deposited near the mouth of the river as a delta or alluvial fan. In any segment of a river, one or more of these processes can be observed.

In the entire river system, the net result of these processes is erosion, transportation, and deposition of sediment. Erosion occurs in a systematic way, and the resulting landscape evolves in a predictable manner through a sequence of stages. Although we cannot observe the evolution of a landscape from start to finish, we can construct several conceptual models which explain landscape development. In the next chapter, we will consider some of these models to give us some insight into the changing surface of the earth.

Figure 8.33 *Alluvial fans in Nevada. As alluvial fans grow they merge into a broad alluvial slope which covers much of the adjacent basin.*

Additional Readings

Crandell, D. R., and H. H. Waldron. 1956. "A Recent Volcanic Mudflow of Exceptional Dimensions from Mt. Rainier, Washington." Amer. J. Sci. 254: 359-362.

Morisawa, M. 1968. Streams—Their Dynamics and Morphology. New York: McGraw-Hill Book Company.

Sharpe, C. F. S. 1938. Landslides and Related Phenomena. New York: Columbia University Press.

Young, A. 1972. Slopes. Edinburgh: Oliver and Boyd.

9 Evolution of Landforms

R iver systems are not simply channels through which water and sediment flow from the continents to the sea; they are the major agents by which the surface is sculptured into an almost infinite variety of erosional and depositional landforms. The processes of stream erosion show that a landscape evolves through a series of predictable stages from the time erosion begins until it is modified or stopped. Through study of regions in various stages of development, geologists have constructed conceptual models of landscape evolution. No two regions are exactly alike, and there is almost an infinite variety of detail, even within similar areas. These idealized models, however, permit one to recognize more fully the series of steps in landscape development and the importance of river systems in forming the surface features of the earth.

In this chapter, we consider, first, several general models of landscape development, each illustrated by a series of block diagrams. Second, we consider how the structure and nature of the bedrock cause erosion to occur in different ways and at different rates to produce distinctive landscape types. Third, we deal with the problem of rates of erosion: how they are measured and what the measurements signify in terms of the earth's history.

Major Concepts

1. Erosion of a landscape by running water evolves through a series of predictable stages.
2. Climate and rock structure greatly influence the nature of the landscape.
3. Evolution of landscapes by running water may be interrupted by (a) earth movements, (b) glaciation, or (c) vulcanism.
4. Rock formations erode at different rates so that resistant and nonresistant units are etched out into relief. This produces alternating cliffs and slopes on horizontal rocks and alternating ridges and valleys on folded rocks.
5. Rates of erosion show that the continents are eroded rapidly and must have been repeatedly uplifted in the geologic past to remain above sea level.

Model of Erosion in Humid Climates

Statement

A newly uplifted area theoretically evolves through a series of stages until it is ultimately eroded to near sea level. In the early stages stream erosion begins to dissect the area. Undrained divides are broad and wide. With time the area becomes completely dissected into rolling hills. Further erosion reduces the area to a surface of low relief near sea level.

Discussion

The block diagrams (*figure 9.1*) illustrate the sequence of changes resulting from a newly uplifted area in a humid climate such as the continental shelf or adjacent coastal plain. In this model, it is assumed that uplift was rapid and that the surface was elevated more than a thousand meters above sea level.

Initial Stage. The newly exposed surface shown in *block a* is at first poorly drained, but eventually an integrated drainage system develops by headward erosion. Streams begin to erode V-shaped valleys into the uplifted landmass; downcutting and headward erosion are the dominant erosional processes. The valley walls extend right down to the stream bank. Tributary development proceeds rapidly as the drainage net is extended up the regional slope by headward erosion. New tributaries develop on the slopes of the valley walls produced by the major rivers. Stream capture is common in the headwater region. The stream gradients are steep, and the channels are marked by waterfalls and rapids. But broad areas of the original surface remain uncut, and marshes or lakes occupy original depressions in the newly uplifted surface.

As erosion continues, the local relief of the area (difference in elevation between valley floor and divide summit) increases and valleys are cut deeper and wider. Tributary valleys branch out from the larger streams and further dissect the initial surface. Slope retreat is more evident as an important process.

Intermediate Stage. As tributaries increase in number and length, the original surface is completely destroyed, and the landscape becomes marked by steep slopes leading down to the deeply incised stream channels (*block b*). The drainage network becomes so well developed that little or no area remains undrained. Local relief steadily increases. Streams also begin to erode laterally during this stage and develop flood plains. Slope retreat becomes more prominent. The processes of erosion ultimately cease to produce greater local relief and, thereafter, erosion and deposition tend to reduce the local relief, as well as the general elevation of the area.

With the land reduced to a lower elevation, the rate of erosion proceeds more slowly. Relief decreases, but slopes continue to retreat from the river channels and are worn down by weathering and mass movement. The rivers widen their valleys by lateral erosion as they meander across the flood plain.

Evolution of Landforms

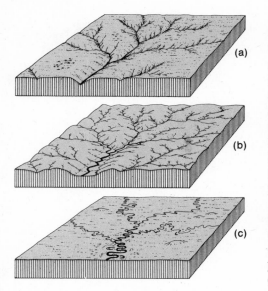

Figure 9.1 *Evolution of a landscape in a humid climate. (a)* Initial Stage. *Large areas are undissected. Downcutting and headward erosion are the dominant processes. The streams have steep gradients, with rapids and waterfalls. Only a small percent of the area consists of valley slopes. The streams have not developed flood plains nor an extensive meandering pattern. (b)* Intermediate Stage. *The area is completely dissected so that most of the surface consists of valley slopes. Relief is at a maximum, and a well-integrated drainage system is established. The main streams meander, and a flood plain begins to develop. The topography is characterized by smooth rolling hills. (c)* Late Stage. *The landscape has been eroded to a nearly featureless surface near sea level. Rivers meander over a broad flood plain. Some isolated erosional remnants remain, but deposition is nearly as important as erosion.*

Figure 9.2 *An area in the intermediate stage of dissection. Valley slopes make up a large percent of the surface; relief is at a maximum; and the drainage system is well integrated, with a large number of streams per unit area.*

Late Stages. The landscape is ultimately reduced to a low, nearly featureless erosional surface near sea level (*block C*). This surface is referred to as a **peneplain** over which the streams meander slowly, depositing almost as much sediment as they erode. The peneplain is the ultimate result of erosion in this model; it is near sea level and no further effective stream erosion can occur upon it. Subsequent modification, if conditions remain stable, are deep weathering and deposition of sediment on the flood plain. Oxbow lakes, meander scars, and natural levees are common, and locally an isolated erosional remnant of resistant rock, a **monadnock**, may protrude above the peneplain surface.

Examples. Erosion of a landscape is much more complex than the model described above. It is doubtful if a large area evolves through all of these stages, but there are numerous examples of regions in various stages of dissection. The coastal plains of the Atlantic and Gulf regions have only recently emerged from below sea level, so vigorous erosion has not occurred.

The Allegheny Plateau (Pennsylvania) with adjacent areas is an example of a region near the intermediate stages of erosion in a humid climate, as most of the area is dissected to some extent and, with continued erosion, relief will be diminished (*figure 9.2*). The best example of a surface eroded to the late stage is the broad low region of eastern Canada (*figure 9.3*). This area is eroded to within a few hundred meters of sea level and has remained low for many millions of years. Its erosional his-

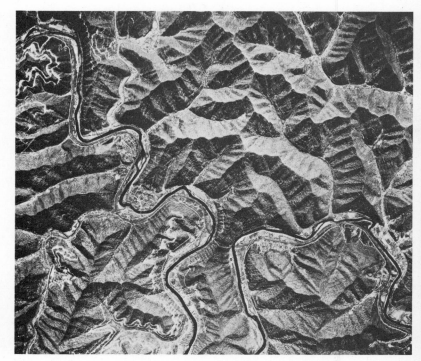

tory is much more complex than the model described above because it has been partly covered by the sea at various times and has been subjected to recent glaciation.

Model of Erosion in an Arid Climate with Through-Flowing Streams

Statement

The evolution of a landscape in an arid or semiarid climate in which there is an integrated drainage system with major streams flowing to the sea is similar to that which occurs in a humid climate: the surface is systematically dissected through a series of stages until it is worn down to a peneplain. However, there are several significant differences in the details of the landscape. In an arid climate, the topography is much more angular, as resistant limestones and sandstones form steep cliffs and nonresistant shales form broad slopes. Slope retreat is more evident in an arid region and, in some areas, wind deposits significantly alter the landscape. Many tributary streams are dry except during major rain or spring runoff.

Discussion

The series of diagrams shown in *figure 9.4* illustrates the evolution of an idealized landscape in an arid region with through-flowing streams.

Initial Stage (block a). The profile of canyons developed in the early stages of dissection is characterized by alternating cliffs and slopes, whereas the topography in humid regions is characteristically rounded. Slope retreat is a more obvious process and, with a minimum of precipitation, an intricate network

Figure 9.4 *Evolution of a landscape in an arid climate with through-flowing streams. (a)* Initial Stage. *Alternating cliffs and slopes characterize the typically angular topography. A network of stream channels begins to develop, and cliffs retreat laterally. Much of the original surface is undissected. (b)* Intermediate Stage. *Erosion continues to dissect the area into a network of deep canyons. Local relief reaches a maximum, and resistant rock layers form cap rock for plateaus, mesas, and buttes. (b)* Late Stage. *Erosion shrinks the mesas and buttes and forms a peneplain near sea level. Only a few resistant remnants punctuate the nearly flat region.*

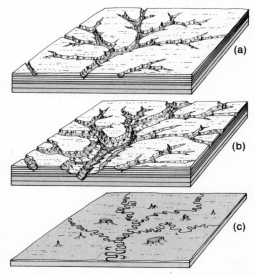

Evolution of Landforms   **189**

Figure 9.5 *The canyonlands of the Colorado Plateau, an area in the intermediate stage of landscape evolution in an arid climate.*

of stream channels is slow to develop. Cliffs retreat laterally by undercutting of the nonresistant shales below them. It is common to find large areas of resistant rock stripped clean of the overlying formation, even though the surface may be over a thousand meters above sea level. As in the initial stages in a humid climate, large regions of the original surface remain undissected.

Intermediate Stage (block b). As erosion continues, the original surface becomes dissected by a network of deep canyons. Local relief reaches a maximum, and alternating hard and soft layers are etched out into cliffs and slopes. Resistant rock layers play an important role in landscape development, as they form the cap rock for plateaus, mesas, and **buttes.**

Late Stage (block c). In the final stages of erosion, the mesas and buttes continue to shrink as slopes retreat from the drainage channels. A peneplain forms near sea level, and the entire surface becomes nearly flat and featureless.

Example. The Colorado Plateau of the western United States may be considered as an example in this type of landscape development, although the region has been subjected to recent earth movements and is somewhat more complicated. In *figure 9.5,* the spectacular landscape of the Canyonlands National Park can be seen to be the result of dissection of an uplifted plateau, with differential slope retreat being responsible for the plateaus and mesas eroded from the resistant rock formations. The patterns of erosion of this region are controlled by dissection of the region by the Colorado River. On a regional basis, all major cliffs recede away from the river and its major tributaries.

The Earth's Dynamic Systems

Model of Erosion in an Arid Climate with Internal Drainage and Block Faulting

Statement

Because of limited rainfall in arid regions, many streams are shorter than the slopes down which they flow. That is, the water in the stream either seeps into the ground or evaporates before it reaches a lake or the ocean. Instead of a well-integrated drainage system with a long, continuous stream system flowing to the sea, streams in desert regions are typically shorter and terminate in alluvial fans or shallow **playa lakes**. This is especially true of regions where block faulting has produced structural depressions of internal drainage. The major landforms in such areas are mountain ranges, resulting from erosion of fault blocks, and basins which become filled with sediment.

Discussion

The idealized stages in landscape development in an arid region with internal drainage and block faulting are illustrated in *figure 9.6*.

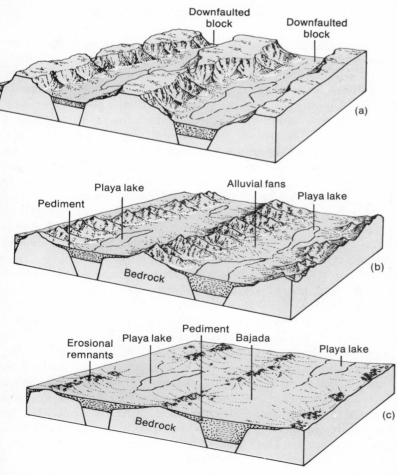

Figure 9.6 *Evolution of a landscape in an arid climate with internal drainage and block faulting. (a) Initial Stage. Faulting produces maximum relief. Some areas are mountainous and undissected. Playa lakes develop in the central part of the basins. (b) Intermediate Stage. The mountain range becomes completely dissected, and the mountain front retreats, developing a pediment. Alluvial fans spread out into the valley. (c) Late Stage. The basins become nearly filled with sediment. Erosion wears down the mountain ranges to small isolated remnants. The pediments expand, and the alluvial fans merge to form bajadas. Most of the surface is an alluvial slope.*

Initial Stage (block a). The initial stage develops by block faulting or folding and upwarps. Maximum relief usually occurs in the initial stage as a result of crustal deformation, not erosion. Relief then continues to diminish throughout subsequent stages, unless major uplift recurs to interrupt the evolutionary trend by producing greater relief during the later stages. Depressions between mountain ranges generally do not completely fill up with water to form large lakes because of the low rainfall and excessive evaporation. Instead, shallow, temporary lakes called playa lakes form in the central part of the basin and fluctuate considerably in size. They may be completely dry for many years and then expand and cover a large part of the valley floor during years of high rainfall. These lakes are commonly saline because they have no outlet.

More sediment is produced by weathering in the uplifted mountain mass than can be carried away by the intermittent streams, which may flow only during spring runoff. The streams are thus commonly overloaded with sediment and deposit much of their load where they emerge from the mountain front. The sediment deposited by the river accumulates in a broad alluvial fan. Throughout the history of an intermontane basin, mountain ranges are eroded and the debris is deposited in the adjacent basin.

Intermediate Stage (block b). As geologic processes continue, the mountain mass becomes dissected into an intricate network of canyons and is worn down to a lower level. At the same time, the mountain shrinks in size as the front recedes through the process of slope retreat. The fans along the mountain front grow and merge to form a large alluvial slope called a **bajada.** As the mountain front retreats, an erosional surface called a **pediment** develops on the underlying bedrock and continues to expand as the mountain shrinks. Pediments are generally covered by a thin veneer of alluvium as they form.

Late Stage (block c). The final stage in this model of arid landscape development is characterized by small, islandlike remnants of the mountains surrounded by an extensive erosional surface, the pediment. The pediment shows little relief and is largely covered with erosional debris.

Example. This model, or variations of it, can be used effectively in explaining the landscape of the Basin and Range province of the western United States (*figure 9.7*). Here, block faulting has occurred over a large area to produce alternating mountain ranges and intervening fault block basins. To the north, throughout much of eastern Utah and all of Nevada, the area is in the initial stages of development. The basins occupy about half of the total area and pediments are small. The relief of the mountain ranges is high, with most fans just beginning to coalesce to form a broad alluvial slope. In Arizona and Mexico, erosion in the Basin and Range has proceeded much farther (*figure 9.8*) than in Utah and Nevada, and the area could be considered to be in the later stages of development. The ranges are

The Earth's Dynamic Systems

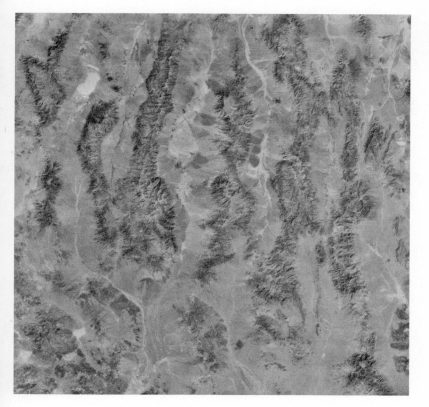

Figure 9.7 The Basin and Range in central Nevada, an area of block faulting and internal drainage in an early to intermediate stage of development.

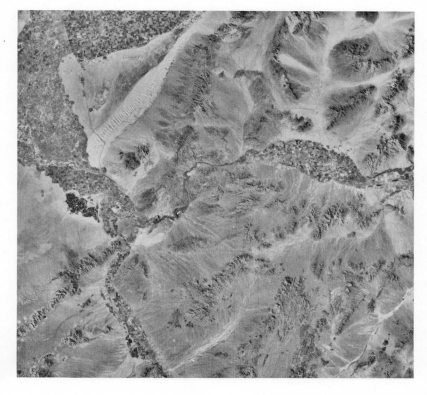

Figure 9.8 The Basin and Range in Arizona, an area of block faulting and internal drainage in the intermediate to late stages of arid landscape evolution.

eroded down to small remnants of their original size and extensive bajadas, extending over approximately four-fifths of the area, cover wide pediments through which isolated remnants of bedrock protrude.

Models of Erosion in Volcanic Terrains

Statement

The extrusion of lava provides a new surface which is attacked by erosion and is modified through a series of stages until the flows are completely destroyed or buried by sediment or younger lava. Lava flows are generally very resistant to erosion. They flow down previously established drainage systems

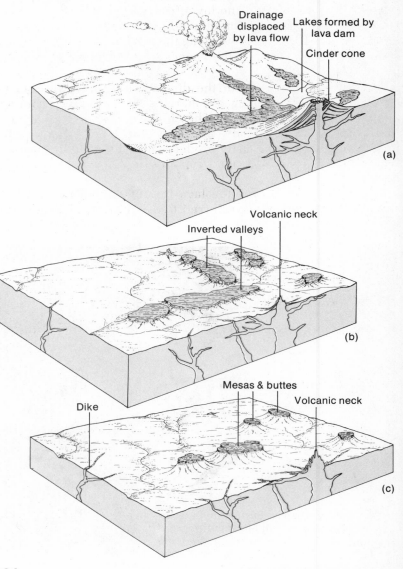

Figure 9.9 Evolution of landforms involving local vulcanism. (a) Initial Stage. Lavas extruded from the volcanic vents flow down the existing rivers and streams and block the normal drainage. Lakes commonly form upstream, and new stream channels develop along the margins of the lava flow. Volcanic cones are fresh and relatively untouched by erosion. (b) Intermediate Stage. The volcanic cones are worn down, leaving only volcanic necks. Erosion along the margins of the lava flow removes the surrounding rock so that the flow forms a sinuous ridge or inverted valley. (c) Late Stage. Inverted valleys are eroded to mesas and buttes. Volcanic necks and dikes commonly form peaks and isolated ridges.

194

and displace the river channel. In areas of small local volcanic eruptions, the lava flows downslope and is confined to the stream valley. Thus, the stream channel is displaced and erosion proceeds along the margin of the flow to develop an **inverted valley**.

Where large volumes of lava are extruded, the flows may completely bury the landscape. New drainage established across or along the margins of the lava plain ultimately dissects it into a plateau.

Discussion

Local Vulcanism. The sequence of landforms resulting from erosion of an area where minor volcanic activity has occurred is illustrated in *figure 9.9*. In the initial stage (*block a*) volcanic flows enter a drainage system, following the river channel and partly filling the valley. The lava flow disrupts the drainage in several ways. Lakes are impounded upstream, and the river is displaced and forced to flow along the margins of the lava. Subsequent stream erosion is then concentrated along the margins of the lava flow.

As erosion proceeds in the displaced drainage (*block b*), new valleys are cut along the margins of the flow, becoming deeper and wider with time. The cinder cones, formed during the initial volcanic activity, are soon obliterated because the unconsolidated ash is easily eroded. Only the conduit through which the lava was extruded remains as a resistant volcanic neck.

As erosion continues, the lava flow is eroded into a long, sinuous ridge, commonly referred to as an inverted valley be-

Figure 9.10 *Inverted valleys in southwestern Utah, a region with local volcanic activity in an intermediate stage of development. The long narrow ridge is an ancient valley which became partly filled with lava extruded from vents in the far background. The drainage was displaced to the flow margins where it cut new valleys, leaving the old valley, filled with lava, stranded high above the present surface.*

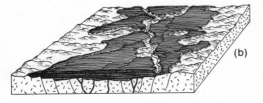

Figure 9.11 *Evolution of a landscape involved with regional vulcanism.* (a) *Initial Stage. Extensive lava flows fill the preexisting drainage and flood the surrounding area. New drainage develops along the margins of the volcanic field, as well as across the new surface of the lava.* (b) *Intermediate Stage. The lava field begins to be dissected by the new drainage system. The igneous rock is usually very resistant so the surrounding area is usually eroded faster, leaving the lava field as a plateau.* (c) *Late Stage. The area surrounding the volcanic flow is reduced to lowland. Remnants of lava form a cap rock on isolated high mesas and buttes.*

cause the previous stream valley beneath the lava is now higher than the surrounding area.

In the final stages of erosion (*block c*), the inverted valley is reduced in size and ultimately becomes isolated mesas and buttes.

Examples of inverted valleys are common along the western and southern margins of the Colorado Plateau (*figure 9.10*) where recent vulcanism has occurred during the erosion of the landscape.

Regional Vulcanism. In regions of extensive volcanic activity, the entire area may be completely buried by lava flows (*figure 9.11*). In the initial stage (*block a*) drainage is usually displaced near the margins of the lava plain, but some rivers may migrate across it. When extrusion ceases (*block b*), stream erosion begins to dissect the lava plain and ultimately cuts it into isolated plateaus and mesas (*block c*). These in turn are eventually consumed.

The Snake River Plain in Idaho is a classic example of an area of extensive basaltic vulcanism which is presently being dissected by the Snake River (*figure 9.12*).

Interruptions in Landscape Development

Statement

The models described above assume rapid uplift followed by stabilization so that erosional processes continue uninterrupted. In reality, this exact sequence rarely, if ever, occurs, for various forces may interrupt the process at any stage of development. Recurrent uplift interrupts the processes by creating greater relief, thus **rejuvenating** the forces of erosion. Glaciation may completely interrupt the evolution of the landscape by modifying the surface with the products of glacial erosion and deposition. Volcanic activity also interrupts the trends of landscape

Figure 9.12 *The Snake River Plains of southern Idaho, an area of recent regional volcanic activity in the early stage of erosion.*

The Earth's Dynamic Systems

development by depositing lava and ash over a terrain previously subjected only to stream erosion. Climatic changes may occur and produce differences in the style of landforms produced. Changes in sea level would also affect the evolutionary trends in landscape development.

Discussion

Rejuvenation. The most significant interruptions in landscape development are usually due to earth movements: broad regional uplift, subsidence, or folding and faulting associated with mountain-building processes. The result is an increase in stream energy and accelerated erosion so that the evolutionary trends are interrupted and many aspects of landscape development return to the condition of the initial stage and begin development again. If uplift occurs in a late stage of development, the gradient of a meandering stream pattern is increased and the stream begins to erode a new valley. The meandering pattern characteristic of the late stage in erosion may become entrenched and form a deep canyon, but the river will have many characteristics of the initial stage of erosion, such as high gradient and rapids.

Classic entrenched meanders are found throughout the Colorado Plateau, as is shown in _figure 9.13._ Another example is the Sierra Nevada Mountains in eastern California which have been rejuvenated repeatedly by westward tilting. Evidence of rejuvenation is also found in uplifted and dissected erosional surfaces both in the Colorado Rockies and in the Appalachian Mountains.

Glaciation. The glacial epoch disrupted the erosional pro-

Figure 9.13 Entrenched meanders in the drainage system of the Colorado Plateau illustrate rejuvenation of stream erosion. The meandering pattern of the river was formed during an earlier period when the river had a low gradient. Recent uplift has increased the gradient and the river's power to erode. Downcutting became the dominant process and the meanders became entrenched.

cesses not only in the area covered with ice but also in other areas throughout most of the continents. The ice sheets completely obliterated the drainage system over which they moved, and along the margins of the great continental ice sheets deposits of glacial debris hundreds of meters thick completely covered the preexisting drainage channels. Thus, much of the surface in Canada and northern Europe shows a strong imprint of glaciation rather than erosion by streams and rivers.

In addition, the melting ice deposited a huge load of debris along the ice margins so that the remaining streams became overloaded with sediment. Instead of eroding, the streams began to fill their valleys with sediment. For example, in the lower Mississippi Valley, more than 60 m (200 ft) of sediment is believed to have accumulated largely from deposition of the great load of glacial sediment. The river now flows over a broad flat-floored valley built up by fluvial sedimentation.

Differential Weathering and Erosion

Statement

Weathering and erosion of different rock types proceed at vastly different rates. Factors such as climate, degree of compaction and cementation, and the dominant minerals within the rock are important but, in general, the resistant rocks are sandstones, quartzites, basalt, conglomerate, and limestones in arid regions. Nonresistant rocks include shale, slate, and limestone in humid regions. The net result is that when a sequence of rocks is exposed to weathering and erosion, the more resistant rocks stand out in relief, whereas the nonresistant rocks recede.

Differential weathering and erosion occur on all scales ranging from steep cliffs separated by gentle slopes to layering etched out in delicate laminae less than 1 mm thick and are responsible for much of the spectacular scenery of our planet.

Discussion

Differential erosion is perhaps best developed in arid regions. In these areas, not only is the type of rock important but jointing and availability of surface and ground water combine to produce fascinating details of the landscape.

Cliffs and Slopes. Probably the most widespread example of differential erosion is the alternating cliffs and slopes developed on a sequence of alternating hard and soft sedimentary rocks. Soft shales typically form slopes, and the more resistant sandstone or limestones produce cliffs (*figure 9.14*). The height of the cliff and width of the slope are largely functions of the thickness of the formations involved. Most of the topography of the Colorado Plateau is in some way a product of differential weathering on sequences of alternating shales and sandstones or limestones.

Columns and Pillars. Buttes, pinnacles, and columns are

198

Figure 9.14 In the Colorado Plateau, slopes form on nonresistant shale formations which erode rapidly and quickly retreat from the river channel. Cliffs form on hard sandstone and limestone which erode very slowly.

Figure 9.15 Eroded columns in a resistant sandstone formation.

Figure 9.16 The eroded columns of Bryce Canyon, Utah. Rapid erosion along joint systems separates the columns from the main cliff. Differential erosion accentuates the difference between rock layers to produce the fluted columns.

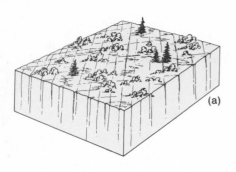

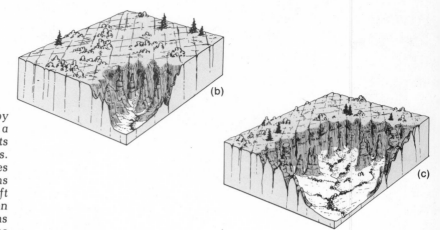

Figure 9.17 *Evolution of columns by differential erosion associated with a retreating cliff. (a) Intersecting joints separate the rocks into columns. (b) Weathering along the joints produces columns which erode into various forms as a result of alternating hard and soft layers. (c) As weathering and erosion proceed, the cliff retreats. Old columns are completely destroyed, but new ones are continually created.*

simply details of differential erosion on a receding cliff. If the cliff is a massive unit, vertical columns such as those shown in *figure 9.15* will develop. If stratification produces alternating layers of hard and soft rock, additional detail may be etched out (*figure 9.16*). Jointing commonly plays an important role, for it permits weathering to attack a rock body from many sides at once. The columns and pillars famous in Bryce Canyon National Park, for example, result from differential weathering along a set of intersecting joints. The joints separate the rock into columns. Differential erosion of the nonresistant shales which separate the more resistant sandstone and limestone forms deep recesses in the columns to produce the fascinating slopes and forms (*figure 9.17*).

Natural Arches. An interesting form of differential weathering is the natural arches formed in massive sandstones. Most natural arches in the Colorado Plateau result from granular disintegration of sandstone formations. Cement which holds the sand grains together is dissolved away where there are maximum concentrations of surface and ground water. Jointing also plays an important role. A model illustrating the process by which natural arches are formed is shown in *figure 9.18*. The sandstones are not tightly cemented and, where water is most abundant, the cement is dissolved and the rock disintegrates. In the arid west, there is little precipitation, and much of that seeps into the porous sandstone instead of flowing down the surface drainage system. However, the subsurface water is most abundant beneath the stream channel and will emerge as a seep in a cliff beneath a dry waterfall. Moisture from the small amount of water flowing over the cliff accelerates granular disintegration and develops an alcove or recess at the base of the dry falls. Ground water percolates through the sandstones, follows the di-

The Earth's Dynamic Systems

rection of surface drainage lines above, and seeps out near the base of the formation below the falls. The presence of this moisture further dissolves the cement, and the disintegrated sand particles are washed or blown away. Thus, the area in a cliff beneath a stream is most susceptible to granular disintegration and soon forms a large alcove.

If the sandstone is cut by a series of joints, a large block may be separated from the cliff, producing an arch (figure 9.19). Weathering then proceeds inward from all surfaces on the arch until it is destroyed, leaving only standing columns.

Erosional Landforms and Structure

Statement

Erosional landforms are greatly influenced by the structure and nature of the rock into which they are cut. As was discussed in the previous section, different rock types weather and erode at different rates, causing the more resistant units to stand out as ridges above the more readily eroded material adjacent to them. Thus, the structure of the rock is commonly expressed in the landforms and exerts a signifcant influence upon their shape and pattern.

Discussion

Inclined Strata. If a sequence of alternating resistant and nonresistant rock is tilted, the resistant formations will be eroded into long, asymmetrical ridges, and the interbedded soft units will form elongate valleys parallel to the ridges (figure 9.20). Ridges formed on gently dipping strata are known as **cuestas**; sharp ridges formed on steeply dipping rocks are referred to as **hogbacks.**

Usually, inclined strata are part of larger flexures such as domes or basins. Hogbacks or cuestas form by rapid headward

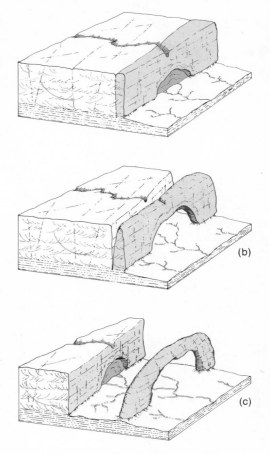

Figure 9.18 *The formation of natural arches in massive sandstones. (a) In arid regions much water seeps into the subsurface below a stream channel. This water moves laterally along an impermeable layer and eventually emerges as a seep near the base of a cliff. The cement which holds the sand grains is soon dissolved in this area of greatest moisture, and the sand grains fall away, forming a recess or alcove beneath the dry falls from the stream above. (b) If a joint system in the sandstone is roughly parallel to the cliff face it may be enlarged by weathering, separating a slab from the main cliff. (c) An arch is produced as the alcove enlarges. Weathering then proceeds inward from all surfaces until the arch collapses.*

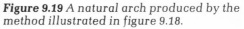

Figure 9.19 *A natural arch produced by the method illustrated in figure 9.18.*

Figure 9.20 *A hogback formed by differential erosion on a resistant layer in an inclined series of sedimentary rocks.*

erosion along the strike or trend of the nonresistant units. They commonly develop by extension of tributaries from several drainage systems. Major superposed streams cut across resistant and nonresistant rocks alike. Tributaries begin to develop along the slopes of the valleys, but those which are established on the weak rock rapidly extend themselves headward, eroding a **subsequent** valley along the nonresistant unit and leaving the inclined resistant beds protruding as a hogback ridge.

Domes and Basins. The initial influence of domes and basins may appear as little more than a modification of drainage but, with differential erosion, a sequence of alternating hard and soft rocks is expressed by a series of circular hogbacks or cuestas in intervening **strike valleys** cut on the less resistant

The Earth's Dynamic Systems

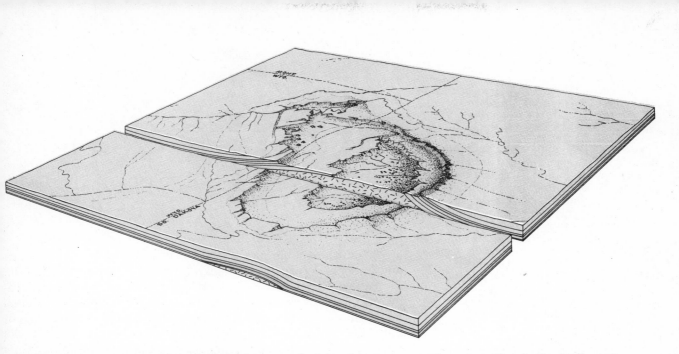

rock (figure 9.21). If the older rocks in the center of the dome are nonresistant, the center of the uplift may be eroded down into a topographic lowland bordered by inward facing escarpments formed on the younger resistant units. If the older rocks are more resistant, the center of the dome will remain high and form a dome-shaped hill or ridge. Large domal structures have outward-facing cliffs, whereas large basins have inward-facing escarpments.

Folded Rocks. As erosion proceeds on a sequence of folded rocks, resistant formations are etched out and form ridges which follow the zigzag outcrop pattern of the limb of the folds. The Jura Mountains of Switzerland and the Appalachian Mountains of the eastern United States illustrate well the variations in topography of folded rocks. The rocks of the Jura Mountains, folded very recently in late Tertiary time, are so young that erosion is just beginning to break through the crests of the folds (figure 9.22). **Anticlines** (uparched rocks) form most of the mountain ridges and **synclines** (downfolded rocks) the valleys. In places,

Figure 9.21 *Circular hogbacks and intervening strike valleys formed by differential erosion on a structural dome. Black Hills, South Dakota.*

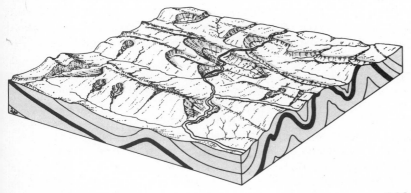

Figure 9.22 *Differential erosion on the folded rocks of the Jura Mountains. The relatively young folds are in the early stages of dissection, and erosion is just beginning to breach the crests.*

Evolution of Landforms

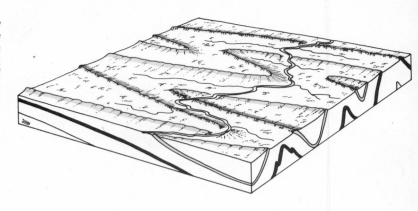

Figure 9.23 *Differential erosion on the resistant sandstone formations in the Appalachian Mountains. Erosion has proceeded to a much later stage than in the Juras. The tops of the folds are eroded off, and only the resistant rock units remain as ridges.*

Figure 9.24 *The evolution of a fault scarp. (a) Original dissected upland. (b) First major period of faulting. Streams erode through the scarp produced by faulting to form the triangular-shaped spurs. (c) Recurrent movement along the fault may produce a series of fresh scarps which are dissected by stream erosion. Older faceted spurs recede and are worn down. (d) Most of the fault scarp and faceted spurs are eroded away and the cliff recedes from the fault line. (e) Continued erosion reduces the mountain mass to a lowland.*

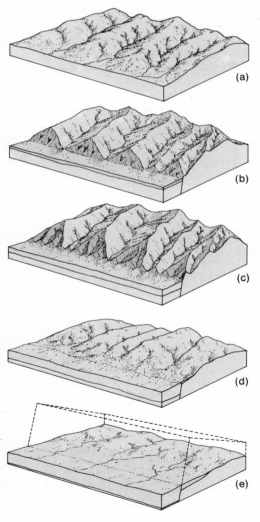

(a)

(b)

(c)

(d)

(e)

a resistant limestone formation is stripped off the crest of some anticlines and forms a hogback along the flanks of the fold.

By contrast, in the Appalachian Mountains, where the folding occurred much earlier (late Paleozoic), the topography bears no relation to the structural relief: ridges are carved on resistant quartzite and conglomerate formations, and the valleys are developed on weak shales and limestones. These mountains have suffered a much greater amount of erosion; therefore, the resistant units form ridges corresponding to the trend of the limbs of the folds. Long strike valleys (formed along the trend of the beds) have developed by headward erosion along the weak formations, and the drainage network has assumed a **trellis pattern** (figure 9.23).

Faults. Many of the major faults of the world which are still active are seen topographically as cliffs. **Normal faulting** will generally produce a cliff which is soon cut by stream erosion to form a series of triangular **faceted spurs** (figure 9.24). Continued erosion forms gullies along the blunt face of the faceted spur so the cliff produced by the fault is considerably modified. Recurrent movement along the fault may produce a fresh cliff at the base of the older modified faceted spurs, but it also is soon modified by gulleying (figure 9.24c). When movement on the fault ceases, the cliff erodes down and back from the fault line.

Strike-slip faults are fractures in which displacement has been horizontal rather than up or down and one block has moved past the other parallel to the strike or trend of the fault plane. Since there is little vertical movement, high cliffs do not form from strike-slip faults; instead, the fracture is expressed at the surface by a low ridge or offset drainage (figure 9.25).

Rates of Erosion

Statement

In many ways slow processes such as erosion are more difficult to measure than extremely rapid events, but various approaches to the problem provide some reliable estimates of the rate at which the continents are being worn away. These

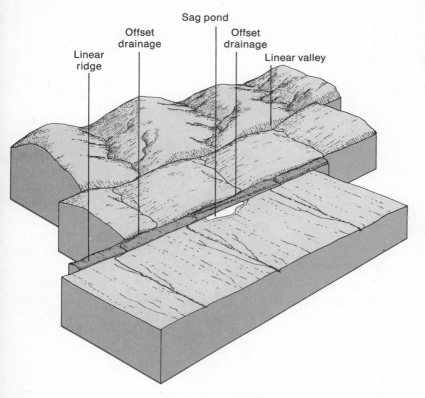

Linear ridge · Offset drainage · Sag pond · Offset drainage · Linear valley

Figure 9.25 *Erosional features along a strike-slip fault. Relatively little relief is produced by strike-slip displacement. Streams are offset, linear ridges and valleys form in sliver blocks, and local sag ponds may form in depressions along the fault line.*

measurements are very significant in the study not only of landforms but also of continental evolution, earth dynamics, and sedimentary rock genesis.

The most significant estimate of erosion rate for large drainage basins or entire continents is based on measurements of the amount of sediment transported by major river systems. In addition, measurements made at specific localities may be based on archeological evidence, datable volcanic extrusions, and various records of slope surface retreat. More complex studies are made by comparing uplift and erosional history of entire mountain ranges.

Discussion

Estimates Based on Volume of Sediment Transported by Rivers. A significant estimate on the rate of erosion for an entire river system can be made by measuring carefully the amount of sediment carried out of the drainage basin each year. Of course, rates of erosion vary greatly, depending on climate, topography, rock composition, and stage of stream evolution, but this method provides a good estimate of the rate at which the surface of the continent is being lowered. Extensive measurements of the amount of material carried in solution and in suspension have been made for most rivers throughout the United States and many of the larger rivers elsewhere in the world.

On the basis of sediment transported by the major rivers, the average rate of erosion in the United States is estimated to

Evolution of Landforms

205

Figure 9.26 Lava cascades in the western Grand Canyon. The remnants of basalts at the base of the cliff are 1.2 million years old. These formed a lava dam across the river 350 m (1400 ft) high. Erosion of the Colorado River has been able to cut down through this dam plus an additional 15 m (50 ft) since it was formed. The average rate of downcutting is 40 cm (8 in.)/1000 years.

be approximately 6 cm (2.4 in.)/1000 years. We should emphasize that the estimate of 6 cm/1000 years is an *average*. The surface is *not* lowered uniformly by this amount; indeed, some areas within the drainage basin may be unaltered or even built up by local accumulations of sediment. Others may be lowered more than 25 cm in only a few hours by the erosive power of a flash flood. Measurement of the amount of sediment removed from an area indicates only how much erosion occurs, not where it occurs. On a regional basis most rapid rates of erosion occur where the relief is greatest and where the total potential energy available from rivers for erosion and transportation is highest.

Estimates Based on Geologic Record. Volcanic extrusions provide another method of determining rates of erosion. If the time of extrusion can be determined by radioactive dating, the amount of erosion since the extrusion can be measured by comparing the difference in elevation between the present drainage and the ancient channel into which the basalt flowed. The average rate of erosion based on studies of inverted valleys in southern Utah is 5.4 cm/1000 years, which compares favorably with the average rate of erosion in the Colorado Plateau based on sediment load carried by the rivers (9 cm/1000 years).

Vulcanism in the western Grand Canyon also provides a means of estimating rates of downcutting by the Colorado River. Recent spectacular lava cascades on the rim of the inner gorge and remnants of lava dams adhere to the walls of the canyon (*figure 9.26*). The age of the dams, based on potassium-argon dating, is 1.2 million years. During this time span, the Colorado River has been able to cut through the 350 m (1400 ft) high dam and incise itself another 15 m (50 ft) into the sedimentary rocks below. The rate of canyon cutting based on these figures is 40 cm (8 in.)/1000 years. The above rates seem very slow but, geologically speaking, they are rapid indeed.

Summary

The evolution of a landscape is complex and is influenced by many factors such as climate, rock type, structure, and elevation. A landscape evolves through a series of stages, and various graphic models may be constructed to show how, under ideal conditions, a landscape will evolve in (1) a humid climate, (2) an arid climate with external drainage, (3) an arid climate with internal drainage, and (4) volcanic terrains. It should be emphasized that these are ideal models and that the "real world" is much more complex. At any time, the evolution of a landscape may be interrupted or changed by such things as earth movements, glaciation, or vulcanism. Much of our scenic landscape is the product of differential erosion in which rock units of varying resistance are etched out into cliffs, slopes, columns, or natural arches. Rates of erosion are slow by human standards, but measurements of the amount of sediment transported by rivers indicate that, on an average, the continental surface is being lowered at a rate of 6 cm/1000 years.

Although running water is by far the most effective agent of erosion, water also seeps into the ground and percolates through the pore spaces of the rock. As it moves it dissolves soluble minerals and erodes through subsurface solution activity. In the next chapter we will consider ground water as part of the hydrologic system.

Additional Readings

Bloom, A. L. 1969. The Surface of the Earth. Englewood Cliffs, N.J.: Prentice-Hall.

Easterbrook, D. J. 1969. Principles of Geomorphology. New York: McGraw-Hill Book Company.

Garner, H. F. 1974. The Origin of the Landscapes. New York: Oxford University Press.

Gordon, R. B. 1972. Physics of the Earth. New York: Holt, Rinehart and Winston.

Hunt, C. B. 1973. Natural Regions of the United States and Canada. San Francisco: W. H. Freeman and Company.

Thornbury, W. D. 1969. Principles of Geomorphology. New York: John Wiley and Sons.

Tuttle, S. D. 1970. Landforms and Landscapes. Dubuque, Iowa: William C. Brown Company, Publishers.

10 Ground Water

W ater moving in the pore spaces of rocks beneath the earth's surface is a geologic process not easily observed and, therefore, not readily appreciated. Yet, ground water is an integral part of the hydrologic system, and it is one of our most important natural resources. Ground water is not a rare or unusual phenomenon; it is distributed everywhere beneath the surface. It occurs not only in humid areas but beneath the desert regions, as well as under the frozen Arctic and high mountain ranges. In many areas, the amount of water seeping into the ground may be equal to or greater than the surface runoff. Ground water rarely flows as distinct underground streams but slowly percolates through the pore spaces of the rocks and is eventually discharged back to the surface. As it moves, it dissolves the more soluble minerals, produces caves and caverns, and is responsible for a special type of topography unlike that formed by streams and rivers.

In this chapter, we will consider the role of ground water as part of the system of moving water on the earth's surface.

Major Concepts

1. The movement of ground water is controlled largely by the porosity and permeability of the rocks.
2. The water table is the surface below which all pore spaces in the rock are saturated with water.
3. Ground water moves slowly (percolates) through the pore spaces in rocks by the pull of gravity, and in artesian systems it is moved by **hydrostatic pressure.**
4. Natural discharge of ground water is generally into streams, marshes, and lakes.
5. **Artesian water** is water confined under pressure like that in a pipe. It occurs in permeable beds bounded by impermeable formations.
6. Erosion by ground water produces **karst topography** characterized by **sink holes,** **caves**, and **disappearing streams.**

Porosity and Permeability

Statement

Water is able to infiltrate into the subsurface because solid bedrock, as well as loose soil, sand, and gravel, contains pore spaces. The pores or voids within a rock may be the space between grains, vesicles, cracks, or solution cavities. Two important physical properties of a rock largely control the amount and movement of ground water. One is porosity, i.e., the percentage of the total volume of the rock consisting of voids. Porosity determines how much water a rock body can hold. The second factor is permeability, the capacity of a rock to transmit fluids. Permeability depends upon such things as size of the voids and the degree to which they are interconnected. Porosity and permeability are not synonymous; some rocks such as shale have a high porosity, but the pore spaces are so small that it is difficult for water to move through them, so the rock, even though it has high porosity, is impermeable.

Discussion

Various types of porosity are shown in *figure 10.1*. In sand or gravel deposits (*diagram a*) pore space may occupy 12 to 45% of the total volume. If several grain sizes are abundant, the space between the large grains is partly occupied by smaller grains, and the porosity is greatly reduced (*diagram b*). In sandstone, the pore space is partly filled with cement (*diagram c*) which significantly reduces the porosity of the rock.

All rocks exposed at the surface are cut by numerous fractures which constitute a significant type of porosity (*diagram d*). As the water moves along joints in limestones, it commonly dissolves passageways (*diagram e*), which may grow to considerable size and become subterranean caves.

Ground Water

Figure 10.1 *Types of porosity in rocks.*
(a) Porosity resulting from space between well-sorted sedimentary grains (sand and gravel) with little or no cement.
(b) Porosity in poorly sorted sedimentary deposits. Much of the space between the large grains is occupied by smaller grains, thus lowering the porosity of the rock.
(c) Porosity in cemented sedimentary rocks (sandstone) or conglomerate. Much of the original porosity is destroyed by cementing material. (d) Porosity due to fractures. (e) Porosity due to fractures which have been enlarged by solution activity.

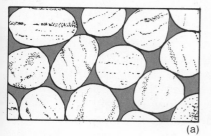
(a)

(b)

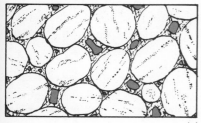
(c)

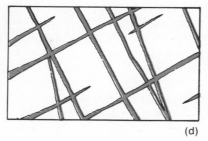
(d)

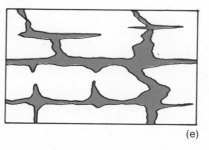
(e)

Porosity in rocks at depths greater than 3 km (2 mi) is very scanty because of the high pressures resulting from the weight of the overlying rock. At depths greater than 16 km (10 mi) rocks yield to slow flowage so that pore space is destroyed.

Permeability is the capacity for transmitting fluids. It varies with viscosity of the fluid, the hydrostatic pressure, the size of openings, and particularly the degree to which the openings are interconnected. Therefore, a rock may have high porosity but low permeability.

Rocks commonly having high permeability are conglomerates, sandstones, basalt, and certain limestones. Permeability in sandstone and conglomerate is high because of the relatively large interconnected pore spaces between the grains. Basalt is permeable because it is often extensively fractured with columnar jointing, and the tops of most flows are vesicular. Fractured limestones are permeable, as are limestones in which solution activity has created many small solution cavities. However, dense unfractured limestone is quite impermeable. Rocks that have low permeability are shale, unfractured granite, quartzite, and other dense crystalline metamorphic rocks.

Regardless of the degree of permeability, the rate at which ground water moves is extremely slow when compared to the turbulent flow of rivers. Whereas the velocity of water in rivers is measured in kilometers per hour, the velocity of ground water flow commonly ranges from 1 m (3.3 ft) per day to 1 m per year. The highest rate of percolating movement measured in the United States in exceptionally permeable material is only 250 m (825 ft) per day. Only in special cases, such as flow of water in caves, does the movement of ground water approach the velocity of slow-moving surface streams.

The Water Table

Statement

As water seeps into the ground, it continues to migrate downward by the pull of gravity through two zones in the rock and soil. The upper zone, known as the **zone of aeration**, is filled partly with air and partly with water; the water in this upper zone is called suspended water. Below a certain level all the openings within the rock are completely filled with water (*figure 10.2*). The upper surface of this **zone of saturation** is called the **water table**. It is generally only a meter or so deep in humid regions, but in the desert it may be hundreds of meters below the surface. In swamps and lakes the water table is, in essence, at the surface.

Discussion

Although the water table cannot be observed directly, it has been studied and mapped from data collected from wells, springs, and surface drainage. In addition, the movement of

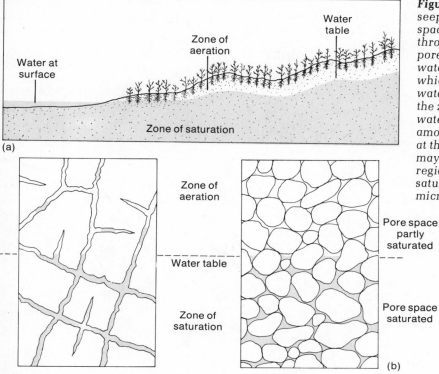

Figure 10.2 *The water table. (a) As water seeps into the ground through the pore space in the rock and soil, it passes first through the zone of aeration, in which the pore space is occupied by both air and water, then into the zone of saturation, in which all the pore space is filled with water. The water table marks the top of the zone of saturation. The depth of the water table varies with the climate and amount of precipitation. It lies essentially at the surface in lakes and swamps, but it may be hundreds of meters deep in desert regions. (b) The zones of aeration and saturation and the water table in microscopic view.*

ground water has been studied by means of dyes or other tracers, so our knowledge of the invisible body of ground water is quite extensive.

Several important generalizations can be made about the water table and its relation to the surface topography and surface drainage. These are shown diagrammatically in *figure 10.3.* In general, the water table roughly parallels the topographic surface, being highest beneath the hills and lowest in the valleys. In flat country the water table is flat; in areas of rolling hills it rises and falls with the surface of the land. The reason for such a configuration is that ground water moves very slowly, so a difference in the level of the water table is built up and maintained in areas of high elevation. During periods of heavy rainfall, the water rises in the areas beneath the hills, and during drought the water table tends to flatten out.

Where impermeable layers such as shale occur within the zone of aeration, the ground water is trapped and cannot migrate down to the main water table, and a local **perched water table** is formed within the zone of aeration. If the perched water table extends to the side of a valley, springs and seeps occur.

The water table is at the surface in lakes, swamps, and most streams, and water moves from the subsurface to these areas. Most streams in arid regions are above the water table and lose much of their water through seepage into the subsurface.

Gravity pulls ground water slowly downward through the zone of aeration to the water table. When the water passes

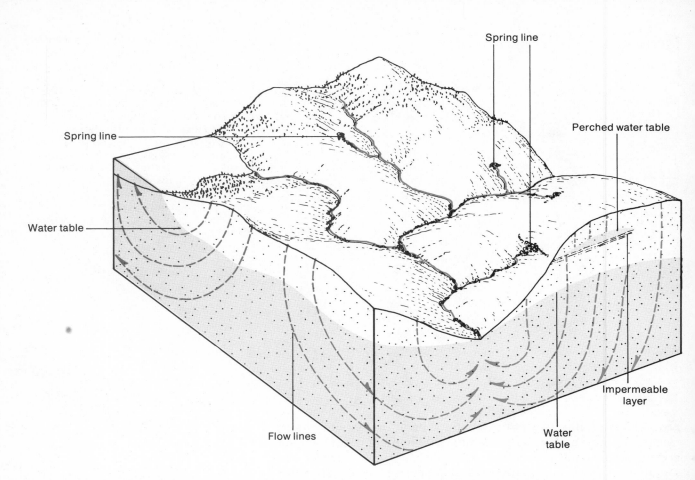

Spring line

Spring line

Perched water table

Water table

Flow lines

Water table

Impermeable layer

Figure 10.3 *The movement of ground water. The configuration of the water table is a subdued replica of the topography. In cross section the water table has hills and valleys much like those on the surface. Differences in the height of the water table causes differences in the pressure on the water in the saturation zone at any particular point. Thus, water moves down beneath the high areas of the water table because of the higher pressure and upward beneath the low areas where pressure is less. It commonly seeps into streams, lakes, or swamps where the water table is at the surface. A line of springs and seeps occurs where an impermeable layer of rock that has caused a perched water table to form is exposed at the surface.*

through the **zone of aeration** and encounters the water table, it continues to move downward through the pull of gravity along curved paths from areas where the water table is high toward areas where it is low—towards lakes, streams, and swamps (*figure 10.3*). The path of ground-water movement is *not* down the slope of the water table as one might first suspect. The explanation for this seemingly indirect flow is that the water table is not a solid surface like the ground surface. It is a surface of a liquid which in some ways resembles the surface of a wave. Water on any given point below the water table is under greater pressure from the higher areas of the water table beneath the hill and, therefore, tends to move directly downward and toward points of least pressure.

Although the paths of ground-water movement may seem indirect, they conform to the laws of fluid physics and have been mapped in many areas by tracing the movement of dye injected into the system. The movement of the dye shows that there is a continual slow circulation of ground water from infiltration at the surface to seepage into streams, rivers, and lakes. Thus, the ground-water body is not stagnant and motionless; rather it is an important part of the hydrologic system and is intimately related to surface drainage.

The Earth's Dynamic Systems

Natural and Artificial Discharge

Statement

Natural discharge of ground-water reservoirs occurs wherever the water table intersects the surface of the ground. Such places are generally in channels of streams and on the floors and banks of marshes and lakes. Springs or seeps occur wherever the water table intersects the valley wall. Natural discharge into streams and lakes is the major link between the ground-water reservoir and other parts of the hydrologic system. Water wells are made simply by drilling holes down into the water table.

Discussion

Natural discharge of the ground-water reservoir into the drainage system introduces a significant volume of water into the surface drainage. Indeed, were it not for ground-water discharge, many permanent streams would be dry during parts of the year. Most natural discharge is near or below the surface of streams or lakes and, therefore, goes largely unnoticed. It is detected and measured by comparing the volume of precipitation to surface runoff.

The most obvious natural discharge consists of seeps and springs. Springs result from a variety of geologic structures and variations in the rock sequence, some of which are shown in figure 10.4. Where a permeable bed is underlain by an impermeable layer, the ground water is forced to move laterally to the outcrop of the bed (*block a*). Conditions such as this are commonly found in mesas and plateaus where interbedded sandstone and shale occur. The spring line is commonly marked by a line of vegetation. In limestone terrain, springs occur where the base of the cavernous limestone outcrops (*block b*); the Mammoth Cave area of Kentucky is a good example. Lava formations which outcrop along the sides of a canyon often develop important springs, as ground water migrates readily through the vesicular and jointed basalt (*block c*). An excellent example is the Thousand Springs area of Idaho where there are copious flows along the side of the Snake River Valley. Other springs develop along faults which produce an avenue of greater permeability along the fracture zone and frequently displace strata so that impermeable beds block the flow from the permeable layers (*block d*).

Wells. Ordinary wells are made simply by drilling holes through the zone of aeration into the zone of saturation as shown in figure 10.5. Water then flows out of the pores into the well and fills it to the level of the water table. When a well is pumped, the water table is drawn down around the well in the shape of a cone, called the **cone of depression**. If water is withdrawn faster than it can be replenished, the cone of depression continues to grow, and ultimately the well may go dry. In cases of large wells such as those used by cities and industrial plants,

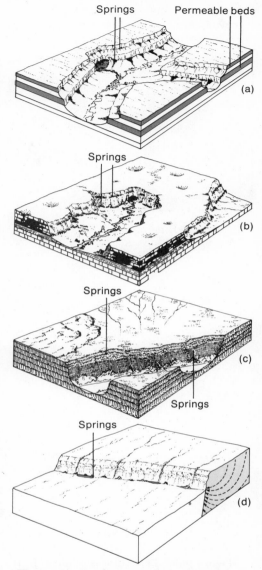

Figure 10.4 *Geologic conditions conducive to the formation of springs. (a) Permeable beds separated by impermeable beds. A spring line develops where permeable beds are exposed in valley walls.*
(b) Area of cavernous limestone. Springs form where cavernous limestone is exposed in valley walls. (c) Porous basalt. Surface water readily seeps into vesicular and jointed basalt flows. It then migrates laterally and forms springs where basalt units are exposed in canyon walls. (d) Fault lines. Many faults displace rocks so that impermeable beds are placed next to permeable beds. A spring line commonly results as ground water migrates along a fault line.

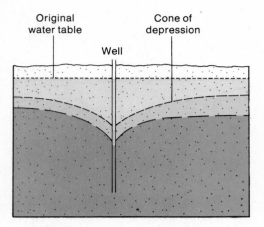

Figure 10.5 *Diagram showing the cone of depression formed around a pumping well.*

the cone of depression may be many hundreds of meters in diameter. All wells within the cone of depression are affected (*figure 10.6*). This undesirable condition has been the cause of "water wars" fought physically as well as in the courts. Inasmuch as ground water moves and is not fixed in a stationary place like mineral deposits, it is difficult to state who owns it. Many disputes are now being arbitrated by using computers that simulate actual subsurface conditions such as permeability, direction of flow, and level of water table. Computers can predict what changes will occur in the ground-water system when given amounts of water are drawn out of a well over specified periods of time.

Extensive pumping lowers the general surface of the water table. This has been done with some serious consequences in some metropolitan areas of the Southwest, where the water table has been lowered hundreds of meters. There is a limited supply of ground water and, although the ground-water reservoir is continually being replenished by precipitation, migration of ground water is so slow that it may take hundreds of years to raise a water table to its former position of balance with the hydrologic system.

Artesian Water

Statement

Artesian water is ground water that is confined in some way so that it builds up an abnormally high hydrostatic pressure. In sedimentary rocks an artesian system commonly results where permeable beds such as sandstone lie between such impermeable beds as shale. The water confined in the sandstone bed behaves much like water in a pipe (*figure 10.7*). Hydrostatic pressure builds up so that if a well or fracture intersects the bed, water will rise in the opening and may produce a flowing well or artesian spring.

Discussion

The geologic conditions necessary for artesian water are illustrated in *figure 10.7* and include the following:
1. The rock sequence must contain interbedded permeable and impermeable strata. The sequence occurs commonly in nature as interbedded sandstone and shale. Permeable beds are usually referred to as **aquifers**.
2. The rocks must be tilted and be exposed in some elevated area where infiltration can occur.
3. Sufficient rainfall must exist in the outcrop area to furnish an adequate supply of water.

The height to which artesian water will rise above the aquifer is indicated by the dotted line in *figure 10.7*. This line is not horizontal because of friction in the aquifer and loss of pressure resulting from fractures which develop leaks. If a well

Figure 10.6 *When water is pumped from a deep well, the cone of depression may extend outward for hundreds of meters and effectively lower the water table over a large area. Shallow wells in the area will then run dry because they will be above the water table.*

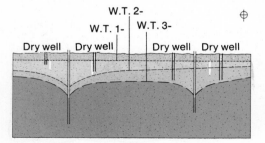

The Earth's Dynamic Systems

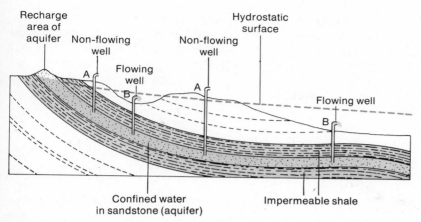

Recharge area of aquifer

Non-flowing well

Hydrostatic surface

Non-flowing well

Flowing well

A

B

A

Flowing well

B

Confined water in sandstone (aquifer)

Impermeable shale

Figure 10.7 *Geologic conditions necessary for an artesian system: (1) A permeable bed (aquifer) must be confined between impermeable layers. (2) Rocks must be tilted so the aquifer can receive infiltration from surface water. (3) There must be adequate infiltration to fill the aquifer and create a hydrostatic pressure.*

is drilled at Location A, water will rise in the well, but will not flow at the surface. At Location B, however, a flowing well results.

Examples. In most areas underlain by sedimentary rocks, artesian water is a commonplace occurrence because the geologic conditions necessary for an artesian system can be produced in sedimentary rocks in a variety of ways. Examples of well-known areas in the United States will help you appreciate how artesian systems result from different geologic settings.

One of the best-known artesian systems underlies the Great Plains states (*figure 10.8a*). The sequence of interbedded sandstones, shales, and limestones is nearly horizontal throughout most of Kansas, Nebraska, and the Dakotas but is warped up along the eastern front of the Rockies. The sandstone formations constitute several important aquifers which receive surface water along the flanks of the Black Hills and the Rocky Mountains. Water confined in these formations is under hydrostatic pressure and gives rise to an extensive artesian system in the Great Plains states.

Figure 10.8b illustrates another regional artesian system in the inclined strata of the Atlantic and Gulf Coast plains. The rock sequence consists of permeable sandstone and limestone beds alternating with impermeable clay. Surface water flowing toward the coast seeps into the beds where they are exposed at the surface and slowly moves down the dip of the permeable strata.

A third example is from the western states where the arid climate makes artesian water a very important resource (*figure 10.8c*). Here the subsurface rocks in the intermontane basin consist of sand and gravel deposited in ancient alluvial fans which grade basinward into playa clay and silt. The playa deposits act as confining layers between the permeable sand and gravel. Water seeping into the fan deposits becomes confined as it moves away from the mountain front.

Artesian systems also underlie some of the great desert regions of the world, and natural discharge from them to the surface is largely responsible for the oases. An example of the

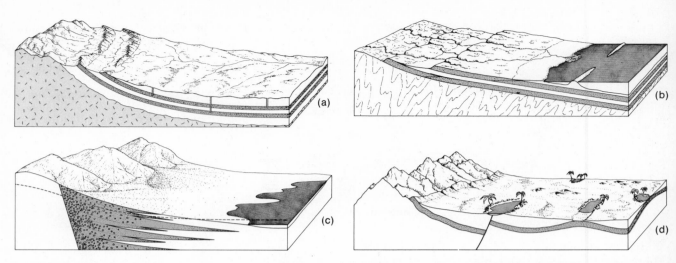

Figure 10.8 *Examples of artesian systems. (a) Great Plains states. Permeable Cretaceous sandstones underlie much of the Great Plains and are warped up along the Rocky Mountains where they receive infiltration. This forms a widespread artesian system. (b) Atlantic and Gulf Coast states. Tertiary and Cretaceous rocks dip uniformly toward the sea. Water enters permeable beds where they are exposed and is confined to form a large artesian system. (c) Intermountain basin in the western United States. Permeable sand and gravel deposited as alluvial fans interfinger with impermeable clay deposited in playa lakes. Water seeping into the lenses of buried fan deposits may be confined by the clay to form an artesian system. (d) Sahara Desert. Gently warped permeable beds underlie much of the Saraha and receive water where they are exposed at the mountain front. The artesian water may find a natural discharge through fractures or where the rock is exposed and form oases.*

system in the Sahara is shown in *figure 10.8d.* Oases occur where artesian water is brought to the surface by fractures or folds or where the desert floor is eroded down to the top of the aquifer.

Thermal Springs and Geysers

Statement

In areas of recent volcanic activity, lava flows and bodies of intrusive rock near the surface may remain very hot over long periods of time. Ground water migrating through these thermal areas becomes heated and, when discharged to the surface, produces thermal springs and **geysers,** a unique phenomenon in which a column of hot water and steam is explosively discharged at regular or irregular intervals.

Discussion

The three most famous regions of hot springs and geysers are Yellowstone National Park, Iceland, and New Zealand. These areas are all regions of recent volcanic activity so that the temperature of the rocks just below the surface is quite high.

The eruption of geysers is caused by pressure building up to a critical point in ground water contained in the fractures, caverns, or layers of porous rock (*figure 10.9*). Since the water at the base of the fracture is under greater pressure than that above, it must be heated to a higher temperature than the water above before it will boil. Eventually, a slight increase in temperature or a decrease in pressure resulting from liberation of dissolved gases causes the deeper water to boil. As steam is produced, it expands, throwing the water from the underground chambers and fractures high into the air. After the pressure is released, the fractures and caverns are refilled with water, and the process is repeated. This process accounts for the periodic eruption of many geysers, the interval between eruptions being that amount of time required for water to percolate into the fracture and be heated to a critical temperature. Geysers such as Old Faithful in Yellowstone National Park erupt at definite intervals because the rocks are permeable and the "plumbing system" can be refilled rapidly. Other geysers require more time for water to percolate into the chambers and, as a result, erupt at more irregular intervals.

Geothermal Energy. The thermal energy involved in ground water offers an attractive source of useable energy for man. Presently, it is utilized in local areas of Italy, Japan, and Iceland in a variety of ways. Present estimates show that 1% to 2% of our current needs for energy could come from a geothermal source.

In Iceland geothermal energy has been used successfully since 1928. The plan is simple. Wells are drilled into geothermal areas, and the steam and hot water is piped to storage tanks and then pumped to homes and municipal buildings where it is used for heating and hot water supply. The cost of this direct heating is only about 60% of that of fuel oil heating and about 75% of the cheapest method of electric heating. Steam from geothermal energy is also used to run electric generators, producing a form of energy somewhat easier to transport. Corrosion is a problem, however, because most thermal waters are acid and contain undesirable dissolved salts.

Erosion by Ground Water

Statement

Slow-moving ground water cannot erode by abrasion in the same manner as a surface stream, but ground water is capable of dissolving great quantities of soluble rock and moving it in solution; in some areas it is the dominant agent of erosion. The process of ground-water erosion starts with the development of a cavern system as ground water percolating through joints, faults, and bedding plains dissolves the soluble rock. In time the fracture enlarges to form a cavern system, some of which extends as a subterranean network for many kilometers. The caves grow larger and ultimately the roof collapses to produce a craterlike depression called a sink hole. Solution activity then enlarges the sink hole to form a solution valley which continues to grow until its soluble rock is completely removed.

Discussion

Rock salt and gypsum are the most soluble rocks and are rapidly eroded by solution activity, but they are relatively rare and are not widely distributed throughout the continent. Limestone is also fairly soluble and, inasmuch as it is a common rock type, solution activity plays an important role in eroding limestone terrains in humid regions.

Perhaps the best way to appreciate the significance of solu-

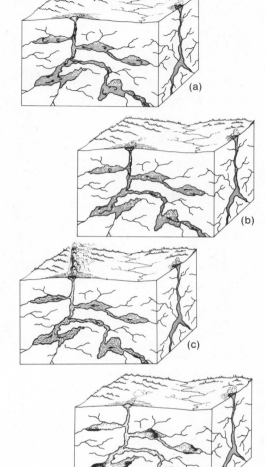

Figure 10.9 *The origin of geysers. (a) Ground water circulating through the hot rocks in an area of recent volcanic activity collects in caverns and fractures. As steam bubbles rise, they grow in size and number and tend to clog in restricted parts of the geyser tube. (b) When this happens, the expanding steam forces water upward, and it is discharged at the surface vent. The deeper part of the geyser system then becomes ready for the major eruption. (c) The preliminary discharge of water reduces the pressure on the water at depth. Water from the side chambers and pore spaces begins to flash into steam, forcing the water in the geyser system to erupt. (d) When the pressure from the steam is spent and the geyser tubes are empty, eruption ceases. The system then begins to fill with water again, and the eruption cycle starts anew.*

tion activity is to consider the nature of a cave system and the amount of rock removed by solution. A map of a cave system may show long, winding corridors, with branched openings that enlarge into chambers, or a maze of interlacing passageways and channels, controlled by intersecting joint systems (figure 10.10). Where a vertical sequence of limestone formations occurs, several levels of cave networks may exist. Mammoth Cave, Kentucky, for example, has over 50 km (30 mi) of continuous subterranean passages. In humid areas where limestone is exposed, collapse of caves develops numerous sink holes, and the sinks grow and increase in number to produce a unique landscape known as karst topography. Instead of a well-integrated surface drainage system with principal valleys and tributaries, karst topography is characterized by a surface pitted with sinks, large closed depressions known as solution valleys, and disappearing streams.

A karst topography evolves through a series of stages which are shown in the block diagrams in figure 10.11. In the initial

Figure 10.10 *A map of Anvil Cave, Alabama, shows the extent to which caverns are controlled by a fracture system. The joints occur in two intersecting sets, one trending nearly north-south and the other east-west. Solution activity along the joints has produced the network of caverns.*

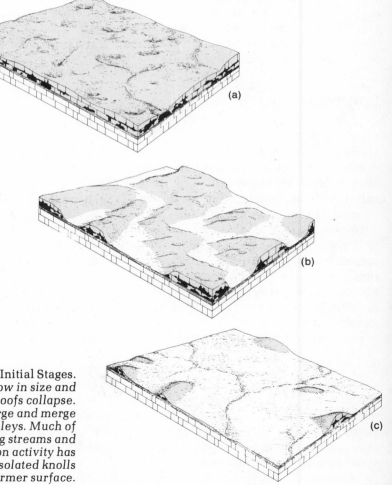

Figure 10.11 *Evolution of karst topography. (a) Initial Stages. Scattered sink holes dot the landscape and grow in size and number as caverns enlarge and their roofs collapse. (b) Intermediate Stage. Individual sinks enlarge and merge with those in adjacent areas to form solution valleys. Much of the original surface is destroyed. Disappearing streams and springs are common. (c) Late Stage. Solution activity has removed most of the limestone formation. Only isolated knolls remain as remnants of the former surface.*

The Earth's Dynamic Systems

stages of development, water follows surface drainage until a large river cuts a deep valley below the limestone layers. Ground water then moves through the joints and bedding surfaces in the limestone and emerges at the river. As time goes on the passageways become larger and caverns develop. Surface waters disappear into solution depressions. The roofs of caves collapse, producing numerous sink holes (*block a*). Springs commonly occur along the margins of valleys of major streams. Sink holes increase in number and continue to grow in size as solution activity dissolves the limestone terrain. The cavernous terrain of central Kentucky, for example, has over 60,000 holes. Sinks ultimately coalesce to form solution valleys (*block b*). Most of the original surface is finally dissolved (*block c*), leaving only scattered mesas and small buttes. When the soluble bedrock has been removed by ground-water solution, surface drainage then reappears.

An example of a karst topography in an intermediate stage of development is shown in *figure 10.12*. Sink holes in this area have been enlarged to such an extent that the dominant features are rounded mounds.

Deposition by Ground Water

Statement

The mineral matter dissolved by ground water may be deposited in a variety of ways. The most spectacular deposits are commonly found in caves as **dripstone**. Less obvious are the deposits in permeable rocks such as sandstone and conglomerates where ground water commonly deposits mineral matter to form a cement between grains. The precipitation of minerals by ground water also is responsible for the formation of certain mineral deposits such as uranium found in the Colorado Plateau.

Discussion

Cave deposits are familiar to almost everyone, and a great variety of forms have been named. However, most originate in a similar way and are collectively referred to as dripstone. The process is shown in *figure 10.13*. As the water enters the cave (usually from a fracture on the ceiling), part of it evaporates so that a small amount of calcium carbonate is left behind. The next drop adds more calcium carbonate and, eventually, a cylindrical or cone-shaped projection is built downward from the ceiling. Many beautiful and strange forms result, some of which are shown in *figures 10.14* and *10.15*. Iciclelike forms growing down from the ceiling are called **stalactites**. These are commonly matched by columns growing up from the floor (**stalagmites**), as the water dripping from the stalactite precipitates additional calcium carbonate on the floor directly below. Many stalactites and stalagmites unite to form columns. Water percolating from a fracture in the roof may form **drip curtains,** and

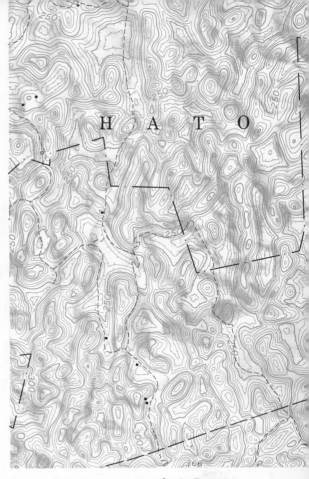

Figure 10.12 *Karst topography in Puerto Rico. Solution activity has removed much of the limestone formation which formed the original surface, leaving haystacklike mounds separated by solution valleys.*

Figure 10.13 Formation of dripstone. *Dripstone originates on the ceiling of a cave when a water drop seeps through a crack and deposits a small ring of calcite as it evaporates. The ring grows into a tube which commonly acquires a tapering shape, as water may seep from adjacent areas and flow down its outer surface.*

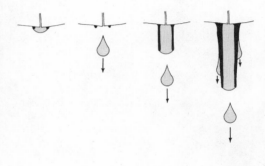

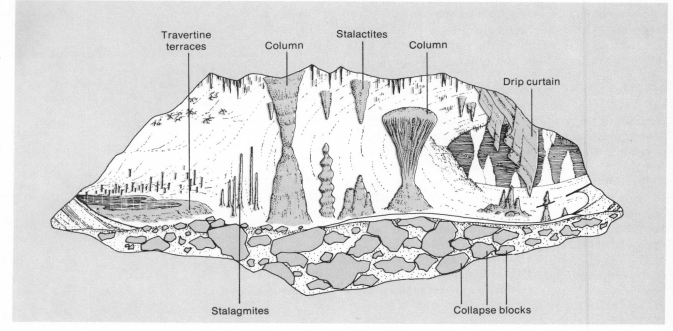

Travertine terraces

Column

Stalactites

Column

Drip curtain

Stalagmites

Collapse blocks

Figure 10.14 (above) *Varieties of cave deposits. The many interesting forms of cave deposits are all composed of calcite and originate when water seeping into the cave evaporates and deposits the dissolved mineral.*

Figure 10.15 (below) *Onadaga Cave, Crawford County, Missouri, shows many of the forms of dripstone illustrated in figure 10.14.*

pools of water on the cave floor flow from one place to another, making **travertine terraces.**

Alteration of the Ground-Water System

Statement

In previous sections of this chapter, we have seen that ground water is part of the hydrologic system and is intimately related to precipitation, surface drainage, and discharge. As in many other natural physical systems, there is a tendency for a balance or a condition of equilibrium to be established. For example, the level of the water table represents an equilibrium between the amount of precipitation and infiltration, surface drainage, porosity, permeability of the rock, and the volume of ground water discharged back to the surface. When one of these factors is modified, the others respond to reestablish equilibrium.

A variety of ground-water problems result from man's activities because ground water constitutes a most valuable resource and is being exploited at an ever-increasing rate. Some of the more important problems resulting from altering the ground-water system are (1) changes in chemical composition of ground water, (2) invasion by salt water, (3) changes in position of the water table, and (4) subsidence.

Discussion

Changes in Composition. The composition of ground water may be changed by increasing the concentration of dissolved

Figure 10.16 The effects of solid waste disposal sites on the ground-water system. (a) A permeable layer of sand and gravel overlying an impermeable shale creates a potential pollution problem because the contaminants are free to move with the ground water. (b) An impermeable clay or shale confines the pollutants and prevents significant infiltration into the ground-water system in the limestone below. (c) A fractured rock body provides a zone where pollutants may move readily in the general direction of ground-water flow. (d) An inclined permeable aquifer below a disposal site permits pollutants to enter an artesian system and move down the dip of the beds, contaminating the artesian system.

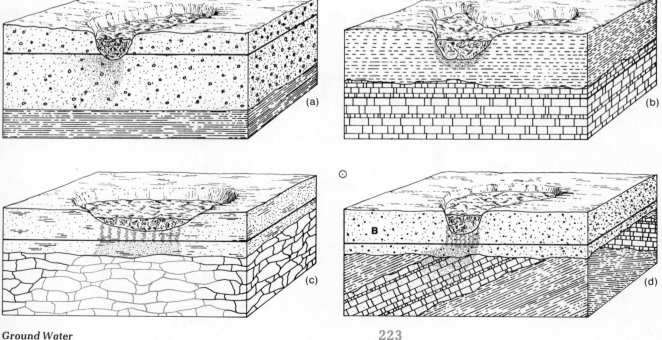

(a)

(b)

(c)

(d)

solids in surface water. When the soil mantle of the earth is regarded as an infiltration system through which ground water moves, it is evident that any concentration of chemicals or waste creates local pockets which could potentially contaminate the ground-water reservoir. Material **leached** (dissolved by percolating ground water) from waste disposal sites includes both **chemical and biological contaminants. Upon entering the ground-water flow system, the contaminants move according to the hydraulics of that system. The character and strength of the leachates** depends partly upon the length of time the infiltrated water is in contact with the waste deposit and partly upon the volume of infiltrated water. In humid areas where the water table is shallow and in constant contact with refuse, leaching is constantly producing maximum potential for pollution.

Figure 10.16 illustrates four geologic environments in which waste disposal affects the ground-water system.

In *diagram a* the geological setting is one in which the near-surface material is permeable and essentially homogeneous. Leachate percolates downward through the zone of aeration and, upon reaching the water table, becomes part of the ground-water flow system. As part of the flow system, the leachate moves in the direction of the slope of the water table and ultimately becomes part of the surface drainage system.

In *diagram b* an impermeable layer of shale confines pollutants and prevents their free movement in the ground-water system.

Diagram c illustrates a disposal site above a fractured rock body. Upon reaching the fractured rock, the contaminants may move more readily in the general direction of ground-water flow. Also, dispersion of the contaminants is limited because of the restriction of flow to the fractures.

Diagram d illustrates a critical condition where a waste disposal site is located in permeable sand and gravel above an inclined aquifer. Here leachate moves down past the water table and enters the aquifer as recharge. If the waste disposal site is located directly above the aquifer, as shown in the diagram, most leachate will enter the aquifer, but some will move down the gradient of the impermeable shale. If the site were located at position B, very little leachate would penetrate the aquifer, but it would move downgradient over the impermeable shale.

Salt-Water Encroachment. Where permeable rocks are in contact with the sea, as on an island or peninsula, a lens-shaped body of fresh ground water will be buoyed up by the denser salt water below in a manner illustrated in *figure 10.17a*. In a very real sense, the fresh water floats upon the salt water and is in a state of balance with it. The ratio of the density of fresh water to salt water is 40:41. Thus, the elevation of the water table above sea level is one-fortieth as great as the depth of the base of the fresh-water lens below sea level. For example, if the water table is 3 m (10 ft) above sea level, the contact between fresh and salt

Figure 10.17 *Relation between fresh and salt water on an island or peninsula. (a) A lens of fresh ground water beneath the land is buoyed up by denser salt water. (b) Excessive pumping causes a cone of depression in the water table and a cone of salt-water encroachment at the base of the fresh-water lens. (c) Fresh water pumped down an adjacent well would raise the water table around the well and lower the interface between the fresh and salt water.*

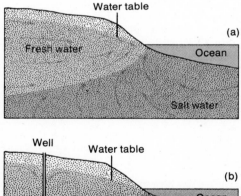

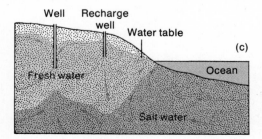

The Earth's Dynamic Systems

water would be 120 m (400 ft) below sea level. This is a delicate balance, easily upset by excessive pumping from wells because, as a cone of depression develops in the water table around a well when it is pumped, a much larger cone of salt-water intrusion develops at the base of the fresh-water lens below the well, as shown in *figure 10.17b*. With excessive pumping the cone of salt-water intrusion would extend up to the well and contaminate the fresh water. It would then be necessary to stop pumping for a long period of time to allow for the water table to rise to its former position and eliminate the salt-water encroachment. Restoration of the balance between the fresh-water lens and the underlying sea water may be hastened if fresh water is pumped down into an adjacent well (*figure 10.17c*).

Changes in the Position of the Water Table. The water table is intimately related to surface runoff, configuration of the landscape, and ecologic conditions at the surface. The balance between the water table and surface conditions, established over thousands or millions of years, may be completely upset by causing the water table to change its position. Two examples will serve to illustrate some of the many potential ecological problems.

In southern Florida, water from Lake Okeechobee flowed for the past 5000 years as an almost imperceptible river, only a few centimeters deep and 64 km (40 mi) wide, to create the Everglades. The movement of the water was not confined to channels but flowed as a sheet in a great curving swath for more than 160 km (100 mi) (*figure 10.18*). The surface slope of the Everglades is only 2 cm/km to the south; therefore, the water moved slowly to the coast, preventing salt water from invading the Everglades and subsurface aquifers along the coast. In effect, the water table in the swamp was at the surface, and the ecology of the Everglades was in balance with the water table.

Today, many canals have been constructed to drain swamp areas for farm land, to help control flooding, and to supply fresh water to the coastal megalopolis (*figure 10.19*). The canals have diverted the natural flow of water across the swamps and have effectively lowered the water table, in some places as much as a half meter below sea level. This has produced many unforeseen and often unfortunate results. As the water table lowered, salt-water encroachment occurred all along the coast, intruding public and private wells. This has forced some cities to move their water wells far inland in order to obtain fresh water.

The most visible effects, however, are in the ecology of the swamp. In the past during periods of natural drought, the high water table was sufficient to maintain a marsh condition. Now during droughts, the surface is dry. Forest fires ignite the dry organic muck, which burns like peat and smolders for long periods after the surface fires are out. This effectively destroys the ecology of the swamp. Lowering of the water table also causes the muck to compact and subside, in places as much as 2 m (8 ft).

Figure 10.18 *Natural drainage of southern Florida in 1871. The surface water spread southward from Lake Okeechobee in a broad sheet only a few centimeters deep. This maintained a swamp condition in the Everglades and established a water table very close to the surface.*

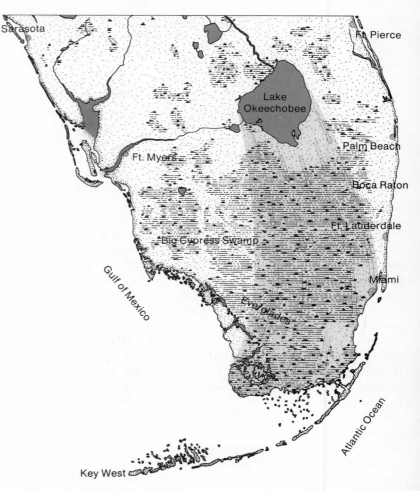

In addition, when the muck is exposed to the air, it oxidizes and disappears at a rate of about 2.5 cm (1 in.) per year. Once the muck is gone from the swamp, it can be replaced only by nature.

In contrast to problems produced by lowering the water table, raising the water table may modify many surface processes. An example is found in the environmental changes caused by irrigation in the Pasco Basin of Washington. This area lies in the rain shadow of the Cascade Mountains and receives only 15 to 25 cm (6 to 10 in.) of precipitation a year. The surface conditions of the basin have developed in response to an overall increase in aridity over the last several million years, and the surface material, slope, and vegetation are in balance with an arid climate and low water table.

In recent years, extensive irrigation has caused the water table to rise, introducing many changes in surface conditions. Today, 100 to 150 cm (40 to 60 in.) of water is applied to the ground by irrigation, which produces the effect of a simulated climatic change of considerable magnitude. The higher water table has rapidly developed large springs along the sides of river valleys. The springs are now permanent, reflecting satura-

The Earth's Dynamic Systems

Figure 10.19 *Modification of the natural drainage in southern Florida. The system of canals diverts the natural flow of surface water across swamps; the water table is lowered; swamp conditions are destroyed; and salt water encroaches in the wells along the coast.*

tion of much of the ground. Erosion is accelerated, and many farms and roads have been severely damaged. Landslides present the most serious problems, as the slopes which were stable under arid conditions are now unstable because they are partly saturated from the greatly raised water table and from the formation of perched water bodies.

In many areas, it is imperative that we modify our environment by reclaiming land or by irrigation. But unless we are careful, the detrimental effects may outweigh the advantages. Before an environment is modified seriously, we must attempt to understand as many consequences of altering the natural systems as possible.

~~Subsidence.~~ **Subsidence** of the surface related to ground water may be the result of natural earth processes, such as the development of sinks in a karst area, or it may be man-made as a result of fluid withdrawal.

Collapse into sink holes is an ever-present hazard in limestone terrains. Numerous examples exist where buildings and roads have been damaged by sudden collapse into previously undiscovered caverns below. In the United States important

Ground Water

227

karst regions occur in central Tennessee, Kentucky, southern Indiana, Alabama, Florida, and Texas (*figure 10.20*). The problem of collapse is difficult to solve. Important construction in karst regions should be preceded by test borings to determine the possible presence of subterranean cavernous zones. Concrete slurries may be pumped into solution cavities, but such remedies can be very expensive.

Compaction and subsidence present serious problems in many areas of newly deposited sediments. In New Orleans, for example, large areas of the city are now 4 m (13 ft) below sea level due largely to pumping ground water. As a result, the Mississippi River flows some 5 m (17 ft) above the city, and rainwater must be pumped out of the city at considerable cost. In addition, as the earth subsides, water lines and sewers are damaged.

When ground water, oil, or gas is withdrawn from the subsurface, significant subsidence may also occur, causing damage to construction, water supply lines, sewers, and roads. Long Beach, California, has subsided 9 m (29 ft) as the result of 34 years of oil production from the Wilmington Oil Field. This has resulted in almost $100 million damage to wells, pipe lines, transportation facilities, and harbor installations. Houston, Texas, has subsided as much as 1.5 m (5 ft) from withdrawal of ground water.

Figure 10.20 *Major areas of karst topography in the United States.*

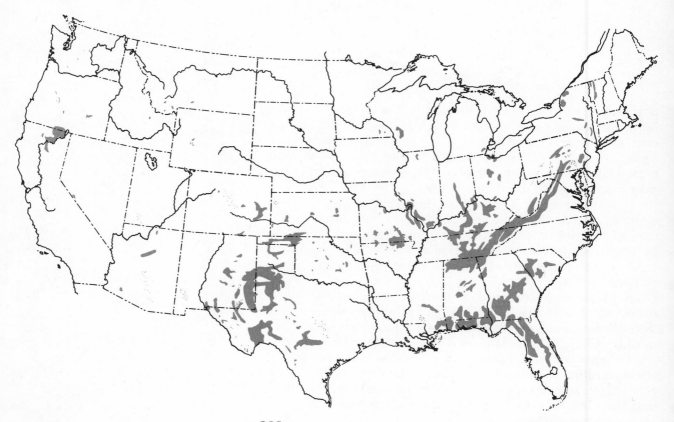

Probably the most spectacular example is Mexico City, which is built on a former lake bed. The subsurface formations consist of water-saturated clay, sand, and volcanic ash. As ground water is pumped for domestic and industrial use, the sediment compacts, and slow subsidence is widespread. Large structures such as the Opera House (weighing 54,000 metric tons [60,000 short tons]) have settled more than 3 m (10 ft), and half of the first floor is now below ground level. Other buildings are noticeably tilted, as illustrated in *figure 10.21*.

Figure 10.21 *Subsidence of buildings in Mexico City resulted from compaction after ground water was pumped excessively from unconsolidated beds beneath the city. Subsidence has tilted this building more than 2 m so that the windows in the background of the photograph are now at the level of the sidewalk.*

Ground Water

229

Summary

Ground water is an integral part of the hydrologic system and is intimately related to surface water. The movement of ground water is very slow and controlled largely by porosity and permeability of the rock. At some depth below the surface all pore spaces are filled with water. The upper surface of this saturated zone is called the water table. Movement of ground water below the water table is closely associated with surface runoff, and the ground-water reservoir discharges into streams, lakes, and swamps.

Artesian water is water that is confined between impermeable beds and is under hydrostatic pressure.

Ground water erodes by solution activity in soluble rocks such as limestone, salt, and gypsum and produces karst topography characterized by sink holes, solution valleys, and disappearing streams.

Under normal conditions ground water is in some degree of balance with surface runoff, surface topography, and salt water in the ocean. Alteration of the ground-water system may produce many unforeseen problems, as does any alteration of the hydrologic system.

In the next chapter, we shall consider a natural alteration, in fact a complete disruption, of the hydrologic system. When the earth's climate changes sufficiently to produce widespread and prolonged cold, the water in the hydrologic system is frozen and glaciers form. While an ice age is rare in the earth's history, this phenomenon significantly modifies or interrupts the geologic processes on the earth's surface.

Additional Readings

Davis, S. N., and R. J. M. deWiest. 1966. Hydrogeology. New York: John Wiley and Sons.

Sayre, A. N. 1950. "Ground Water." Sci. Amer. 183(5): 14-19 (Offprint No. 818).

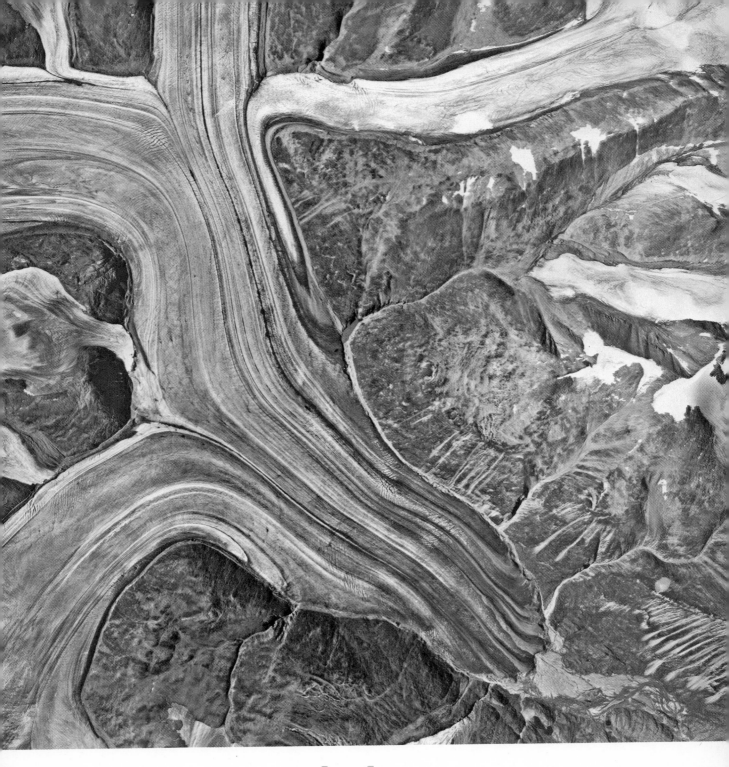

11 Glaciation

Throughout most of geologic time the climate has been relatively mild, and the hydrologic system has operated without interruption. However, we are just emerging from an ice age—a rare event in the history of the earth, for widespread glaciation has occurred only a few times since the earth's creation.

When glaciation does occur it completely disrupts the hydrologic system, and most geologic processes are interrupted or significantly modified. During an ice age, much precipitation becomes trapped in the glaciers and does not flow immediately back to the sea. Consequently, sea level drops and the hydrology of streams is greatly modified. As the great ice sheets advance over the continents, they disrupt and obliterate the preexisting drainage networks. The moving ice scours and erodes the landscape and transports eroded rock debris to its margins. There it is deposited as sheets or ridges of **moraine** (unconsolidated, unsorted debris) which covers features of the preexisting topography. The crust of the earth is pushed down by the weight of the ice, and melt water commonly flows toward the ice margins to establish lakes. New drainage systems carrying large volumes of water and sediment are established along the ice front. The area beneath the ice is completely depopulated, as both plants and animals are forced to migrate in front of the advancing ice. This causes great stress on animal populations and, because many species fail to adapt, they become extinct. The number of individuals in surviving species is significantly reduced. Far beyond the margins of the glaciers, many stream systems are modified by the large runoff produced by melting ice. Even in arid regions changes in climatic conditions associated with glaciation are reflected in the development of large lakes which form in closed basins.

Today in mountain areas tongues of ice hundreds of meters thick move down high stream valleys. As they move, they sculpture broad U-shaped valleys, form sharp glacial peaks, and deposit eroded debris as moraines at the front of the ice.

The causes of an ice age remain a tantalizing unanswered question. Many hypotheses explaining worldwide temperature changes have been proposed, but recently attention is being focused on continental drift and the shifting ocean currents as major controlling factors.

Major Concepts

1. Glacial ice forms where more snow accumulates each year than melts. With sufficient thickness ice will flow under its own weight to form a glacier.
2. As ice flows, it erodes the surface by abrasion and **plucking**. Sediment is transported by the glacier and deposited where the ice melts.
3. The two major types of glaciers, continental and valley, produce distinctive erosional and depositional landforms.
4. The major effects of the "Ice Age" include rise and fall of sea level, isostatic adjustments, modification of drainage systems, creation of numerous lakes, and migration and selected extinction of plants and animals.
5. The cause of glaciation is not clearly understood but may be related to oceanic currents and continental drift.

Glacier Systems *fields of flowing ice*

Statement

A glacier is a system of flowing ice originating on land through the process of accumulation and metamorphism of snow. Water enters the system as snow in the upper part of the glacier and is recrystallized and transformed into ice. Here, the rate of snow accumulation is greater than the rate at which it is lost by melting and evaporation during the summer, so each year there is a net gain in volume of ice. When the ice reaches a critical thickness, it will flow out from the zone of accumulation at rates of a few centimeters per day. At the lower end or periphery of the glacier, ice leaves the system by a combination of melting and evaporation. Perennial snow fields that do not move are not considered to be glaciers; nor is the ice of the frozen Arctic Ocean.

There are two principal types of glaciers: (1) valley, or alpine glaciers, which are streams of ice that originate in high snow fields of mountain ranges and flow down valleys previously carved by streams, and (2) continental glaciers, which are enormous sheets of ice thousands of meters thick that spread out and cover large parts of continents.

Discussion

Figure 11.1 is a highly idealized diagram showing the major parts of a glacial system and comparing the size and form of a continental glacier with that of an alpine glacier. The continental glacier is roughly circular or elliptical and spreads out radially from the zone of accumulation. The ice is several thousand meters thick and covers thousands of square kilometers. The alpine glacier is much smaller and is largely confined to the valleys of a stream system. It has branching tributaries, and flow direction is controlled by the stream valley which the glacier occupies. The essential parts of both glacial systems are (1) the zone of accumulation where there is a net gain of ice because more snow accumulates each year than is lost by melting and evaporation and (2) the zone of **ablation** where ice leaves the system by melting and evaporation.

In the zone of accumulation, snow is transformed into glacial ice through a process identical to rock metamorphism. Freshly fallen snow consists of delicate hexagonal ice crystals or needles with as much as 90% void space (*figure 11.2*). As snow accumulates, the ice at the points of the snowflake melts and migrates toward the center, eventually forming an elliptical granule of recrystallized ice approximately 1 mm thick called a **firn** or **névé**. As the snow accumulates and deepens with repeated annual deposits, the loosely packed névé granules are compressed by the weight of the overlying snow. Melt water, resulting from daily temperature fluctuations and pressure of the overlying snow, seeps through the pore space between the grains and freezes, aiding in the process of recrystallization. Air

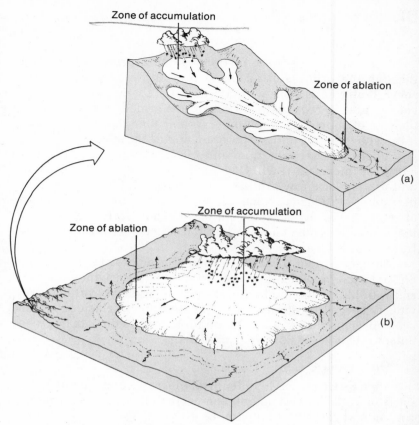

Figure 11.1 *Glacier systems. (a) Valley Glacier. The valley or alpine glacier constitutes an open system of ice flowing under the pull of gravity. The valley glacier is much smaller than a continental glacier and is confined to a river valley with a sloping floor. The movement of ice is therefore controlled by the topography. Snow enters the system in the headwaters, is transformed into ice, and flows down the valley to the end of the system, where it melts or evaporates. (b) Continental Glacier. Snow enters the system by precipitation and is transformed into ice. The ice flows outward from the zone of accumulation under pressure of its own weight. The ice leaves the system by evaporation and melting in the zone of ablation. The balance between rate of accumulation and rate of melting determines the size of the glacier system.*

Figure 11.2 *Transformation of a snow flake into granular ice. The process involves melting of the arms of the snow crystal and refreezing the water near the center to produce a more compact ice grain.*

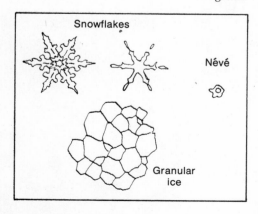

in the pore space is driven out and, when the ice reaches approximately 50 m (160 ft) in thickness, the weight is sufficient to initiate plastic flow. The ice then flows from the zone of accumulation.

At the lower end of the glacier, loss of ice by combined melting and evaporation exceeds the rate of accumulation. This area of net loss is known as the zone of ablation and is the exit boundary of the system.

It is important to understand that the margins of a glacier constitute the boundaries of a system of flowing ice in much the same way as the banks and mouth of a river constitute the boundaries of a river system. If more snow is added in the zone of accumulation than is lost by melting or evaporation at the end of the glacier, the mass of ice increases and the form of the glacier expands. If the accumulation of ice is less than ablation, there is a net loss of mass and the form of the glacier recedes. If there is a balance between accumulation and ablation, the mass of ice remains constant, and the *front of the ice remains stationary. However, ice within the glacier flows continually toward the terminal margins regardless of whether the end of the glacier is advancing, retreating, or stationary.* This might be compared to slowly advancing a straw into a fire. The straw is continually moving, but is always being consumed by the fire as it advances. As long as the temperature is low enough and adequate precipi-

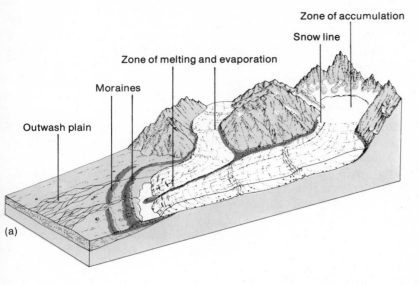

(a)

Outwash plain

Moraines

Zone of melting and evaporation

Snow line

Zone of accumulation

Figure 11.3 *Structure and flowage of valley glaciers. (a) Sketch showing the major parts of a simple valley glacier and some of the important landforms resulting from valley glaciation. (b) Valley glaciers on Baffin Island, Canada, showing many of the features illustrated in diagram (a). Note the snow-covered zone of accumulation, the rough ice in the zone of ablation, the flow structures, and the moraines.*

(b)

tation is available, the glacier will flow. The length of the glacier in no way represents the amount of ice which has moved through the system, just as the length of a river does not represent the volume of water which has flowed through it.

With this brief introduction, let us now look at the two types of glaciers in greater detail.

Valley Glaciers. The diagram and photograph in *figure 11.3* show the characteristics of typical valley glaciers. As we have mentioned before, valley glaciers are long, narrow tongues of ice that originate in snow fields of high mountain ranges and flow down preexisting valleys cut by streams. They range in length from a few hundred meters to streams more than 100 km long. In many ways, they closely resemble a fluvial drainage system. They are nourished by precipitation in the higher reaches of the mountains; they flow downgrade through a system of tributaries leading to a main trunk stream; and they receive and transport considerable amounts of debris from mass movement along the sides of the valleys through which they flow. As can be seen on the diagram and photograph, the boundary between the zone of accumulation and zone of ablation is approximated by the snow line. Above the snow line, the surface of the glacier is smooth and white, as more snow accumulates than is lost by melting and evaporation. Below the snow line, melting and evaporation exceed snowfall. Therefore, the surface of the ice in this area is rough and pitted and perhaps even broken by open crevasses. Above the snow line the crevasses are usually hidden beneath the snow cover. At the end of the glacier, sediment transported by the ice is simply dropped as the ice melts.

Continental Glaciers. In terms of their effect upon the landscape and hydrologic system of the earth, continental glaciers (ice sheets) are by far the most important type of glacier, but they are extremely large and difficult to study. Studies of the Antarctic ice sheet by teams of scientists from various countries during the last 15 years have provided much new information which permits us to construct a reasonably accurate model of the continental glacial system. The diagram in *figure 11.4* illustrates the major features of a continental glacier and some of the important effects it produces.

A continental glacier is a roughly circular, elliptical plate of ice approximately 3000 m (10,000 ft) thick which moves from the center of accumulation radially outward. It is not confined to a system of valleys but is capable of moving up and over relatively large topographical obstructions. Continental glaciers can, however, move more rapidly down broad preexisting valleys or lowlands, and as a result the margins of the glacier are commonly lobate. The thickness of a continental glacier may be as much as 3.3 km (2 mi), but rarely more. Ice does not have the strength to support the weight of a thicker accumulation. If appreciably more ice is added, the glacier will simply flow faster away from the center of accumulation.

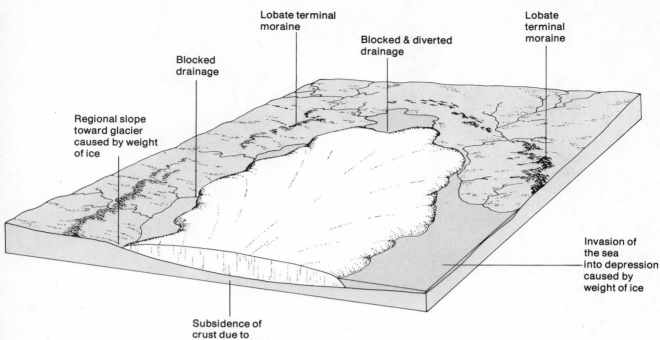

Blocked
drainage

Lobate terminal
moraine

Blocked & diverted
drainage

Lobate
terminal
moraine

Regional slope
toward glacier
caused by weight
of ice

Invasion of
the sea
into depression
caused by
weight of ice

Subsidence of
crust due to
weight of the ice

An important result of continental glaciation is the effect the weight of the ice has upon the crust. The huge weight of ice depresses the continent so that the surface of the land commonly slopes down toward the center of the ice. This depression acts as a trap for melt water and, during the retreat of the glacier, large lakes form along the ice margins; in some areas, an arm of the sea may invade this depression (figure 11.4). Previous drainage systems beneath the ice are completely obliterated and rivers flowing toward the ice margins are impounded to form lakes which may overflow and develop a river channel parallel to the ice margins.

The Barnes Ice Cap of Baffin Island, shown in figure 11.5, provides an excellent example of many features of a continental glacier. Note how the ice has blocked the drainage from the north to form the group of lakes along the ice margin. A photograph of the southern margin of the ice cap (figure 11.6) shows the large gently arched surface of the glacier, sediment deposited along the ice contact, and stream channels formed by melt water on the glacier surface.

The ice cap of Greenland is perhaps a better example because the ice covers nearly 80% of the island. In cross section, the glacier appears as a lens or a low, broad dome. The upper surface is a broad, almost flat-topped arch and is typically smooth and featureless. The base of the glacier is also curved but is considerably more irregular than the top because the shape of the ice is governed partially by the topography of the land over which it flows. The Greenland ice cap is nearly 3030 m (9900 ft) thick in the central part of Greenland but thins

Figure 11.4 Schematic diagram of a continental glacier and some of the major features it produces. The weight of the ice depresses the continent, so the land slopes toward the glacier. This produces glacial lakes along the ice margin or permits an arm of the sea to invade the depression. The original drainage is greatly modified as some streams flow toward the ice margins and form lakes. The glacier advances more rapidly into lowlands, so the margins are typically lobate. As the system expands and contracts, lobate moraines are deposited along the margins, and a variety of erosional and depositional landforms are formed beneath the ice.

CONN LAKE

BIELER LAKE

I S L A N D

B A R N E S

I C E   C A P

BLANCHFIELD LAKE

B A R N E S   I C E   C A P

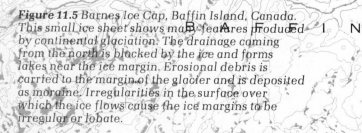

Figure 11.5 Barnes Ice Cap, Baffin Island, Canada. This small ice sheet shows many features produced by continental glaciation. The drainage coming from the north is blocked by the ice and forms lakes near the ice margin. Erosional debris is carried to the margin of the glacier and is deposited as moraine. Irregularities in the surface over which the ice flows cause the ice margins to be irregular or lobate.

B A F F I N

I S L A N D

toward the margins. The zone of accumulation is in the central part of the island, where the ice sheet is nourished by snow from storms moving from west to east. The snow line varies from 50 to 250 km (30 to 150 mi) inland, so the area of wastage constitutes only a narrow belt along the margins. In rugged terrain, especially near the margins of the glaciér, the direction of ice movement is greatly influenced by topographic features of the land (*figure 11.7*). Measurements in Greenland show that the main ice mass moves forward at approximately 10 to 30 cm (4 to 12 in.) per day, but the speed of the outlet glaciers through the mountain passes may be extreme, with velocities as great as 1 m per hour. In some places, the ice can be seen moving.

Erosion, Transportation, and Deposition by Glaciers

Statement

Glaciers transport and deposit sediment in a specific way, quite distinct from running water. Beneath the moving ice, melt water seeps into joints or fractures, freezes, and expands,

Figure 11.6 *The margins of Barnes Ice Cap. The upper surface of the glacier is gently arched and melt water has formed meandering streams on the glacier surface. Erosional debris carried by the moving ice is deposited at the glacier margins to form a terminal moraine.*

Figure 11.7 *Schematic diagram showing outlet glaciers advancing through a mountain pass as the continental glacier moves upon a mountain range.*

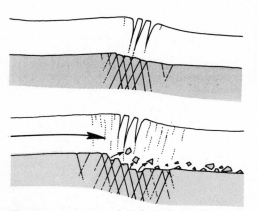

Figure 11.8 *Erosion by glacial plucking. Blocks of rock separated by joints freeze to the base or sides of the glacier and are lifted from the outcrop. They then act as abrasives and wear down the surface over which they move.*

wedging blocks of rock loose. The loosened blocks freeze to the bottom of the glacier and are plucked or quarried from the bedrock below and become incorporated into the moving ice (*figure 11.8*). The angular blocks are firmly gripped by the glacier as they become frozen to the ice and act as tools which grind and scrape the bedrock like a file. Aided by the pressure of the overlying ice, the angular blocks become very effective agents of erosion, capable of wearing away large quantities of bedrock. The rock fragments incorporated in the glacial ice are also abraded and worn down as they grind against the walls and floor of the valley and usually develop flat surfaces which are deeply scratched. The bedrock surface eroded by a glacier commonly shows grooves and striations from abrasion.

The rock material incorporated in the ice moves "in suspension" or is pushed like soil in front of a bulldozer. Deposition occurs as the ice melts, so **till** (glacial sediment) is characteristically unsorted and unstratified.

Discussion

Evidence of the abrasive and quarrying action of glaciers can be seen on most bedrock surfaces over which glacial ice has moved. Small hills of bedrock are commonly streamlined by glacial abrasion, and their upstream side is typically rounded

Figure 11.9 *Roche Moutonnee. Ice moving over bedrock commonly erodes the surface into streamline shapes, with glacial plucking producing a ragged ridge on the downcurrent side.*

The Earth's Dynamic Systems

Figure 11.10 *Glacial striations in the High Sierras, California.*

off while the downstream side is made steep and rugged by glacial plucking (*figure 11.9*). Glacial striations such as those illustrated in *figure 11.10* may range from a hairline scratch to large furrows over a kilometer long.

The inclusions of rock fragments in the glacier are collectively referred to as the load of the glacier. The load is concentrated near the contact between the ice and the bedrock from which it was derived. Near the terminus, much of the load is carried upward along shear planes and concentrated on the surface of the ice (*figure 11.11*). This results where a narrow zone of ice at the terminus is so thin that the flow slows down, forcing the ice upstream to shear up and over it.

A part of the glacial load in valley glaciers consists of rock fragments which avalanche down the steep valley sides and accumulate along the glacier margins. Frost action is especially active in the cold climate of valley glaciers and produces large quantities of angular rock fragments. This material is transported along the surface of the margins of the glacier, forming conspicuous dark bands called **lateral moraines** (*figure 11.12*). Where a tributary glacier enters the main valley, the lateral moraines join to form a long dark band down the center of the main glacier called a **medial moraine**. Thus, in addition to transporting the load near its base, a valley glacier acts like a conveyor belt and transports a large quantity of surface sediment to the terminus.

At the terminus of a glacier, the ice leaves the system through melting and evaporation, and the load is deposited as a **terminal** or **end moraine** (*figure 11.13*).

Rates of Flow and Glacial Erosion. The rate of ice flow in a glacier may be considered extremely slow when compared to the

Figure 11.11 *Movement of sediment near the margins of a glacier. The ice near the ends of a glacier is commonly stagnant, so that much of the sediment load is carried upward along shear planes and concentrates on the surface of the ice. Melting causes the form of the glacier to recede from profile 1 to profile 2, but the ice continues to move forward and sediment is deposited as end moraines.*

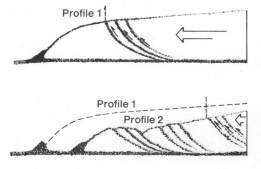

Figure 11.12 *Lateral and medial moraines on a valley glacier.*

flow of water in rivers, but the movement is continuous and, over the years, vast quantities of ice may move over the landscape. Measurements show that some of the large valley glaciers of Switzerland move as much as 180 m (600 ft) per year; smaller glaciers move 90 to 150 m (300 to 500 ft) per year. Some of the most rapid rates have been measured on the outlet glaciers of Greenland where ice is funneled through mountain passes at the rate of 8 km (5 mi) per year. From these and other measure-

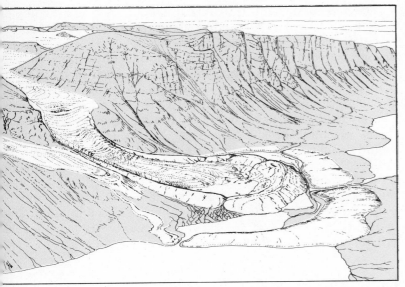

Figure 11.13 Sketch of terminal moraines formed by a glacier on Baffin Island, Canada. The moraine accumulates as the sediment load is dropped where the ice melts and leaves the glacier system.

ments it appears that flow rates of a few centimeters per day are common and that velocities of 3 m (10 ft) per day are exceptional.

Recent close observation of glacial movement shows that, at times, the entire glacier may surge forward more than several hundred meters per day. These surges are now known to be fairly common because the movement of ice can be monitored by orbiting satellite photography. In the past, only a few surges were seen because the flow is so short-lived. Some of the best-known examples include a glacier in the Himalayas, which moved 11 km (7 mi) in three months, and the Steele glacier in the Yukon Territory where a surge in 1966 had a maximum surface displacement of approximately 8 km (5 mi). Glacial surges are apparently the result of sudden slippage along the base of the glacier caused by the buildup of extreme stress upstream. Stagnant or slow-moving ice near the end of a glacier may act as a dam for the faster-moving ice upstream. When this happens, stress builds up behind the slow-moving ice and may cause a sudden surge in flow when a critical point is reached. Surges may also be caused by a sudden addition of mass to the glacier such as a large avalanche or landslide onto the glacier surface.

In considering the flow of ice in a glacier and the erosion it may cause, it is important to remember that a glacier is a system of moving ice with material entering the system in the zone of accumulation, flowing through, and then leaving the system at the distal margins. The ice within the glacier continually flows through the system regardless of whether the front of the glacier is advancing, retreating, or stationary. The length of a valley glacier is, therefore, no indication of the amount of erosion it can accomplish or has accomplished. *The erosion of a glacier is a function of the time the glacier exists and the velocity of the flowing ice within the system.*

Two examples will illustrate this point. A glacial valley 2₀ km (12 mi) long may be eroded 600 m (2000 ft) deeper than the original stream valley. This large amount of erosion was not accomplished by movement of 20 km of ice down the valley. It was the result of many thousands of kilometers of ice flowing through the valley. If the ice occupied the valley during each glacial epoch and moved 0.3 m (1 ft) per day, a total of approximately 72,000 km (45,000 mi) of ice would move down the valley in the glacier. Abrasion caused by such a long stream of ice would be enormous and completely adequate to wear down the valley 600 m (2000 ft) deep. Likewise, the continental glacier moving southward from the Hudson Bay area did not simply move a distance of 600 to 800 km (400 to 500 mi). The system expanded out that distance from the center of accumulation and over the total period of 800,000 years, roughly 72,000 km (45,000 mi) of ice moved through the system.

Landforms Developed by Valley Glaciers

Statement

Valley glaciers modify the system of river valleys through which they move, both by erosion and deposition. They characteristically form a broad, deep U-shaped valley which is relatively straight. The head of the valley is sculptured into a large "amphitheater" called a **cirque**. Where several cirques approach a summit from different directions, a sharp peak, called a **horn**, is formed. The most significant landforms produced by deposition are moraines and **outwash plains**.

Discussion

The series of idealized diagrams in *figure 11.14* provides an illustration of some of the major landforms resulting from valley glaciation and permits a comparison and contrast of landscape formed only by running water with that which has been modified by valley glaciers.

Block a shows a mountain region being eroded by streams. A relatively thick mantle of soil and weathered rock debris covers the slopes. The valleys have a typical V-shaped cross section and have many bends at tributary junctions so that ridges or spurs appear to overlap when one looks up the valley.

The area is shown occupied by glaciers in *block b*. The glaciers grow and flow down the tributary valleys and merge to form a major glacial stream. During epoch b several thousand kilometers of ice may flow down the valley through the glacier system and melt or evaporate at the end of the glacier. This enormous quantity of ice moving along through the valley may erode it as much as 610 m (2000 ft) below the original level of the stream valley.

Frost action is a major process which effectively sharpens the mountain summits. Cirques are enlarged by a glacial pluck-

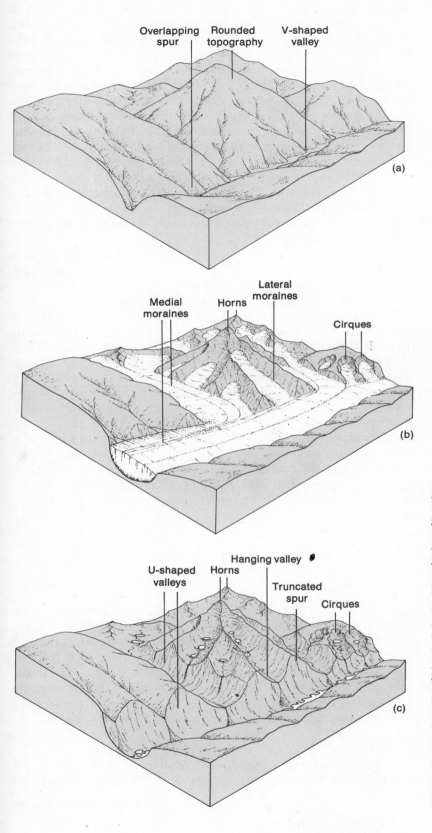

Overlapping spur Rounded topography V-shaped valley

(a)

Medial moraines Horns Lateral moraines Cirques

(b)

U-shaped valleys Horns Hanging valley Truncated spur Cirques

(c)

Figure 11.14 *Landforms developed by valley glaciation. (a) Topography before glaciation is shaped by running water. Valleys are typically V-shaped and have many curves and irregularities in plane view. Hills are rounded. (b) Valley glaciers form in high areas and move down major stream valleys. The major glacier has a network of tributaries which deliver ice to the main glacier. Frost action produces abundant rock fragments which accumulate as lateral moraines on the flanks of the glacier. (c) When the glacier recedes, the topography has been significantly modified. Sharp, angular landforms dominate. The valleys previously occupied by glaciers are deep and U-shaped. Bowl-shaped depressions called cirques develop at the head of the valley. Where several cirques meet, a sharp pyramidlike peak called a horn is formed. Tributaries form hanging valleys which may have spectacular waterfalls.*

ing and grow headward toward the mountain crest to form a horn. It should be noted that where tributaries enter the main glacier, the surface of each glacier is at an even level. However, the main glacier is much thicker and therefore erodes its valley to greater depths than the tributaries. Thus, when the glaciers recede from the area, the floors of the tributary valleys are left higher than that of the main valley and **hanging valleys** result.

Block c shows the region after the glaciers have disappeared. The most conspicuous and magnificent landforms developed by valley glaciers are the long, straight U-shaped valleys or troughs. Many are several hundred meters deep and tens of kilometers long. The heads of glacial valleys terminate in large amphitheater or bowllike cirques which commonly contain small lakes. The general topography of the peaks and ridges adjacent to the glacier is rugged and sharp because of the severe effects of frost action.

The block diagrams in *figure 11.14* do not show the landforms developed at the terminus of a valley glacier. These are illustrated in the photographs in *figure 11.15*. In *figure 11.15a*, three types of moraine are immediately obvious: (1) lateral moraines along the margins of the glacier, (2) medial moraines formed where two lateral moraines join, and (3) end moraine. The end moraine characteristically conforms to the shape of the terminus of the ice and forms a broad arc which commonly traps melt water to form a temporary lake (*figure 11.15b*). If during the recession of the ice there are periods of stabilization, recessional end moraines may form behind the terminal end moraine.

The great volume of melt water released at the terminus of a glacier reworks large amounts of the previously deposited moraine and redeposits the material in an outwash plain beyond the glacier. Outwash sediments have all the characteristics of stream deposits and are typically rounded, sorted, and stratified.

Landforms Developed by Continental Glaciers

Statement

As a continental glacier flows over the land, it removes pre-existing soil and erodes the surface of bedrock as much as several meters. This material is transported toward the margin of the ice where it is deposited, producing a variety of distinctive depositional landforms such as moraines, **kettles**, **drumlins**, **eskers**, lake sediments, and **outwash** deposits. Till deposited by continental glaciers may be more than 300 m (990 ft) thick and, thus, blanket most of the topography upon which it rests.

Discussion

Most landforms developed by valley glaciers are small enough to be seen and appreciated from various points on the

The Earth's Dynamic Systems

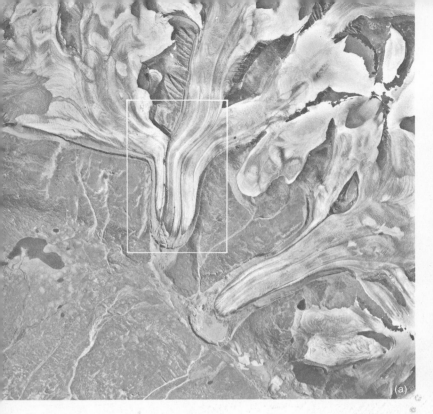
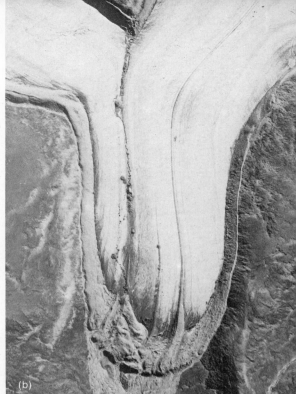

ground. By contrast, landforms produced by continental glaciers are usually regional in extent. The distinctive landforms produced by continental glaciation can be explained best by reference to the block diagrams shown in *Figure 11.16*, followed by examples shown on aerial photographs and maps. *Block a* shows the margins of an ice sheet at its maximum advance. Debris transported by the glacier accumulates at the glacial margin as a terminal end moraine. Beneath the glacier is a variable thickness of debris overridden by the glacier called ground moraine. This may be reshaped by subsequent advances of ice to produce streamlined hills called drumlins. Melt water from the ice forms braided streams which flow over the outwash plain and deposit glacial sediment. During retreat of the glaciers, melt water forms subglacial channels or tunnels. Temporary lakes may develop where melt water is trapped along edges of the glacier, and deltas and other shoreline features develop along the lake margins. Deposits on the lake bottom are typically stratified into a series of layers (**varves**). Ice blocks left behind by the retreating ice front may be partly or completely buried in the outwash plain. When an isolated block of debris-covered ice melts, a depression called a kettle is formed.

Block b shows the area after the glacier has completely disappeared. The end moraine appears as a belt of hummocky hills, the topography of which may be several kilometers wide with a local relief of 100 to 200 m. From the ground, the moraine would probably not be recognized as anything more than a

Figure 11.15 *Valley glaciers and their landforms. (a) Regional view of glacial systems on Baffin Island, Canada. Note the snow field in zone of accumulation, lateral moraines which merge into medial moraines, terminal moraines, and the outwash plain. (b) Details of the terminus of the central glacier are shown in figure a.*

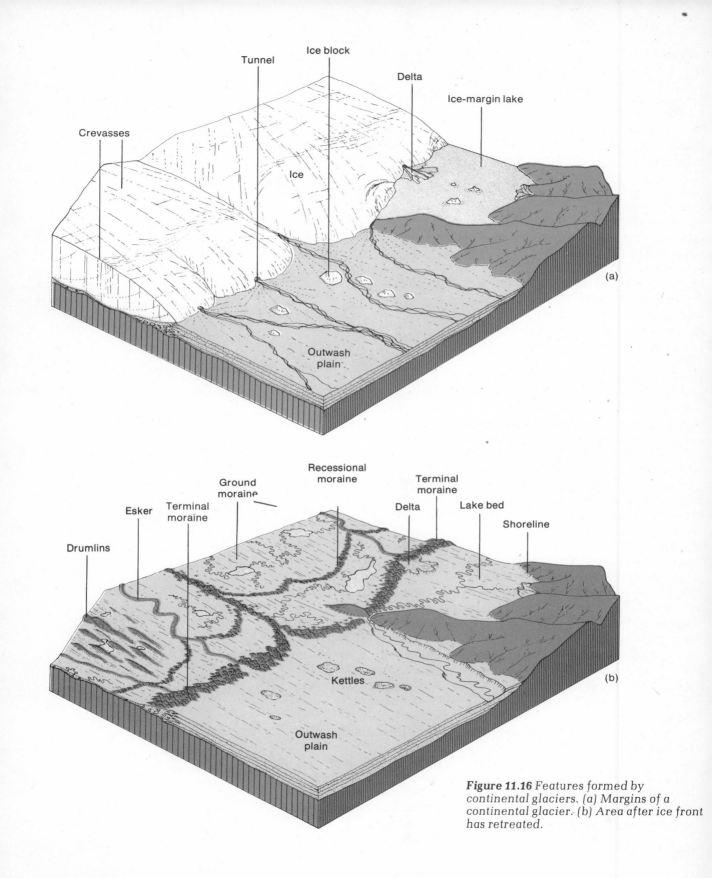

Figure 11.16 *Features formed by continental glaciers. (a) Margins of a continental glacier. (b) Area after ice front has retreated.*

series of hills by the untrained observer but, when mapped, it is arcuate in pattern, conforming to the lobate margin of the glacier. Many small depressions occur throughout the moraine, some of which may be filled with water to form small kettle lakes and ponds.

Additional landforms become visible upon retreat of the ice. Several *recessional end moraines* behind the terminal end moraine are left as the retreating ice front becomes stagnant. Behind the end moraines, the surface is covered by a variable thickness of ground moraine which buries much of the pre-glacial drainage. Ground moraine is typically poorly drained, and many lakes and marshes form. Locally, fields of drumlins rise above the general level of ground moraine in swarms of streamlined hills of unstratified glacial debris. These are typically 6 to 15 m (20 to 50 ft) high and 1 to 3 km (.6 to 1.9 mi) long. Typical drumlins are asymmetrical, with the steep slope pointing in the direction from which the ice advanced. They apparently form in regions where ground moraine deposited by one glacial advance is overridden by a subsequent advance of the ice, an action which molds the moraine into clusters of elliptical hills.

Sinuous ridges composed of coarse sand and gravel may extend for miles across glaciated regions. These features, called eskers, roughly parallel the direction of ice movement and form by deposition of sediment by streams flowing in tunnels within and beneath the glacier.

The outwash plain is relatively unmodified as the glacier recedes but continues to be built up by melt-water stream deposition. Glacial lakes are soon drained by the establishment of some outlet, leaving a flat plain over which new streams meander. The previous position of the lake is marked by stranded beaches, wave-cut terraces, and deltas.

Excellent examples of the features described above are found throughout the area of the United States north of the Missouri and Ohio rivers, which roughly mark the margins of the Pleistocene glaciers. Northward into Canada where the centers of outflow were located, the effects of glacial erosion dominate the landscape. Examples of typical features formed by continental glaciation are shown in *figure 11.17.*

Pleistocene Glaciation

Statement

The glacial and interglacial periods which occurred during the last 2 to 3 million years constitute one of the most significant events in recent earth history. During this time the normal hydrologic system was completely interrupted throughout large areas of the world and considerably modified elsewhere. Evidence for such a recent event is overwhelmingly abundant, and over the last hundred years field observations by thousands

Figure 11.17 *Features formed by continental glaciers. (a) Eskers—Canadian Shield. (b) Drumlins—Canadian Shield. (c) Terminal moraine—Alberta, Canada. (d) Till deposited on bedrock—central Iowa.*

of geologists have provided incontestable evidence that continental glaciers covered large parts of Europe, North America, and Siberia. These ice sheets disappeared only within the last 10,000 to 15,000 years.

The general extent of glaciation in the northern hemisphere is shown in *figure 11.18*, and a more detailed map showing glacial features in North America is shown in *figure 11.19*. These maps were compiled after many years of work by hundreds of geologists during which the location and orientation of drumlins, eskers, moraines, striations, and glacial stream channels were mapped.

Perhaps a dozen major periods of Pleistocene glaciation are recorded in the United States by deposits of broad sheets of till and complex moraines separated by ancient soils and layers of windblown silt. Striations, drumlins, eskers, and other glacial features show that all of Canada, the mountain areas of Alaska, and eastern and central United States down to the Missouri and Ohio rivers were covered with ice (*figure 11.19*). There were three main centers of accumulation. The largest was centered over Hudson Bay. Ice from it advanced radially northward to the Arctic islands and southward into the Great Lakes area. A smaller center was located in the Labrador peninsula. Ice spread southward from this center into the New England states. In the Canadian Rockies in the west, valley glaciers coalesced into ice caps which merged into a single ice sheet and then moved westward to the Pacific shores and eastward down the Rocky Mountain foothills.

Throughout much of central Canada, the glaciers eroded the soil, unconsolidated overburden, and solid bedrock to an average depth of 15 to 25 m (50 to 75 ft). This material was transported to the margins of the glaciers and accumulated as ground moraine, end moraines, and outwash in a broad belt from Ohio to Montana. In places, the glacial debris is over 300 m (1000 ft thick) but average thickness is about 15 m (50 ft). Melt water from the ice transported much debris down the Mississippi River and a large amount of the fine-grained debris was transported and redeposited by wind.

The presence of so much ice upon the continent had a profound effect upon the entire hydrologic system of the earth. Sea level dropped more than 100 m (330 ft). River patterns were obliterated or modified. Glacial processes of erosion and deposition replaced running water in many areas as the dominant geologic processes. Even areas far removed from the ice were affected in many unsuspected ways.

Discussion

Rhythms of Climatic Changes (Substages of the Ice Age). Even before the theory of worldwide glaciation was generally accepted, many observers recognized that the Pleistocene epoch consisted of more than a single advance and retreat of the ice. Extensive evidence now shows that there were many pe-

Figure 11.18 *The extent of Pleistocene glaciation in the northern hemisphere.*

Figure 11.19 Physiographic map showing glacial features in the central and eastern United States.

riods of growth and disappearance of continental glaciers. The interglacial periods of warm climates are represented by buried soil profiles, peat beds, and lake and stream deposits which separate deposits of unsorted, unstratified glacial debris.

Potassium-argon dating methods show that the first advance of the ice began about 2 million years ago and radiocarbon dating indicates the last glaciers began to retreat 11,000 years ago. Remnants of these last glaciers, occupying about 10% of the world's land surface, still exist in Greenland and Antarctica. We may thus be living in an as yet unfinished Ice Age.

Studies of details of advance and retreat of the ice are significant because they provide a surprising amount of climatic data helpful in determining whether we are living in a minor interglacial substage or in the beginning of an ice-free period typical of most of geologic history. Although we are unable to predict at present whether the climate will be warmer or colder, based on the established pattern of climatic change in the past, we can say that significant climatic changes are likely to occur. We are near the end of a glacial cycle. A return to normal (warmer) climates would lead to the melting of existing ice caps and the submergence of all major sea ports and lowland cities of the world. A return to an Ice Age with another major advance of continental glaciers would drive mankind from much presently habitable land. Whatever does happen, it is unlikely that stability will be established with the present conditions of ice covering approximately 10% of the earth's land surface. If the past is indicative of normal events, this ice will disappear before climatic conditions can be considered normal.

Isostatic Adjustment. Isostatic adjustment of the earth's crust resulted when the weight of the ice depressed the continent. In Canada a large area around Hudson Bay was depressed below sea level. A rebound from this depression has been in effect ever since the ice melted, and the former sea floor has been elevated almost 300 m (1000 ft). The rate of rising in this area is still about 2 cm (1 in.) per year. It is calculated that an additional rise of 80 m (265 ft) is needed before the land regains its preglacial level and before isostatic balance is reestablished.

Tilting of the earth's crust as it rebounds from the weight of the ice can be measured by mapping the elevation of shorelines of ancient lakes (*figure 11.20*). The shorelines were level when they were formed but were tilted as the crust rebounded from the unloading of the ice. In the Great Lakes region, old shorelines slope downward to the south, away from the centers of maximum accumulation of ice, indicating a rebound of 400 m (1320 ft) or more.

Changes of Sea Level. One of the most important effects of Pleistocene glaciation was the repeated rise and fall of sea level, a phenomenon which corresponded to the retreat and advance of the glaciers. During a glacial period, water became locked upon the land as ice, and sea level was lowered. As the glaciers melted, sea level rose. The amount of change in sea

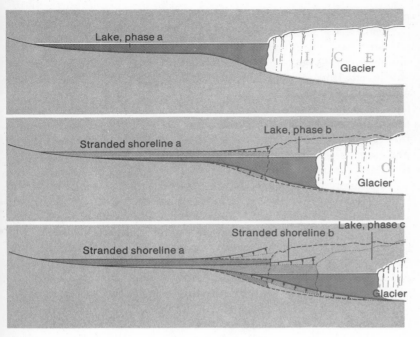

Figure 11.20 Tilted shorelines produced by isostatic adjustment of the crust after a glacier recedes. (a) A lake develops along glacier margins. Shoreline features such as beaches and bars are horizontal. (b) As the ice recedes, isostatic rebound occurs. Shoreline features formed during phase a are tilted away from the ice. Younger horizontal shoreline features are formed in lake b. (c) Continued retreat of ice permits further isostatic rebound and tilting of both shorelines a and b, which converge away from the glacier.

level can be calculated because the area of maximum ice coverage is known in considerable detail, and the thickness of ice can be estimated by comparison with known volumes of ice in the glaciers of Antarctica and Greenland. The Antarctic ice sheet alone contains enough water to raise sea level throughout the world by about 70 m (230 ft).

The dates of sea level changes are well documented by radiocarbon samples of terrestrial organic matter and near-shore marine organisms obtained from drilling and dredging off the continental shelf. These data show that about 35,000 years ago the sea level was near its present position. Gradually it receded; by 18,000 years ago it had dropped nearly 137 m (450 ft) (figure 11.21). It then rose rather rapidly to within 6 m (20 ft) of its present level. These fluctuations caused the Atlantic shoreline to recede 100 to 200 km (60 to 125 mi), exposing vast areas of the shelf. Early man probably inhabited large parts of the continental shelf now more than a hundred meters below sea level.

The glaciers extended far out across the exposed shelf of the New England coast, as evidenced by unsorted morainal debris and even mastodon remains dredged from these areas. In the central and southern Atlantic states, drainage systems extended across the shelf and eroded stream valleys now revealed by depth soundings.

Modification of Drainage Systems. Prior to glaciation, the landscape of North America was carved by running water, and well-integrated drainage systems were established which collected runoff and transported it to the sea. Much of North America was drained by rivers flowing northward into Canada because the regional slope throughout the north central part of

Figure 11.21 Changes in sea level based on radiocarbon ages of nearshore marine organisms obtained from samples from the Atlantic and Gulf coasts.

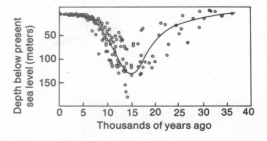

the continent was to the northeast. The preglacial drainage patterns are not known in detail, but various features of the present drainage together with segments of ancient stream channels now buried by glacial sediments suggest a pattern similar to that shown in *figure 11.22*. Prior to glaciation, the major tributaries of the upper Missouri and Ohio rivers were part of a northward-flowing drainage system which included the major rivers which drained the Canadian Rockies, such as the Saskatchewan, Athabasca, Peace, and Liard rivers. This drainage system emptied into the Arctic Ocean, probably through Lancaster Sound.

As the glaciers overrode the northern part of the continent, they effectively buried the trunk stream of the major drainage and caused the north-flowing tributaries to be dammed along the ice front. This created a series of lakes along the ice front. As the lakes overflowed, the water drained along the ice front and established the present course of the Missouri and Ohio rivers. A similar situation created Lake Athabasca, Great Slave Lake, and Great Bear Lake and their drainage through the MacKenzie

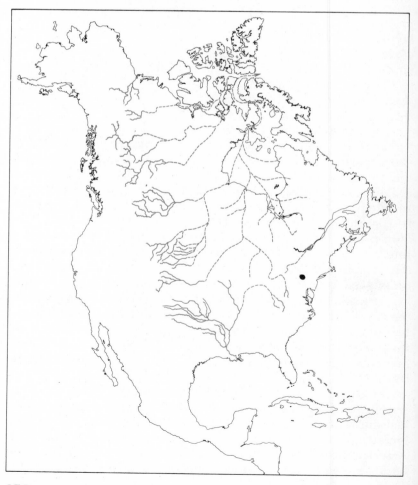

Figure 11.22 *Preglacial drainage of North America. Before glaciation, the regional drainage from the central and northern Rockies was to the north across the shield through Hudson Bay to the Arctic Sea.*

The Earth's Dynamic Systems

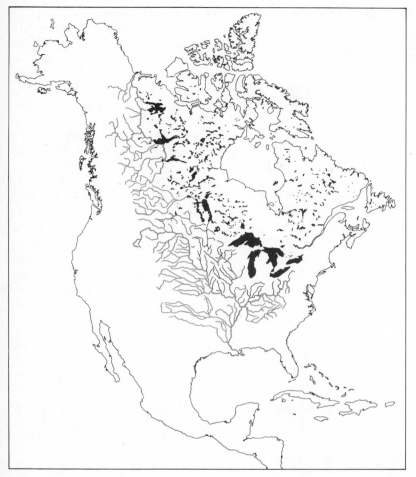

Figure 11.23 *Present drainage of central North America. The preglacial drainage was largely impounded against the glacial margins and developed new outlets to the sea through the Missouri, Ohio, and Mackenzie rivers.*

Deranged Drainage Pattern.

River (*figure 11.23*). Compare this diagram with *figure 11.5*, which shows a drainage system undergoing similar modifications at the present time as a result of the Barnes Ice Cap.

Extensive and convincing evidence for these changes is most clearly seen in South Dakota where the Missouri River flows in a deep trenchlike valley roughly parallel to the regional contours. The channel of the Missouri River is cut at right angles to the regional slope. All important tributaries enter from the west. East of the Missouri River abandoned preglacial valleys are now filled with glacial debris and mark the remnants of preglacial drainage. This hypothesis is also strongly supported by recent discoveries of huge, thick deltaic deposits in the mouth of Lancaster Sound, deposits difficult or impossible to explain by the present drainage pattern.

Beyond the margins of the ice, the hydrology of many streams and rivers was affected profoundly by increased flow resulting from melt water or by greater precipitation associated with the glacial epoch. With the appearance of the modern Ohio and Missouri rivers, significant Arctic and Atlantic drainage was diverted to the Gulf of Mexico through the Mississippi River. Many streams became overloaded and partly filled their valleys with sediment. Others became more effective agents of downcutting and deepened their valleys. Although the history of each river is complex, the general effect of glaciation on many rivers was to produce thick alluvial fill in their valleys, now being eroded to form stream terraces.

Lakes. The Ice Age created more lakes than those produced by all other geologic processes combined. The reason for this becomes obvious in light of the vast amount of water released during the retreat of the glaciers. Throughout central Canada and southern Scandinavia, the area covered by glaciers was eroded and scoured by the moving ice which completely disrupted the preexisting drainage system. This left a myriad of closed, undrained depressions in the bedrock which filled with water and became lakes (*figure 11.24*).

Southward in the north central United States, the drainage system was modified in a different manner. Here, the glaciers did not erode the surface but deposited ground and end moraines over extensive areas. These deposits formed many closed depressions throughout Michigan, Wisconsin, and Minnesota which were soon filled with water to form tens of thousands of lakes. Many of the lakes still exist; many others have been drained or filled with sediment, leaving a record of their existence in peat bogs, lake silts, and abandoned shorelines.

Figure 11.24 *Lakes created by continental glaciation in North America.*

258

During the retreat of the ice sheets of both Europe and North America, several conditions combined to create large lakes adjacent to the glacier. Their formation can be envisioned with the help of the basic model of continental glaciation shown in *figure 11.4*. The ice on both the American and European continents was more than 2400 m (8000 ft) thick near the centers of maximum accumulation but tapered toward its margins. Subsidence of the crust was greatest beneath the thickest accumulation of ice. In parts of Canada and Scandinavia, the crust was depressed over 600 m (2000 ft). As the ice melted, rebound of the crust naturally lagged behind, producing a regional slope toward the ice. This slope formed basins which lasted for thousands of years and became lakes or were invaded by the sea. The Great Lakes of North America and the Baltic Sea of northern Europe were formed basically in this way.

Although the formation of the Great Lakes is extremely complicated, the major elements of their history are illustrated in the four diagrams of *figure 11.25*. The preglacial topography of the Great Lakes region was greatly influenced by the structure and character of the rocks exposed at the surface. Reference to a geologic map of this region will show that the major structural feature is the Michigan Basin, which exposed a broad circular belt of weak Devonian shale surrounded by a more resistant Silurian limestone. Preglacial erosion undoubtedly formed a wide valley or lowland along the shale, while escarpments developed on the resistant limestone. As the glaciers moved southward into this area, large lobes of ice advanced down the great valleys to erode them into broad, deep basins. Lakes Michigan, Huron, Erie, and Ontario all lie along this belt of weak Devonian shale scoured by large lobes of the glaciers. As the glaciers receded, melt water accumulated in the basins to form the ancestors of the Great Lakes. Drainage was initially to the south through an outlet near present-day Chicago as ice dammed the northward-flowing drainage. Continued recession of the ice enlarged these lakes and exposed the basin of Lake Superior which ultimately developed drainage south from Duluth to the Mississippi. When Lake Superior was again overrun by ice, Lake Michigan drained through the Chicago outlet, and the eastern lakes found an outlet through the Mohawk Valley and down the Hudson River. As the ice receded farther, a new outlet was found through the St. Lawrence estuary. The opening of this channel drained much of the lake water and caused a low level in the lakes, but crustal rebound closed the Duluth outlet and caused the lake basins to fill again. The three upper lakes were fused into a single large body of water leaving three outlets: one into Lake Erie, another to the south past Chicago, and a third by way of the Ottawa River. With additional crustal rebound, the present drainage system was established.

Niagara Falls came into existence when the level of the lake fell below the resistant limestone formation at Lewiston, with

Figure 11.25 *The evolution of the Great Lakes.*

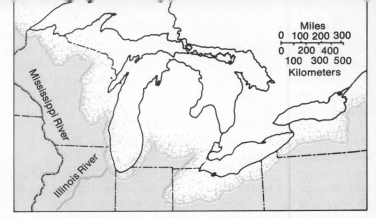

(a) Approximate position of the ice front 16,000 years ago. The ice advanced into lowlands surrounding the Michigan Basin, with large lobes extending down from the present sites of Lake Erie and Lake Michigan.

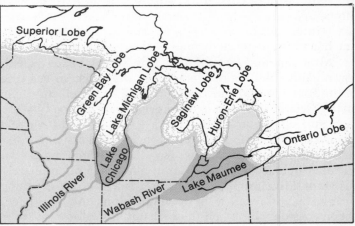

(b) The ancestral Great Lakes appeared about 14,000 years ago as the ice receded. The northern margins of the lakes were against the retreating ice. Drainage was to the south, to the Mississippi River.

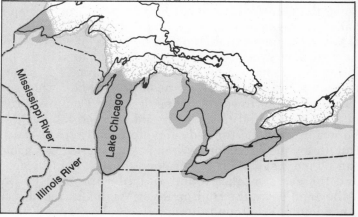

(c) As the ice front continued to retreat, an eastern outlet developed to the Hudson River, but the western lakes still drained into the Mississippi River. The lakes began to assume their present outline about 10,500 years ago.

(d) Niagara Falls originated about 8000 years ago when the glacier receded back past the Lake Ontario Basin and water from Lake Erie flowed over the Niagaran escarpment into Lake Ontario.

drainage from Lake Erie to Lake Ontario through the Niagara River. Undercutting of the weak shale below the limestone caused the falls to retreat upstream about 1.2 m (4 ft) per year. Since their inception, the falls have retreated 11.2 km (7 mi) upstream (see *figure 8.3*).

To the northwest, another group of lakes formed in much the same way, but they have since been reduced to only a small remnant of their former selves (*figure 11.26*). The largest of these marginal lakes, known as Lake Agassiz, covered the broad flat region of Manitoba, northwestern Minnesota, and the eastern part of North Dakota. It drained first into the Mississippi River, and at lower stages developed outlets into Lake Superior. Later, when the ice dam retreated, it drained into Hudson Bay. Remnants of this vast lake include Lakes Winnipeg and Manitoba, and Lake of the Woods. The vast expanse of lake sediments deposited on the floor of Lake Agassiz provides much of the rich soil for the wheat lands of North Dakota and Manitoba. Even now ancient shorelines of Lake Agassiz remain clearly etched along its margins.

Figure 11.26 *Large lakes which originated at or near the margins of the glacier in North America.*

Figure 11.27 *Stages in the evolution of the Baltic Sea.*

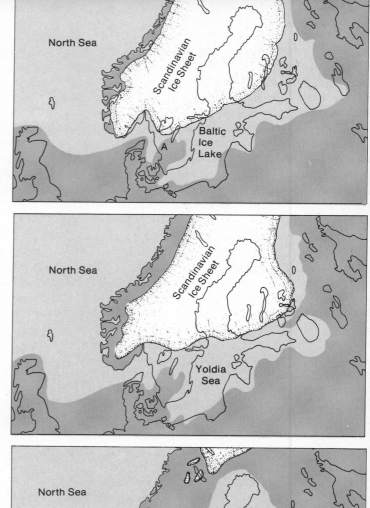

(a) Approximate position of the Scandinavian ice sheet 10,200 years ago. As the glacier receded, melt waters formed a large lake in roughly the present site of the Baltic Sea. An outlet developed at point A, and the lake began to drain through an ice-margin river into the North Sea. Sea level was rising rapidly from the melt water.

(b) By 9800 years ago the outlet widened, and sea water invaded the Baltic Lake to form a large embayment of marine water marginal to the ice front.

(c) The land in front of the ice was rising more rapidly than the sea as a result of isostatic rebound from the removal of the ice sheet. The broad connection between the Baltic Sea and the North Sea was narrowed and reduced to a short river. The Baltic then became another fresh-water lake which lasted about 2000 years.

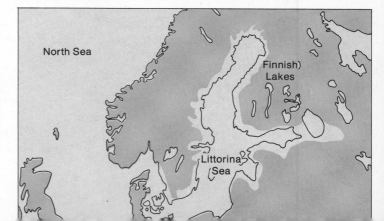

(d) The land, sea level, and the lake were all rising at different rates, and eventually isostatic rebound began to lag behind the rise of sea level. The large lake was again connected to the ocean and became the ancestral Baltic Sea, which was considerably larger than the present-day descendant, which is still shrinking as the land rises to the north.

Northward, along the margin of the Canadian Shield, Lake Athabasca, Great Slave Lake, and Great Bear Lake are remnants of other great ice-margin lakes.

The recession of the Scandinavian ice sheet caused similar depressions along the ice margins and large lakes were formed which ultimately connected to the ocean to form the Baltic Sea (figure 11.27). Thus, the origin of the lakes which bordered the Canadian ice sheet is very similar to that of Lakes Ladoga and Onega in the USSR, the Gulf of Finland, and the Baltic Sea. Both areas have a similar history, reflecting isostatic depression and rebound of the earth's crust in response to the weight of the glacial ice.

Pluvial Lakes. The climatic conditions which brought about the growth and development of glaciers also had profound effects in arid and semiarid regions far removed from the large ice sheets. The increase in precipitation which fed the glaciers also increased the runoff of major rivers and intermittent streams, resulting in the growth and development of large **pluvial** (L. *pluvia*, "rain") **lakes** in numerous isolated basins throughout the world. Most pluvial lakes were restricted to the relatively arid regions where, prior to the glacial epoch, there was insufficient rain to establish an integrated through-flowing drainage system. Instead, stream runoff flowed into closed basins and formed playa lakes. With increased rainfall, the playa lakes enlarged and partly or completely filled the basins and ceased to be ephemeral. These lakes developed a variety of shoreline features such as wave-built terraces, bars, spits, and deltas now recognized as high-water marks in many desert basins. Pluvial intervals during which the lakes covered their greatest area usually corresponded with glacial intervals. During interglacial stages the pluvial lakes shrank to small saline remnants or dry dusty playas.

The greatest concentration of pluvial lakes in North America was in the northern part of the Basin and Range province of western Utah and Nevada. Here, the fault block structure has produced more than 140 closed basins, most of which show evidence of former lakes or former high-water levels of existing lakes. The distribution of the former lakes is shown in figure 11.28. Lake Bonneville was the largest by far and occupied a number of coalescent intermontane basins. Remnants of this great body of fresh water are the Great Salt Lake, Utah Lake, and Sevier Lake. At its maximum extent, Lake Bonneville was approximately the size of Lake Michigan, covering an area of 50,000 km² (20,000 mi²) and was 300 m (1000 ft) deep. The principal rivers entered the lake from the high Wasatch Mountains to the east and built large deltas, shoreline terraces, and other coastal features which are now high above the valley floors (figure 11.29). The most conspicuous feature is a horizontal terrace high on the Wasatch Mountain front. As the level of the lake rose to 300 m (1000 ft) above the floor of the valley, it overflowed to the north into the Snake River and then to the sea. The outlet

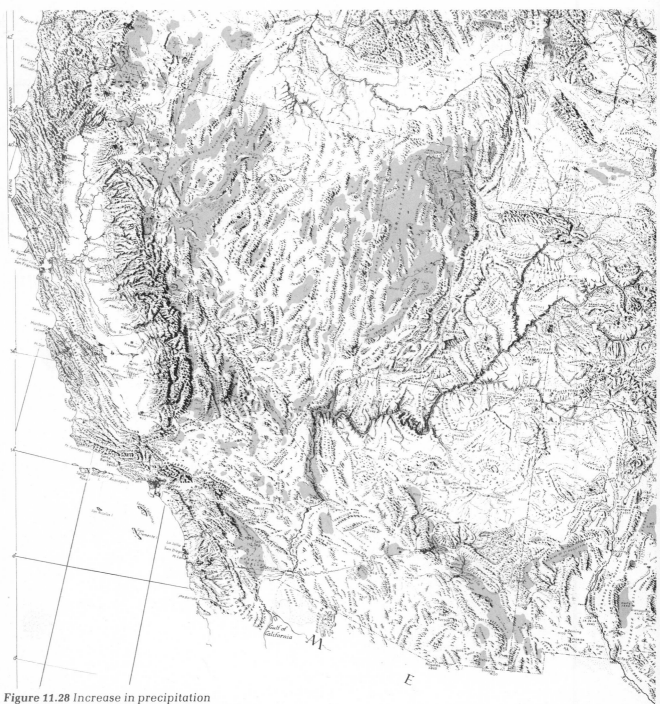

Figure 11.28 *Increase in precipitation associated with the glacial periods created numerous large pluvial lakes in the closed basins of the western United States. Lake Bonneville, in western Utah, was the largest, with present remnants being the Great Salt Lake, Utah Lake, and Sevier Lake.*

was established on unconsolidated alluvium and rapidly eroded down to bedrock, 100 m (350 ft) below the original pass. The level of the lake was then stabilized and fluctuated with the pluvial epochs associated with glaciation. Some valley glaciers from the Wasatch Mountains extended down to the shorelines of the old lake and some of their moraines have been carved by wave action. This indicates clearly that glaciation was contemporaneous with the high level of the lake. As the climate became dryer the lakes evaporated, leaving faint shorelines at lower levels.

The Channelled Scablands. The continental glacier in western North America moved southward only a short distance into Washington but played an important role in producing a strange complex of interlaced deep channels, a type of topography which is perhaps unique. This area, known as the Channelled Scablands, covers much of eastern Washington and consists of a network of braided channels 15 to 30 m (50 to 100 ft) deep. The term "scabland" is appropriately descriptive because, when viewed from the air, the surface of the basalt plateau has the appearance of great wounds or scars (figure 11.30). Many of the channels have steep walls, abandoned "dry" waterfalls, and cataracts. In addition, there are deposits with giant ripple marks and huge bars of sand and gravel. These features attest to erosion by running water of exceptional kind and degree which would be considered catastrophic by normal standards; yet today there is not enough rainfall to maintain a single permanent stream across the area.

Figure 11.29 Shoreline features of Lake Bonneville at the southern margins of Salt Lake Valley. The horizontal terrace along the base of the range marks the high-water mark of the lake.

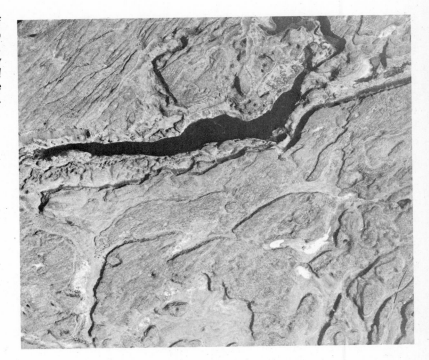

Figure 11.30 *The Channelled Scablands of Washington consist of a complex of deep channels cut into the basalt bedrock. The Scabland topography is completely different from that produced by a normal drainage system and is believed to have been produced by catastrophic flooding.*

The process by which the Scablands was eroded is outlined as follows: A large lobe of ice advanced southward across the Columbia Plateau and temporarily blocked the Clark Fork River, one of the major north-flowing tributaries of the Columbia River (figure 11.31). The impounded water backed up to form glacial Lake Missoula, a long narrow lake extending diagonally across part of western Montana. Sediments deposited in this lake now partly fill the long narrow valley. As the glacier receded, the ice dam failed, releasing a tremendous flood over the southwestward-sloping Columbia Plateau. The enormous discharge was but little diverted by the preexisting shallow valleys and spread over the basalt surface, scouring out channels and forming giant ripple marks, bars, and other sediment deposits. Estimates suggest that during the flood, as much as 40 km³ (9.5 mi³) of water per hour might have discharged from glacial Lake Missoula. Since the glaciers advanced several times into the region, catastrophic flooding likely occurred many times. Glacial Lake Missoula formed each time the ice front advanced past the Clark Fork River and then flooded the Scablands with each ice recession and subsequent dam failure.

Effect of the Ice Age on Life. The severe climatic changes occurring during the Ice Age had a drastic impact on most life forms. With each advance of the ice, large areas of continents (the area beneath the ice) became totally depopulated. In addition, a tremendous stress was put upon plants and animals retreating southward in front of the advancing glacier. The most severe stresses resulted from drastic climatic changes, reduction of living space, and curtailment of total food supply.

Most species were displaced, along with their environment, across distances of approximately 3200 km (2000 mi) as the glacier advanced. As the ice retreated, new living space became available in the deglaciated areas, whereas the formerly exposed continental shelves were inundated by the rising sea level. During the major glacial advances, new migration routes opened from Asia to North America and from Southeast Asia to the islands of Indonesia. Land plants, of course, were forced to migrate with the climatic zones in front of the glacier. Displaced storm tracts and changes in precipitation affected even the tropics as the glacier pushed cold weather belts southward.

The repeated and overwhelming changes in the environment brought about by the cycles of advancing and retreating ice were too great for many life forms, and a large number of

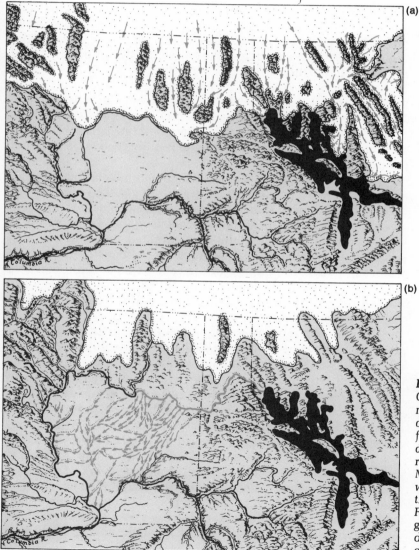

Figure 11.31 Postulated origin of the Channelled Scablands. (a) The ice sheet in northern Washington blocked the drainage of the north-flowing Clark Fork River to form a long deep lake in northern Idaho and western Montana. (b) As the glacier receded, the ice dam which formed Lake Missoula failed catastrophically, and water from the lake quickly drained across the Scabland, eroding deep channels. Repeated advance and retreat of the glacier probably produced several ice dams which failed as the ice melted each, causing catastrophic flooding.

species, particularly giant mammals, became extinct. The Imperial Mammoth, 4.2 m (14 ft) high at the shoulders, once roamed much of North America. The sabertooth "tiger" became extinct about 14,000 years ago. Fossils of the giant beaver, as large as a black bear, and the giant ground sloth which measured 6 m (20 ft) tall when standing on its hind legs, have been found in Pleistocene sediments. In Africa fossil sheep 2 m (6 ft) tall have been found, in addition to pigs as big as the present-day rhinoceros. Even giant kangaroos and other marsupials in Australia existed during the Pleistocene.

Effects of Winds. The presence of ice over so much of the continent clearly modified patterns of atmospheric circulation. Winds near the glacial margins were strong and unusually persistent because of the abundance of dense cold air coming off the glacier fields. These winds picked up and transported large quantities of loose fine-grained sediment brought down by the glacier. The dust deposited by the winds accumulated as **loess** (windblown silt) hundreds of meters thick, forming an irregular blanket over much of the Missouri River Valley, central Europe, and northern China.

Sand dunes were much more widespread and active in many areas during the Pleistocene. A good example is the Sand Hills region of western Nebraska, which covers an area of about 60,000 km² (24,000 mi²). This region was a large active dune field during the Pleistocene, but today it is stabilized by a cover of grass.

The Oceans. The effect of Pleistocene glaciation was felt even in the deep ocean basins. In addition to the changing sea level which altered shorelines and exposed much of the continental shelves, the glacial periods caused the ocean waters to cool. The cold water undoubtedly affected the kind and distribution of marine life, as well as the chemistry of the sea water. In addition, patterns and strengths of oceanic currents were significantly changed. Circulation was restricted significantly by features such as the Bering Strait, extensive pack ice, and exposed shelves.

Even the deep ocean basins did not escape the influence of glaciation. Where glaciers entered the sea, icebergs broke off and rafted their enclosed load of sediment out into the sea. As the ice melted, the debris, ranging in size from huge boulders to fine clay, settled out on the deep ocean floor, resulting in an unusual accumulation of coarse glacial boulders in fine oceanic mud. Ice-rafted sediment is most common in the Arctic, Antarctic, North Atlantic, and northeastern Pacific.

In the warmer reaches of the ocean, the glacial and interglacial periods are recorded by alternating layers of red clay and small calcareous shells of microscopic organisms. The red mud accumulated during the cold periods when fewer organisms inhabited the cold water. During the warm interglacial periods, life flourished and layers of shells mixed with mud were deposited.

The Earth's Dynamic Systems

Records of Pre-Pleistocene Glaciation

Statement

Glaciation has been a very rare event in the history of the earth. Prior to the last Ice Age, which began 2 to 3 million years ago, the climate of the earth was typically mild and uniform throughout long periods of time. This is indicated by the type of fossil plants and animals preserved in the stratigraphic record as well as by the characteristics of sediments themselves. However, there are widespread glacial deposits, consisting of unsorted, unstratified debris containing striated and faceted cobbles and boulders, which clearly record several major periods of ancient glaciation prior to the last Ice Age. This rock commonly rests upon striated and polished bedrock and is intimately associated with varved (representing seasonal sedimentation) shales and deposits of sandstone and conglomerate typical of former outwash deposits. Such evidence implies several periods of glaciation in the remote history of the earth.

Discussion

Several periods of widespread pre-Pleistocene glaciation have been recognized. The best documented record is found in late Paleozoic rocks of South Africa, India, South America, and Australia, indicating widespread continental glaciation 200 to 300 million years ago. In these areas numerous exposures of ancient glacial deposits are found, many of which rest upon a striated surface of older rock.

Deposits of older glacial sediment found on every continent but South America indicate that two other very widespread periods of glaciation occurred during late Precambrian time.

Small bodies of glacial sediment of other geologic periods have been found in local areas, but they are not nearly so well documented. Glaciation, therefore, has been a relatively rare phenomenon and appears not to have occurred in any regular cycles. The only feature that seems to be common to all periods of glaciation is that the continents appear to have been relatively high and were undergoing periods of mountain-building. Mountain-building has occurred many times in the geologic past but often without accompanying glaciation, a fact that suggests that an elevated landmass is not the primary cause of glaciation, only one of several prerequisites. A period of glaciation must require a special combination of conditions which have occurred only several times in the last 4 billion years.

Causes of Glaciation

Statement

Although the history of Pleistocene glaciation is well established and the many and varied effects of glaciation are clearly recognized, it is difficult to explain why glaciation takes place. For over a century geologists and climatologists have struggled with the problem, but it still remains unsolved.

Any theory which adequately explains glaciation must explain the following facts.

1. During the last Ice Age, which lasted 2.5 to 3 million years, there were repeated advances of the ice, separated by interglacial periods of warm climate. Therefore, glaciation is not related to a slow process involving long-term cooling.

2. Glaciation is a rare and unusual event in earth's history. There was widespread glaciation 200 to 300 million years ago and during late Precambrian time, approximately 700 million years ago. Other episodes of glaciation were apparently not widespread. Therefore, glaciation does not appear to be a result of some normal cyclic event; it seems to stem from a rare combination of conditions on the earth.

3. Throughout most of the earth's history, the climate has been milder and more uniform than it is now. A period of glaciation would require a lowering of the earth's average surface temperature by about 5° C and perhaps an increase in precipitation.

4. In order for continental glaciers to grow, an elevated landmass must be situated in such a position that storms would bring moist cold air to them. Glaciers may move into lowlands in lower latitudes, but they originate in highlands in high latitudes. Greenland and Antarctica are favorable topographic conditions today, as are the Labrador Peninsula, the northern Rocky Mountains, Scandinavia, and the Andes. It is doubtful that ice sheets would originate in low-lying continental plains even if the climate were sufficiently cold.

5. Adequate effective precipitation is required for glacial development. There are a number of areas sufficiently cold to produce glaciers even today, but these areas are too dry and have insufficient snowfall for a glacier system to develop.

Discussion

Astronomical Hypotheses. A number of hypotheses have been advanced to explain glacial periods by extraterrestrial or astronomical influences. These may be grouped into categories involving (1) variations in solar radiation and (2) variations in the earth's orbital path.

Variations in solar radiation with short-term fluctuations of as much as 3% have been observed, and some meteorologists believe larger changes may be possible. If minimum output in radiation occurred simultaneously with favorable topographic conditions on earth, a glacial period could result. Unfortunately, there is presently no way to determine variations in solar radiation in the past, so this hypothesis cannot be verified.

Variations in the earth's orbital path are measurable. They alter the length of seasons, perhaps to an amount which could produce alternating colder and warmer cycles. When the cooling produced by rotational wobble is synchronous with cooling of orbital deviation and minimal radiation from the sun, the earth (at least one hemisphere) may be cooled sufficiently to initiate glaciation. These astronomical variations occur in cycles repeated every few thousand years and apparently have existed throughout geologic time. However, there seems to have been no corresponding cyclic repetition of glaciation in the geologic record. Unless astronomical variations are supported by "geologic accidents" on the earth's surface, they fail to explain the erratic occurrence of glaciation throughout geologic time.

Atmospheric Changes. Changes in the composition of the earth's atmosphere sufficient to insulate the earth from the sun's radiation have been suggested to explain the glacial periods. The most widely supported hypothesis of this kind attributes the temperature fluctuation necessary for growth of ice sheets to a decrease in the carbon-dioxide content of the atmosphere. Carbon dioxide plus water produces an important "greenhouse" effect. With a high carbon-dioxide content in the atmosphere, short-wave solar radiation can reach the earth, but the long-wave thermal radiation cannot escape, so the earth's climate remains moderate. A drop in air temperature of 4° C may be sufficient to initiate an ice age. Extensive growth of vegetation could remove carbon dioxide from the atmosphere and bring about such a drop in temperature. However, the drop in temperature would retard plant growth and rebalance the carbon-dioxide content!

Another proposal is that fine volcanic dust injected into the atmosphere would insulate the earth by reflecting back into space part of the solar radiation, thus causing a drop in temperature. However, periods of exceptional volcanic activity do not correspond with glaciation, so this hypothesis is generally discarded.

Oceanic Controls. Oceanic currents are intimately involved in the temperature and moisture balance in climate and in the past may have played an important role in the growth and decay of ice sheets.

One hypothesis is based on circulation of oceanic currents in the Atlantic and Pacific. Prior to the Pleistocene, North and South America were not connected by the Isthmus of Panama. The warm equatorial currents of the Atlantic moved westward into the Pacific instead of northward in the Gulf Stream. The Arctic Ocean, being frozen, could not supply the moisture needed to feed the glaciers, although temperature was sufficiently low.

As the Isthmus of Panama grew and connected the Americas, the equatorial currents were deflected northward, providing enough heat to melt the Arctic sea ice. The Arctic waters were then free to evaporate and provide the moisture for the glaciers. As the ice sheets grew, sea level dropped, and the effectiveness of the Gulf Stream was reduced. The Arctic waters then froze, and the glaciers decayed and disappeared from lack of nourishment. As the glaciers receded, the Gulf Stream became more effective, and melt from the Arctic sea ice provided precipitation necessary for another advance of the ice sheets.

An important aspect of this hypothesis is that it explains the repeated advance and retreat of the glaciers during the Pleistocene epoch under an essentially stable heat budget.

Continental Drift. The theory of continental drift has recently been utilized to help explain the growth and decay of ice sheets during the Pleistocene, and earlier glacial periods as well. Throughout most of geologic time, the polar regions appear to have been occupied by a broad, open sea which allowed major ocean currents to move unrestricted to the polar regions. Equatorial waters were spread over the polar regions, warming them with water from more temperate latitudes. This unrestricted circulation produced mild, uniform climates which persisted throughout most of geologic time.

Movement of the large American continental plate westward from the Eurasian plate throughout Tertiary time culminated in the development of the north-south-trending Atlantic Ocean with the North Pole in the small, nearly land-locked basin of the Arctic Ocean. Meanwhile, the Antarctic continent drifted over the South Pole by late Miocene time. Continental glaciation probably began in the Antarctic as the high landmass moved over the cold polar region where a supply of precipitation from the surrounding oceans sufficient to nourish the glaciers was available. Evidence from deep-sea cores in the southern oceans strongly suggests that glaciation in the Antarctic began long before the Pleistocene epoch and has continued there ever since. The presence of a continental glacier on Antarctica probably caused further global cooling of both hemispheres. By the beginning of Pleistocene time, the present location and configuration of the continents and ocean basins had been established. With the present belts of climatic zones, areas near the polar regions were at the glacial threshold.

According to a theory proposed by Donn and Ewing, precipitation derived from the ice-free Arctic Ocean initiated continental glaciation in the high latitudes of the northern hemisphere. Channels connecting the Arctic Ocean with the Atlantic occurred in the shallow seaways on either side of Greenland. The cold polar air originating over the snow-covered continents has a very low percentage of water vapor, but the air masses from the open oceans contain substantial amounts of water. Ice caps grew into ice sheets which spread over the northern parts of Canada and Eurasia. Sea level dropped, restricting the circulation through the shallow channels connecting the Atlantic and Arctic Oceans. This restriction reduced the amount of warm water carried to the Arctic, thus lowering the temperature of the water. The glaciers themselves also lowered the average atmospheric temperature, and the ice-free Arctic Ocean became frozen over, probably by the middle of the glacial stage. Evaporation from the Arctic, now essentially nil, cut off the supply of snow to the major centers of the ice sheets. The southern extent of glaciation was determined by the amount of moisture available from the North Atlantic and North Pacific Oceans.

An interglacial period was initiated when the temperatures of the ocean surface water, particularly the North Atlantic, were lowered sufficiently to decrease the rate of evaporation and, therefore, of precipitation over the ice sheets. The glaciers, now starved for moisture, began to decay and retreated toward their centers of spreading.

As the ice sheets receded the ocean surface waters became warmer, and the Arctic ice eventually began to melt. This initiated another ice age, but a time lag of about 5000 years between the melting of the ice and warming of oceanic waters permitted the ice to disappear completely before the new glacial cycle began.

Greenland and Antarctica sustain continental glaciers because they are surrounded by the open sea, which supplies the moisture necessary for glaciers to exist.

This hypothesis accounts for the complete short-term cycle of glacial and interglacial stages which recurred throughout the Pleistocene. It also agrees with the fact that glaciation is a rare event requiring special geographic conditions involving continental plates, polar regions, and ocean currents. In addition, it explains the glacial period in the southern hemisphere during late Paleozoic time because the supercontinent at that time was located in the area of the South Pole and surrounded by oceans to supply the necessary moisture. A major ice age thus developed but terminated as the continents drifted apart. Moreover, this theory emphasizes precipitation as the major controlling factor, whereas other theories are based on almost inexplicable changes in temperature. However, preliminary exploration of the sediment on the Arctic sea floor does not show traces of alternating cold and warm water fauna which should be found if the Arctic was alternately frozen and unfrozen.

Summary

Several times in the geologic past systems of flowing ice called glaciers have completely interrupted the normal hydrologic system over much of the world. Water enters a glacier system as snow, is recrystallized into solid ice, flows through the glacier, and ultimately leaves the system near its outer margins by melting or evaporation. Two main types of glaciers are recognized: (1) valley glaciers and (2) continental glaciers. Each type produces distinctive landforms.

The last "Ice Age" began to terminate 15,000 to 20,000 years ago. The period of glaciation lasted for more than a million years and had a profound effect on many aspects of our physical environment. The most significant results of glaciation include (1) rise and fall of sea level, (2) isostatic adjustment of the crust,

(3) modification of drainage systems, (4) creation of numerous lakes, and (5) migration and selective extinction of many plants and animals.

The causes of glaciation are not completely clear, but a glacial period may be related to continental drift and modification of oceanic currents.

Melting of the glaciers raised the level of the oceans and inundated many coastlines. Today these shores are continuously modified by movement of the water in the oceans. In the next chapter we shall discuss these shoreline processes.

Additional Readings

Flint, Richard F. 1971. Glacial and Quaternary Geology. New York: John Wiley and Sons.

Paterson, W. J. B. 1969. The Physics of Glaciers. London: Pergamon Press.

Post, Austin, and E. R. LaChapelle. 1971. Glacier Ice. Seattle: University of Washington Press.

Price, R. J. 1973. Glacial and Fluvioglacial Landforms. New York: Hagner Publishing Company.

Sharp, R. P. 1960. Glaciers. Eugene: University of Oregon Press.

Turekian, K. K., ed. 1971. The Late Cenozoic Glacial Ages. New Haven: Yale University Press.

12 Shoreline Processes

Water in the ocean is in constant motion. It is moved by wind-generated waves, tides, **tsunamis** (seismic sea waves), **longshore currents,** and a variety of **density currents.** Many of these are active along the coasts where they expend their energy eroding the headlands and depositing sediment along beaches and bars. The processes operating along the shoreline move in a direction which tends to establish a degree of equilibrium or balance in which the configuration of the shoreline is such that the least amount of energy will be spent upon it.

However, the present shorelines of the world are not the result of modern shoreline processes alone. Nearly all coasts have been affected profoundly by the rise in sea level associated with the melting of the glaciers which began 10,000 to 15,000 years ago. With a rise in sea level, large parts of the continents were flooded, and the shorelines were placed against a landscape resulting from continental processes. Therefore, the configuration and other characteristics of a given shoreline may be largely the result of stream erosion or deposition, glaciation, vulcanism, earth movements, marine erosion or deposition, or growth of organisms.

The great concentration of population on or near shorelines indicates that they are of major importance to man. To live properly in this environment, man must understand its processes and work within the framework of shoreline dynamics. In this chapter we describe the origin of wind-generated waves, tides, tsunamis, and longshore currents. We then consider how erosional and depositional features along the coast are formed and how they evolve with time. A brief analysis of shorelines, based on the geologic agents responsible for their development, concludes the chapter.

Major Concepts

1. Waves generated by the wind provide most of the energy for shoreline processes.
2. Longshore currents are generated by waves advancing obliquely toward the shore and are one of the most important processes operating on the coast.
3. Erosion along a coast tends to develop a **sea cliff** by undercutting action of waves and longshore currents. As the cliff recedes, a **wave-cut platform** develops and enlarges until equilibrium is established.
4. Sediment transported by waves and longshore currents is deposited in areas of low energy, forming **beaches**, **spits,** and **barrier islands.**
5. Processes of erosion and deposition along a coast tend to develop a straight or gently curving shoreline on which there is a minimum of energy expended.
6. **Reefs** grow in a very special environment and form coasts which may evolve into an **atoll**.
7. The worldwide rise in sea level associated with the melting of glaciers has drowned many coasts. As a result many coasts have been only slightly modified by modern marine processes.

Waves

Statement

Oscillatory wind-generated waves are formed in the open ocean by the transfer of energy from the wind to the water as turbulence and gusts cause variations in pressure on the water surface. Wind-generated waves are oscillatory, in that a given water particle in the wave moves in a circular orbit and returns to approximately its original position; only the form of the wave moves continually forward. As a wave approaches shallow water, the column of water affected by wave motion encounters the sea floor. The resulting friction causes the gradual slowdown so that a series of waves becomes crowded together. Wave height increases as energy from a deep column of water is concentrated into a shallower one. As wave height increases and velocity decreases, the wave crest extends beyond the support of the underlying column of water, causing the wave to collapse or break. All water in the wave then moves as turbulent flow up the beach.

Discussion

Most shoreline processes are either directly or indirectly the result of wave action; therefore, an understanding of wave phenomena is fundamental to the study of shoreline processes. All waves are a means of moving some form of energy from one place to another. This is true of sound waves, radio waves, or water waves. All waves must be started by some force or energy. By far the most important types of waves in the ocean are those generated by the wind. Other types of waves in the ocean are quite rare and include seismic sea waves, or tsunamis (waves generated by earth movements), **tidal bores,** and deep-water waves. Although the transformation of wind energy into water wave energy is not completely understood, it appears that pressure fluctuations and frictional drag associated with the turbulent flow of the wind play an important role. Turbulent wind depresses the water surface where it moves downward and elevates the surface where it moves upward. This produces an irregular wavy surface on the water.

The description of ocean waves utilizes the same terms as those applied to other wave phenomena. These are illustrated in *figure 12.1*. The **wave length** is the horizontal distance between adjacent **crests** or adjacent **troughs**. **Wave height** is the vertical distance between crest and trough. The time between the passage of two successive crests is referred to as the **wave period**.

Waves tend to be choppy and irregular in the area where they are generated because a number of wave systems may be superposed on each other. The size and orientation of the various wave systems are different, but as waves move out from their place of origin, the shorter waves are left behind, and wave patterns develop some measure of order. Most large waves

Shoreline Processes

Figure 12.1 *Nomenclature of wave morphology.*

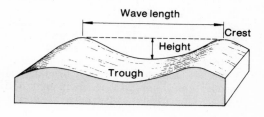

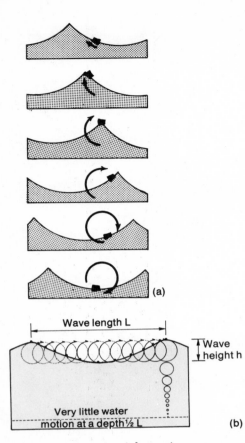

Figure 12.2 (a) Particle motion as a wave advances from left to right can be seen by movement of a floating object. As the wave advances, the object is lifted up to the crest and returns to the trough. The wave form advances, but the water moves in an orbit, returning to its original position. (b) The orbital motion of water in a wave decreases with depth and dies out at a depth equal to about half the wave length.

Figure 12.3 As a wave approaches the shore, several significant changes result as the orbital motion of the water interacts with the sea floor. (a) The wave length decreases due to frictional drag, and the waves become crowded together closer to shore, (b) the wave height increases as the column of water moving in an orbit stacks up on the shallow sea floor, and (c) the wave becomes asymmetrical due to increasing height and frictional drag on the sea floor and ultimately breaks. The water then ceases to move in an orbit and rushes forward to the shore.

generated in the oceans may travel great distances, some as much as 10,000 km (6000 mi), before breaking on the shore.

The height of a wave is a function of three factors: (1) the velocity of the wind, (2) the distance over which the wind blows, referred to as the **fetch**, and (3) the wind duration. The importance of fetch is apparent in small lakes, for no matter how long and hard the wind blows, large waves cannot be produced.

Water Motion in Waves. Wind-generated waves are described as oscillatory; that is, as each wave form passes a given point, water particles move only in a circular orbit and return to approximately their original position. The wave form, but not the individual water particles, travels out from the area where the waves are generated. This motion can be observed in any floating object as it appears to bob up and down with the passage of each wave *(figure 12.2a)*. As the wave form approaches, a particle of water in the trough will move upward and toward the crest. As the crest passes, the particle moves downward and back to its original position. Under strong wind, a water particle may advance slightly more than it recedes, but the water does not advance forward with the form of the wave.

Beneath the surface, the orbital motions of the water decrease rapidly to a level known as **wave base** where the motion completely dies out *(figure 12.2b)*. For practical purposes, water can be considered motionless at a depth equal to about one-half the wave length. In currents, water moves forward in a definite direction and does not return to its original position.

Breakers. The energy of a wave depends upon its length and height. The greater the height of a wave, the greater the size of the orbit in which the water moves. The total energy of a wave may be represented by a column of water in orbital motion. As a wave approaches shallow water, some very important changes occur *(figure 12.3)*: (1) The wave base encounters the bottom; friction causes the wave to slow down gradually so that each wave progresses more slowly than the one behind. Therefore, the waves are crowded together. (2) The wave height increases as the column of water affected by wave motion encounters the sea floor. Thus, energy from a deep column of water is concentrated into a shallower one. Each wave, therefore, becomes higher as it passes into shallow water. As the

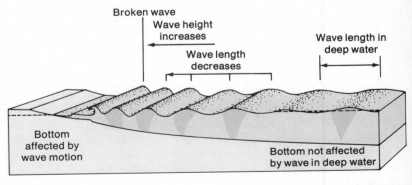

wave form becomes progressively higher and the velocity decreases, a critical point is reached where the forward velocity of the orbit distorts the wave form. The wave crest then extends beyond the support of the underlying column of water, and the wave collapses or breaks. (3) At this point, all the water moves forward. The wave form is largely lost, and the energy is released as a wall of moving turbulent water. After the breaker collapses, the **swash** (turbulent sheet of water) flows up the beach slope. The swash is a powerful surge causing a landward movement of sand and gravel on the beach. As the force of the swash is dissipated against the slope of the beach, the **backwash** flows down the beach slope, but much of it seeps into the permeable sand and gravel of the beach.

Tides

Statement

A tide is the alternating rising and falling of the ocean's surface as it responds to the gravitational attraction of the moon and the sun. Tides are also generated in the atmosphere and in the solid body of the earth, but our main concern here is the effect of ocean tides upon the coasts. Tides affect beach processes in two ways: (1) by initiating a rise and fall of the water level and (2) by generating currents.

Discussion

Since the first century, it has been known that the moon controls the tides in some manner, but it was not until Sir Isaac Newton published the Law of Gravitation in 1687 that the true explanation of tide became known. The Law of Gravitation states that two bodies attract each other with a force directly proportional to the product of their masses and inversely proportional to the square of the distance between them.

The forces involved in tides include: (1) the gravitational attraction of the moon on the earth and (2) the centrifugal force on the earth resulting from the earth-moon system revolving about a common center of gravity. The centrifugal force caused by the earth's revolving about the earth-moon system is constant and in opposition to the gravitational force exerted by the moon. The water in the oceans is free to move and so is deformed by these forces. As can be seen in *figure 12.4*, on the side of the earth facing the moon water is pulled toward the moon, causing a high tide. As the moon revolves around the earth, the tidal bulge in the ocean follows the moon. There is also a tidal bulge on the opposite side of the earth caused by centrifugal force. The balance between gravitational attraction and centrifugal force is exact at the center of the two bodies, but at the earth's surface the two forces are not equal. On the moon side of the earth the gravitational force is greater, while on the opposite side of the earth the centrifugal force is greater and causes a

Figure 12.4 *The origin of tides is related to gravitational attraction of the moon and centrifugal force of the earth-moon system. On the side of the earth facing the moon, gravitational attraction is greater, forming a tidal bulge in the ocean's water. On the other side of the earth, centrifugal force is greater, causing another tidal bulge. The two forces balance at the center of the earth.*

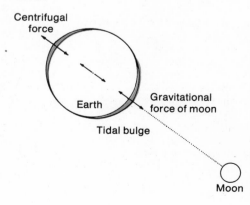

bulge in the ocean surface. There are thus two high tides produced as the moon orbits the earth.

The movement of tides produces a significant effect on many coasts. Although the difference between high and low tide in the open ocean is small, exceptionally high tides are produced in shallow seas where the rising water is concentrated in bays and estuaries. In the Bay of Fundy between New Brunswick and Nova Scotia, the range between high and low tide is as much as 21 m (70 ft). Off the coast of Brittany, tides of 12 m (40 ft) with current velocities as high as 20 km (12 mi) per hour are produced. In some rivers, high tide may reverse the flow of water as it rushes upstream in a breaking wave called a tidal bore. Thus, tidal currents are capable of transporting sediment and eroding the sea floor near shore.

Tsunami

Statement

Waves produced by the wind are the most significant in the development of coastal landforms, but unusual waves called tsunamis (seismic sea waves), produced by earthquakes, volcanic eruptions, or submarine landslides, may have disastrous effects upon coasts.

Discussion

A tsunami wave differs from a wind wave in that the energy is transferred to the water from the sea floor so that the entire depth of water is involved in the wave motion. The wave front travels out from its point of origin at very high speeds, ranging from 480 to 800 km (300 to 500 mi) per hour, and may traverse the entire ocean. The wave height is only 30 to 60 cm (1 to 2 ft), and the wave length ranges from 55 to 200 km (60 to 120 mi) so that in an open ocean a tsunami may pass unnoticed. As the wave approaches the shore, the energy distributed in the deep column of water becomes concentrated into an increasingly shorter column which results in a rapid increase in wave height. The waves which were less than 60 cm (2 ft) high in the deep ocean build rapidly to heights of more than 15 m (50 ft) in many cases and well over 30 m (100 ft) in rare instances. Thus, they exert an enormous force against the shore and may inflict serious damage and great loss of life.

A number of tsunamis are well documented by records from seismic stations and coastal observations. For example, the Hawaii tsunami of April 1, 1946, originated in the Aleutian Trench off the island of Unimak. The waves moved across the open ocean, imperceptible to ships which lay in their path because the wave height was only 30 cm (1 ft). Moving at an average speed of 760 km (470 mi) per hour, they reached the Hawaiian Islands, a distance of 3200 km (2000 mi) away, in less than five hours. Since the wave length was 145 km (90 mi), each

The Earth's Dynamic Systems

wave crest arrived about twelve minutes apart. As the wave approached the island, its height increased to a minimum of 17 m (55 ft), thus producing an extremely destructive surf which swept inland, completely demolishing houses, trees, and almost anything in its path.

Wave Refraction

Statement

As waves approach shallow water they are commonly bent or refracted so that the crest of the wave becomes almost parallel to the shore. Bending occurs because part of the wave begins to drag bottom in shallow water and slows down, while the remainder of it in deeper water moves forward at normal velocity. Therefore, the wave actually changes its direction and is curved or refracted. As a result, **wave refraction** concentrates energy on the **headland** and disperses the wave energy in a **bay**.

Discussion

The effect that wave refraction has on concentrating and dispersing energy can be appreciated by carefully analyzing the energy in a single wave. In *figure 12.5* the unrefracted wave is divided into several equal parts each of which has an equal

Figure 12.5 *Wave refraction concentrates energy on the headlands. Each segment of the wave A-B, B-C, and C-D has the same amount of energy. The segment B-C encounters the sea floor sooner than A-B or C-D and moves more slowly. This causes the wave to bend so that the energy contained in segment B-C is concentrated on the headland, while that contained in A-B and C-D is dispersed along the beach.*

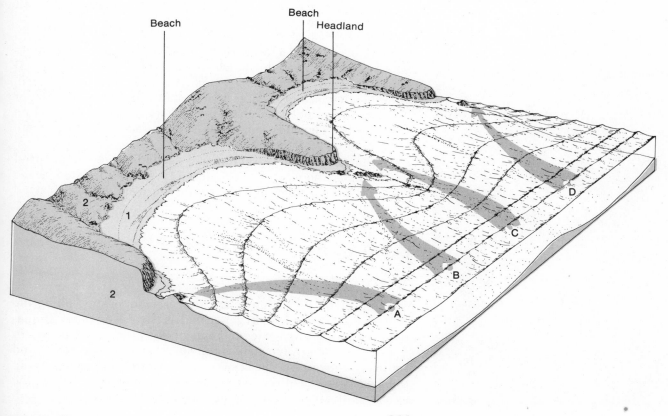

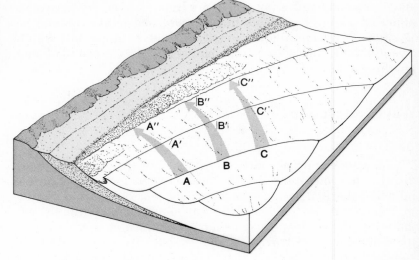

Figure 12.6 Wave refraction also occurs where a wave approaches the shore at an angle. The wave at point A encounters the sea floor before B or C and is slowed down by frictional drag, forcing the wave to bend.

Figure 12.7 Aerial photograph showing wave refraction along a coast line.

amount of energy. As the waves move toward the shore and are refracted, the energy in segment B-C is concentrated along the headlands throughout a relatively short distance. However, an equal amount of energy in segments A-B and C-D is distributed throughout a much greater length of shore and the wave height is reduced. Because of the concentration and dispersal of energy, breaking waves are powerful erosional agents on the headlands but are weak in bays, where they generally deposit sediment to form beaches.

Waves advancing obliquely toward the shore are also refracted. In *figure 12.6* the crest A advances to A' as it encounters bottom and is slowed down. During the same interval of time, B advances to B' because its velocity is only slightly reduced. Point C advances to C', a greater distance because this part of the wave does not encounter the bottom. Thus, the crest line of the wave becomes curved as shown and tends to become more parallel with the shore.

Patterns of wave refraction can easily be seen on an aerial photograph. The example in *figure 12.7* shows the major wave front advancing obliquely toward the shore and being refracted around the headlands and points.

Longshore Drift

Statement

One of the most important processes operating along the coast is the longshore drift generated as waves advance obliquely to the shore. The basic elements of this process are shown in *figure 12.8*. As a wave strikes the shore at an angle, water and sediment moved by the breaker are transported obliquely up the beach in the direction the wave is advancing. When the energy of the wave is spent, the water and sediment return with

The Earth's Dynamic Systems

the backwash directly down the beach perpendicular to the shore. The next wave moves the material obliquely up the shore again, and the backwash returns it directly down the beach slope. If we follow the path of a single grain of sand, we see that it is moved in a zigzag pattern with each wave, with a resulting net transport parallel to the shore.

There are two zones in which longshore movement occurs: One is along the upper limits of wave action and is related to the surge and backwash of the waves described above. The other is in the surf and breaker zone in which material is transported in suspension.

Discussion

With longshore drift, the beach may be thought of as a "river of sand." With time, this process results in the movement of an enormous volume of sediment along the coast. Longshore currents move parallel to the coast in the direction the waves strike the shore. If wave direction is constant, longshore drift will be in one direction. If the waves approach the shore at different angles during different seasons, longshore currents may be reversed.

One of the best ways to appreciate the process of longshore drift is to consider examples in which the process has influenced the affairs of man. A good example is in the area of Santa Barbara along the southern coast of California, where data have been collected over a considerable period of time.

Santa Barbara is a picturesque coastal town at the base of the Santa Ynez Mountains. It is an important educational, agricultural, and recreational area, and the local people desired a harbor capable of accommodating deep-water vessels (figure 12.9a). Studies by the U.S. Army Corps of Engineers indicated that the project was unfavorable because of the strong longshore currents which transported large volumes of sand to the south. But in 1925 $750,000 was raised locally to construct a

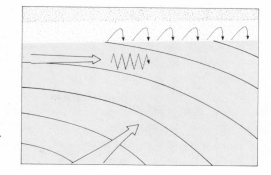

Figure 12.8 *The origin of longshore currents. When a wave moves toward the beach at an oblique angle, sediment is lifted by the surf and moved diagonally up the beach slope. The backwash then carries the particles back down the beach at right angles to the shoreline. This action transports the sediment along the coast in a zigzag pattern. Particles are also moved under water in the breaking zone by this action.*

Figure 12.9 *The effect of a breakwater on longshore drift at Santa Barbara, California. (a) The Santa Barbara coast before the breakwater was built. (b) After the initial breakwater was constructed the harbor became filled with sand because longshore currents could not be generated in the protected area behind the breakwater. (c) After the breakwater was connected to the shore, longshore currents moved sand around the breakwater and filled the mouth of the harbor. Sand is now dredged from the harbor and pumped down the coast.*

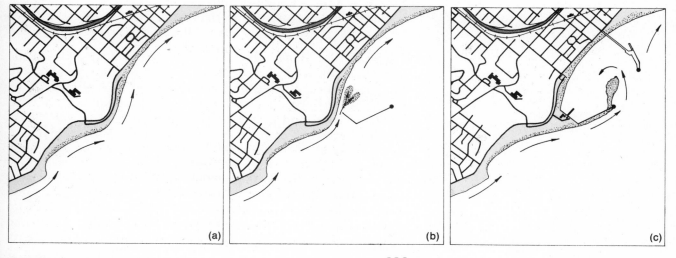

(a) (b) (c)

breakwater 460 m (1500 ft) long. In this area longshore drift transports sand supplied by rivers draining the mountains and moves its southward to Monterey. Sediment on the beach may appear to be stable but, in the continual motion of longshore currents, sand is moved southward from beach to beach. Near Monterey, the head of a deep submarine canyon is close to shore, and sand transported by the longshore currents is delivered to the head of the canyon. It then moves down the canyon to the deep ocean floor where it ultimately comes to rest and builds a deep-sea fan. It has been estimated that new sand reaches this area at the rate of 592 m³ (775 yd³) per day or about 214,000 m³ (280,000 yd³) per year. The currents are so strong that boulders .6 m (2 ft) in diameter may be transported.

In spite of reports advising against the project, an initial breakwater was built and a deep-water harbor constructed (figure 12.9b). The initial breakwater was not tied to the shore, but sand moved by longshore currents began to pour through the gap and fill the harbor which was protected from wave refraction and longshore drift by the breakwater (diagram b). This necessitated connecting the breakwater to the shore. Sand then accumulated behind the breakwater and built a smooth curving beach (diagram c). Longshore currents then moved to the sand around the end of the breakwater and deposited it inside the harbor. This produced two disastrous effects: (1) The harbor became so choked with sand that it was not usable except for very shallow draft vessels. (2) The beaches downcoast were deprived of their source of sand and began to erode. Within twelve years, the property damage exceeded $2 million as the beach in some areas was cut back 75 m (245 ft). To solve the problem, a dredge was installed in the Santa Barbara harbor to pump out the sand and return it into the system of longshore currents. Most of the beaches have been partly replenished, but the cost of dredging exceeds $30,000 per year.

Erosion Along Coasts

Statement

Solid rock along coasts is eroded by chemical action of the sea water, direct hydraulic action of waves and currents, and abrasive action of sand and gravel moved by the waves. The most effective process in wave erosion is probably the abrasive action of sand and gravel moved about by the waves. These "tools of erosion" operate in a narrow zone between high and low tide as they are transported by wave action and by longshore currents. The abrasive action of sand and gravel along the shore is similar to that of the traction load in rivers; however, instead of cutting a vertical channel, the sand and gravel along the shore cut horizontally and form a wave-cut cliff and a wave-cut platform (figure 12.10). It should be emphasized that, in the erosion of a shoreline, marine and subaerial (on the ex-

Figure 12.10 *Wave erosion along a coast undercuts the base of a sea cliff by abrasion. This causes the cliff to retreat inland.*

posed land surface) agents operate together to produce erosion above wave level. Seepage of ground water, frost action, wind, and mass movement, all combine with the undercutting action of waves to erode back the coast.

Solution activity also takes place along the coast and is especially effective in eroding limestone. Even noncalcareous rocks may be weathered rapidly because the chemical action of sea water is greater than that of fresh water alone. It has been demonstrated that many rocks weather three to fourteen times faster in sea water than in fresh water.

The hydraulic action of waves, particularly storm waves, constitutes another effective agent of marine erosion. When a wave breaks against a cliff, the sheer impact of the water alone is capable of exerting pressures of over 254 kg/m^2 (6000 lb/ft^2). Thus, water is driven into every crack and crevice and tightly compresses the air within the fractures of the rocks. The compressed air acts as a wedge which widens the cracks and loosens the blocks. As the wave recedes, the compressed air suddenly expands, producing an effective quarrying process for removing blocks of rock. Compressed air resulting from the breaking of large waves has provided enough force to move blocks of rock weighing as much as 6.3 metric tons (7 short tons).

Figure 12.11 *Evolution of a shoreline by wave erosion. (a) Initial stage. Wave action begins to develop a notch at sea level which collapses to form a wave-cut cliff. (b) Intermediate Stage. Continued wave erosion causes the cliff to recede, developing a wave-cut platform. Sea stacks, arches, and caves result from differential erosion along zones of weakness. (c) Late Stage. The wave-cut platform grows so large that wave energy is dissipated across it. Erosion along the shore is greatly reduced. Beaches develop, and the sea cliff retreats through mass movement.*

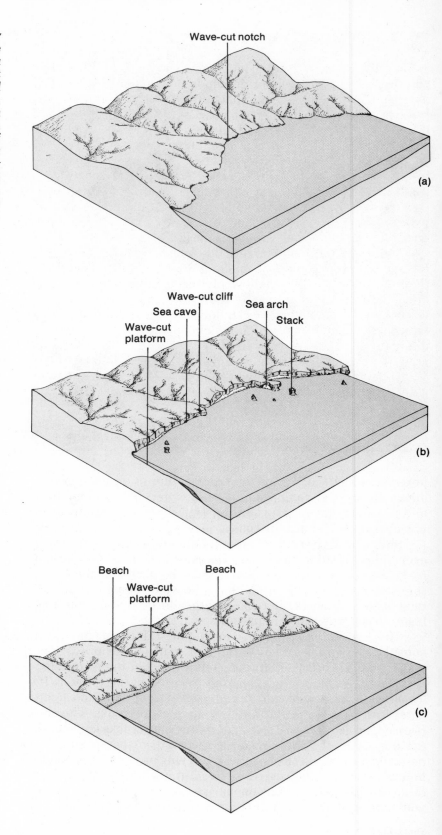

Discussion

Wave-cut sea cliffs are common coastal features, especially where steeply sloping land descends beneath the water. The development of a sea cliff is shown in *figure 12.11*. In the early stage *(figure 12.11a)*, wave action carves out a small notch at sea level. As the notch is enlarged, the overhanging cliff ultimately collapses. The fallen debris is soon broken up and removed by wave action, and the process is repeated on the fresh surface of the new face on the cliff *(figure 12.11b)*. As the sea cliff retreats, a wave-cut platform is produced at its base, the upper part of which is commonly visible near shore at low tide. Stream valleys which formerly reached the coast at sea level are shortened and left as hanging valleys as the cliffs recede.

As the platform is enlarged, the waves break progressively farther from shore and lose much of their energy by friction as they travel across the shallow platform. Consequently, wave attack upon the cliff is greatly reduced. Beaches may now develop and the cliff face is worn down by weathering and mass movement *(figure 12.11c)*. The wave-cut platforms effectively dissipate wave energy, thus limiting the size to which they may grow. Some volcanic islands, however, appear to have been completely truncated by wave action and slope retreat, leaving a flat-topped platform.

Erosion of a sea cliff proceeds at different rates depending on the durability of the rock and the degree of exposure of the coast to wave attack. Zones of weakness such as outcrops with joint systems, fault planes, or vertical beds of shale between harder sandstones, provide loci for accelerated erosion. If a joint extends across a promontory, wave action may first hollow out an alcove which later enlarges to a **sea cave**. As the headland is commonly subjected to erosion from two sides, caves excavated along a zone of weakness may join to form a **sea arch** *(figures 12.12 and 12.13)*. Eventually, the arch collapses, leaving an isolated pinnacle known as a **sea stack** in front of the cliff.

Erosion along shores tends to develop a profile of equilibrium in much the same way that a river system approaches a state of equilibrium with its volume, gradient, and sediment load. In a shoreline at equilibrium, energy available in the waves is sufficient to transport available sediment but not great enough to erode back the shore. If more sediment is available than can be transported, the shoreline will be built out into the sea. If wave action is strong enough to erode back the sea cliffs, it will do so until the wave-cut platform becomes large enough to limit further growth. Naturally, the rate at which a sea cliff retreats, leaving a wave-cut platform, depends upon the physical characteristics of the bedrock and strength of the waves. In poorly consolidated material the average rate of cliff retreat may be 1.5 to 2 m (5 to 6 ft) per year. Since Roman times, parts of the British coast have been worn back more than 5 km (3 mi), and many villages and ancient landmarks have been swept

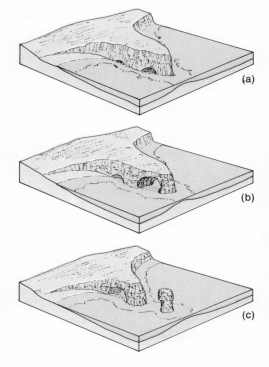

Figure 12.12 *Evolution of sea caves, arches, and stacks. (a) Wave energy is concentrated on the headlands as a result of wave refraction. Zones of weakness such as joints, faults, or nonresistant beds erode faster and develop sea caves. (b) Sea caves enlarge to form an arch. (c) Eventually the arch collapses to form a sea stack. A new arch may develop from the remaining headland.*

(a)

(b)

(c)

Figure 12.13 Sea stacks and an arch along the coast of Iceland.

away. New volcanic islands such as Surtsey (near Iceland), composed almost entirely of volcanic ash, are soon completely planed off by wave action.

Deposition Along Coasts

Statement

Sediment transported by waves and longshore currents is deposited along the coast in areas where energy is low (bays and inlets). Deposition features include beaches, spits, and barrier islands. These deposits are greatly influenced by longshore currents and tend to develop straight or gently curving coastlines.

Discussion

Beaches. A beach may be defined as a shore built of unconsolidated sediment. Sand is the most common material, but some beaches are composed of cobbles and boulders and others

Figure 12.14 Major features along a beach.

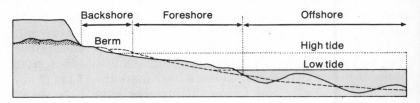

Figure 12.15 *A curved spit developed by longshore drift moving sand beyond the mainland.*

Figure 12.16 *Development of a tombolo. (a) An offshore island acts as a breakwater to incoming waves and creates a "wave shadow" along the coast behind it where longshore drift cannot occur. (b) Sediment moved by longshore drift is trapped in the "shadow" zone. (c) The deposit of sediment ultimately grows until it connects with the island. Longshore currents are then able to move sediment along the shore and around the island.*

of fine silt and mud. The type of material composing a beach, as well as the slope and other physical characteristics, is dependent largely upon wave energy but also on the particle sizes that are available. Beaches composed of fine-grained material are flatter than those composed of gravel—a result of low wave energy. Rivers draining the continents are the prime source of material for beaches although sediment from the adjacent sea floor and erosion of headlands is also important. In the tropical areas, the greatest source of sand is often shell debris derived from erosion of nearshore coral reefs.

The beach is normally subdivided into three major zones, each of which closely parallels the coast (figure 12.14).

1. *Offshore* (from low tide to the breaker zone). In many coasts this area is marked by features such as barrier **bars** and adjacent parallel troughs.
2. *Foreshore* (from low tide to the highest level reached by high-tide storms).
3. *Backshore* (from the point marking the highest area reached by storm waves, known as the **berm** crest, to the base of the sea cliff or the area of first vegetation). Movement of sediment on the beach is largely by longshore drift, but where sand is abundant and longshore drift is weak, the beach may build outward into the sea by accumulation of layer upon layer of new sediment on the beach face.

Spits. In areas where a straight shoreline is indented by bays or estuaries, longshore drift may extend the beach from the mainland to form a spit. A spit may continue to grow far out into the bay as material is deposited at its end and it migrates; it tends to become curved as a result of wave refraction (figure 12.15). With continued growth, the spit may extend completely across the front of the bay to form a baymouth bar.

Spits may also grow outward and connect with an offshore

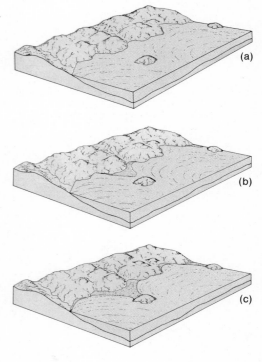

Figure 12.17 *A barrier island off the Atlantic Coast.*

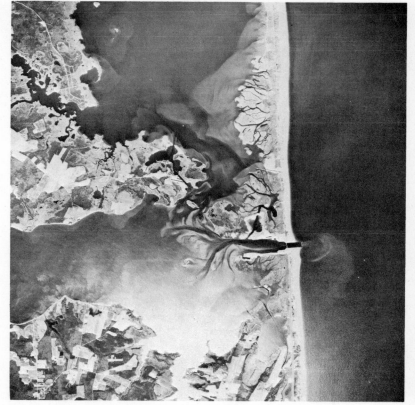

Figure 12.18 *The origin of a barrier island from longshore migration of a spit. (a) Sediment moves along the shore and is deposited as a spit in the deeper water near a bay. (b) The spit grows parallel to the shore by longshore drift. (c) Tidal inlets cut the spit, which is now long enough to be considered a barrier island.*

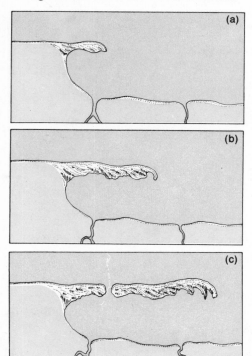

island to form a **tombolo**. This feature is commonly produced by the effect the island has on wave refraction and longshore drift (*figure 12.16*). An island near shore may cause wave refraction around it so that little or no wave energy strikes the shore behind it. Longshore currents are not generated in this "shadow zone," so deposition of sediment carried along the coast can occur behind the island. Ultimately, the sediment carried by long-shore currents can build up a deposit which becomes connected to the island to form the tombolo. Longshore currents then move uninterrupted along the shore and around the island.

Barrier Islands. Barrier islands are long, low islands of sediment trending parallel to the shore (*figure 12.17*). They are almost invariably present along shorelines adjacent to gently sloping coastal plains and are typically separated from the mainland by a **lagoon.** Most barrier islands are cut by one or more tidal inlets.

The origin of barrier islands has been a controversial subject for many years. One theory contends that they result from the growth of spits across irregularities in the shoreline such as that illustrated in *figure 12.18*. Certainly some barrier islands develop in this manner. Others may form by the shoreward migration of offshore bars. Another hypothesis contends that barrier islands are the result of the worldwide rise of sea level which drowned early-formed beaches and isolated them off

The Earth's Dynamic Systems

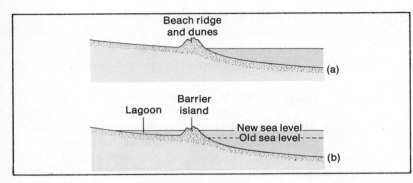

Figure 12.19 *Development of a barrier island by sea level rising over former beach ridges. (a) On a gently inclined coast, wave action builds a beach ridge at sea level. The ridge may grow over 32 m high as sand is piled up by wind action. (b) A rise in sea level during melting of glaciers drowns the former beach ridge and produces a barrier island which is then modified by wave action.*

shore. The process is illustrated in *figure 12.19*. During a lower stand of sea level, such as was produced during the Ice Age, a beach would form along a low coastal plain. Sand from the beach would be piled up into a dune ridge parallel to the shore. As sea level rose (with the melting ice) the dune ridge would be drowned and become a barrier island modified by wave action and longshore currents.

Evolution of Shorelines

Statement

Processes of erosion and deposition along the shore tend to develop a straight, or gently curving coast in which there is a minimum of energy expended. Headlands are eroded back, bays are filled in with sediment and, ultimately, if there is no change in sea level, a shoreline will reach a state of equilibrium with energy expended upon it so that neither erosion nor deposition dominates.

Discussion

The series of stages in the evolution of a shoreline is shown in *figure 12.20*. The area was originally shaped by stream erosion and subsequently partly flooded by rising sea level. In the initial stage (*diagram a*) bays are formed by drowned river valleys, and some hilltops form peninsulas and islands. In *diagram b* marine erosion begins to attack the shore, and the headlands of the islands and peninsulas are eroded to produce sea cliffs. As erosion proceeds (*diagram c*), the islands and headlands are worn back, and the cliffs increase in height. A wave-cut platform develops, and sediment begins to accumulate as spits and beaches. The wave-cut platform enlarges, reducing wave energy so that a beach forms at the base of the cliff. In a more advanced stage of development (*diagram d*), the islands are completely eroded away and bays become partly sealed off by growth of spits, forming a lagoon. The shoreline then becomes straight and simple.

In the final stages of marine development (*diagram e*), the shoreline is cut back beyond the limits of the bay. Sediment moves along the coast by longshore currents, but the wave-cut

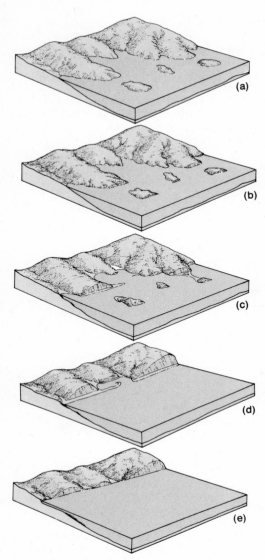

platform is now so wide that it effectively eliminates further erosion of the cliff by wave action. The shoreline is straight and approaches a state of equilibrium with the energy acting upon it. Further modification of the cliffs results from subaerial weathering and erosion.

Naturally, the development of the idealized shoreline outlined above would be modified by special conditions of structure and topography and would be adjusted with fluctuations of sea level, but the general process of erosion of the headlands by wave action and the straightening of the shoreline by both erosion and deposition would basically follow the generalized sequence. An area rarely proceeds through all of these stages, however, because sea level fluctuations upset the previously established balance. Many areas along the coast of the United States have reached a stage equivalent to b, c, or d.

Reefs

Statement

Coral reefs are a unique type of coastal phenomenon because they are of organic origin. The framework of the reef is a ridged structure built by a complex community of marine organisms such as corals, algae, sponges, and other invertebrates. Reefs grow only in certain parts of the world because they cannot tolerate temperatures below 15.5° C and require shallow, clear marine water. Shallow continental shelves and the shores of volcanic islands in tropical areas are ideal reef environments.

The growth of reefs along the shores of volcanic islands begins with fringing reefs attached to the land. If the island subsides or sea level rises, the reef grows upward to form a barrier reef separated from the mainland by a lagoon. As the landmass erodes away or if subsidence continues, the barrier reef will continue to grow upward and develop into an atoll, which encloses a lagoon in which there is no central landmass.

Discussion

Reef Ecology. The marine life which forms a reef can flourish only under strict conditions of temperature, salinity, and water depth. Modern coral reefs occur in warm tropical waters between the limits of 25° S lat. and 30° N lat. The optimum water temperature is between 25° C and 30° C. Colonial corals need sunlight and cannot live in water deeper than about 76 m (250 ft). They grow most luxuriantly just a few meters below sea level. Dirty water inhibits rapid, healthy growth of corals because it cuts off sunlight. Therefore, corals are absent or are stunted near the mouths of large rivers which bring in muddy water. Salinity conditions must range from 27 to 40 parts per thousand. As a result, a reef may be killed by a flood of fresh water from the land which appreciably alters the degree of salinity. Coral reefs are remarkably flat on top, the upper

Figure 12.20 *Evolution of an embayed coastline. (a) A rise in sea level floods a topography eroded by a river system and forms bays, headlands, and islands. (b) Wave erosion cuts cliffs on the islands and peninsulas. (c) The wave-cut cliffs recede and grow higher, and headlands are eroded back to a sea cliff. Sediment begins to accumulate as beaches and spits. (d) The islands are eventually completely eroded away, beaches and spits enlarge, and lagoons form in the bays. (e) A straight shoreline is produced by additional retreat of the cliffs and by sedimentation in bays and lagoons. The large wave-cut platform now limits further erosion by wave action.*

The Earth's Dynamic Systems

surface being positioned at the level of the upper third of the tidal range. They are usually exposed at low tide, but must be covered at high tide. In summary, corals thrive only in clear, warm, shallow seas where wave action brings sufficient oxygen and food.

Reef Types. The most common types of reefs in the present oceans are fringing reefs, barrier reefs, and atolls (*figure 12.21*).

1. **Fringing reefs,** generally ranging from 0.5 to 1 km (.3 to .6 mi) wide, are attached to such landmasses as the shores of volcanic islands. The corals grow seaward toward their food supply and are usually absent near the deltas and the mouths of streams where the waters are muddy. Heavy sedimentation and high runoff make some tropical coasts of continents less attractive for fringing reefs than the small oceanic islands.

2. **Barrier reefs** are separated from the mainland by a lagoon which may be more than 20 km (12 mi) wide. As seen from the air, the barrier reefs of the South Pacific islands are marked by a zone of white breakers. At intervals, narrow gaps occur through which excess shore and tidal water may exit.

   The finest example of this type of reef is the Great Barrier Reef of Australia which stretches 800 km (500 mi) along the northern shore of Australia at a distance of 30 to 160 km (20 to 100 mi) off the Queensland coast.

3. **Atolls** are more or less circular reefs which rise from deep water and enclose a shallow lagoon in which there is no exposed central landmass. The outer margin of an atoll is naturally the site of most vigorous coral growth and commonly forms an overhanging rim from which pieces of coral rock break off and accumulate as talus on the slopes below. A cross-sectional view of a typical atoll shows that the lagoon floor is shallow and is composed of calcareous sand and silt and rubble derived from erosion of the outer side. (See front of *diagram c, figure 12.21.*)

Atolls are by far the most common type of coral reef. Over 330 are known, of which all but ten lie within the Indo-Pacific tropical area. Drilling indicates that coral reef material in some atolls extends down as much as 1400 m (4610 ft) where it rests on a basalt platform carved on an ancient volcanic island. Reefs this thick presumably represent accumulations over a period of 40 to 50 million years. Upward growth of the reef to form such an atoll requires a gradual and continuous relative subsidence.

Origin of Atolls. In 1842, Charles Darwin first proposed a theory to explain the origin of atolls. As indicated in *figure 12.21* the theory is based on continued relative subsidence of a volcanic island. The theory suggests that coral reefs are originally established as fringing reefs along the shores of new volcanic islands. As the island gradually subsides, the coral reef grows

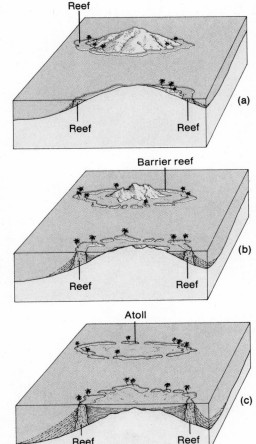

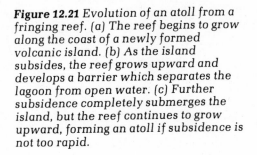

Figure 12.21 *Evolution of an atoll from a fringing reef. (a) The reef begins to grow along the coast of a newly formed volcanic island. (b) As the island subsides, the reef grows upward and develops a barrier which separates the lagoon from open water. (c) Further subsidence completely submerges the island, but the reef continues to grow upward, forming an atoll if subsidence is not too rapid.*

upward along the outer margins; the rate of upward growth essentially keeps pace with subsidence. With continued subsidence the area of the island becomes smaller, and the reef becomes a barrier reef. Ultimately, the island is completely submerged and the upward growth of the reef forms an atoll. Erosional debris from the reef fills the enclosed area of the atoll to form a shallow lagoon. The depth of the reef material in atolls tends to confirm Darwin's theory.

Platform Reefs. Reefs also grow in isolated oval patches in warm, shallow water on the continental shelf and were apparently more abundant during various past geologic periods. Modern patch reefs appear to be somewhat randomly distributed, although some appear to be oriented in belts—a feature suggesting control from submarine topographic highs, such as drowned shorelines.

Types of Coasts

Statement

In previous sections of this chapter, we have seen that erosion and deposition by waves and currents operate to varying degrees along all coasts; therefore, it might seem logical to classify a coast according to the stage to which it has evolved. However, nearly all coastlines have been profoundly affected by the worldwide rise of sea level associated with the melting of glaciers. As a result, many coasts have been modified only slightly by marine processes and are essentially the same as they were when the sea first came to rest against them. These coasts are called **primary coasts**. Coastlines shaped by marine erosion or deposition or the growth of organisms are called **secondary**. Many subtypes may be recognized; some of the most important are described below.

Discussion

Primary Coasts. The configuration of primary coasts is a result of the rise in sea level which floods a landscape previously shaped by such subaerial agents as streams, glaciers, vulcanism, or earth movements. These agents produce a highly irregular coastline characterized by bays, estuaries, fiords, headlands, **peninsulas,** and **offshore** islands. The landforms may be either erosional or depositional, but very little modification has resulted from marine processes. Some of the more common types are illustrated in *figure 12.22.*

1. Stream Erosion. When an area is eroded by running water and is flooded subsequently by the rise of sea level, the landscape becomes partly drowned. Stream valleys become bays or estuaries, and former hills become islands. The bays following the tributary valley system inland result in a coastline having a dendritic pattern. Chesapeake Bay is a well-known example.

Figure 12.22 *Primary Shorelines—Shape of coast produced from non-marine processes. (a) Stream erosion. (b) Stream deposition. (c) Glacial erosion (valley).*

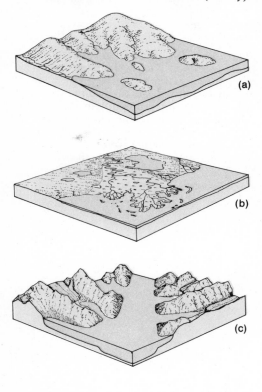

The Earth's Dynamic Systems

2. Stream Deposition—Deltaic Coasts. At the mouths of major rivers, fluvial deposition results in the buildup of deltas out into the sea and dominates the configuration of the coast. Deltas may assume a variety of shapes and are locally modified by marine erosion and deposition.

3. Glacial Erosion. Drowned glacial valleys are usually called fiords and form some of the most rugged and scenic shorelines in the world. Fiords are characterized by long trough-like bays which cut into mountainous coasts and may extend inland as much as a hundred kilometers. In polar areas, glaciers still remain at the heads of many fiords. Walls of fiords are steep and straight. Hanging valleys with spectacular waterfalls are common. Usually the water is deep, sometimes in excess of several hundred meters.

4. Glacial Deposition. Glacial deposition dominates coasts in the northern latitudes where continental glaciation once extended out beyond the present shoreline onto the continental shelf. The ice sheet left drumlins or moraines to be drowned by subsequent rise in sea level. One example, Long Island, is partly submerged moraine, and in Boston Harbor partly submerged drumlins form elliptical islands.

Secondary Shorelines. Secondary coasts are shaped by marine erosion and deposition and are characterized by wave-cut cliffs, beaches, barrier bars, and spits, or they are produced by biologic agents such as marsh grass, mangroves, or coral reefs. Marine erosion and deposition tend to smooth out and straighten the shoreline and establish a balance between the energy of the waves and the configuration of the shore. The most common types are illustrated in *figure 12.23.*

1. Wave Erosion Coasts. Wave erosion begins to modify primary coasts as soon as the landscape produced by other agents is submerged. The energy of waves is concentrated on the headlands and develops a wave-cut platform slightly below water level. Ultimately, a straight cliff is formed with hanging stream valleys and a large wave-cut platform. The cliffs of Dover, England, are an excellent example.

2. Marine Deposition Coasts. In areas where abundant sediment is supplied by streams or marine currents, various marine deposits determine the characteristics of the coast. Barrier islands and beaches are the dominant features. The shoreline is modified as waves break across the barriers and drive them inland. The barrier also increases in both length and width as sand is added. The lagoons behind the barrier receive sediment and fresh water from the streams and thus are often capable of supporting dense marsh vegetation. Gradually, the lagoon fills up through deposition of stream sediment or from the invasion of sand from the barrier bar through tidal deltas. The barrier coasts of the south Atlantic and Gulf states are excellent examples.

3. Coasts Built by Organisms. Coral reefs develop another type of coast which is very prominent in the islands of the south-

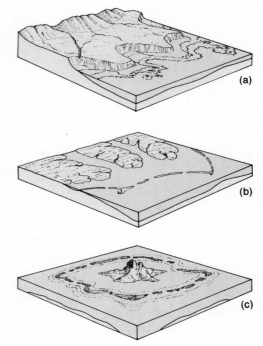

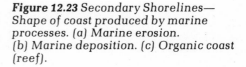

Figure 12.23 *Secondary Shorelines— Shape of coast produced by marine processes. (a) Marine erosion. (b) Marine deposition. (c) Organic coast (reef).*

western Pacific. The reefs are built up to the surface by corals and algae and may ultimately evolve into an atoll. Another type of organic coast prevalent in the tropics results from the growth of mangrove trees out into the water, particularly in shallow bays. Mangroves are plants whose roots must extend into water of high salinity. The root systems form an interlocking mesh, an impressive barrier which acts as a breakwater. Sediment is trapped in the space between the roots, and the mangrove barrier grows seaward.

Summary

The coasts are continually being modified by erosional and depositional processes operating on the shore. Wind-generated waves provide most of the energy for shoreline processes, but tides and tsunamis may be locally important. Where waves strike the shore at an angle, longshore currents are generated and move both water and sediment parallel to the shore. The sediment may accumulate as beaches, bars, or spits.

Erosion along the shore tends to develop a sea cliff by the undercutting action of waves. As the cliff recedes, a wave-cut platform develops just below sea level.

The worldwide rise in sea level associated with melting of the glaciers has drowned many coasts. As a result, coasts are classified on the basis of the process which has been most significant in developing their configuration.

Coastal processes, both erosional and depositional, tend to develop a straight, smooth shoreline in which there is a balance or equilibrium between the shape of the shoreline and the amount of energy spent upon it.

We have emphasized that wind is the major force in developing waves which in turn modify the coast. Wind is also an effective agent in transporting and depositing loose sand and dust on the continents and is responsible for the major landforms in the great desert regions of the world. In the next chapter we will consider the geologic work of wind which is part of the moving surface fluids (gas and liquid) on our planet.

Additional Readings

Bascom, W. 1959. "Ocean Waves." Sci. Amer. 201(2):74-84 (Offprint No. 828).

Bascom, W. 1964. Waves and Beaches. New York: Anchor Books (Doubleday and Company, Inc.)

Oceanography. 1971. A Scientific American Book. San Francisco: W. H. Freeman and Company.

Weyl, P. K. 1970. Oceanography: An Introduction to the Marine Environment. New York: John Wiley and Sons.

13 Wind

Where loose sediment is abundant and climatic conditions are favorable, wind will transport sand and dust, producing depositional features which locally dominate the landscape. Foremost among these are the great dune fields in the large deserts, sand dunes along many of the world's coasts, and blankets of windblown dust called loess which covers millions of square kilometers in parts of the middle latitude continents.

In spite of this, wind is probably the least effective agent of erosion, although many curious erosional landforms are mistakenly attributed to it. Even in the desert most erosional landforms are the product of running water, and the greatest effects produced by wind are expressed in shifting sand dunes. The most significant process involving the wind is the transportation and deposition of loose fine-grained sediment eroded by other geologic agents.

In this chapter we are concerned primarily with how wind transports and deposits sediments to form distinctive landforms.

Major Concepts

1. Wind is not an effective agent of erosion but is capable of transporting loose, unconsolidated fragments of sand and dust.
2. Sand is transported by saltation and surface creep. Dust is transported in suspension and may remain high in the atmosphere for long periods of time.
3. Sand dunes migrate as sand grains are blown up and over the windward side of the dune and accumulate on the lee slope. The internal structure of a dune consists of stratification inclined in a downwind direction.
4. Various types of dunes form in response to wind velocity, sand supply, constancy of wind direction, and the characteristics of the surface over which the sand moves.
5. Windblown dust (loess) forms blanket deposits which may mask the former landscape. It originates from deserts or from fine rock debris deposited by glaciers.

Wind as a Geologic Agent

Statement

Wind is an effective geologic agent locally because it is capable of lifting and transporting loose sand and dust, but its ability to erode solid rock is very limited. The main action of wind as a geologic agent is in transportation and deposition in arid regions.

Discussion

Formerly, it was thought that the wind, like running water and glaciers, had great erosional power—power to abrade and wear down the earth's surface—but it has become increasingly apparent that few major erosional topographical features are formed by wind erosion. Even in the desert, where water is not an obvious geologic agent, wind action is not the major erosional process; its major function is to transport loose sediment. Most erosional landforms in deserts were produced by running water in times of wetter climate and are, in a sense, "fossil" landscapes formed under processes no longer active. Sand may be transported by the wind with considerable velocity and, locally, wind abrasion may aid in eroding and shaping some details of landforms. But wind erosion is hardly capable of producing erosional features of great areal extent. Even minor alcoves and niches, or wind caves, as well as certain topographic features called "pedestal rocks" (often thought to be caused by wind erosion), are produced by differential weathering, not wind abrasion.

Though relatively insignificant as an erosional agent, the wind is effective in transporting loose, unconsolidated sand, silt, and dust and is responsible for the formation of great "seas of sand" in the Sahara, Arabian, and other deserts, as well as the blankets of windblown dust covering millions of square kilometers in China, the central United States, and parts of Europe.

Prevailing wind patterns are the result of the earth's rotation and the resulting **Coriolis effect** (deflection due to rotation), plus the effect of the configuration of continents and oceans, the location of mountain ranges, and variations in solar radiation with latitude (figure 13.1). The great deserts of the world, such as the Sahara and others in northern Africa and the deserts of Asia, are located largely in belts 15° to 35° north and south of the equator. Air rises at the equator and descends about 30° to 35° north and south of the equator. As it descends, it is heated by compression. The air is further heated as it moves along the ground toward the equator. The increased temperature of the air increases its ability to hold water. As a result, evaporation of the surface moisture occurs rather than precipitation. When the air reaches the equator it again rises, cools, and thus releases this moisture as rain.

Other deserts such as the Kalahari in South Africa and the

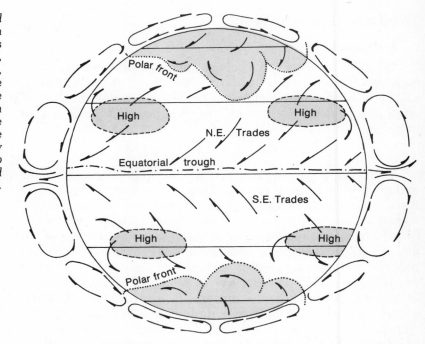

Figure 13.1 *Atmospheric circulation and prevailing wind patterns. Heated air from the equatorial regions rises and moves toward the poles, where it is cooled, compressed, and forced to descend, forming the subtropical high pressure belts. The air then moves toward the equator as trade winds. In the northern hemisphere this air is deflected by the earth's rotation to flow southwest. In the southern hemisphere it is deflected to flow northwest. The cold polar air tends to wedge itself toward the lower latitude and form the polar fronts.*

Figure 13.2 *Deflation basins in the Great Plains are produced where solution activity has dissolved the cement that binds the sand grains together in the horizontal rocks. The loose sand is removed by the wind to form a basin. Water trapped in the basin further dissolves cement, and the basin is enlarged.*

Basin and Range in the western United States lie in the interior of continents or where high mountain ranges intercept the moisture-laden air, forcing it to rise and the moisture to condense and fall as rain before reaching the desert region.

Wind action is most significant in these desert areas but is not confined to them. Most coasts are modified by winds which pick up loose sand on the beach and transport it inland.

It is interesting to note the significance of wind action on the planet Mars, as was so dramatically demonstrated by the orbiting satellite studies of 1971 and 1972, for without water and vegetation, wind action can be the dominant process on the surface of Mars. In former geologic periods of the earth when climatic conditions were much different than now, the work of the wind may have been more significant in various regions of our planet.

Wind Erosion

Statement

Wind erosion is manifested in two ways: (1) by **deflation**, the lifting and removing of loose sand and dust particles from the earth's surface, and (2) by abrasion, the sandblast action of windblown sand.

Discussion

Deflation. Deflation is responsible for many depressions called **blowouts** in areas where weak unconsolidated sediment is exposed at the surface. In the Great Plains area of the United

States, tens of thousands of small deflation basins dot the landscape (figure 13.2). They may be shallow depressions several meters in diameter or large ones more than 50 m (165 ft) deep and several kilometers across. The basins develop where calcium carbonate cement in the bedrock is dissolved by ground water, leaving loose material to be picked up and transported by the wind.

Wind is capable of moving only sand- and dust-size particles so that deflation leaves concentrations of coarser material called **lag deposits**, or **desert pavements** (figure 13.3). Locally, these are very striking desert features and stand out in contrast to dune fields or playa lake deposits.

Deflation occurs only where unconsolidated material is exposed at the surface; it does not occur where there are thick covers of vegetation or layers of gravel. Therefore, the process is limited to areas such as deserts, beaches, and barren fields.

Abrasion. The effects of wind abrasion can be seen on the surfaces of bedrock in most desert regions. The process, which is the same as that used in the artificial sandblasting of building stone, produces a delicate etching of details of the rock's internal structure as resistant laminae, hard pebbles, resistant minerals, or fossils are left protruding from the softer surrounding rock. Some pebbles called **ventifacts** are shaped by the wind (figure 13.4). Since the movement of sand by the wind is concentrated within 2 m (6.5 ft) of the ground, sandblasting is most effective near the surface.

Figure 13.4 Ventifacts are pebbles shaped and polished by wind action.

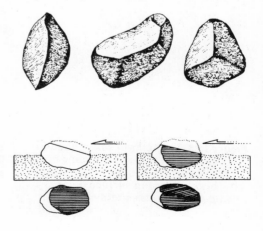

Transportation of Sediment by Wind

Statement

Field observations and wind-tunnel experiments indicate that windblown sand moves by skipping or bounding into the air (a process called saltation) and by rolling or sliding along the surface (called surface creep). Fine silt and dust are carried in suspension over great distances and settle back to the ground only after turbulent wind ceases.

Discussion

Movement of Sand. Although both wind and water transport sand by saltation, the mechanics of motion involved are quite different because the viscosity of water is so much greater than that of air. In water, saltation involves a hydraulic lift; in air, saltation results in a series of elastic bounces (*figure 13.5*). The initial energy which lifts the grains into the air comes from impact with other grains. When the wind reaches a critical velocity, the sand grains begin to move by rolling or sliding. As a grain moves along the surface, it strikes other grains. The impact may cause one or both of the grains to bounce into the air, where they are forced forward by the stronger wind above the ground surface. The force of gravity soon pulls the grain back to earth at an angle of impact generally ranging from 10° to 16°. If the surface over which the sand is moving is solid rock, the grain will bounce back into the air. If the surface consists of loose sand, the impact of the falling grain may eject other grains into the air; these are then moved forward by the wind in a parabolic

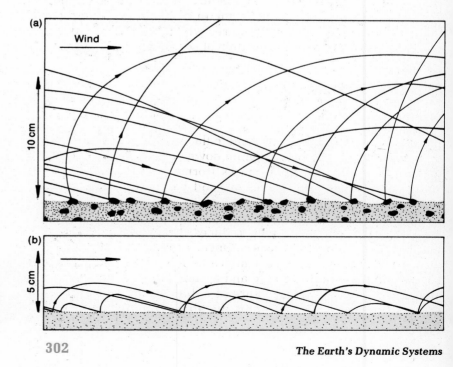

Figure 13.5 *The paths of saltating sand grains. (a) Grains impacting on a pebble surface bounce high into the air. (b) A loose sand surface dampens impact so grains saltate close to the ground.*

The Earth's Dynamic Systems

path. In this way, a sandy surface is set in motion by individual saltating grains.

Some sand grains, too large to be ejected into the air by impact from other grains, move by creep (rolling and sliding). Their momentum comes by impact from saltating grains, not directly from the wind. Approximately one-fifth to one-fourth of the sand moved by a sand storm travels by rolling and sliding. Particles over 1 cm in diameter are rarely moved by the wind.

Movement of Dust. Dust-sized particles are light enough to be lifted by an eddy of turbulent air, but it is difficult for turbulent air alone to initially lift dust particles from the surface. This is because dust lies within a thin zero velocity layer close to the ground. Laboratory experiments show that particles less than .03 mm in diameter cannot be lifted from the surface by the wind once they have settled to the ground, regardless of wind velocity. Dust is initially ejected into the air by impact from saltating sand grains or other disturbances. Once in the air, dust may be lifted high into the atmosphere and transported many kilometers.

Estimates on the quantity of material transported by the wind are based on numerous samples throughout many parts of the world. For example, some dust-falls in the eastern United States originated 3200 km (2000 mi) to the west, and dust from the Sahara is known to have eventually fallen on England. In a single dust storm originating in the Sahara as much as 1.8 metric tons (2 million short tons) of dust fell on Europe and much more on northern Africa. During the "dust bowl" years of the 1930s, a dust cloud extending 3650 m (12,000 ft) above the ground near Wichita, Kansas, was estimated to contain 50,000 metric tons (55,200 short tons) per cubic kilometer. Thus, locally the atmosphere may be the dominant factor in developing the surface features of an area, surface features which are largely the result of transportation and deposition of sediment.

Migration of Sand Dunes

Statement

Wind commonly deposits sand in the form of dunes (mounds or ridges), which generally migrate downwind. In many respects, dunes are similar to ripple marks (formed either in air or in water) and larger sand waves or sand bars common in many streams and in shallow marine water.

Many dunes originate where an obstacle creates a protected area, or wind shadow, which reduces the wind velocity to a point where deposition occurs. Once formed, the dune itself acts as a wind shadow and migrates downwind as sand is carried up the windward side and accumulates on the leeward side. Thus, the internal structure of a dune is characterized by cross-stratification inclined in the direction the wind was blowing.

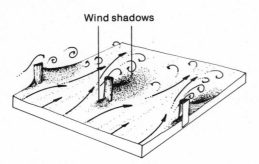

Wind shadows

Figure 13.6 *Wind shadows form behind an obstacle which diverts the wind. Sand accumulates in the protected areas, and the accumulation may grow large enough to form a dune.*

Sand dunes are probably the most familiar type of wind deposit, but the factors which control their development are far from simple. Commonly, a sand dune originates when an obstacle such as a bush or large rock deflects the wind and causes a pocket of quieter air to form behind it (*figure 13.6*). As sand is blown over or around the obstruction into the wind shadow, a mound of sand accumulates. Once a small dune is formed, it acts as a barrier itself and disrupts the flow of air, causing continued deposition downwind (*figure 13.7*). Dunes range in size from a few meters high to huge deposits as much as 200 m (660 ft) high and 1 km wide.

A profile showing movement of sand in a typical dune is diagrammed in *figure 13.8*. Dunes are asymmetrical, with a gently inclined windward slope and a steeper downwind slope called the **lee slope** or **slip face.** The steep slip face of the dune indicates the direction of the prevailing wind. As the wind blows over the dune, it transports the sand by saltation and by surface creep up the windward slope to the crest of the dune. The wind continues upward past the crest, creating divergent air flow and eddies just over the lee slope. Beyond the crest of the dune the sand drops out of the wind stream and accumulates on the slip face. The maximum slope of the slip face is 34°, the angle of repose for dry, well-sorted sand. Therefore, slumping is common on this unstable slope. As more sand is transported from the windward slope and accumulates on the lee slope, the dune migrates downwind. The internal structure of a migrating dune consists of cross-stratification which forms on the lee slope

Figure 13.7 *A wind shadow from a fence post and the resulting sand accumulation on the Great Salt Lake Desert.*

The Earth's Dynamic Systems

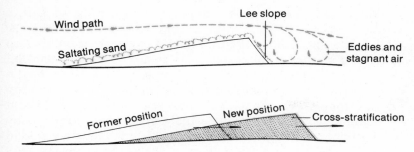

Figure 13.8 *Figure 13.8 A sand dune migrates as sand grains are moved up the slope of the dune and accumulate in a protected area on the downwind face. The dune slowly moves downwind, grain by grain. As the grains accumulate in inclined layers on the downwind slope they produce a series of layers (cross-beds) inclined in a downwind direction.*

and is, therefore, inclined in a downwind direction. This structure enables the geologist to map the direction of ancient wind systems in windblown sandstone simply by measuring the direction in which the cross-strata are inclined.

Types of Sand Dunes

Statement

Sand dunes may assume a variety of fascinating shapes and patterns depending upon a number of factors such as (1) sand supply, (2) wind velocity, (3) variability of wind direction, and (4) the characteristics of the surface over which the sand moves. The most common dune varieties are **transverse dunes**, **barchan dunes**, **longitudinal dunes**, **star dunes**, and **parabolic dunes**.

Discussion

Transverse Dunes (figure 13.9a). Transverse dunes tend to form in areas where there is a large supply of sand and a constant wind direction. These dunes completely cover large areas and tend to develop a wavelike form, with sinuous ridges and troughs perpendicular to the prevailing wind. Transverse dunes commonly form in desert regions which have exposed ancient sandstone formations that provide an ample supply of sand or along beaches where sand is transported landward by strong onshore winds. These dunes commonly cover large areas known as sand seas, so called because the wavelike form of the dune produces a surface resembling a stormy sea. Typically, the dune develops a gentle windward slope and a steep leeward slope characteristic of other dunes.

Barchan Dunes (figure 13.9b). Where the supply of sand is limited and winds of moderate velocity blow in a constant direction, crescent-shaped dunes, called barchans, tend to develop. Typically, the barchan is a small isolated dune from 1 m (3.3 ft) to 50 m (165 ft) high. The tips or horns of the barchan point downwind, and sand grains are swept around the barchan as well as up and over the crest. With a constant wind direction a beautifully symmetrical crescent may be formed, but with shifts in wind direction one horn may become much larger than the other. Although barchans are typically isolated dunes, they often are arranged in a chainlike fashion extending downwind

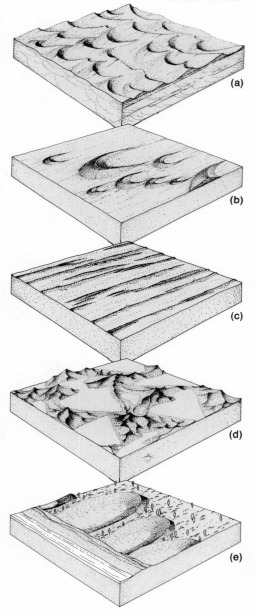

Figure 13.9 *Schematic diagrams showing some of the major dune types. (a) Transverse dunes develop when the wind direction is constant and the sand supply is large. (b) Barchan dunes develop when the wind direction is constant but the sand supply is limited. (c) Longitudinal dunes are formed by converging wind directions in an area where there is a limited sand supply. (d) Star dunes develop where the wind direction is variable. (e) Parabolic or blowout dunes are formed by strong onshore winds.*

(a)

(b)

(c)

(d)

(e)

from the source of sand. They migrate 7.5 to 15 m (25 to 50 ft) a year.

Longitudinal Dunes (figure 13.9c). Longitudinal dunes, also called **seif** dunes (Arabian word for "sword"), are long parallel ridges of sand, elongate in a direction parallel to the vector resulting from two converging wind directions. They develop where strong prevailing winds converge and blow in a constant direction over an area having a limited supply of sand. Many longitudinal dunes are less than 4 m (13 ft) high but may extend downwind for several kilometers. In larger desert areas, they may become 100 m (330 ft) high and 120 km (75 mi) in length and are usually spaced 0.5 km (0.3 mi) to 3 km (1.8 mi) apart. Longitudinal dunes occupy the vast area of central Australia, called the sand ridge desert, and are especially well developed in some desert regions of north Africa and the Arabian peninsula.

Star Dunes (figure 13.9d). A star dune is a mound of sand having a high central point from which three or four arms or ridges radiate in various directions. This dune type is typical of parts of north Africa and Saudi Arabia. The internal structure of these dunes suggests that they were formed by winds blowing in three or more directions.

Parabolic Dunes or Blowout Dunes (figure 13.9e). Parabolic dunes develop along coastlines where vegetation partly covers the sand. In spots where vegetation is absent, small deflation basins are produced by strong onshore winds. The blowout depression grows larger as more sand is exposed and removed. Usually, the sand piles up on the lee slope of the shallow deflation hollow and forms a crescent-shaped ridge. In map view, the parabolic dune is similar to a barchan, but the tips of the parabolic dune point upwind. Because the form resembles a hairpin, it is called a hairpin dune.

Loess

Statement

Loess is a deposit of windblown dust which slowly accumulates over large areas as a blanket deposit, often masking former landforms. The particles are very similar, if not identical, to the dust in the air at the present time. They consist mostly of angular fragments of quartz with lesser amounts of feldspar, micas, and calcite. Normally, loess deposits lack stratification and weather in vertical slabs, forming perpendicular cliffs.

Discussion

Desert Loess. In northern China extensive loess deposits consist primarily of disintegrated rock material brought by the prevailing westerly winds from the Gobi Desert in central Mongolia. The yellow-colored loess reaches a thickness of more than 60 m (200 ft) and blankets a large area. It is responsible for the characteristic color of the Yellow River and the Yellow Sea, as

The Earth's Dynamic Systems

the material is easily eroded and transported in suspension by running water.

Loess in the eastern Sudan of North Africa is also considered to have originated from a desert, probably the Sahara to the west. Similarly, the loess deposits of Argentina have been derived from the arid regions to the west rather than from glacial deposits to the south.

Glacial Loess. Much of the loess of North America and Europe appears to have originated from rock debris ground by glaciers and deposited as outwash. This sediment, commonly called rock flour, is readily transported by wind. In the central United States, the great bulk of loess occurs in the bluffs and uplands of the Mississippi Valley, the major drainage system through which the melt waters of the glaciers drained. Near the rivers, the loess is 30 m (100 ft) or more thick but becomes thinner away from the river channel. The most extensive accumulations are east of the flood plains, and in many areas loess lies directly upon glacial or glacial-fluvial deposits. Apparently, as the glaciers melted, the fine rock flour ground up by the glacier was deposited so rapidly as glacial outwash that a protective plant cover could not be established—a situation typical of many outwash plains. Winds then picked up the dust-sized particles and deposited them east (downwind) from the flood plain. In Europe, the loess deposits also appear to have been derived from the adjacent glaciated areas.

Summary

Wind is capable of transporting huge quantities of sand and dust which, when deposited, form distinctive landforms of sand dunes and thick layers of loess. It is an important geologic agent of sediment transport and deposition in the arid regions of the world, and it may be the most important surface process presently active on Mars.

Sand is transported by the wind by saltation and surface creep; dust is transported in suspension. Windblown sand typically accumulates in dunes which migrate downwind, as sand is transported particle by particle up the windward slope and accumulates in the relatively quiet area of the lee slope of the dune. This produces cross-bedding which is inclined downwind in the dune.

A variety of dune types result from variations in sand supply, wind direction and velocity, and the characteristics of the desert surface. The most significant include (1) transverse dunes, (2) barchan dunes, (3) longitudinal dunes, (4) star dunes, and (5) parabolic dunes.

Loess accumulates as a blanket deposit which may completely cover the preexisting surface. The dust is derived from rock flour near glacial margins or from desert regions.

In the previous chapters, our study of surface processes

associated with the hydrologic system has emphasized the development of surface features of the earth through erosion and deposition. In the following chapter, we will see how the erosional debris (sedimentary rocks) constitutes a record of erosion on our planet throughout time and how paleogeography (ancient geography) can be studied from sedimentary rocks.

Additional Reading

Bagnold, R. A. 1941. The Physics of Blown Sand and Desert Dunes. London: Methuen Publishing Company.

14 Sedimentary Rocks and Paleogeography

The geologic processes described in previous chapters generally produce only subtle changes in the landscape during a lifetime but, over a period of tens of thousands, or millions, of years, the effect of these processes on the landscape is considerable. Given enough time, the erosive power of the hydrologic system is capable of reducing entire mountain ranges to featureless lowlands over which the sea may expand and deposit new sedimentary layers. The record of erosion through time and the changing landscape of the continents is preserved in the sequence of sedimentary rocks. Each bedding plane is a remnant of a former surface of the earth, and the rock type, together with its internal structure, is a product of a previous period of erosion.

To interpret the sedimentary record correctly, one must first understand something about such modern sedimentary environments as deltas, beaches, and rivers. Sediment deposited today in these areas provides an insight into how ancient sedimentary rocks originated. From the sedimentary record the geologist can find the trends of ancient shorelines, map the position of former mountain ranges, and determine drainage patterns of ancient river systems. With this information, he is able to make paleogeographic maps of ancient landscapes and, looking back through time, see our planet as it was millions of years ago.

Many lines of evidence are important in paleogeographic studies, including the composition of the rocks, lateral variations in rock bodies, and the nature of fossils contained in the rocks. Statistical studies of cross-bedding in sandstones commonly provide the regional framework for the study of ancient landscapes. Paleogeography is a subject of more than academic interest because it constitutes a major exploration tool by which petroleum, coal, and other mineral resources in sedimentary rocks are found.

In this chapter we attempt to show how clues from modern sedimentary environments can be used to determine the nature of past events and the growth and decay of ancient landscapes. The results of such studies form the basis for interpreting a detailed history of the earth.

Major Concepts

1. Today, sediment is being deposited in a wide variety of environments. An understanding of the processes and products of these environments provides a basis for interpreting ancient sedimentary rocks.
2. Stratification, cross-bedding, and other primary sedimentary features are records of processes which change during sedimentation.
3. **Unconformities** are records of reversals in geologic processes, most of which indicate (a) a termination of sedimentation, (b) uplift and erosion, and (c) subsidence and a new period of sedimentation.
4. From details of rock type, sedimentary structures, fossils, and sequences of rocks, it is possible to make a paleogeographic synthesis and determine the conditions which prevailed during a selected interval of the geologic past.

310

Environments of Sedimentation

Statement

The term **environment of sedimentation** refers to the place where sediments are deposited and to the physical, chemical, and biological conditions that exist in that place. The idealized diagram in *figure 14.1* shows in a general way the regional setting of the major sedimentary environments. Continental environments include those areas of sedimentation which occur exclusively on the land surface, unaffected by tidal water. Most important are the fluvial environments associated with the major river systems, alluvial fans, desert dunes, lakes, and the margins of glaciers. Marine environments are exclusively beneath sea level and include the shallow seas which cover parts of the continental platform plus the floor of the deep ocean basins. Between these two are transitional or mixed environments which occur along the coast and are influenced by both marine and nonmarine processes. These include deltas, beaches, tidal flats, barrier islands, reefs, and lagoons. If you will carefully study the physiographic map, you will soon note examples of most of these environments as they exist today.

Each of these environments is characterized by certain physical, chemical, and biological conditions which leave an imprint on the type of sediment deposited. The characteristics of the rock, its texture, composition, and internal structure, together with preserved fossil remains are all records of past sedimentary environments and, when correctly interpreted, permit

Sedimentary Rocks and Paleo-geography

Figure 14.1 *Schematic diagram showing the major environments of sedimentation.*

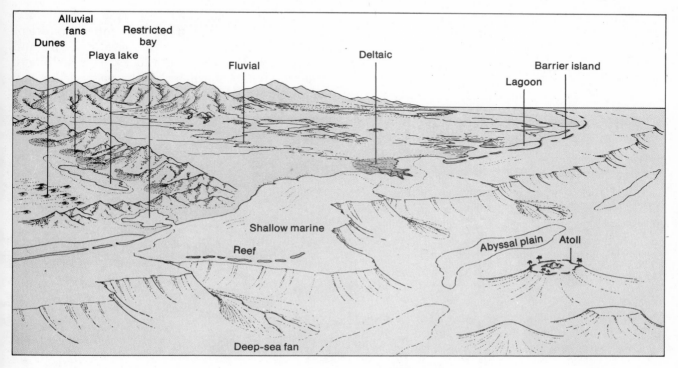

311

the geologist to reconstruct paleogeographic conditions which once existed upon the earth.

Discussion

The Fluvial Environment. The great rivers of the world act as the major channel by which erosional debris is transported from the continent to the sea. However, before reaching the sea, most rivers meander across flat alluvial plains and deposit much sediment. Within this environment sedimentation occurs in the stream channels, in abandoned meanders, and on the flood plain.

The resulting rocks consist of channels or lenses of conglomerate or sandstone enclosed in deposits of fine silt and shale (figure 14.2).

Alluvial-Fan Environment. In many arid regions of the world, thick deposits of sedimentary rock are accumulating in a series of alluvial fans at the base of mountain ranges. An excellent example of this sedimentary environment is in the Basin and Range province of the western United States (figure 14.3). The fans occur in a belt at the base of the mountains. In many areas, the deposits are hundreds of meters thick and consist of unweathered coarse sand and gravel of the rock types which form the mountain range. The size of the fragments varies greatly. The coarsest fragments, which may be blocks the size of a house, accumulate near the mountain front, and finer fragments are deposited out in the valley.

Flash floods and mudflows, which may occur only once in several decades, dominate the processes of sediment transport and deposition in an arid region. The torrent from a cloud burst sweeps away loose debris which soon develops into a mixture of water, mud, and sand, much denser than water alone and capable of moving large boulders and blocks. This sediment, soon deposited on the fan, is characterized by poor sorting and crude stratification. Playa sediments in the central part of the

Figure 14.2 *The fluvial environment. (a) Point bar sedimentation occurs on the inside of a meander bend. Erosion on the outside of the meander cuts a channel in the horizontal flood plain deposits which is subsequently filled with sand deposited on the point bar. (b) An ancient fluvial channel in the Tertiary sediments of central Utah.*

(a) (b)

basin consist of fine silt and clay, carried across the fan, and various salts precipitated from solution. They are normally well bedded in thin continuous layers of silt, mud, gypsum, and salt. These deposits interfinger with the coarse fan material toward the source area.

Eolian (Wind) Environment. Wind is a very effective sorting agent. Silt and dust are lifted high into the air and may be transported thousands of kilometers before being deposited as a blanket of loess. Sand is winnowed out and is transported by surface creep and saltation and accumulates in dunes, but gravel cannot be moved effectively by the wind.

The most significant ancient wind deposits are sandstones that accumulated in large dune fields comparable to the present Sahara, Arabian, and Australian deserts. These form widespread deposits of clean sand which preserve to an unusual degree the large-scale cross-bedding developed by migrating dunes (figure 14.4). Wind deposits are distinctly different from those sediments formed in most continental environments but may be confused with some beach deposits.

Glacial Environment. Glacial deposits consist of unsorted, chemically unweathered rock debris transported and deposited directly by the glacier. Stratification of any kind is practically nonexistent (figure 14.5). Typically, the deposits of continental glaciers are widespread sheets of unsorted debris which may rest on the polished and striated floor of the underlying rock. In many glacial deposits fine-grained particles may dominate, but larger boulders and pebbles are typically present.

Streams from the melt waters of glaciers typically rework the unsorted glacial debris and redeposit it as stratified, sorted

Figure 14.3 *The alluvial fan environment. (a) Alluvial fans are formed where streams from a mountain range enter a dry basin and deposit their load of sediment. The sediment is characteristically coarse grained, with conglomerate being the most abundant rock type. In the central part of the basin, fine silt and mud may be deposited in a playa lake adjacent to the fan. (b) Ancient alluvial fan deposits consisting of coarse gravel.*

 (a)

 (b)

Figure 14.4 *The eolian environment. (a) Sand transported by the wind accumulates in dunes. The sand is blown up and over the dune and accumulates in the wind shadow of the steep dune face. This forms large-scale cross-strata which dip in a downwind direction (see figure 13.8). (b) Ancient dune deposits consist of well-sorted, well-rounded sand deposited in large-scale cross-strata.*

Figure 14.5 *The glacial environment. (a) A glacier transports large boulders, sand, and fine silt and deposits them as the ice melts. The resulting sediment is unsorted and unstratified, with the individual particles being angular. (b) Ancient glacial deposits consisting of unsorted, unstratified, angular sediment.*

stream deposits beyond the glacier. Thus, the unsorted glacial deposits are commonly directly associated with sorted stream deposits from the melt water. The glacial environment is normally accompanied by lakes whose deposits are typically varved.

The Delta Environment. One of the most significant environments of sedimentation occurs where the major rivers of the world enter the sea and deposit most of their sediment in large marine deltas. The delta environment may be very large, covering areas of more than 36,000 km² (20,000 mi²); it is commonly very complex and involves as many as twenty distinct subenvironments such as beaches, bars, lagoons, swamps, stream channels, and lakes (figure 14.6).

Because deltas are large features and include a number of both marine and nonmarine subenvironments, a large variety of sediment types accumulate in them. Sand, silt, and mud dominate, but these may accumulate in a variety of other places:

(a)

(b)

(a) (b)

flood plains, river channels, tidal flats, beaches, barrier bars, etc. Therefore, a deltaic deposit is recognized only after considerable study of the size and shape of the various rock bodies and their relationships to each other. Within a delta both marine and nonmarine fossils may be preserved.

The Shoreline Environment. Much sediment accumulates in the zone where the land meets the sea. Within this zone, a variety of subenvironments occur as beaches, bars, spits, lagoons, and tidal flats, each with its own characteristic sediment. Where wave action is strong, mud is winnowed out, and only sand or gravel accumulate as beaches or bars. The beach sediment is constantly moved back and forth by wave action and moves parallel to the shore by longshore currents. Thus, the material is sorted and abraded by transport so that typical beach deposits consist of clean, well-sorted sand. Where wave energy is high, gravels accumulate (*figure 14.7*).

The sediment of a beach generally consists of materials which are stable under long periods of transportation. Quartz dominates in most beaches where the primary supply of sediment is from the continent. In the oceanic islands, basalt is typically the only rock type, and some beaches are composed of fragments of basalt and are therefore black. Coral growth along the margins of tropical islands supplies much material for the beach because fragments of shells are broken from the reef, reworked by wave action, and deposited as abraded grains of calcium carbonate. Most white beaches of these islands are composed of this material.

The development of offshore bars often produces quiet lagoons behind them. Here, fine sediment, rich in organic matter brought in by tides, accumulates as black mud (*figure 14.8*).

Other shoreline environments include tidal flats where fine-grained silt and mud accumulate in thin laminae. Ripple marks are prominent and are commonly preserved in the rock

Figure 14.6 The delta environment. *(a) The delta environment includes a number of both marine and nonmarine subenvironments, such as stream channels, flood plain, lakes, swamps, tidal flats, beaches, bars, and shallow marine areas. (b) Ancient deltaic deposits cover many hundreds of square kilometers. The sediment is deposited in a systematic way, with stream-channel deposits cut into the offshore mud and silt.*

Figure 14.7 *The gravel beach environment. (a) Beach gravels accumulate along shorelines where high wave energy is expended along the shore. The gravels are well sorted, well rounded, and commonly stratified in low, dipping cross-strata. (b) Ancient gravel beaches are relatively thin and widespread and commonly associated with clean, well-sorted sand deposited offshore.*

Figure 14.8 *The lagoon environment. (a) The lagoon is protected from the high energy of waves by a barrier island or reef. Sediment which is transported into a lagoon is therefore restricted to fine silt and mud. (b) Ancient lagoonal sediments are characterized by well-stratified silt and mud, usually high in organic matter.*

record (figure 14.9). Burrowing organisms such as shrimp, clams, and worms living on the tidal flat rework the sediment, partly destroying the laminae. Many organisms leave a distinctive track or burrow in the sediment, indicating the environment of deposition.

Organic Reef Environment. An organic reef is a solid structure of calcium carbonate constructed of shells and secretions of marine organisms. The framework of most typical reefs is built by corals and algae, but many other organisms thrive in the reef environment, and their shells and hard parts contribute to the reef sediment. Reef-building corals live in colonies in which thousands of individuals inhabit small cup-shaped chambers in a calcareous framework. As successive generations grow upward and outward, the hard calcareous framework also branches upward and outward like a branching plant (figure 14.10). The interspace in the dead structure collects fragments of shells

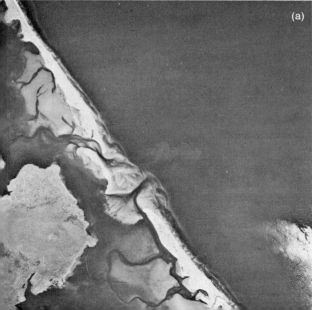

and is cemented by algae. The entire structure forms a white, porous limestone.

The framework of most typical reefs, consisting of the mass of colonial corals, forms a wall which slopes steeply seaward. Wave action continually breaks up part of the seaward face, and the blocks and fragments of the reefs accumulate as debris on the seaward slope. Behind the reef, toward the shore or toward the interior of an atoll, a lagoon forms in which lime mud and evaporite salts are deposited.

Gradual subsidence of the sea floor permits continuous upward growth of reef material to a thickness of as much as 1210 m (4000 ft). Because of their limited ecological tolerance (warm, shallow areas), fossil reefs are excellent indicators of ancient environments.

Shallow Marine Environment. If you study the physiographic map it will soon become apparent that shallow seas border

Figure 14.9 *The tidal flat environment. (a) The tidal flat is alternately covered with a shallow sheet of water and then exposed to the air. The tidal currents produce ripple marks on the silt and mud which accumulate in this low-energy environment. Mud cracks commonly form at low tide and are subsequently covered and preserved. (b) Ancient tidal flat deposits.*

Figure 14.10 *The organic reef environment. (a) A reef consists of a complex community of corals, algae, and other shallow-marine organisms. A rock is formed from the calcareous shells of these organisms. (b) Fossil reefs are found in sequences in shallow marine limestone.*

(a) (b)

Figure 14.11 *The shallow marine environment. (a) Limestone and mud are the most abundant sediment types to accumulate in a shallow marine environment. In this view of the Great Bahama Banks off the southern coast of Florida, lime mud pellets are transported by the currents in a field of submarine sand waves or dunes. (b) Thin interbedded layers of limestone, shale, and sandstone are commonly formed in a shallow marine environment as the shallow seas expand and contract, causing the various subenvironments to shift back and forth.*

most of the land areas of the world and extend to the interior of the continent in areas such as Hudson Bay, the Baltic Sea, and the Gulf of Carpentaria (north of Australia). These shallow seas constitute an extremely important sedimentary environment because sediment eroded from the continent is transported toward these areas by river systems. Shallow seas were more widespread during the geologic past, and much of the continent is mantled with sediment which accumulated within this environment.

The characteristics of the sediment deposited in a shallow sea environment depend to a considerable degree on the supply of sediment from the land and the local conditions of climate, wave energy, circulation of water, and temperature. If there is a large supply of land-derived sediment, sand and mud accumulate to ultimately form sandstone and shale, respectively. If sediment from the land is not abundant, limestone is generally precipitated or deposited by biologic means. An excellent example of limestone being deposited in a shallow sea today is the area of the Great Bahama Bank southeast of Florida (*figure 14.11a*). Evaporite deposits such as salt and gypsum are also produced in restricted bays of shallow seas.

Shallow seas will expand over a lowland or withdraw as sea level fluctuates with changing climatic conditions or earth movements which cause the continents to be depressed or uplifted. As the sea expands deposits of mud typically cover a thin sheet of sand laid down by the advancing beach, and limestone will commonly be deposited over the mud. The stratification of these three rock types is a common product of the shallow marine environment (*figure 14.11b; see also figure 14.24*).

An additional feature characteristic of this environment is the vast and varied marine life which flourishes on the shallow sea floor. The abundance of life forms is made possible because sunlight can penetrate and permit the growth of plants which

The Earth's Dynamic Systems

constitute the base of the food chain in the sea. Therefore, fossils of shallow marine organisms are characteristic of rocks formed in shallow seas.

The Deep Ocean. The deep ocean floor adjacent to the continents is for the most part extremely smooth because of deposition of sediment brought down across the continental shelf by turbidity currents. In the marine environment a turbidity current is an infrequent event in which mud and coarser sediment are thrown into suspension by an earthquake or submarine landslide. This forms a slurry of sediment and water which, being denser than the surrounding clear water, sinks beneath it and moves out rapidly down the submarine slope (*figure 14.12*). The principle of turbidity current flow can easily be demonstrated in the laboratory by pouring muddy water down the side of a tank filled with clear water. The mass of muddy water moves down the slope of the tank and across the bottom at a relatively high speed without mixing with the clear water. Turbidity currents are also observed where streams discharge muddy water into a clear lake or reservoir. The denser muddy water continues to move out along the bottom of the basin and may flow for a considerable distance. For example, the turbidity currents produced by the Colorado River flowing into Lake Mead continue to flow beneath the clear water of the lake for more than 200 km (125 mi) and occasionally boil up into the outlet pipe at Boulder Dam.

A turbidity current "triggered" by an earthquake near Grand Bank of Newfoundland broke a series of transatlantic cables as it swept across the continental slope. The speed at which the current moved was determined to be 80 to 95 km (50 to 60 mi) an hour. This mass of muddy water swiftly spread a distance of 150,000 km (100,000 mi) on the floor of the Atlantic.

As a turbidity current moves across the flat floor of a basin, its velocity gradually decreases, and the coarse sediment is deposited. After the turbid water ceases to move, the remaining sediment held in suspension gradually settles out, depositing a layer of sediment in which there is a continuous gradation from coarse material at the base to fine material at the top without a break or sudden change in grain size. Apparently, deposition is continuous with no internal lamination. A subsequent turbidity flow may deposit a succeeding layer of graded sediment with a sharp contact between layers. Graded bedding can originate

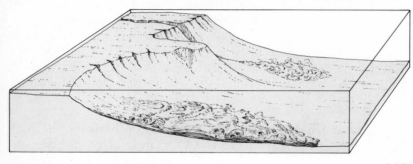

Figure 14.12 *Schematic diagram showing movement of turbidity currents down the slope of the continental shelf. Sediment is moved largely in suspension and, as the current slows down, the coarse grains are deposited first, followed by successively finer-grained sediment.*

Sedimentary Rocks and Paleogeography

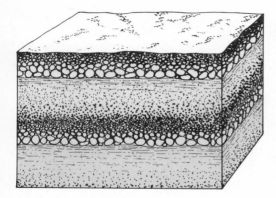

Figure 14.13 Graded bedding produced by turbidity currents occurs in widespread layers, generally less than a meter thick. The deep marine environment commonly produces a great thickness of graded units which are easily distinguished from sediment formed in most other environments.

from other processes, but turbidity currents constitute the major process by which this structure is formed in the sea (*figure 14.13*).

Deposits of turbidity currents form **deep-sea fans** and abyssal plains adjacent to the continental margins. They are characterized by a sequence of graded beds with each layer extending over large areas and are, thus, easily distinguished from sediment deposited in most other environments.

Sediment also accumulates on the floor of the open ocean far from the continents. This material consists of fine particles of mud which are carried in suspension and gradually sink. The most abundant sediment is a fine-grained brown or red clay, the primary origin of which is still subject to conjecture. It may be derived from windblown dust, volcanic ash, or possibly from mud dispersed from the continents. The reddish color is a result of oxidation of iron, for free oxygen in the cold water oxidizes the fine particles as they slowly settle.

Some sediment in the open ocean apparently crystallized directly from sea water. In large parts of the South Pacific, these deposits form the bulk of the sediment. Most abundant are manganese nodules, composed of oxides of manganese and iron and many rarer elements.

Calcium carbonate in the form of microscopic shells of marine organisms also accumulates as mud to form an organic **ooze** which mantles much of the shallower sea floor in the open ocean. In cold, deep water with a higher content of dissolved carbon dioxide, the shells begin to dissolve and are an insignificant component of the sediment accumulating at depths greater than 4550 m (15,000 ft).

Unconformities

Statement

In a sequence of sedimentary rocks, there are many major discontinuities indicating significant interruptions in the rock-forming processes. If sedimentation ceases and the area is subjected to erosion, an unconformity (a physical discontinuity in the succession of strata) is produced and is of paramount importance in the interpretation of geologic events.

Unconformities are classified in three main categories on the basis of the structural relations between the underlying and overlying rocks: (1) **angular unconformities**, (2) **disconformities**, and (3) **nonconformities**. Each records three important geologic events: (1) formation of the older rocks, (2) erosion of the older rocks to a surface independent of its structure, and (3) burial of this ancient erosional surface beneath younger strata.

Discussion

Angular Unconformity. The relationship between the rock bodies shown in *figure 14.14* is an example of an angular un-

conformity, the significance of which can be appreciated only by careful consideration of what the angular discordance implies. The older strata were deposited, folded or tilted, then eroded before the deposition of the overlying beds. A complete reversal of geologic processes occurred: the area ceased to be the site of deposition and was subjected to erosion and, subsequently, became the site of deposition again.

Disconformity. A disconformity consists of an erosional surface carved on horizontal rocks without tilting or folding. The rocks above and below the erosional surface are parallel. The erosional surface may strip off the top of the older sequence and cut channels into the older beds, but there is no structural discordance between the two rock bodies.

Disconformities are formed by broad, gentle upwarps in which the layered sediments on the sea floor are elevated above sea level without folding. Subsequent subsidence then lowers the rocks below sea level again so that deposition of the new sequence of beds is essentially parallel to the stratification of the older. Disconformities are generally more difficult to recognize inasmuch as the beds above and below the erosional surface are parallel (*figure 14.15*).

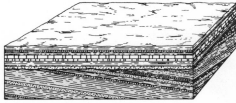

Figure 14.14 *An angular unconformity in the eastern Grand Canyon is easily recognized, as the lower sequence of rocks is inclined to the right at an angle of about 15°, whereas the overlying sequence is horizontal. The unconformity separating these two major groups of rocks is about midway up the canyon wall.*

Sedimentary Rocks and Paleogeography

Figure 14.15 *Disconformities do not show an angular discordance, but an erosional surface separates the two rock bodies. In this exposure it is clear, from the channel in the lower part of the photograph, that the lower limestone units were deposited and eroded before the upper units were deposited.*

Nonconformity. A nonconformity is an unconformity separating plutonic igneous rocks or metamorphic rocks from overlying sedimentary strata. A structure of this sort is shown in *figure 14.16.* Clearly, the light Colorado granite dike in the center of the photo did not invade the overlying horizontal sandstone, because the basal layers of the sandstone contain pebbles of the granite and coarse sand of quartz and feldspar derived from the weathering and disintegration of the granite below. Moreover, the sandstone is unaltered by heat, and locally a soil profile can be seen above the granite and below the sandstone.

Why does sandstone deposited on a granite or metamorphic rock indicate a major discontinuity in rock-forming processes? The answer becomes readily apparent when we consider the environment in which these rocks form. Both originate deep within the earth's crust, the granite cooling slowly at great depth, and the metamorphic rock recrystallizing at high temperatures and pressures far below the surface. In order for these rocks to be weathered and eroded, the overlying cover of rock must be removed. Thus, sedimentary rocks deposited upon

The Earth's Dynamic Systems

a granitic or metamorphic rock imply (1) formation of an ancient sequence of rocks into which the granite was intruded or which covered the metamorphic rocks, (2) uplift and erosion to remove the cover and expose the granites or metamorphic rocks to the surface, and (3) deposition of the younger sedimentary rocks on the old erosional surface developed on the granite.

Characteristics of Buried Erosional Surfaces. Most unconformities are best seen in a vertical section exposed in a canyon wall, road cut, or quarry, where they appear as an irregular line. An unconformity is not a line, however, but a buried erosional surface. The present surface of the earth, especially the coastal plains, is an example of what an unconformable surface is like. Channels cut by streams are responsible for many irregularities, and resistant rocks protruding up above the surrounding surface can cause local relief of several hundred meters.

It can be assumed that a soil profile develops over most ancient erosional surfaces, but much is eroded away as the area slowly subsides beneath the sea. Locally, however, soil profiles and minor irregularities in former landscapes are preserved in their entirety beneath the younger sediments. Erosional sur-

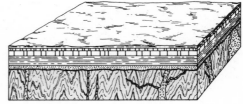

Figure 14.16 *The metamorphic rocks and the igneous dikes shown in this figure were formed at great depths in the crust. The rocks which covered them when they were formed were subsequently eroded away so that the igneous and metamorphic rocks were exposed to the surface. The horizontal sedimentary rocks, lying on the igneous and metamorphic, form this nonconformity.*

Sedimentary Rocks and Paleogeography

faces beneath volcanic ash or lava flows surrounding tree stumps in growth position are commonly preserved in minute detail. A rare example of a buried surface preserving remarkable detail is the ancient town of Pompeii, Italy, which was covered by ash from Mt. Vesuvius in 79 A.D.

Paleocurrents and Paleogeography

Statement

Paleogeography is the study of ancient landscapes and the characteristics of past conditions on the earth's surface. In this study a geologist attempts to determine the location of ancient mountain chains, drainage systems, and areas once covered by inland seas. He attempts to delineate the trends of ancient shorelines, determine ancient climatic conditions, and establish the direction of paleowinds. Although a seemingly impossible task, there are many clues in the sedimentary rocks, some of which we discussed in the previous sections of this chapter.

One approach found to be successful is the study of paleocurrents. Fortunately, there are a number of sedimentary structures which clearly show current direction. Ripple marks form at right angles to current flow; cross-bedding is inclined in a downcurrent direction because the sediment is deposited on the lee slope of a sand wave or sand dune. When mapped and treated statistically, these structures show the patterns of sediment movement. This information may then be combined with studies of the regional relations of rock bodies, to show ancient drainage directions, trends of ancient shorelines, or patterns of paleowinds.

Discussion

The paleogeographic interpretation of current structure depends upon the environment of sedimentation. In river deposits the flow is unidirectional from the highland to the sea (or lake basin) and, although local variations result as the river meanders, the average direction of current structures formed in a river is down the regional slope. This clearly is illustrated by the flow direction in the Mississippi River and its relation to major geographic elements. As shown in *figure 14.17*, the flow of the river may be in any direction locally, but the average is downslope. From the measurements of average flow direction of the rivers alone, we can conclude that the regional slope was to the south and that the shoreline was oriented in an east-west direction. A line drawn perpendicular to the average paleocurrent direction is a general contour line and is parallel to the ancient shoreline.

By such procedures it is possible to map current structures in a formation of river deposits and from them deduce the location of old highlands, the direction of the regional slope down

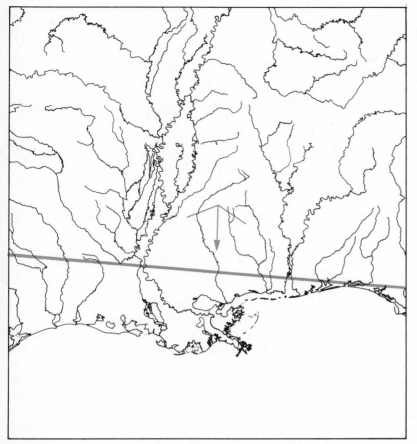

Figure 14.17 *Map of the Mississippi River system showing that, although the river meanders greatly, the average direction of flow is down the regional slope. Current structures such as cross-bedding in river sediments provide an important insight into paleogeography.*

which the sediments flowed, and the general trend of the ancient shoreline.

The problem is more complicated in marine sediments, for there are a variety of currents not directly related to paleogeography. A study of current structures in marine sediments does show the current systems in the ancient sea and the direction in which the sediment was moving at the time it was deposited.

Wind deposits are similar to marine sediments in that the paleocurrent direction, as determined from cross-bedding, may not be related to the slope of the land and trend of shoreline, but studies of paleowinds may yield broader conclusions such as evidence of paleolatitudes, significant in the study of continental drift.

Two examples will serve to illustrate the types of paleogeographic interpretations made possible through studies of cross-bedding and ripple marks in sedimentary rocks.

The first example is a series of Precambrian rocks exposed along the shores of Lake Superior. These rocks are interbedded with lava flows and consist of conglomerate and sandstone, which typically occur in channels. The composition and association with ancient lava flows leaves little doubt that these rocks

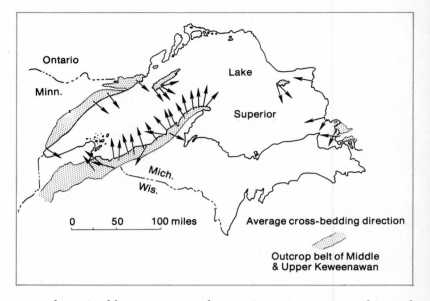

were deposited by streams on the continent in an area subjected to vulcanism. The average direction of cross-bedding in the conglomerates and sandstones is shown in *figure 14.18*. From this map it is apparent that the sediment was derived from both the southeast and the northwest and was deposited in a basin somewhat larger, but in approximately the same position as the present Lake Superior. An outline of the basin into which the sediment was transported (*figure 14.19*) was made by drawing lines perpendicular to the average cross-bedding direction, considered to be roughly parallel to the ancient basin contour lines. Highlands must have existed in Michigan and parts of Wisconsin during this time, as well as in parts of Minnesota and Ontario.

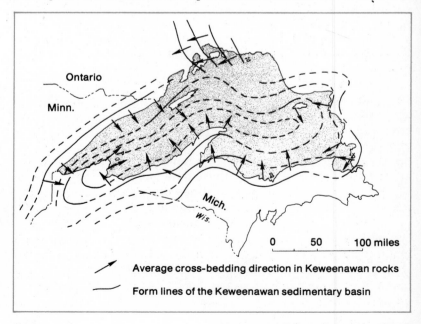

Figure 14.19 On the basis of cross-bedding directions, it is possible to reconstruct the size and shape of the ancient basin in which the Keweenawan sediments were deposited. The arrows show the average cross-bedding direction and the lines represent contour lines, showing the interpreted form of the ancient sedimentary basin.

The Earth's Dynamic Systems

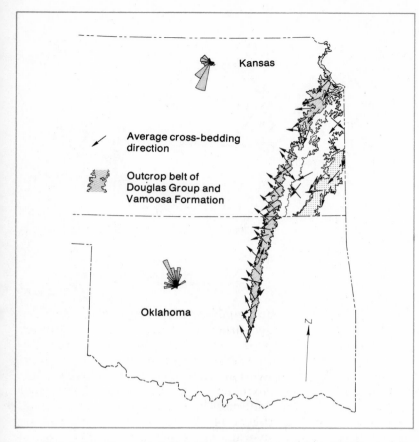

Figure 14.20 *Cross-bedding directions in the fluvial sandstones of eastern Kansas and Oklahoma show a source from the northeast, east, and southeast.*

Our second example is taken from the Pennsylvanian rocks of Kansas and Oklahoma. These rocks consist of interbedded sandstones, limestones, and shales. The sandstones are mostly continental in origin, as indicated by fossil wood and minor coal seams. The limestones consist of fragments of shells and oolites and are obviously marine in origin. The interstratification of these deposits must have resulted from expansion and contraction of the sea.

Cross-bedding directions in the fluvial sandstones (figure 14.20) indicate a regional slope to the southwest in Kansas and north and northeast in Oklahoma. However, the marine limestones show two opposing directions 180° apart at essentially every outcrop (figure 14.21). These directions were up and down the regional slope indicated by the nonmarine sandstones. Paleogeographic interpretation of these data is shown in figure 14.22. The trends of the ancient shorelines during the time the sediments were deposited formed a large arc with the slope of the land towards a low near the Kansas-Oklahoma border. The cross-bedding in the limestones is interpreted as having been produced by tidal currents which moved perpendicular to the shore. From these data we may conclude that an arm of the sea extended across much of Kansas and Oklahoma and expanded and contracted with time.

Figure 14.21 *Cross-bedding directions in the marine limestones interbedded with the fluvial sandstones show two opposing directions of sediment movement 180° apart.*

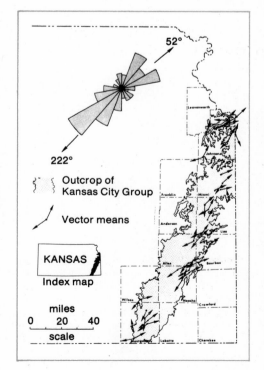

Sedimentary Rocks and Paleogeography

Figure 14.22 *The paleogeographic synthesis of the data is presented in figures 14.20 and 14.21. The fluvial sandstones show that the regional slope was towards the Kansas-Oklahoma line and that the ancient shoreline formed a broad arc. Hypothetical expansion and contraction of the sea which caused the marine and nonmarine sediments to be interbedded are shown by dashed and dotted lines. The opposing paleocurrent directions in the marine limestones are interpreted to be the result of tidal currents.*

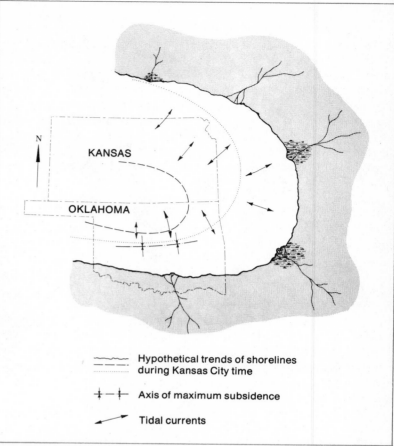

Hypothetical trends of shorelines during Kansas City time

Axis of maximum subsidence

Tidal currents

Paleogeographic Synthesis

Statement

Mapping the direction of cross-bedding and ripple marks is only one way to approach the problem of interpreting paleogeographic features from sedimentary rocks. Other data such as the characteristics of the rock units, their relationship to one another (stratigraphy), and their contained fossils, constitutes important material from which knowledge of paleogeography can be obtained.

A paleogeographic synthesis usually concerns a relatively large area, and the conditions that prevailed there for a selected period of time—for example, the Atlantic coast during upper Cretaceous time, or the mid-continent during early Pennsylvanian time. Geologists measure the thickness of strata in the field and determine the vertical and lateral relationships between rock bodies. Cross sections and maps are constructed to enable the geologist to visualize the relationship of regional scale. Additional helpful data include studies of fossils, mineralogy, and sedimentary structures.

Discussion

In previous sections of this chapter we have studied the relationship between sedimentary rocks and such sedimentary environments as river systems, deltas, beaches, and shallow seas. These environments are related in a very specific way to geographic features; as physical geographic conditions evolve in a systematic way, so do the sedimentary environments and the sequence of rocks produced in them. To illustrate this point let us consider an example of a shallow sea, such as Hudson Bay or the Baltic Sea, which expands progressively over the surrounding low country and then recedes. The event is shown in a series of diagrams in *figure 14.23*.

In the first diagram the sea begins to expand over a lowland drained by a river system. Sand accumulates along the shore, mud is transported in suspension offshore, and limestone is precipitated farther offshore beyond the zone of mud. All three types of sediment are deposited simultaneously but each in a different environment. Stream deposits (not shown in the diagram) accumulate on the flood plain of the river system.

As the sea expands over the lowland, each environment is displaced with each major movement of the shoreline (*diagrams a to c*). This results in the beach sands being deposited over the stream sediments, the offshore mud being deposited over the previous beaches, and the limestone over the mud. As the sea continues to expand inland, layers of sand, mud, and limestone are deposited.

As the sea withdraws (*diagram d*), the mud is progressively deposited over the limestone and the nearshore sand over the mud. The net result is a wedge of limestone encased in a wedge of shale which, in turn, is encased in a wedge of sandstone. Below and above the marine deposits are fluvial sediments deposited by the river system. Subsequent uplift and erosion of the area will produce a definite pattern of rock (*diagram e*), beginning at the base with a beach sand overlain by shale and limestone deposited offshore which is, in turn, overlain by mud and beach sand. A sequence such as this indicates a transgression-regression of the sea. Cross-bedding and ripple marks in the underlying and overlying fluvial sediments show the regional slope and location of source area plus the trend of the shoreline and the direction in which the sea advanced.

An excellent example of regional paleogeographic synthesis is found in the upper Cretaceous rocks of western North America. These rocks, exposed throughout the Rocky Mountains and in the Great Plains as far east as Minnesota and Iowa, are mainly clastic, with sandstones and conglomerate dominating in the west. Thick coal beds also abound in the west and pinch out when traced eastward. Shales with limestone units are most abundant in the east. From hundreds of outcrops geologists have measured the thickness of each rock type and have compiled the cross section shown in *figure 14.24*. Details of the major rock types are shown in *figure 14.25*.

Figure 14.23 *Schematic diagrams showing how expansion and contraction of a shallow sea would produce a cycle of sand-shale-lime-shale-sand. (See text for explanation.)*

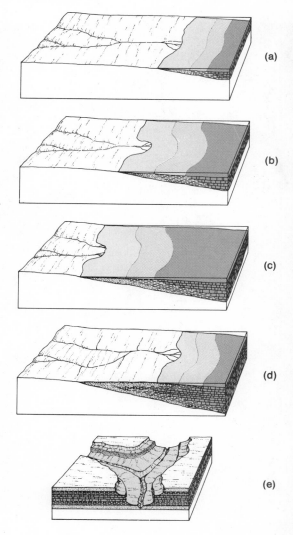

(a)

(b)

(c)

(d)

(e)

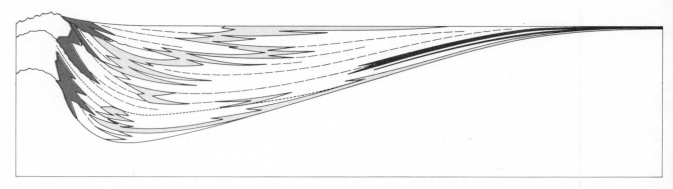

Figure 14.24 (above) *Schematic diagram showing the vertical and lateral variations in Cretaceous rocks of the western United States.*

Figure 14.25 (below) *The vertical sequence of Cretaceous sandstone (beach deposits) and shale (lagoon and offshore marine deposits) in central Utah. This is part of the sequence in the left part of figure 14.24.*

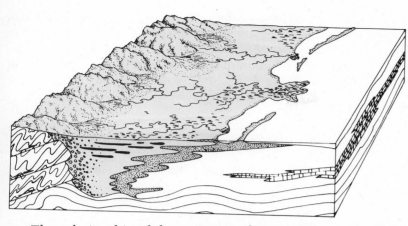

Figure 14.26 *Interpretation of the environments and paleogeography which produced in the Cretaceous rocks in the western United States. (See text for further explanation.)*

The relationship of the various rock types indicates a series of former sedimentary environments which extended in a north-south direction. The coarse gravels are interpreted as alluvial fan deposits derived from a mountain range which extended in a north-south direction through western Utah. These grade eastward into the coals and sandstone, deposited in nearshore environments such as barrier islands, beaches, and swamps. Eastward in the Great Plains states, mud and calcium carbonate accumulated in a broad, shallow sea to form the shale and limestones. The paleogeographic conditions under which these rocks accumulated are shown in the diagram in *figure 14.26*.

As the seas expanded and contracted, the shoreline environments of beaches and swamps moved back and forth to produce the interfingering relations of the sand and coal with the marine shale. This basic geographic pattern existed throughout upper Cretaceous time until finally the sea withdrew completely. Subsequently, the entire region was uplifted and much of the sedimentary sequence was eroded.

Summary

Sedimentary rocks constitute a record of the operation of the hydrologic system upon the earth throughout time. They accumulate in a variety of depositional environments (alluvial fan, fluvial, deltaic, shoreline, shallow marine, etc.). Each sedimentary environment is characterized by certain physical, chemical, and biological conditions which produce distinctive features in the rock and permit the geologist to interpret certain paleogeographic conditions which once existed. Some of the more important features of a rock which permit paleogeographic interpretations are: (1) the geometry of the rock body (size and shape), (2) texture and composition of the rock, (3) vertical and lateral variations in the rock body, (4) fossils, (5) unconformities, and (6) paleocurrent structures (cross-bedding and ripple marks). By studying these features on a regional basis the geologist is commonly able to determine

(1) the location of ancient mountain ranges (source of sediment), (2) the nature of the geologic processes operating in the source area, (3) trends of ancient shorelines, (4) characteristics of the sedimentary environment, and (5) changes in the environment with time.

Additional Readings

Coleman, J. M., and S. M. Gagliano. 1964. Cyclic Sedimentation in the Mississippi River Deltaic Plain. Gulf Coast Ass. Geol. Soc. Trans. 14:67-80.

Dunbar, C. O., and J. Rodgers. 1957. Principles of Stratigraphy. New York: John Wiley and Sons.

Glennie, K. W. 1970. Desert Sedimentary Environments. New York: Elsevier Publishing Company.

Laporte, L. F. 1968. Ancient Environments. Englewood Cliffs, N.J.: Prentice-Hall.

Morgan, J. P., ed. 1970. Deltaic Sedimentation, Modern and Ancient. Society of Economic Paleontologists and Mineralogists Special Paper 15. Tulsa, Oklahoma.

Morisawa, Marie. 1968. Streams. New York: McGraw-Hill Book Company.

Reineck, Hans E., and Singh, J. B. 1973. Depositional Sedimentary Environments. New York: Springer-Verlag.

15 Plate Tectonics

The theory of plate tectonics is accepted by most geologists today; yet, only a few years ago, the continents and ocean basins were considered to be permanent features which had existed since the beginning of the earth. We now believe that the lithosphere is in constant motion and that continents have drifted thousands of kilometers across the earth's surface. Movement of lithospheric plates causes earthquakes, mountain-building, metamorphism, and igneous activity. What is the nature of the mobile plates? What causes them to move? If the lithosphere is in motion, is it possible to measure rates and direction of movement? Have the plates always been in motion? Will they continue to move in the future? Why have we discarded the older concepts of permanent continents and ocean basins? These are some of the questions we will try to answer in this and subsequent chapters.

Although continental drift was proposed more than fifty years ago, the theory of plate tectonics was not developed until the early 1960s, when oceanographic surveys had provided enough data to make accurate regional topographic maps of the ocean floors. These data show that the ocean floors are not flat, featureless areas covered with sediment; nor are they like the continents. Instead, there is a worldwide rift system along the crest of the mid-oceanic ridge and a system of deep trenches along some of the ocean margins. These two features are the most seismically active areas in the world.

In this chapter, we will briefly review the development of the theory of plate tectonics and examine the evidence upon which it is based. We will then consider the nature of the lithospheric plates and their boundaries and the way in which they move relative to each other in order to produce the tectonic features of our planet.

Major Concepts

1. The theory of continental drift was proposed in the early 1900s and was supported by a variety of geologic evidences. However, without a knowledge of the nature of the oceanic crust, a complete theory of earth dynamics could not be developed.
2. A major breakthrough in the development of the plate tectonic theory occurred in the early 1960s, when the topography of the ocean floors was mapped and magnetic and seismic characteristics were determined.
3. The lithosphere may be divided into a series of plates bounded by the mid-oceanic ridge, trenches, mountain ranges, or transform faults.
4. The plates move apart where the convecting mantle rises and spread laterally beneath the mid-oceanic ridge. The plates descend back into the mantle beneath the trenches and are consumed.
5. The energy for plate tectonics is internal heat, probably generated by radioactivity in the asthenosphere.

Plate Tectonics

Development of the Theory

Statement

The theory of plate tectonics has brought about a sweeping change in our understanding of the earth and the forces which shape it. Some scientists have considered this change in thought to be as profound as that which occurred when Darwin reorganized biology in the nineteenth century or when Copernicus determined that the earth is not the center of the universe. Yet, the idea of continental drift was initially developed many years ago and was supported by some very convincing evidence. The evidence came only from the continents because prior to the 1950s we had no effective means of studying the ocean floor. The major breakthrough in development of the plate tectonic theory occurred only after the ocean floors were mapped and their physical characteristics studied. Only then could geologists consider facts concerning the entire planet and develop a theory of its dynamics.

Discussion

Soon after the first reliable world maps were made, geographers noticed that the continents, particularly Africa and South America, would fit together like a jigsaw puzzle if they could be moved, but the idea of a mobile crust with moving continents was not seriously considered until the early 1900s. The concept of continental drift was first put forth by an American geologist, Frank B. Taylor, in 1908 but was perhaps best explained by Alfred Wegener, a German meteorologist, in 1915 when he published a book entitled *The Origins of Continents and Oceans.* Wegener based his theory not only on the shape of the continents but on geologic evidence such as similarities in the fossils found in Brazil and Africa. He drew a series of maps showing three stages in the drift process and called the original large landmass **Pangaea** (meaning "all lands"). Wegener believed that the continents of light granitic rock somehow plowed through the denser basalts of the ocean floor, driven by forces related to the rotation of the earth (*figure 15.1*).

Most conservative geologists and geophysicists rejected Wegener's theories, although many scientific observations supporting continental drift were known in Wegener's time. A few scholars, however, seriously considered continental drift, especially Arthur Holmes of England, who developed the theory in his textbook, *Principles of Physical Geology,* and a South African, Alex L. Du Toit, who compared the landforms and fossils of Africa and South America and further expounded on the theory in his book *Our Wandering Continents.*

Some of the evidence used in the early arguments for continental drift is as follows:

Paleontologic Evidence. The striking similarity of certain fossils found on the continents on both sides of the Atlantic is difficult to explain without considering that the continents were

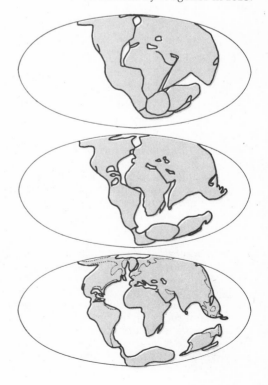

Figure 15.1 *Continental drift as visualized by Wegener in 1915.*

once connected. The record of life indicates that new species appear at one point and disperse outward from that point. Floating and swimming organisms could migrate in the sea from the shore of one continent to another, but the Atlantic Ocean would present an insurmountable obstacle for migration of land-dwelling organisms such as reptiles, land-dwelling mammals, insects, and certain land plants. Let us consider two examples.

Fossils of a seed fern *Glossopteris* have been found in rocks from South America, South Africa, Australia, India, and within 480 km (300 mi) of the South Pole in Antactica. Mature seeds of this plant were several millimeters in diameter, too large to be dispersed across the ocean by winds. The presence of *Glossopteris* in rocks of the same age in the southern continents is therefore considered strong supporting evidence for continental drift.

The distribution of Paleozoic and Mesozoic reptiles provides similar arguments in favor of continental drift because fossils of several species have been found in the now separated southern continents. An example is the mammallike reptile belonging to the genus *Lystorsaurus*. This creature was strictly a land dweller, and its fossils are found in abundance in South Africa and South America, as well as Asia. In 1969, a U.S. expedition to Antarctica discovered fossils of this animal in rocks of similar age, so members of the family are found on all of the southern continents. It is clear that this reptile could not swim thousands of kilometers across the Atlantic and Antarctic Oceans, so some previous connection must be postulated. Continental

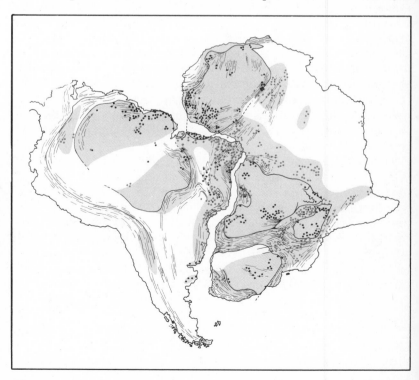

Figure 15.2 *South America and Africa fit together, not only in outline but in rock types and geologic structure. The brown areas represent the shields of metamorphic and igneous rocks formed at least 2 billion years ago. The gray areas represent younger rock, much of which has been deformed by mountain-building. Structural trends such as fold axes are shown by dashed lines. Most of the deformation occurred 450 to 650 million years ago. Dots show sites of rocks dated by radiometric means. Black dots represent rocks older than 2 billion years. Brown dots denote younger rocks. Note that several fragments of the African shields are left stranded along the coast of Brazil.*

The Earth's Dynamic Systems

drift supplies one possible explanation. The existence of a land bridge, similar to Central America but now submerged, is another possibility; however, surveys of the ocean floor show no evidence of such a submerged land bridge.

Evidence from Structure and Rock Type. There are a number of geologic features which abruptly terminate at the coast of one continent and appear on the facing continent across the Atlantic (*figure 15.2*). Folded mountain ranges at the Cape of Good Hope trend east-west and terminate abruptly at the coast, but an equivalent of this structure recognized by age and style of deformation is found in Buenos Aires. Other structures and rock types, including igneous, sedimentary, and metamorphic rocks, match between the two continents. The agreement in the kimberlite pipes (the host rock of diamonds) is especially impressive.

The folded Appalachian Mountains, which extend northeastward across the eastern United States and up through Newfoundland, terminate at the ocean but reappear at the coast of Ireland and Brittany. There are other examples which could be cited, but the important point is that the continents on either side of the Atlantic fit together, not only in outline but in rock type and structure. This fit is similar to matching pieces of a torn newspaper (*figure 15.2*). Not only do the jagged edges fit, but the printed lines come together to form a single sheet. In this analogy, the structure and the rock types of the continents correspond to the printed lines. One important point should be emphasized. The geologic similarities between continents are found only in rocks older than the Cretaceous period—the time when drifting was believed to have taken place.

Evidence from Glaciation. During the latter part of the Paleozoic era (about 300 million years ago), glaciation occurred throughout a large part of the continents in the southern hemisphere. The deposits left by these ancient glaciers are readily recognized, and striations and grooves on the underlying rocks show the direction in which the ice moved (*figure 15.4*). All the continents in this hemisphere except Antarctica now lie close to the equator. On the other hand, continents in the northern hemi-

Figure 15.3 *The continents fit together like a jigsaw puzzle or pieces of a torn newspaper. Not only does the outline of the pieces fit together, but the picture of printing on the pieces (analogous to age and structural features of the continents) also matches and fits together across the edges of the separate pieces.*

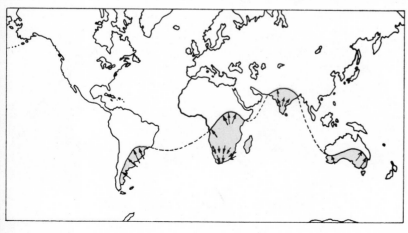

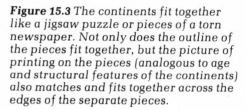

Figure 15.4 *The distribution of late Paleozoic glacial deposits is restricted to the southern hemisphere (except for India). Arrows show the direction of ice movement. These areas are now close to the tropics. The present-day cold latitudes in the northern hemisphere show no evidence of glaciation during this period.*

Figure 15.5 *If the continents were restored to their former positions according to Wegener's theory of continental drift and the former south pole was located approximately where South Africa and Antarctica meet, the location of late Paleozoic glacial deposits and flow directions of the ice would be explained very nicely.*

sphere show no trace of glaciation during this time, but fossil plants indicate a tropical climate in that area. These facts are difficult to explain in light of fixed continents and the climatic belts which are determined by latitude. Even more difficult to explain is the direction in which the glaciers moved, as indicated from regional mapping of the striation and grooves. In South America, India, and Australia, the ice moved inland from the oceans, a situation quite impossible unless there was a landmass where the sea now exists.

If the continents were grouped together as Wegener proposed, the glaciated areas would be grouped in a neat package near the south pole (figure 15.5) and would explain Paleozoic glaciation very nicely. The pattern of glaciation was considered to be strong evidence for continental drift, and many geologists in the southern hemisphere became ardent supporters of the theory because they could see the evidence with their own eyes.

Evidence from Paleoclimates. Other evidence of striking climatic changes tends to support the drift theory. Great coal deposits in Antarctica show that the area now mostly covered with ice once flourished with abundant plant life.

Rock salt deposits and formations of windblown sandstone on other continents provide additional paleoclimatic indications which permit a reconstruction of climatic zones in the past. When the continents are grouped together (figure 15.6), it is clear that the pattern is easily explained. In their present positions, the patterns are quite baffling.

The evidence cited above was considered and debated for years. While this evidence was convincing to many, most geophysicists objected to a theory of continental drift because it was simply impossible for continents to drift through solid rocks of the oceanic crust. In the absence of a reasonable mechanism for the drift process, there was little further development in the theory until after World War II. Then an explosion of knowledge took place which not only provided renewed support for the drift hypothesis but also led to the discovery of a possible mechanism. Scientists were able to study the topography and geophysical characteristics of the ocean floor, in part because of new and sophisticated instruments capable of recording a con-

Figure 15.6 *Paleoclimatic evidence for continental drift. Deposits of coal (C), desert sandstone (D), rock salt (S), windblown sand (dotted area), gypsum (GY), and glacial deposits (GL) each indicate a specific climatic condition at the time of their formation. The distribution of these deposits is best explained if the continents were once grouped together as shown in this diagram.*

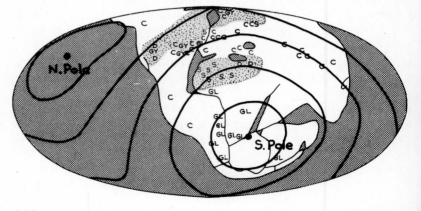

The Earth's Dynamic Systems

tinuous profile of the sea floor or measuring various geophysical properties of the rocks on the deep ocean floor.

Evidence from Paleomagnetics

Statement

The two most powerful arguments for large-scale plate motion are derived from studies of two different aspects of the earth's magnetic field. These are the "paths of polar wandering" and patterns of magnetic reversals on the sea floor.

Discussion

Polar Wandering. We are able to study the history of the magnetic field because the lines of magnetic force running between the north and south poles influence the magnetic properties of a rock as it forms. Volcanic rocks cooling from a lava become magnetized in the direction of the earth's magnetic field when they cool. You will remember from Chapter 2 that the lines of force in the magnetic field occur at different angles at different latitudes, being vertical at the poles and horizontal at the equator (see *figure 2.3*). By using a sensitive instrument called a magnetometer it is possible to determine the direction of the magnetic poles and the magnetic latitude at the time the rock was formed.

Studies of magnetic properties in rocks show that the continents have changed their positions relative to the earth's magnetic field during geologic time. But that's not all; the continents followed different paths relative to the magnetic poles, indicating that they have moved relative to each other (*figure 15.7*). This discovery led to much renewed interest in the theory of continental drift and the conclusion that the Atlantic Ocean opened in relatively recent geologic time.

Magnetic Reversals. Recent studies of the magnetic properties of numerous samples of basalt from many parts of the world demonstrate that the earth's magnetic field has been reversed many times throughout the last 70 to 80 million years. That is, there have been periods of 1 to 3 million years during which magnetic poles have been close to their present location, followed by similar periods during which the north and south poles have been reversed. At least nine reversals have occurred in the last 4.5 million years. The present period of "normal" polarity began about 700,000 years ago and was preceded by a period of reversed polarity beginning 2.5 million years ago which contained two short periods of normal polarity. The major intervals of alternating polarity (about a million years apart) are termed **polarity epochs**, and intervals of shorter duration are termed **polarity events**. The pattern of alternating polarities has been clearly defined, and the polarity epochs and events are known to have occurred in widely spaced parts of the earth. The sequence of magnetic reversals has been well

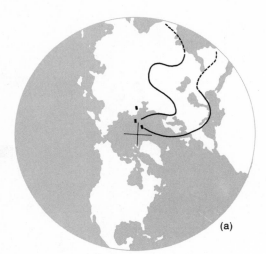

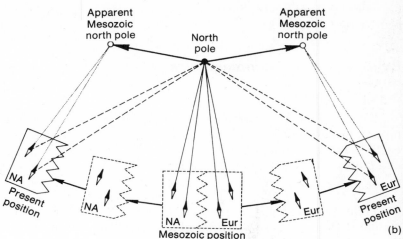

Figure 15.7 *The magnetic properties in the rocks of North America suggest that the magnetic pole has migrated in a sinuous path over the last several hundred million years. Evidence from other continents shows similar migration but along different paths. If the continents have remained fixed, how could different continents have different magnetic poles at the same time? (b) The problem can be answered if the pole has remained fixed while the continents have drifted. If, for example, Europe and North America were joined, the paleomagnetic field preserved in their rocks would indicate a single pole location until they drifted apart. The sequence of rocks on each continent would show different paths of migration to the present pole position.*

documented, and their radiometric ages have been determined so that a reliable chronology of magnetic reversals for the most recent 4 million years has been established (*figure 15.8*). In addition, extrapolation as far back as 76 million years reveals a sequence of at least 171 reversals.

In 1963 Fred Vine and D. H. Matthews suggested that if sea-floor spreading had occurred, it might be recorded in the magnetism of the basalts on the ocean's crust. They postulated that if the earth's magnetic field had been reversed intermittently new basalt forming at the crest of the oceanic ridge would have been magnetized according to the polarity at the time. As spreading of the ocean floor continued, a series of normal and reversed magnetic strips of oceanic crust should have been produced symmetrically from the center of the ridge. Each band of new material moving out across the ocean floor should have retained its original magnetic pattern, a pattern shared by a corresponding band on the other side of the ridge. The result should have been a matching set of parallel bands of normal and reversed magnetized rocks on either side of the mid-oceanic ridge. Subsequent investigations have proven the Vine and Matthews postulate conclusively.

The origin of magnetic patterns on the sea floor is perhaps best understood by considering how the sea floor would evolve during the last few million years according to the theory of plate tectonics and sea-floor spreading. In *figure 15.9a* the sea floor is shown as it is considered to have been during the Gauss normal polarity epoch, named for German mathematician K. F. Gauss. As basalt was injected into the fractures of the mid-oceanic ridge to form dikes or was extruded over the sea floor as submarine flows, it solidified and became magnetized in the direction of the existing (normal) magnetic field. This basalt produced a zone of new crust with normal magnetic polarity along the mid-oceanic ridge. As sea-floor spreading continued, this zone of new crust migrated away from the rift zone but remained parallel to it. About 2.5 million years ago the mag-

The Earth's Dynamic Systems

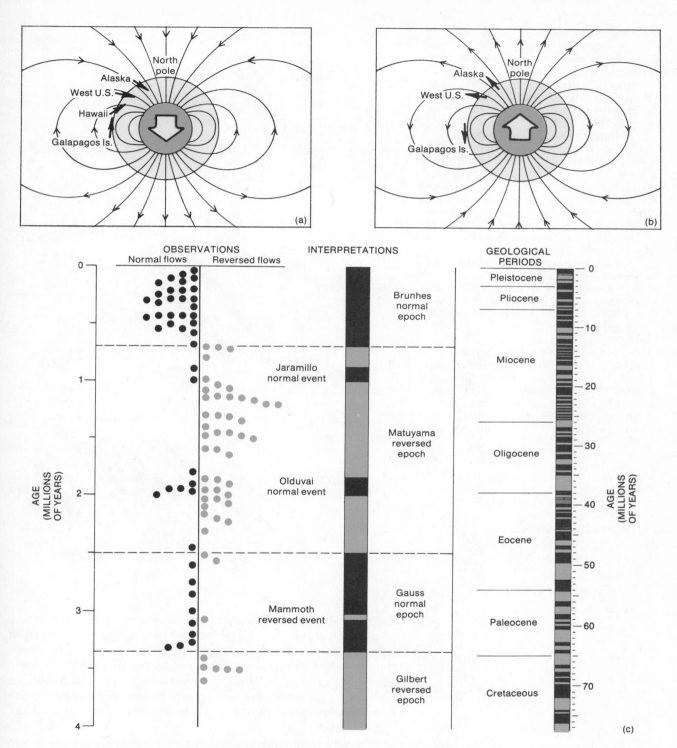

Figure 15.8 *Reversals in lines of force in the earth's magnetic field. (a) With normal polarity. (b) With reverse polarity. (c) Changes in the earth's magnetic field with time. Many rocks preserve a "fossil" record of the earth's magnetic field at the time they were formed. Some show normal polarity, whereas others show reversed polarity. By dating the rocks radiometrically, it is possible to construct a geomagnetic time scale for polarity reversals. In this diagram the patterns of alternating polarities are shown.*

Figure 15.9 *Schematic representation of the magnetisms preserved in new crust generated at oceanic ridges as the lithosphere is transported laterally away from the ridge. (a) The oceanic ridges during the Gauss normal polarity epoch, 2.75 million years ago. As lava cooled and solidified along the ridge in dikes and flows, it became magnetized in the direction of the magnetic field existing at the time (normal). (b) The oceanic ridge during the Matuyama reversed polarity epoch, 2.25 million years ago. As sea-floor spreading occurred, the magnetized crust formed during the Gauss epoch separated into two blocks. Each was transported laterally away from the ridge as if on a conveyor belt. New crust being formed at the ridge became magnetized in the opposite direction. (c) The oceanic ridge at the present time, with polarity reversals of the past 3 million years. The alternating directions of magnetism in the sea floor produced a symmetrical sequence of positive and negative anomalies on each side of the ridge. Note that the pattern of magnetic reversals away from the ridge is the same as the pattern produced in a sequence of basalt flows on the continent. This permits one to correlate rocks on the continent with rocks of equivalent age on the ocean floor.*

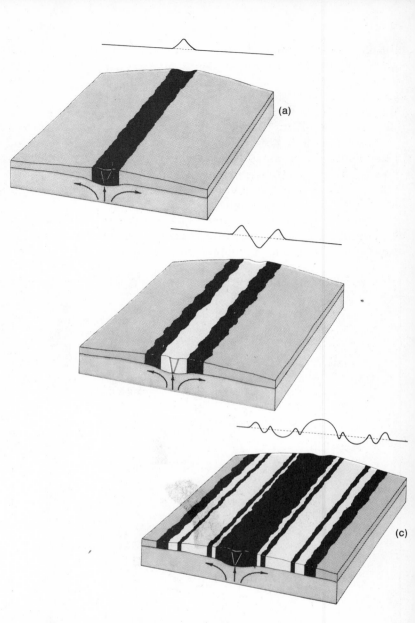

netic polarity was reversed so that new crust generated at the mid-oceanic ridge became polarized in the opposite direction (figure 15.9b). This produced a zone of crust with a reversed polarity. When polarity was changed to normal again, the new crust being created at the ridge was magnetized in a normal direction. In this way, the sequence of polarity reversals for the whole earth became imprinted as magnetic stripes on the ocean crust.

You will note that the pattern of anomalies on the ocean floor on either side of the ridge is the same as that found in a sequence of recent basalts on the continent (figure 15.9c; see also figure 15.8c). That is, the crest of the ridge shows normal polarity and is flanked on either side by a reversed epoch con-

The Earth's Dynamic Systems

taining two short normal events. This is followed by a normal epoch with one brief reversal event, and so on. In brief, the pattern of magnetic reversals away from the crest of the ridge is the same as that found in a vertical sequence of rocks on the continent going from youngest to oldest. These data provide compelling evidence that the sea floor is spreading.

Many magnetic surveys of the ocean floor have been made, and the patterns of anomalies have been determined. *Figure 15.10* shows one of the first areas surveyed along the mid-Atlantic Ridge south of Iceland. Other patterns covering more of the ocean floor are shown in *figure 15.11.*

An important aspect of these surveys is that it is possible to determine the interval of time represented by the anomalies and the rates of plate movement. Studies of magnetic reversals

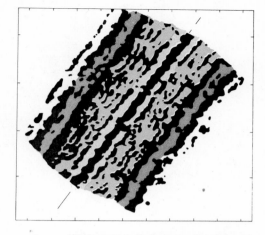

Figure 15.10 *Patterns of magnetic reversal across the mid-oceanic ridge south of Iceland show symmetry on either side of the Mid-Atlantic ridge. The pattern of reversals away from the ridge is similar to the pattern of reversals in a sequence of rocks of similar age on the continent.*

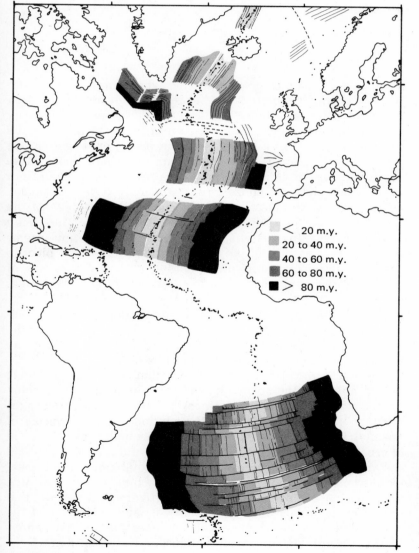

< 20 m.y.
20 to 40 m.y.
40 to 60 m.y.
60 to 80 m.y.
> 80 m.y.

Figure 15.11 *Summary of magnetic patterns and age information for rocks in the Atlantic Ocean. The brown and gray patterns show regions of anomaly trends of a specific age. The trends themselves are shown by lines parallel to the banding. Small dots show the location of earthquakes along the crest of the mid-Atlantic Ridge.*

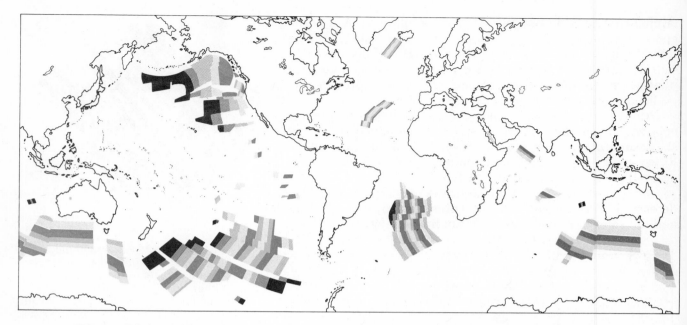

Figure 15.12 *Patterns of sea-floor spreading based on ages of magnetic polarity belts on the ocean floor. The youngest oceanic crust is along the crest of the mid-oceanic ridge. Away from the ridge the crust becomes progressively older. Note that the ridges are offset by transverse faults.*

of rock sequences on the continents where radiometric dates have been determined show that the present normal polarity has existed for the last 0.7 million years. This normal polarity was preceded by the pattern shown in *figure 15.12*. Since the same pattern exists on the ocean floor, provisional ages can be assigned to the magnetic anomalies on the ocean floor. These studies show that almost the total area of the deep-ocean floor was formed during Cenozoic time (during the last 65 million years). These estimates lead to the conclusion that 50% of the deep-sea floor (one-third of the earth's surface) has been created during the last 1.5% of geologic time. It now seems probable that no part of the oceanic crust was formed prior to the Jurassic period 180 million years ago. The average rate of sea-floor spreading, according to these measurements, ranges from 1 to 16 cm (.4 to 6 in.) per year.

Iceland provides an especially interesting example of sea-floor spreading because it is, in essence, a large exposure of the mid-Atlantic ridge and offers a unique opportunity to study the physical mechanism of sea-floor spreading.

Geologic work in Iceland shows that the island is being pulled apart by the spreading sea floor beneath. The tension causes normal faults and fissures parallel to the ridge axis. Through these fissures volcanic eruptions occur. Swarms of parallel dikes are injected into the fissures with each increment of crustal extension. The aggregate width of these dikes is about 400 km (250 mi) and corresponds to the total amount of crustal extension since the beginning of Tertiary time, about 65 million years ago. A geologic map of Iceland (*figure 15.13*) shows that the rocks are oldest ih the extreme east and west ends of the island and become progressively younger toward the center, where present-day vulcanism is almost entirely confined.

Figure 15.13 *A geologic map of Iceland shows that the oldest rocks are along the east and west margins and the youngest rocks are near the center of the island.*

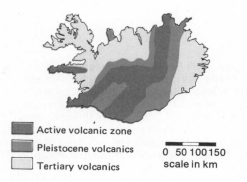

■ Active volcanic zone

■ Pleistocene volcanics

□ Tertiary volcanics

0 50 100 150
scale in km

Evidence of Seismicity

Statement

The establishment of a worldwide network of sensitive new seismic stations in 1966 to monitor nuclear testing has enabled seismologists to amass an amazing amount of data concerning earth dynamics. Earthquakes can be pinpointed with great precision, and the depth and direction of movement of the rocks can be established. This new information indicates that earthquakes are restricted largely to the margins of lithospheric plates (young folded mountain belts, deep oceanic trenches, and the crest of the mid-oceanic ridges). In addition, disturbances in such areas are characterized by specific types and depths of movement. Shallow-focus quakes with vertical movement occur most often at the crest of the mid-oceanic ridges. Shallow-focus quakes with horizontal movement occur along transverse faults across the ridge and along such well-known strike-slip faults of the continents as the San Andreas fault in California. Deep-focus earthquakes are found only under the deep marine trench systems.

These contrasting seismic provinces reflect very different but complementary processes. The mid-oceanic ridges and fracture systems are extensional features produced as the crust is pulled apart. The trenches are **subduction** zones, areas where the rocks are moving at considerable depth below the surface, generating deep-focus earthquakes. Extensive areas in the interior of lithospheric plates are virtually free from earthquakes.

Discussion

Earthquakes are earth vibrations caused by ruptures which result when rocks are strained beyond their elastic limits. This concept can be illustrated by a simple experiment with a brittle plate of steel which bends until it snaps. Energy is stored in elastic bending and is released when rupture occurs, causing the fractured ends to vibrate and send out sound waves.

Detailed studies of active faults show that this model, known as the **elastic rebound theory**, appears to hold for all major earthquakes (figure 15.14). Precision surveys across the San Andreas fault in California show that railroads, fence lines, and roads are first slowly deformed and then offset when movement along the fault occurs as the elastic strain is released.

The focus of an earthquake is important in the study of plate tectonics because it indicates the depth at which rupture and movement occur. Shallow-focus earthquakes are less than 55 km (34 mi) deep, intermediate-focus quakes are 55 to 240 km (34 to 149 mi) deep, and deep-focus quakes are from 240 to 720 km (149 to 447 mi) below the surface. Below 720 km the mantle yields to plastic flow and does not produce earthquakes.

The map in figure 15.15 shows the location of earthquakes and, with dramatic clarity, outlines the plate margins. Shallow earthquakes develop where the mantle material rises and pulls

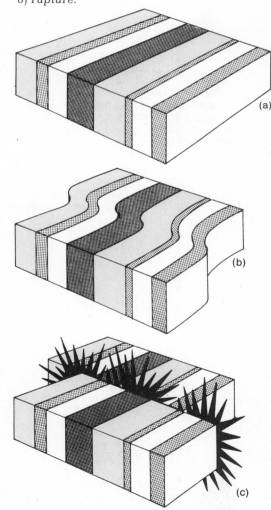

Figure 15.14 The origin of earthquakes. Strain is built up in the rocks until they rupture. Energy is then released and seismic waves move out from the point of rupture.

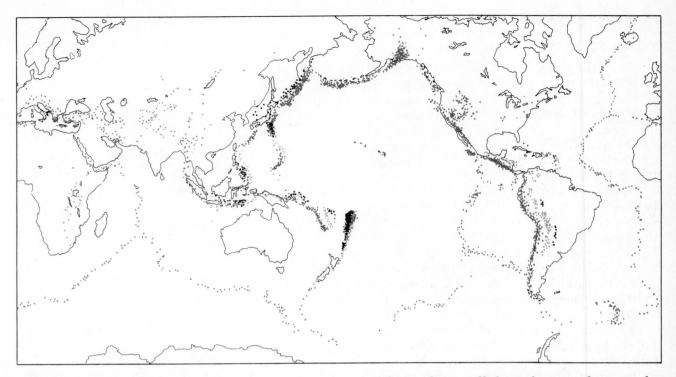

Figure 15.15 The location and focus of earthquakes occurring between 1962 and 1969. Brown dots represent shallow earthquakes, gray dots intermediate earthquakes, and black dots deep-focus earthquakes. Shallow earthquakes are located along the crest of the mid-oceanic ridge. Intermediate and deep earthquakes are restricted to the areas adjacent to deep-sea trenches.

the plates apart. Where plates collide and one is thrust under the other, a zone of deeper earthquakes occurs as the oceanic plate moves down into the mantle. In the central part of the plates, there are no differential movements and, therefore, no earthquakes.

Evidence from Sediment on the Ocean Floor

Statement

To many, the most convincing evidence of all comes from recent drilling into the sediments which form a discontinuous veneer over the basaltic crust of the ocean floor. As predicted by the plate tectonic theory, the youngest sediments were found near the mid-oceanic ridge, where new crust is being created. Away from the ridge the sediment which lies directly above the basalt becomes progressively older, with the oldest sediment found near the continental borders. The deep-sea drilling project proves that the oceanic crust is far younger than that of the continents. The oldest sediments yet found on any ocean floor are only 160 million years old. In contrast, the metamorphic rocks of the continental shields are as much as 3.76 billion years old.

Discussion

The deep-sea drilling project is a truly remarkable expedition in scientific exploration. It began in 1968 with a special ship named the *Glomar Challenger*, designed by a California offshore drilling company. The *Challenger* can lower more than 6100 m (20,000 ft) of drilling pipe in the open ocean, bore a hole in the sea floor, and bring up bottom cores or samples. The project was funded by the National Science Foundation and was planned by JOIDES—Joint Oceanographic Institutions for Deep Earth Sampling—under the direction of Scripps Institution of Oceanography. Since 1968, the *Challenger* has drilled more than 400 holes into the sea floor and has provided much data in support of plate tectonics.

Not only do the thickness and age of sediment increase away from the crest of the mid-oceanic ridge but certain types of sediment also indicate sea-floor spreading. For example, upwelling nutrient-rich water from the Pacific equatorial zone permits planktonic life to thrive there. As the creatures die, their tiny skeletons rain down unceasingly to build a layer of soft white chalk on the sea floor. The chalk marks an environmental zone restricted to the equatorial belt, one that cannot form in the colder waters of higher latitudes. Yet, drilling by the *Challenger* has shown that the "chalk line" in the Pacific occurs at varying distances north of today's equator. The only logical conclusion is that the sea floor moved and that the Pacific floor has been migrating northward for at least 100 million years.

The Nature of Lithospheric Plates

Statement

Now that we have examined some of the evidence for plate tectonics, let us consider some of the details concerning the nature of lithospheric plates, their movement, and some of the results of this process. On the basis of seismic evidence, the plates of the rigid lithosphere are believed to be 100 to 150 km (60 to 90 mi) thick and include both the upper mantle (above the asthenosphere) and the crust. Three types of plate margins are recognized: (1) constructive margins where new crust is generated, (2) destructive margins where the plate moves down into the mantle, and (3) passive margins where the plates simply slip past each other along transform faults.

New crust is created along the constructive margins, where the plates move apart along the crest of the great mid-oceanic ridges and ascending convection currents bring partly melted mantle material to the surface. The plates move because of convection in the underlying asthenosphere. The movement of the plates by the convecting asthenosphere is like the movement of a conveyor belt. As the lithosphere moves slowly away from the ridge, it carries with it the embedded continents. Ultimately, the plates containing oceanic crust are pulled down by the descending currents at the oceanic trenches and are destroyed as they are carried into the mantle. Thus, the oceanic crust is considered, in a sense, to be a surface expression of the mantle, derived from it at the ridge crest and returned to it beneath the trenches. The plates bounded by ridges and trenches are in constant motion but maintain a constant surface area because a rough balance exists between the volume of lithosphere created at the ridges and that destroyed at the trenches (figure 15.16).

Discussion

Seismic studies indicate that the lithosphere (figure 15.16) is relatively strong and rigid, as the seismic-wave velocities in this part of the earth are high. The asthenosphere beneath transmits seismic waves at a much lower velocity, suggesting that it is relatively hot and near the melting point of rock. Many scientists believe that the asthenosphere contains a small amount of liquid that lowers the shear strength, rendering the material soft and mechanically weak. The boundary between the rigid lithosphere and the soft asthenosphere is considered to be the base of the moving plates. Therefore, the plates include the oceanic crust of basalt and sediment, the granitic continental crust where it is present, plus about 100 km of rigid upper mantle.

Figure 15.16 *The major tectonic features of the earth and their relation to the theory of plate tectonics.*

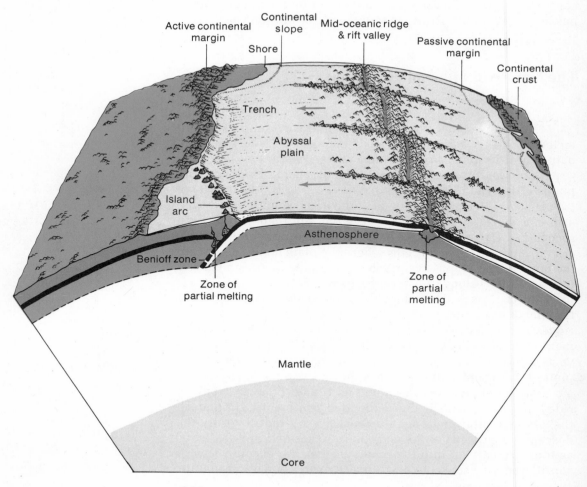

The Earth's Dynamic Systems

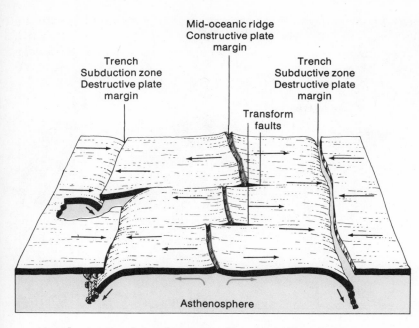

Trench
Subduction zone
Destructive plate
margin

Mid-oceanic ridge
Constructive plate
margin

Trench
Subductive zone
Destructive plate
margin

Transform
faults

Asthenosphere

Figure 15.17 *Types of plate margins. Constructive margins (divergent plate junctions) occur along the oceanic ridges where plates move apart. Destructive margins (convergent plate junctions) occur along the deep trenches. Passive plate margins are along the fracture zones which offset the ridges.*

As shown in the schematic diagram (*figure 15.17*), a constructive margin occurs along the oceanic ridges where the plates are moving away from each other. New crust created along the boundaries is effectively welded to the spreading crust and is being added continuously to the trailing edge of each plate. Faults and joints, resulting from tension as the crust is pulled apart, are the dominant structures, seismic disturbances are shallow, and basaltic eruptions play an important part in generating new crust.

Destructive margins occur along the deep oceanic trenches. Here, two plates approach each other; one is thrust under the margins of the other at an angle of about 45° and is forced back down into the mantle where it is heated and ultimately absorbed into the mantle. Seismic activity along destructive margins includes both shallow earthquakes and deep-focus earthquakes. Andesitic eruptions occur above the descending plate and compressive stresses are commonly produced along the upper plate margins.

Passive plate margins are those in which the plates are neither created nor destroyed but simply slip past each other. These are the great fracture zones of the ocean basins. The direction of relative spreading is parallel to these margins. Shear stresses dominate, shallow seismic activity occurs, and some basaltic extrusion may develop.

At present, six major lithospheric plates and several smaller ones can be recognized (*figure 15.18*). Within this basic framework several variations in the nature of plates may be produced by the distribution of continental masses.

Continental Rifting. Sea-floor spreading and continental drift result from convection or boiling in the asthenosphere which pulls the rigid lithosphere apart, carries it laterally, and

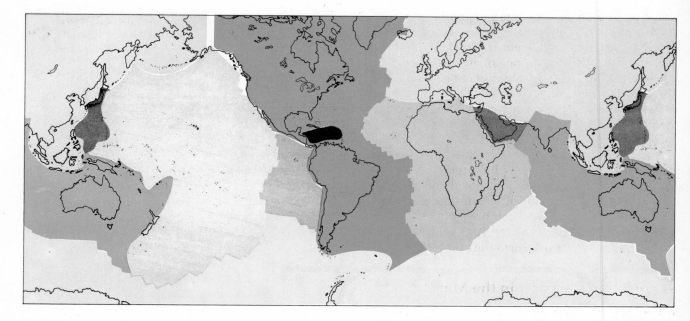

Figure 15.18 Six major lithospheric plates and several smaller ones are delineated by the major tectonic features of the globe plus patterns of seismicity and magnetic anomalies on the sea floor. Some plates such as the American and Indian plates contain continental and oceanic crust. The Pacific plate contains only oceanic crust.

Figure 15.19 Continental rifting. If ascending currents rise beneath a continent, the continent will split and drift apart, creating a new ocean basin.

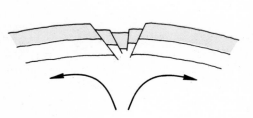

drags part of it down into the mantle again. If convection currents rise beneath a continent, the continent will split, and the two segments will drift apart with the underlying lithosphere (figure 15.19). Several examples showing various stages of continental rifting may be cited. The initial stage of rifting may exist in the system of rift valleys of East Africa, where tension is causing the eastern part of the continent to separate from the rest. The fault block basins, which are partly occupied by large lakes, are huge troughs resulting from the initial separation. The great volcanic mountains along the rifts, such as Mt. Kenya and Mt. Kilimanjaro, grew as magma was ejected and formed new crust along the rift zone. As the continents drift apart, new ocean basins are created along the rift. Examples are the Red Sea (just beginning to form) and the Atlantic Ocean (well developed).

The Atlantic Ocean represents a much more advanced stage in which North and South America have been separated from Europe and Africa. The mid-Atlantic ridge is the zone of spreading, with the American plate moving relatively westward as a single unit. The eastern margins of the American continents are passive, experiencing little, if any, important tectonic activity. In contrast, the western margin of South America constitutes the western margin of the plate at its juncture with the East Pacific plate. Here, deformation and igneous activity have built the Andes Mountains.

As is common in a boiling liquid, the boil, or hot upwelling material, may change its position due to temperature variations which occur during the boiling process. Ocean basins which spread under one set of convection cells (boils) may subsequently be closed by a new set. Thus, we may find in an ocean basin old or dead rifts (mid-oceanic ridges), along which move-

The Earth's Dynamic Systems

ment no longer occurs. Some examples are found in the North Pacific and Arctic Oceans.

Plates with Only Oceanic Crust. If the ascending currents are initiated in former ocean basins, a new oceanic ridge is formed and, in time, the older oceanic crust will be consumed in the trench system (figure 15.20). The Pacific plate is an example consisting totally of oceanic crust formed along the East Pacific rise and consumed along the trench systems of the western Pacific (see physiographic map).

Collision of Continents. Plates carrying continents could ultimately collide at a trench system. Since continental crust is of low density, it cannot subside deep into the mantle; therefore, the margins of colliding continental masses are subjected to extensive horizontal compression and are deformed into mountain belts (figure 15.21). An example is the collision of the India plate with Asia to form the Himalaya Mountains.

Figure 15.20 *If ascending currents in the asthenosphere rise beneath an ocean basin containing no continents, the sea floor will split and spread apart. New crust will be generated at the spreading center and old crust will be consumed at the subduction zone beneath the deep-sea trenches.*

Convection Currents in the Mantle

Statement

Various sources of energy have been proposed to explain convection motion of the mantle. Some believe that the energy is associated with the earth's rotation, earth tides, or heat from the core. Most advocates of the plate tectonic theory, however, favor radiogenic heat produced in the upper mantle. In Chapter 2 we have seen that the heavy elements of the earth are concentrated in the core. Less-dense elements, forming iron-magnesium silicate minerals, are concentrated in the mantle; and the light silica-rich minerals are concentrated in the continental crust. Radioactive elements are probably concentrated in the asthenosphere in the upper part of the mantle. As these elements generate heat, the rocks within the asthenosphere yield to plastic flow and rise in a convection system. Thus, the internal heat is carried to the surface and escapes into space.

The energy for this system was inherited from the early phases of the formation of the earth and is steadily diminishing. The system is irreversible or one-way. Energy within the earth, like that in the sun, will ultimately be spent and the lithospheric plates will cease to move. When this happens in the far future, mountains will no longer form, and all types of volcanic activity and earthquakes will ultimately end. The earth will then be a dead planet, changed only by energy from external sources.

Discussion

Little is known about the actual movement of material in the mantle, but several models of convection systems have been proposed to try to explain the observed facts. Some of these are shown in figure 15.22. In *diagram a* the entire thickness of the mantle is considered to be involved in a series of convection

Figure 15.21 *Collision of continents. The continental crust is too light to move back down into the mantle, so extensive horizontal compression of continental margins results where plates collide.*

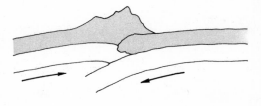

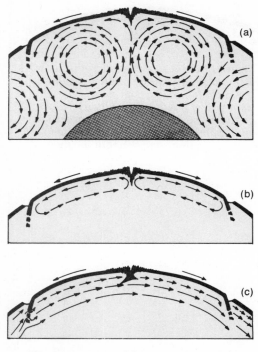

Figure 15.22 *Possible convection systems in the mantle. (a) Large, deep convection cells involving most of the mantle. (b) Shallow convection cells largely restricted to the asthenosphere. (c) Movement of the mantle influenced by rotation of the earth.*

cells. Where the currents rise and move apart horizontally, they initiate the horizontal motion of the lithospheric plates, which move away from the mid-oceanic ridge. Where convection cells descend, they drag the lithosphere down into the mantle. *Diagram b* shows a convection system limited to the asthenosphere. Here, heat is considered to originate from concentration of radioactive elements near the outer part of the mantle. A completely different model is shown in *diagram c* in which movement of mantle material is related to rotation of the earth.

Summary

The theory of plate tectonics was developed during the 1960s as a result of new knowledge obtained concerning the topography, magnetic properties, and seismic characteristics of the oceanic crust. It has subsequently been supported by a wide variety of other geologic and geophysical data from both the continents and the oceans. The theory explains earth dynamics as a result of a series of rigid lithospheric plates which move by convection in the asthenosphere. New crust is created at the mid-oceanic ridge where the plates spread apart, and oceanic crust is consumed in the mantle where the lithosphere descends along the deep oceanic trenches. Mountains form along the margins of continents where they collide with other plates. Recent data from deep-sea drilling substantiate the theory of plate tectonics, as do seismic records of earthquakes.

In the following chapter, we will consider some of the recently discovered facts about the oceanic crust and study some of the details of the evolution of the oceanic crust according to the plate tectonic theory.

Additional Readings

Anderson, D. L. 1971. "The San Andreas Fault." Sci. Amer. 225(5): 52-68 (Offprint No. 896).

Barazangi, M., and J. Dorman. 1969. World Seismicity Maps Compiled from ESSA, Coast and Geodetic Survey, Epicenter Data 1961-1967. Bull. Seismol. Soc. Amer. 59(1):369-380 (Precise plots).

Bullard, E., J. E. Everett, and A. G. Smith. 1965. The Fit of the Continents around the Atlantic. Phil. Trans. Roy. Soc. London Ser. A Math. Phy. Sci. 258:27-41.

Cox, A., ed. 1973. Plate Tectonics and Geomagnetic Reversals. San Francisco: W. H. Freeman and Company.

Cox, A., G. B. Dalrymple, and R. R. Doell. 1967. "Reversals of the Earth's Magnetic Field." Sci. Amer. 216(2):44-54.

Dewey, J. F. 1972. "Plate Tectonics." Sci. Amer. 226(5):56-68 (Offprint No. 900).

Dickinson, W. R. 1971. "Plate Tectonics in Geologic History." Science 174:107-13.

Dietz, R. S., and J. C. Holden. 1970. "The Breakup of Pangaea." Sci. Amer. 223(4):30-41.

Heirtzler, J. R. 1968. "Sea-floor Spreading." Sci. Amer. 219(6):60-70 (Offprint No. 875).

Hess, H. H. 1962. History of Ocean Basins. In A. E. J. Engel, et al., eds. Petrologic Studies: A Volume in Honor of A. F. Buddington, pp. 599-620. Geological Society of America.

Hurley, P. M. 1968. "The Confirmation of Continental Drift." Sci. Amer. 218(4):52-64 (Offprint No. 874).

Isacks, B., J. Oliver, and L. R. Sykes. 1968. "Seismology and the New Global Tectonics." J. Geophys. Res. 73:5855-99.

Le Pichon, X. 1968. Sea-floor Spreading and Continental Drift. J. Geophys. Res. 73:3661-3697.

Marvin, U. B. 1973. Continental Drift. Washington, D.C.: The Smithsonian Institution.

McKenzie, D. P. 1972. "Plate Tectonics and Sea Floor Spreading," Amer. Scientist 60:425-35.

Takeuchi, H., S. Uyeda, and H. Kanamori. 1970. Debate about the Earth. San Francisco: Freeman, Cooper and Company.

Vine, F. J. 1966. Spreading of the Ocean Floor: New Evidence. Science 154 (3744):1405-1415.

Vine, F. J., and D. H. Matthews. 1963. Magnetic Anomalies over Oceanic Ridges. Nature 199:947-949.

Wilson, J. Tuzo (compiler). 1970. Continents Adrift: Readings from Scientific American. San Francisco: W. H. Freeman and Company.

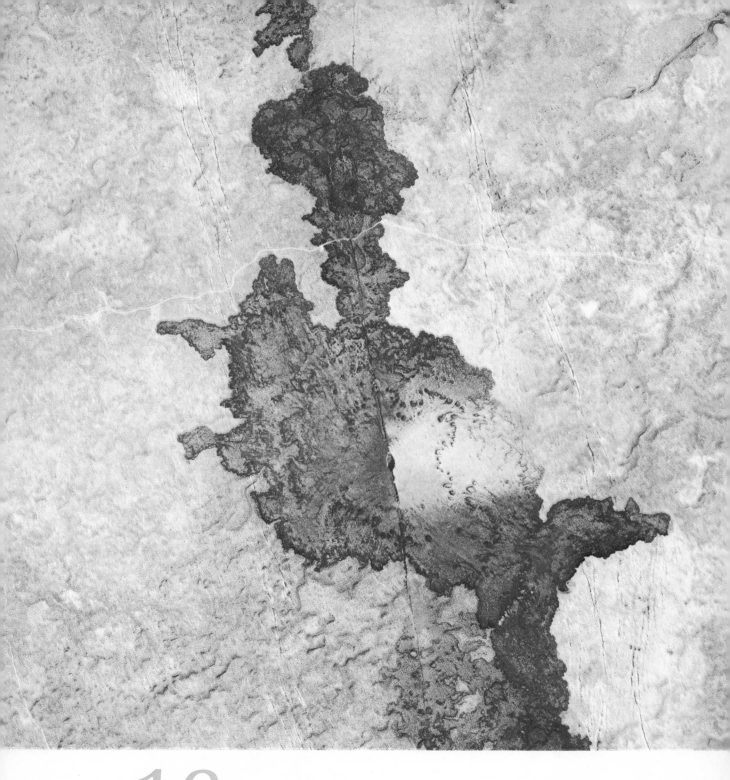

16 The Oceanic Crust

During the past two decades, while space exploration has held the focus of scientific attention, oceanographers have been at work quietly mapping the topography of the ocean floors and studying the nature of the earth's crust beneath the sea. Through this work, in which many nations have participated, a great amount of bathymetric data has been compiled into regional physiographic maps showing in perspective the surface features of the ocean floor. Although man will never see the regional landscape of the sea floor with his own eyes, remote-sensing devices provide him with a picture of the submarine topography just as accurate and true as preliminary topographic maps of the land. The results are fascinating and truly spectacular. A great mountain range, comparable to the most impressive mountain ranges on the continents, extends as a continuous unit over a distance of nearly 64,000 km (40,000 mi). Volcanic seamounts rise as isolated peaks above the surrounding ocean floor. Great fracture zones mark areas of active displacement of the crust, and deep trenches extend more than 9000 m (30,000 ft) below sea level. As we have seen, this knowledge has revolutionized geologic thinking because for the first time it permits us to analyze the regional features of the ocean floors and to relate them to geologic features of the continents. Thus, we are able to consider for the first time all the parts of the planet. No amount of research on the continents alone could reveal what we now know from studies of both the ocean floor and continents.

Major Concepts

1. The oceanic crust is composed of basalt and a thin veneer of marine sediment. It is much thinner than the continental crust and is not deformed by folding and metamorphism. In comparison with the continents it is very young, apparently having been created largely since Cretaceous time.

2. The mid-oceanic ridge is a broad swell offset by great fracture systems in which horizontal displacement has been thousands of kilometers. A rift valley near the crest of the ridge marks the site of the greatest seismic activity and greatest heat flow. It also marks the site of the youngest parts of the oceanic crust. Partial melting of the upper mantle produces basaltic magma which is extruded along the rift valley to form new oceanic crust.

3. Trenches and island arcs result where two plates meet and one is forced to subside into the mantle. Partial melting of the descending plate generates andesitic or granitic magma.

4. The abyssal floor consists of abyssal hills adjacent to the mid-oceanic ridge and the abyssal plains adjacent to the continents.

5. Volcanic activity over local "hot spots" in the upper mantle produce great **shield volcanoes** which form chains of seamounts, guyots, and islands as the plate moves over the source of magma.

6. The continental margins are submerged and form a smooth continental shelf and a steep continental slope which is dissected by submarine canyons.

The Composition and Structure of the Oceanic Crust

Statement

The difference between the oceanic and continental crust is a reflection of two fundamentally different processes involved in differentiation of mantle material and concentration of minerals with low density in the outer zone of the solid earth. The oceanic crust consists of a relatively thin layer of basalt 5 to 8 km (3 to 5 mi) thick overlain by a veneer of marine sediments averaging 0.5 km (0.3 mi) thick. The oceanic crust is not deformed by folding, thrusting, or regional metamorphism. Therefore, the oceanic crust is young, most of it having been created during the past 50 million years. This difference is of fundamental importance in understanding the earth and its dynamics.

Discussion

Questions concerning the nature and origin of ocean basins remained largely unanswered until the 1950s, when many nations made a concentrated effort to map the topography of the sea floor and study its seismicity, magnetism, heat flow, and structure. Although we cannot study and map the oceanic crust by direct observation, as we can the surface of the continents, by a variety of remote-sensing techniques we can determine much about its nature. The principal methods used are (1) precision depth recording, by means of which the topography of the ocean floor can be mapped, (2) seismic studies, from which we can determine the structure and characteristics of the rocks of the crust beneath the oceans, and (3) magnetic survey and heat flow studies, which reveal certain physical properties of the crust and upper mantle.

Three layers of regional extent can be recognized in all ocean basins (*figure 16.1*). Layer 1 is a veneer of soft marine sediment (0.5 to 1 km [0.3 to 0.6 mi] thick) which buries many irregularities in the surface of the bedrock beneath and in places produces a smooth, featureless abyssal plain. The sedimentary layer is thickest near the continental margins and thins toward the crest of the mid-oceanic ridge, where it is completely absent or occurs only in thin, small patches. Seismic velocities of the first layer vary between 1.5 and 3.4 km (.9 to 2.1 mi) per second. Layer 2 is composed of basalt (samples have been obtained from it), but it may also contain some discontinuous lenses of consolidated sedimentary rocks. Layer 3 is commonly called the oceanic layer, which, according to seismic evidence, is entirely basalt. It is 4.7 km (2.9 mi) thick and has an average seismic velocity of 6.8 km (4.2 mi) per second. The mantle, with a seismic velocity of 8.1 km (5 mi) per second, occurs at a uniform depth below the oceanic layer.

The crust in the region of the mid-oceanic ridge differs with respect to its seismic velocity and the depth to the upper mantle from that in the rest of the ocean basin. As is shown in *figure*

The Oceanic Crust

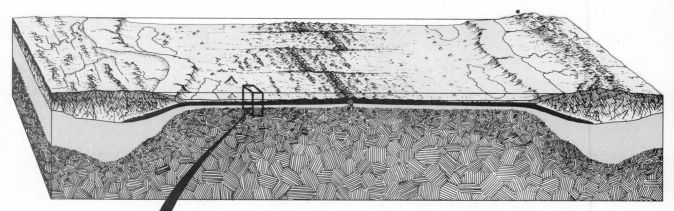

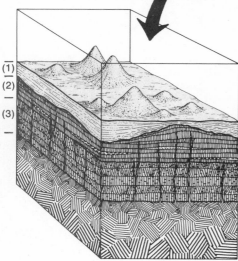

Figure 16.1 *Schematic cross section across the Atlantic Ocean showing the major units of the crust based upon P wave velocities. The oceanic crust is much simpler than that of the continents. The main rock type is basalt which has not been folded nor compressed. Three layers of regional extent are recognized: (1) a veneer of soft sediment 0.5 to 1.0 km thick, (2) a basalt layer with discontinuous lenses of sediment, and (3) an oceanic layer 4.7 km thick, composed entirely of basalt.*

16.1, layer 2 directly overlies the mantle material which extends through layer 3. This unique structural feature of the oceanic crust appears to represent mantle material rising beneath the axis of the ridge.

One of the most significant facts about the oceanic crust is that the samples thus far brought to the surface are all much younger than the rocks which form the bulk of the continents. The oldest known basalts retrieved from the ocean floor are only 50 million years old, and there is little reason to suspect that any basalt on the ocean floor is older than 160 million years. Extensive drilling in recent years has produced sediments ranging only from 125 to 150 million years old.

The age of ocean-floor basalts is related to their position with respect to the mid-oceanic ridge. The youngest basalts are along the crest of the ridge (except for certain island chains like Hawaii) and, away from the ridge, they become progressively older. This fact stands in marked contrast to the continental shields, where rocks range from 1 to 3.5 billion years old and generally become progressively younger toward the continental margin.

The Oceanic Ridges

Statement

The mid-oceanic ridge (*figure 16.2*) is the most pronounced tectonic feature on earth and, if it were not covered with water, it would certainly be visible from observation points on the moon. We have seen in previous chapters that the ridge is a broad fractured swell in which the highest and most rugged topography is located along its axis. The crust of the ridge is marked by a rift valley which extends throughout most of its length. The rift valley is the location of a zone of shallow earthquakes and high heat flow. The most extensive vulcanism on earth occurs along the oceanic ridges where new crust is created by fissure eruptions. This process is responsible for almost all of the crust beneath the world's oceans, but it is largely invisible and unspectacular because it occurs mostly below sea level and

The Earth's Dynamic Systems

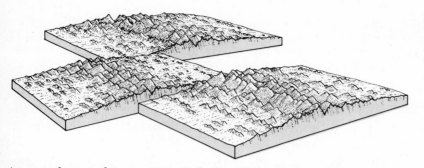

Figure 16.2 *The mid-oceanic ridge is a broad fractured swell offset by great fractures. It marks the site where new crust is generated from the upper mantle.*

is not observed in action. Only locally, where submarine volcanoes build up above sea level (e.g., the recent extrusion of the island Surtsey near Iceland and the older islands of the Azores and St. Helena) is the volcanic activity of the oceanic ridge available for direct observation. Apparently, fissure eruptions along spreading centers are continuous, in a geologic sense, and have been so throughout most of geologic time. It is the magma extruded at the ridge which forms new crust as lava is injected into the tensional fractures along the rift zone or is extruded to form submarine flows and volcanic cones. The origin of the ridge is therefore involved with the origin of basaltic magma.

Discussion

The nature of magma where it originates in the upper mantle is, of course, not known from direct observations, so our understanding of the chemistry and physics of liquid rocks is based on studies of igneous rocks, volcanic products, and synthetic magmas made in the laboratory. Although much remains uncertain, we do know something about the boundaries or limitations under which magma is generated and about some aspects of the mechanism by which it is emplaced in the earth's crust.

Seismic evidence tells us that the crust and the mantle are solid bodies because of the way they transmit S waves (seismic waves which will not go through a liquid). This evidence denies the existence of a permanent worldwide reservoir of magma beneath the solid crust or lithosphere. The outer core of the earth responds as a liquid to seismic waves, but density calculations indicate that the material of the core is much too high in iron and magnesium and, thus, far too dense to produce any magma that forms igneous rocks in the crust. The only source of material with a composition sufficient to produce a magma is the upper mantle. We believe that magma is produced by partial melting of the upper mantle where convection brings hot mantle material toward the surface.

Partial Melting and Composition of Magma. An important fact in considering the origin and composition of magma is that each mineral in a rock has its own melting point. Therefore, when a rock is heated, certain minerals melt first and others follow in a definite sequence. As a result, partial melting may produce a magma quite different in composition than the original rock. Consider, for example, the results of partial melting

The Oceanic Crust

359

of a basalt which is composed of olivine, pyroxene, and plagioclase. As temperature increases to about 800° C, the Na plagioclase melts first, then the pyroxene at 1000° C; olivine does not melt until temperatures of more than 1100° C are reached. Thus, if a basalt were heated to about 1000° C, only part of the minerals would melt and the resulting magma would be rich in the constituent elements of sodium plagioclase and pyroxene. By this process the magma would be enriched in silica and, if cooled, would produce a rock of andesitic rather than basaltic composition.

Various models of partial melting are considered to be very important in plate tectonics. Partial melting of peridotite (a rock from the upper mantle rich in iron and magnesium) is believed to be important in the origin of basalt extruded along the mid-oceanic ridge.

The heat necessary to generate a magma probably comes from radioactivity in the upper mantle. During the early history of the earth, elements were concentrated according to density, the heavy elements gathering in the core and lighter elements in the mantle and crust. Apparently, certain radioactive elements were concentrated in silicate minerals in the asthenosphere and have produced radiogenic heat ever since.

Where the rise in temperature exceeds the melting point of a given mineral, a change in state occurs and the mineral in the liquid phase begins to move upward. Thus, rocks in the upper mantle are partly melted and a magma is generated. It then moves upward and is extruded at the rift system in the oceanic ridge.

Plateau Basalts. Where the rift system passes beneath a continent, the continental crust is split and great volumes of basalt are extruded and spread out over large areas of the continent. These great floods of lava fill lowlands and depressions in the existing topography and, with time, are eroded into basalt plateaus (figure 16.3). These deposits are, therefore, referred to as **flood basalts** or **plateau basalts.** For example, in southern Brazil over a million cubic kilometers of basalt were extruded in a relatively short period of geologic time (10 million years). Similar floods have occurred in the Deccan Plateau of India, the Ethiopian Plateau of Africa, the Columbia River Plateau of the western United States, as well as in large areas of Siberia, Greenland, Antarctica, and northern Ireland. Much older flood basalts are found in northern Michigan, Quebec, the piedmont of the eastern United States, and many other areas.

Plateau basalts are believed to be one evidence of the initial phases of continental drift. Evidence for this can be seen in figure 16.3, a map of the southern continents prior to continental drift, which began in late Mesozoic time. The plateau basalts lie along the present-day continental margins but were originally extruded along the incipient rift system which later developed into the oceanic ridge. The physiographic map shows that the basalts of the Columbia River Plateau lie along the northward

Figure 16.3 *The major plateau basalts are located near the continental margins and are believed to represent extrusions associated with the initial phases of continental rifting.*

The Earth's Dynamic Systems

extension of the east Pacific ridge and may well represent the initial stages of the breakup of the western part of North America.

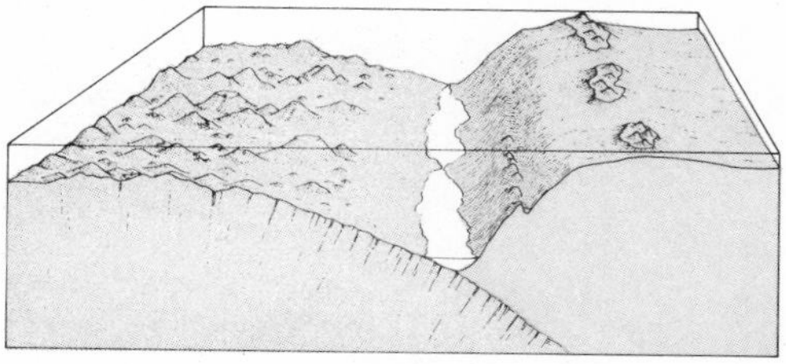

Trenches, Island Arcs, and Disappearing Ocean Floor

Statement

The deep oceanic trenches and associated island arcs are the most active geologic features on the surface of the earth. They are intimately associated with the most intensive active vulcanism, are parallel to the zones of deep-seated earthquakes, and are directly related to the largest gravity anomalies on earth. It is not surprising, therefore, that the trenches have attracted considerable scientific interest, for with their varied manifestations of tectonic activity they have long been considered to be a surface expression of a fundamental process of earth dynamics (figure 16.4).

According to the plate tectonic theory (see figure 15.16), trenches result where two lithospheric plates meet and one is forced to subside back into the mantle (figure 16.4). The descending plate is believed to be inclined at an angle of about 45°. As it moves downward, it creates a zone of earthquakes by sliding past the adjacent rocks. As the descending plate moves into the asthenosphere, it is partly melted by heat from the hotter mantle and from friction. The magma resulting from partial melting of basaltic oceanic crust would be richer in silica and would produce andesite, the rocks typical of **island arcs** and young mountain belts.

Discussion

To explain what is happening as the plate descends down the trenches and under the island arcs, we must say more about the seismicity of this zone. Accurate location of foci of earthquakes in the trench-island arc system shows that the most shallow earthquakes occur near the landward margins of the trench and that the deeper earthquakes occur in a narrow zone sloping at an angle of 45° under the adjacent island arc or continent (figure 16.5). This inclined zone of earthquakes dipping away from the trench goes down as far as 700 km (435 mi) into the mantle and has been known for many years. The zone is named the **Benioff seismic zone** after the seismologist who discovered it.

A relatively recent development in seismology has added to our knowledge of plate movement in the trenches. By studying the direction of the first displacements of the rocks by the P waves arriving at different seismograph stations, seismologists can determine the direction of the stresses in the lithosphere and upper mantle which caused the earthquakes. This type of study confirms the conclusion that the shallow earthquakes in the mid-oceanic ridge are caused by tension and that, under the

Figure 16.4 The deep-sea trenches mark the site of subduction zones where lithospheric plates collide and one is moved under the other and back into the mantle. They are the lowest parts of the earth's surface, associated with zones of deep earthquakes, and are adjacent to intense volcanic activity.

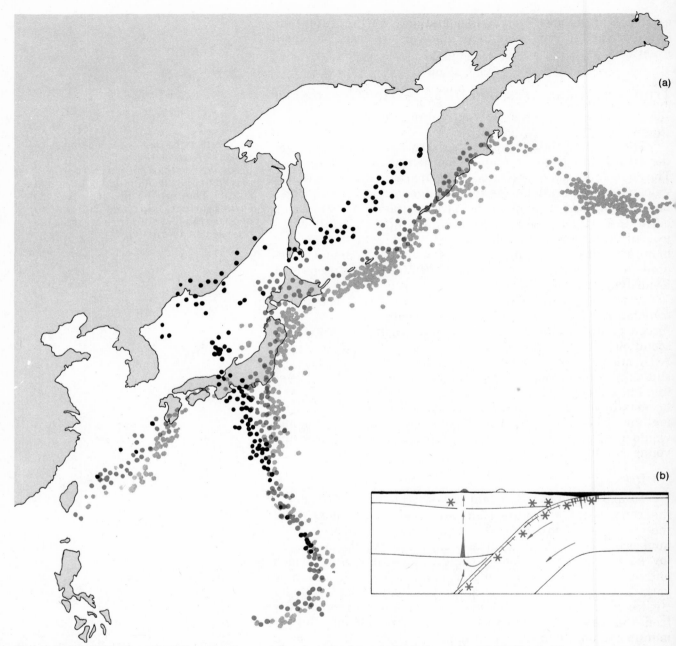

Figure 16.5 (a) Location of earthquakes in the Japanese trench area, 1962-1969. The earthquakes are shallow near the trench but become progressively deeper toward the continent. Gray dots mark shallow earthquakes, black intermediate, and brown deep. (b) Cross section across the trench and island arc showing the seismicity produced along a descending plate.

island arcs, the earthquakes along the Benioff zone are caused by rigid blocks sliding against each other.

It is this type of seismic evidence which confirms so nicely the theory of plate tectonics. The ocean plate is being forced below the island arc down into the asthenosphere.

As the plate descends along the Benioff zone, it encounters progressively higher temperatures at depth. Here, a magma quite different from the basaltic magma characteristic of the extrusions at the mid-oceanic ridge is created. The process, however, is somewhat similar as both result from partial melting. As the oceanic plate is heated, low-temperature minerals begin to melt first. In the basaltic oceanic crust, the first minerals to melt are the sodium-rich plagioclase and amphibole and the sediments. The resulting magma is richer in silica than basalt and, when it cools, it produces an andesitic or granitic rock. Excellent examples are the andesitic volcanoes of the western Pacific, Aleutian Islands, and the Andes Mountains of South America.

The partial melting of basalt and the rise of relatively-low-density silica-rich magma are part of the process by which materials of the earth are differentiated and segregated. We will have more to say about this in the next chapter because the igneous activity associated with the descending plates is a major process by which continents grow and acquire their average composition.

The Abyssal Floor

Statement

Vast areas of the deep ocean consist of broad, relatively smooth surfaces known as the abyssal floor. This type of sea-floor topography was discovered in 1947 by oceanographic expeditions surveying the mid-Atlantic ridge and has subsequently been mapped in detail, using precision depth recorders capable of measuring elevations on the ocean floor with relief as little as 2 m. As we have seen in Chapter 1, the abyssal floor may be divided into two parts, the abyssal hills and the abyssal plains (figure 16.6). You will note from the physiographic map that abyssal hills are located near the flanks of the oceanic ridge and the abyssal plains are located near the continental margins.

The abyssal hills are thought to form from laccolithic intrusions and submarine volcanic activity near the crest of the mid-oceanic ridge. As the plates move away from the ridge, the hills migrate with the moving oceanic crust in the direction of sea-floor spreading and become progressively buried by sediment derived from the continent. Abyssal plains form adjacent to the continents where the abyssal hills are completely buried with sediment.

Discussion

The distribution of abyssal plains and abyssal hills throughout the oceans further substantiates the above explanation of their origin. As shown on the physiographic map, abyssal plains occur only where the topography of the sea floor does not inhibit turbidity currents from spreading out from the continents over the sea floor. Abyssal plains are widespread throughout the Atlantic Ocean and occur in a line on the flat part of the ocean floor at the base of the continents of North America, South America, Africa, and Europe. By contrast, in the Pacific Ocean, there are few abyssal plains because turbidity currents cannot

Figure 16.6 The abyssal floor of the ocean contains two types of landforms. The abyssal hills are believed to be submarine volcanic mounds or laccolithic intrusions. Near the continents the abyssal hills are commonly buried with sediment which forms the flat, featureless abyssal plains.

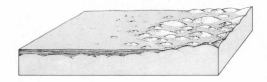

flow past the deep trenches which occur along much of the continental margins. The trenches trap the inflowing sediment so most of the Pacific floor lacks abyssal plains and is covered with abyssal hills instead. The largest abyssal plains in the Pacific are found in the northeast off the coast of Alaska and western Canada. This is the only significant segment of a continental mass bordering the Pacific which is not marked by deep trenches. In the Indian Ocean, abyssal plains occur only along the margins of Africa and India and are absent in the eastern part of the ocean where the deep Java trench along the continental margin acts as a sediment trap. Another major area of abyssal plains lies off the northern shore of Antarctica. Between the Aleutian Trench and the continental shelf, the Bering Abyssal Plain covers most of the deep-sea floor north of the Aleutian Islands. This deep basin is underlain by an abnormally thick section of layered sediment which, in places, may be as thick as 2 km (1.2 mi). As the physiographic map shows, the Aleutian Ridge has cut off this corner of the Pacific basin and has acted as a dam behind which sediments have accumulated rapidly.

Volcanic Islands

Statement

Volcanic islands and seamounts are found in all oceanic basins but not upon the continental shelves. Some obviously originated as extrusions at the oceanic ridge; as the plate drifted away from the rift system where magma is generated, the volcanoes moved away from their source of lava and became extinct. Other seamounts occur in chains removed from the rift and probably originated from local "hot spots" in the mantle.

Discussion

The islands and seamounts of the ocean (figure 16.7) are great volcanic mountains called shield volcanoes built up from the ocean floor by repeated eruptions of basaltic lava. A shield volcano is the largest type of volcanic cone. Unlike other types characterized by steep flanks, shield volcanoes have very gentle slopes (2 to 5°) and are built of innumerable basaltic lava flows which are extruded from a central caldera or through fissure eruptions along the flanks of the cone.

The island of Hawaii, the largest and youngest island in the Hawaiian chain is our best-known example. It consists of five coalescing volcanic mountains which have grown in less than a million years. The island rises 4000 m (13,000 ft) above sea level and more than 9000 m (30,000 ft) above the sea floor. Almost the entire volume, which may be as much as 42,000 km^3 (10,000 mi^3), is made up of separate lava flows, each consisting of less than 0.4 km^3 (0.1 mi^3). Most of the lava is extruded from a vertical fracture system which reflects major weaknesses in the foundation of the volcano and underlying oceanic crust.

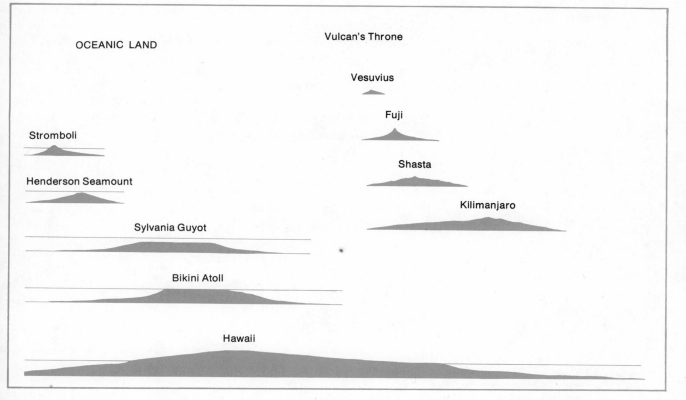

OCEANIC LAND

Vulcan's Throne

Vesuvius

Fuji

Stromboli

Shasta

Henderson Seamount

Kilimanjaro

Sylvania Guyot

Bikini Atoll

Hawaii

Guyots are flat-topped seamounts, some of which may be capped by thousands of meters of coral reef. They range in depth from 180 to 4300 m (600 to 14,000 ft). Guyots originally grew to be islands on the uparched segment of the sea floor near the crest of the oceanic ridge. Here, wave action would be able to erode the top of the island to form a flat surface, and coral reefs could grow on this platform when the island was near sea level. As sea-floor spreading continued, the flat-top island would migrate off the ridge and become submerged. If environmental factors were satisfactory, coral reefs would continue to grow upward as the seamount moved off the flanks of the ridge into deeper water. Thus, atolls such as the Bikini Atoll could develop a reef rock over a thousand meters thick by continued upward growth of corals as the island slowly moved into deeper water.

One of the most striking characteristics of islands and seamounts is that they occur in remarkable chains, stretching for more than 6000 km (3700 mi) across the ocean floor. The Hawaiian archipelago and the Emperor seamounts in the western Pacific are good examples.

The theory of plate tectonics explains the development of guyots, thick reefs, and chains of seamounts by movement of the sea floor over a "hot spot" in the upper mantle where a huge column of upwelling lava known as a "plume" lies in a fixed position under the lithosphere. Vulcanism occurs over the "hot

Figure 16.7 *Volcanic islands and seamounts are huge shield volcanoes built up by extensive extrusion of basaltic lava and are much larger than the volcanic cones built on continents. Flat-topped seamounts called guyots have been eroded and subsequently submerged.*

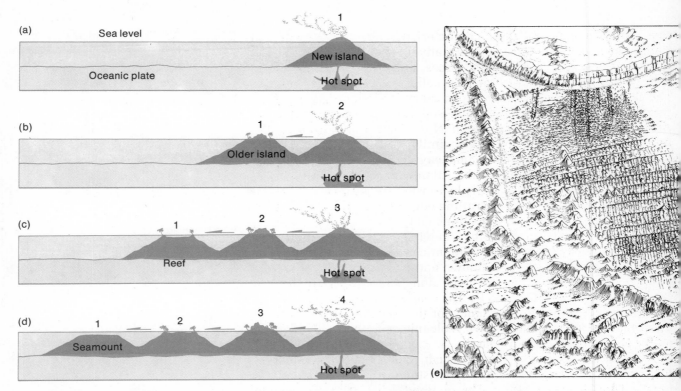

Figure 16.8 *The origin of linear chains of volcanic islands and seamounts according to the plate tectonic theory. (a) A volcanic island or seamount is built up by extrusions from a fixed "hot spot" or source of magma in the mantle. (b) As the plate moves, the volcano passes over the source of magma and becomes dormant. The surface of the island may then be eroded to sea level, and reefs may grow to form an atoll. A new island is formed over the "hot spot." (c) Continued plate movement produces a chain of islands. (d) The chain of islands become progressively older away from the "hot spot." (e) An abrupt change in the direction of plate movement is indicated by the change in direction of the chain of islands in the Pacific. The Emperor Seamount Chain began to form more than 40 million years ago when the plate was moving northward. About 25 million years ago, the plate moved northwestward and started to form the Midway-Hawaiian Chain.*

spot" and a seamount or island grows. As the sea floor moves at a rate of approximately 10 cm per year, the volcano moves past the "hot spot" and becomes extinct, and a new volcano is created in its former position (figure 16.8). In this manner, a steady succession of new volcanoes is created; these in turn migrate with the plate, forming a straight line from the "hot spot" toward the trench. Supporting this theory, the westernmost seamounts in the Pacific (farthest from the "hot spot") show the greatest age both from the standpoint of erosional development and radioactive age determination on samples of bedrock.

Continental Margins

Statement

The abrupt termination of the continents is a major structural and topographic feature of our planet and is of fundamental importance in understanding the origin and evolution of the continents and ocean basins. According to the theory of plate tectonics, the continental slopes originate by rupture as the continents are split and pulled apart. Normal faulting plays an important role in their early development and, with time, the slope or scarp is modified by erosion and sedimentation.

Discussion

According to drift theory, two types of continental slopes may be recognized; each is subjected to different stresses and, as a result, develops different characteristics. The slopes on the trailing edge of the continent are relatively passive and are not subjected to strong compressive forces, unless the direction of drift changes and the rifted margins collide with another plate. The slopes of the Red Sea provide an example of a newly formed continental margin caused by the separation of Africa from Arabia. The continental slopes of North and South America bordering the Atlantic Ocean are similar and are considered to have originated in much the same way (figure 16.9). The major difference is that the slopes of the Atlantic are much older and have been modified considerably by erosion and sedimentation.

In contrast to the passive trailing edge of the continent, the **leading edge** of the continent is subjected to much more stress as it commonly encounters another moving plate. Deformation generally results. Where the continent collides with an oceanic plate, the oceanic plate moves down and under the lighter continental crust, and a trench develops along the continental margin. Such is the case along the Pacific coast of South and Central America. Here, there is a tendency for the continental slopes to be straight and seismically active. Volcanic activity is prominent on the adjacent continental mass.

Submarine Canyons. Submarine canyons are common along the continental slope and have been studied for many years, long before the ocean floor was mapped. The large canyons with V-shaped profiles and a system of tributaries closely resemble the great canyons cut by rivers on the continents (figure 16.10). Many pioneer researchers, therefore, suggested that the canyons were cut by rivers, but the problem of *how* remained unanswered. Some canyons are 3000 m (10,000 ft) below sea level,

Figure 16.9 *The Atlantic continental slope is typical of the slopes on the trailing edges of continents. It is mantled with sediment and dissected by submarine canyons.*

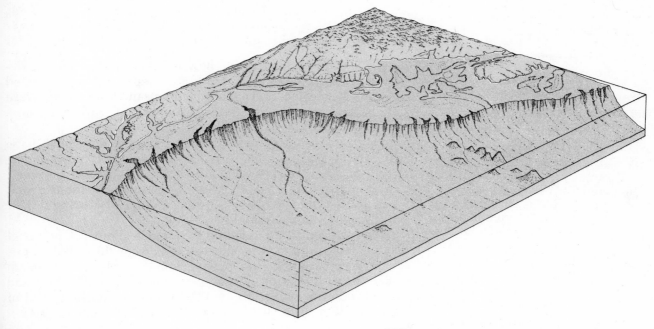

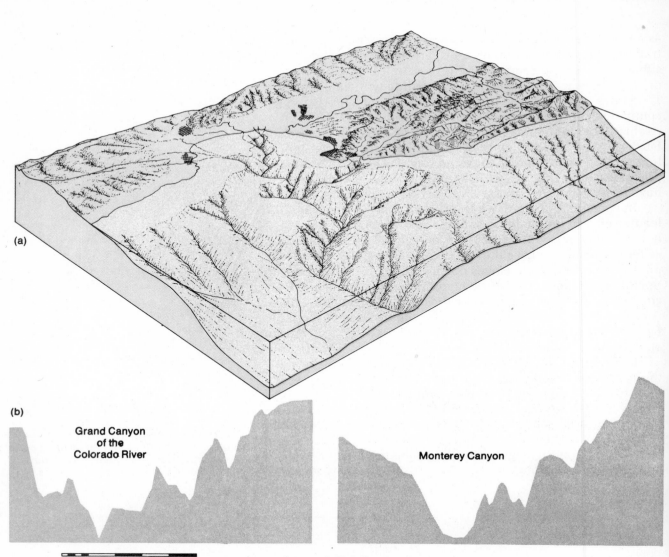

(a)

(b)

Grand Canyon
of the
Colorado River

Monterey Canyon

0 1 2 3 4 5
Scale in miles
Vertical exaggeration x5

*Figure 16.10 (a) The Monterey submarine
canyon off the coast of southern California
shows many characteristics of canyons
cut by rivers. (Constructed from
bathymetric charts.) (b) A profile across
the Monterey canyon is similar to a
profile across the Grand Canyon
constructed with the same number of
control points.*

and it is difficult to conceive how sea level changed to the degree necessary for rivers to cut to that depth. Others suggested
that the continents were uplifted thousands of meters, were
dissected by canyons, and then subsided so the canyons became
submerged.

More recently, submarine canyons were thought to have
been eroded by turbidity currents, but it is not known if turbidity currents are capable of such erosion. It is clear, however,
that much sediment has flowed through many of the canyons,
because of the great volumes of sand and mud occurring at their
lower ends as deep-sea fans.

The theory of plate tectonics could explain the origin of
canyons as a product of stream erosion during the initial phases
of continental rifting. When convection currents occur beneath
a continent, they arch up the overlying continental crust and
begin to pull it apart. The initial response is the formation of
rift valleys such as those in East Africa. The sides of the rift

The Earth's Dynamic Systems

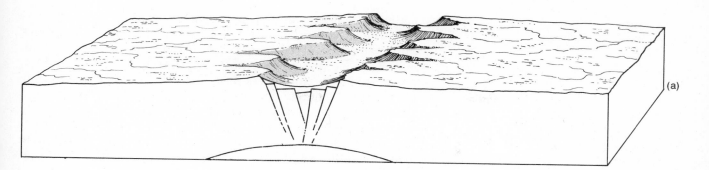

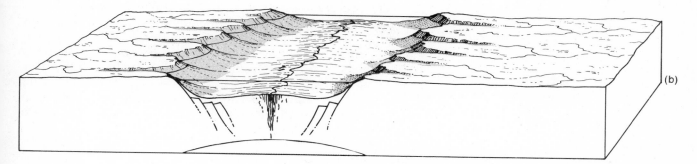

valleys are soon attacked by erosion, and many canyons and ravines are cut into them (figure 16.11). As the continents separate completely and move off the uparched spreading center, the margins become submerged and the canyons drowned. They then act as channels for turbidity flows and could be modified by them.

Figure 16.11 *Formation of submarine canyons by continental erosion along the initial rift zone. (a) The continent is arched and pulled apart by convecting currents in the asthenosphere. Erosion cuts canyons into the sides of the rift valley, the future continent's margins. (b) Continents move off the uparched area, and the margins become partly submerged. The canyons cut by erosion on the sides of the rift valley become submarine canyons. (c) As the continents drift further from the spreading center, the canyons are modified by activity of turbidity currents.*

Summary

The oceanic crust consists of a relatively thin layer of basalt (5 to 8 km thick) overlain by a veneer of marine sediment. It has not been deformed by folding or thrusting or by regional metamorphism and is, therefore, distinctly different from continental crust. The oceanic crust is very young, most having been created during the last 50 million years. According to the plate tectonic theory, the oceanic crust is continually being created by igneous activity along the mid-oceanic ridge, spreads outward away from the ridge, and is ultimately consumed as it descends back into the mantle at the trenches. Abyssal hills form at the mid-oceanic ridge and move with the spreading plates away from the central rift valley. As they move, they become progressively buried with sediment and may be completely covered to form a smooth abyssal plain. Volcanic islands and seamounts originate from extrusions near the rift valley or from local "hot spots" in the mantle beneath the moving plate. Movement of the plate over the "hot spot" produces linear chains of islands and seamounts.

The process of sea-floor spreading and the growth of oceanic basins is part of the more fundamental system by which our planet is being differentiated; the basaltic oceanic crust is separated from the denser peridotite mantle by partial melting. The origin and growth of continents is also involved with differentiation, but the process is quite different, involving partial melting of the oceanic crust, mountain-building, metamorphism, and sedimentary differentiation. We will consider these processes in the next chapter.

Additional Readings

Bullard, E. 1969. "The Origin of the Oceans." Sci. Amer. 221(3):66-75 (Offprint No. 880).

Heezen, B. C. 1960. The Rift in the Ocean Floor. Sci. Amer. 203(4): 98-110.

Heirtzler, J. R. 1968. Sea-floor Spreading. Sci. Amer. 219(6):60-70 (Offprint No. 875).

Keen, M. J. 1968. An Introduction to Marine Geology. Oxford: Pergamon Press.

Menard, H. W. 1969a. "The Deep-Ocean Floor." Sci. Amer. 221(3):126-142 (Offprint No. 883).

———. 1969b. Growth of Drifting Volcanoes. J. Geophys. Res. 74(20): 4827-4837.

Moore J. R., ed. 1971. Oceanography: Readings from Scientific American. San Francisco: W. H. Freeman and Company.

The Ocean. 1969. A Scientific American Book. San Francisco: W. H. Freeman and Company.

Phinney, R. A., ed. 1968. The History of the Earth's Crust. Princeton, N.J.: Princeton University Press.

17 Mountain-Building, Metamorphism, and the Origin of Continents

Although the theory of plate tectonics was developed largely from data obtained from studies of the oceanic crust, a record of plate motion in most of the geologic past is preserved only in the rocks of the continents. The oceanic crust is continually being created along the oceanic ridge and destroyed along the subduction zones. Plate movement can be studied and measured in the oceanic rocks today, but all of the oceanic crust is younger than about 165 million years.

By contrast, a record of ancient plate movement is preserved in the metamorphic rocks of the continental shields. Here geologists find a long and complex history of earth dynamics which can best be explained in the framework of the plate tectonic theory. In this chapter, we will consider how new continental crust is created by a combination of deformation, metamorphism, igneous activity, erosion, sedimentation, and isostatic adjustment. Through plate movement, continents grow by accretion and may also split and drift apart. All of these processes of continental evolution are involved with the fundamental process of segregating and differentiating earth materials according to density.

Major Concepts

1. Mountain ranges are long, narrow belts of folded geosynclinal sediments. Metamorphism and igneous activity occur in the deeper parts of the deformed belt.
2. Erosion with subsequent uplift by isostatic adjustment follow mountain-building so that the deformed belt is ultimately eroded to sea level.
3. The continental shields consist of a series of former mountain belts "welded" together and eroded to sea level. Older deformed belts are near the central part of the shields, and younger mountain ranges are along the continental margins.
4. The theory of plate tectonics provides an explanation for the origin and evolution of continents. Continental growth occurs as **geosynclines** are deformed into mountain ranges which ultimately become part of the shield.

Mountain Belts

Statement

The great linear belts of folded mountains which typically occur along the margins of continents constitute one of the most distinctive tectonic features of our planet. Within these relatively narrow zones, there are many indications that rocks have been deformed by shortening of the earth's crust. Sedimentary rocks which were originally horizontal are now tightly folded, and in many places they are broken and displaced by fractures, with one block thrust over another. In the deeper parts of mountain ranges, now exposed by erosion, it is clear that heat and pressure have been so great that the rocks have crystallized or melted.

For over 150 years, geologists have studied mountain ranges and have mapped the folds, faults, and other structural features in great detail. These studies reveal that in spite of the almost infinite variety of style in topographic form, the following basic structural features are common to the earth's mountain belts.

1. Mountains originate along the margins of continents and are associated with zones of earthquake and volcanic activity.
2. The rocks have been folded and displaced by thrust faults.
3. The sequence of sedimentary rocks in a mountain belt is typically eight to ten times as thick as rocks of equivalent age in the continental interior.
4. Deformation of a mountain belt involves the entire crust, not just the surface layers. A "root" of the mountain, therefore, extends down into the mantle.
5. Intrusion of batholiths and regional metamorphism occur in the deeper parts of the folded belt.
6. Mountain-building has occurred throughout geologic time.

The above observations indicate that folded mountain belts represent the results of a fundamental process operating in the crust and upper mantle.

Discussion

Geologically, a mountain belt is a linear zone in the earth's crust which has been highly deformed by great horizontal forces. In the early stages of its formation, it forms high, rugged ranges, but these are soon eroded down to lowlands and may be smoothed off to form a nearly featureless plain. However, the deformation which characterizes a mountain range involves the whole crust, not just the upper layers which form the highland; therefore, the structure of a mountain range can be studied in the roots of older mountains which now form lowlands. Volcanic peaks and eroded plateaus may form high, rugged topography but, if the rocks are not deformed by compression, they are not considered a mountain belt in the strict geologic sense.

Location. The young active mountain belts on the earth today occur in two long, narrow belts: (1) the Cordilleran belt, which includes the Rockies and Andes, and (2) the Himalayas-

Alpine belt, which extends across Asia, western Europe, and into North Africa. Older mountain ranges, in which deformation has ceased long ago but is still expressed by significant topographic relief, are the Appalachian Mountains of the eastern United States, the mountains of eastern Australia, and the Ural Mountains of Russia.

The location of mountains in long, narrow belts along continental margins is significant because it indicates that they could not result from a uniform worldwide force evenly distributed over the earth. Mountains must be the result of forces concentrated along the margins of continents.

Another important aspect of a number of mountain ranges is that many of them extend to the ocean and abruptly terminate at the continental margin. The northern Appalachians, the Atlas Mountains of Africa, and the mountains of Great Britain are excellent examples. In these systems, the abrupt termination of the folded structures suggests that the mountain systems were once much more continuous and have been separated by continental drift.

Deformation of Mountain Belts. The single most characteristic feature of active mountain chains is the structural deformation of the rocks. The scale of deformation ranges from small deformed grains or fossils in the rock to large-scale folds and faults tens of kilometers wide. In most mountain ranges, the smaller-scale structures, such as the folds seen in *figure 17.1*, are visible and easy to recognize. Larger structural features are

Figure 17.1 *Folded sedimentary rocks constitute a fundamental structure in mountain belts. A fold may range from a small wrinkle to huge structures extending for hundreds of kilometers. Folds in mountain belts clearly indicate compressive forces in the earth's crust. This exposure is in the Wasatch Mountains.*

The Earth's Dynamic Systems

more difficult to visualize, but with the aid of an aerial photograph (figure 17.2) you can begin to grasp the style and magnitude of deformation involved in a mountain belt.

Figure 17.3 shows part of the Appalachian Mountains and the typical surface expression of folded rocks in a mountain belt. The ridges are resistant formations which form a zigzag pattern across the region, similar to that in the mountains seen in the satellite photograph in figure 1.6. In both areas the folded rocks were originally similar to the idealized diagram shown in figure 17.4. In diagram a, the original folded structure is shown in perspective. The uparched folds, called anticlines, and the downarched folds, called synclines, are easily distinguished, and all are inclined or plunge toward the back of the block diagram at an angle of approximately 20°.

Figure 17.2 Large folds are generally eroded, and resistant rock layers along their flanks rise above the surrounding area and clearly express the form of the flexure, as in this view looking north along the east flank of the San Rafael Swell, Utah.

Mountain-Building, Metamorphism, and the Origin of Continents   375

Figure 17.3 *The surface expression of folded rocks typically forms ridges which zigzag back and forth across the fold belt. This picture is a satellite view of the Appalachian Mountains in Pennsylvania.*

Figures *17.4b* and *c* show the fold system after the upper part has been removed by erosion, with the trace of the folds forming a zigzag pattern. If we examine a segment of the folded rocks, such as is shown in the photograph in *figure 17.5*, the subsurface configuration may be interpreted as shown in the cross section across the front of the photograph.

From these illustrations it should be apparent that the resistant ridges in *figure 17.3* which rise over 300 m (1000 ft) above the surrounding surface are simply limbs of huge plunging anticlines and synclines which zigzag back and forth across the Appalachian folded belt. These great folds show that the rocks of the continental crust have been shortened almost 40%.

Intense deformation in some mountain ranges produces more complex folding such as that illustrated in *figure 17.6*. Because the size of the fold structure may be over 160 km (100 mi) wide, it is not generally exposed in a single mountain range; instead, the structure extends through a number of mountain ranges. Details of such intensely deformed structures are extremely difficult to work out because the rocks are generally altered and much evidence is removed by erosion.

The Earth's Dynamic Systems

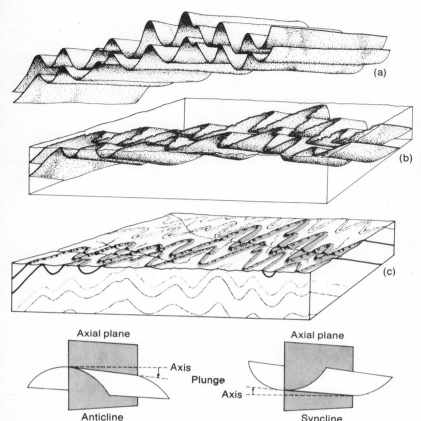

(a)

(b)

(c)

Axial plane

Axis

Plunge

Axis

Anticline

Axial plane

Syncline

Figure 17.4 *The geometry and surface expression of folded rocks. (a) The basic form of folded strata is similar to that of a wrinkled rug. In this diagram, the strata are compressed and inclined to the north. (b) If the top of the folded sequence of rock is eroded away, the trace of the individual layers forms a zigzag pattern at the surface. (c) Rock units which are resistant to erosion form ridges and non-resistant layers erode into linear valleys. Compare this diagram with figure 17.3. (d) Nomenclature of folds. Uparched folds are called anticlines; downwarped folds are termed synclines. A plane dividing the fold into equal parts is the axial plane. The axis of a fold is a line formed by the intersection of the axial plane and a bedding plane. The plunge of a fold is the inclination of the axis from the horizontal.*

Figure 17.5 *The surface expression of a plunging fold is commonly a series of alternating ridges and valleys. In a plunging anticline, such as shown here, the surface trace of the beds forms a V which points in the direction of plunge. The subsurface configureation of a fold may be determined by careful study of the surface layers.*

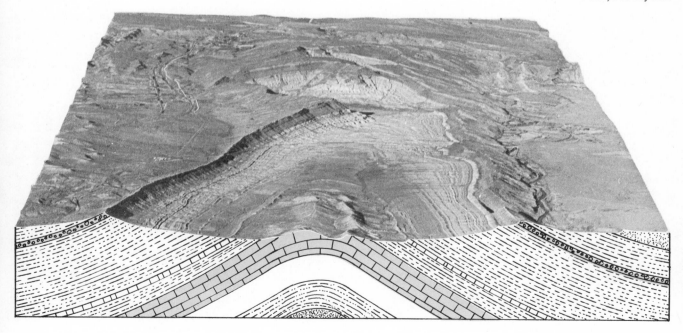

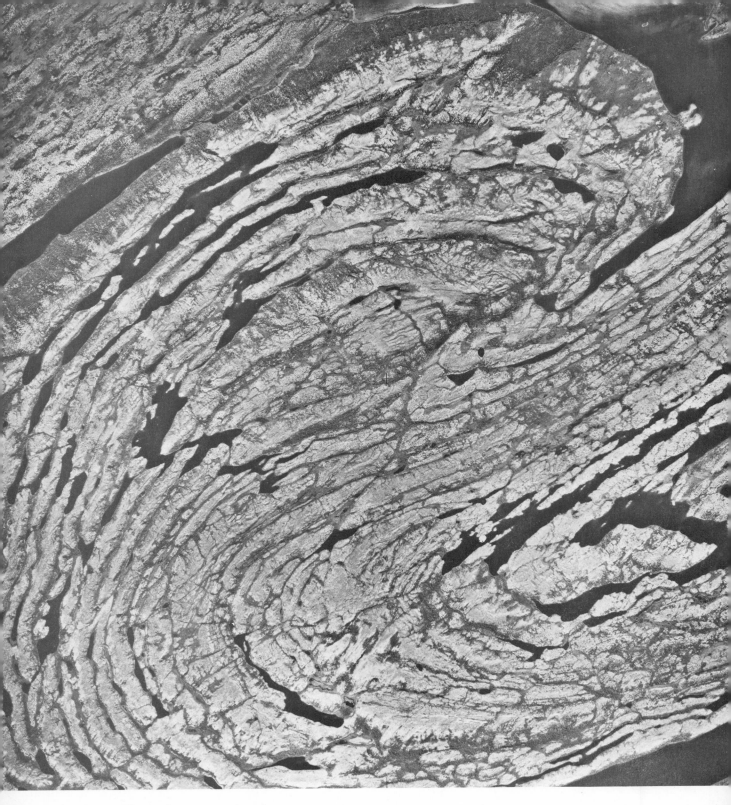

Figure 17.6 *The surface expression of complex folds.*

The geometry and surface expression of complex folds are illustrated in a series of diagrams in *figure 17.7*. In *diagram a* a typical complex fold is shown in perspective. The fold consists of a huge overturned anticlinal structure, with numerous minor anticlines and synclines forming digits on the larger fold. In *diagram b* the fold has been subjected to considerable erosion, and most of the upper limb has been removed. Note the profile of the mountains eroded on the fold. The topographic expression of complex folds (*diagram c*) is variable, usually consisting of a series of mountain ranges unless the area has been eroded to a peneplain. Complex folds are common in the Swiss Alps but were recognized only after detailed geologic studies were conducted in the area for more than half a century.

In many mountain belts the deformation has been so intense that the rocks are not only crumpled and folded but they break along fractures so that one block overrides another. These fractures are referred to as **thrust faults** and commonly evolve in a manner shown in *figure 17.8*. Compressive forces first develop a fold which grows tighter and ultimately fractures, with part of the fold thrust over another. In some areas, sequences of rock several hundred meters thick are turned upside down so that younger rocks are overlain by older ones.

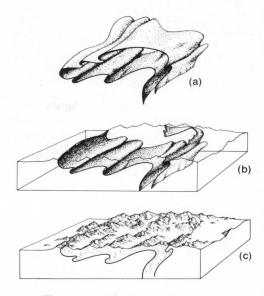

Figure 17.7 *The geometry and surface expression of complex folds. (a) The configuration of a large, tight overturned fold with minor folds on the limbs. (b) The surface outcrop of the fold in diagram a after erosion has removed the upper surface. (c) The topographic expression of the fold in diagram a may be a series of mountains or a flat, eroded surface, such as shown in Figure 17.6.*

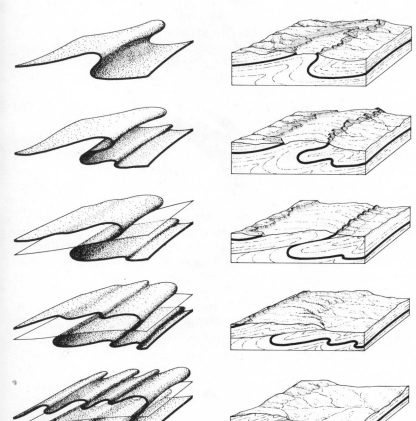

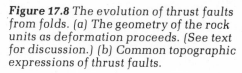

Figure 17.8 *The evolution of thrust faults from folds. (a) The geometry of the rock units as deformation proceeds. (See text for discussion.) (b) Common topographic expressions of thrust faults.*

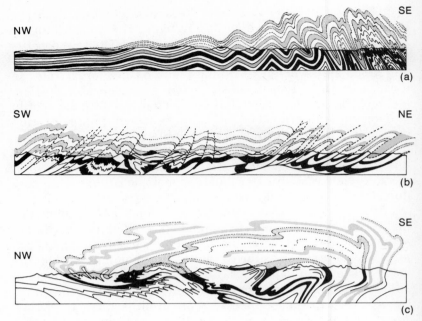

Figure 17.9 *The structure of mountain ranges in cross section. (a) The Appalachian Mountains consist of tight folds which have been eroded down to within two or three thousand meters of sea level. Resistant sandstones form the mountain ridges. (b) The Canadian Rockies contain both folds and thrust faults. (c) The Alps are complex folds, many of which are overturned.*

From detailed field mapping, the structure of an entire mountain range can be determined and illustrated by cross-sectional diagrams. The cross section (*figure 17.9a*) of the regional structure across the Appalachian Mountains shows the magnitude and style of deformation. The mountain belt is asymmetrical, being more intensely deformed to the southeast. Toward the interior of the continent, the deformation gradually dies out. Most of the folds are consistently asymmetrical, with their steeper or overturned flanks to the northwest; rocks toward the southeast are metamorphosed and locally injected by a granitic pluton.

In the Canadian Rockies, the major type of deformation is thrust faulting (*figure 17.9b*), in which large slices of rock have been thrust over one another along a belt 160 km (100 mi) wide.

The structure of the Alps is even more complicated (*figure 17.9c*), with great overturned folds called **nappes** (Fr., "sheet") showing an enormous amount of crustal shortening. The rocks involved are so intensely deformed, that original spherical pebbles are stretched out into rods as much as thirty times longer than the original diameter of the pebble. Most of these structures can be explained only in terms of compressive forces.

The significant fact from these and other examples which could be cited is that in each mountain belt, the internal structures show the results of strong horizontal compressive forces. Similar deformation on older rocks, in which the topographic relief has been eroded down, constitutes one of the basic reasons for concluding that such regions are roots of ancient mountain systems.

Sedimentation and Mountain Belts. One of the more significant facts about mountain belts is that they contain sedimentary

The Earth's Dynamic Systems

rocks which are eight to ten times thicker than rocks of the same age deposited in the interior of the continent. Geologists discovered this over a hundred years ago when they found that the total thickness of sedimentary strata in the Appalachian Mountain region was approximately 15 km (9 mi), whereas strata of equivalent age in Wisconsin and Illinois were only 2 km (1.25 mi) thick. A comparable difference in thickness of strata in folded mountain belts and adjacent platforms was found to be generally true throughout the world. In the Rockies, Himalayas, and Alps, sequences of strata 10 km or even 15 km (6 to 9 mi) thick have been measured; the same rocks covering the adjacent platform do not average more than 1 km (.6 mi) in thickness. This difference indicates that the mountain belts were once segments of the crust which subsided much more than the rest of the continent, permitting the accumulation of great thicknesses of sediment. The areas of abnormally thick accumulations are called geosynclines. After attaining a critical thickness, the geosyncline is subsequently folded and compressed into a mountain range (figure 17.10).

Also significant is the fact that most of the sediment in geosynclines was deposited in shallow water, as indicated by shallow-water fossils, mud cracks, salt casts, and interbedded coal deposits. The great thickness of strata now exposed in mountain ranges never filled the deep ocean basins, even though the cumulative thickness is presently great enough to overfill any existing depth. The only possible explanation is that the crust slowly subsided at about the same rate that sediment was being deposited. The surface of the continental platforms along the margins must have been warped downward to permit the accumulation of an abnormally thick sequence of strata and to allow the entire margins, subsequently deformed, to produce mountain ranges.

For many years, geologists believed there were no existing undeformed geosynclines, but in recent years, with modern techniques of marine geology, they have determined that geosynclines are forming along the passive margins of continents. The continental shelf and slope of the eastern United States are probably the best-known example of a modern geosyncline.

The sediments presently accumulating on the Atlantic continental margins show most sedimentary features found in ancient geosynclines. Thus they become progressively thicker toward the edge of the shelf. (Seismic studies show that the maximum thickness is about 10 km [6 mi].) Sedimentary environments include alluvial plains, deltas, lagoons, beaches, and shallow marine areas.

Metamorphism and Igneous Activity. In the deeper part of folded mountain belts, deformation is so intense that the original sedimentary and volcanic rocks are metamorphosed into **schists** and **gneisses**. Abnormally high heat at certain points within a mountain belt may produce a concentric system of metamorphic zones about a thermal center. In the deeper parts

Figure 17.10 *The evolution of a geosyncline into a mountain range. (a) Sediment accumulates in the shallow seas along the continental margin. (b) Subsidence of the continental margin permits a great thickness of shallow marine sediments to accumulate. (c) and (d) Sedimentary deposits are deformed by compression into a folded mountain belt.*

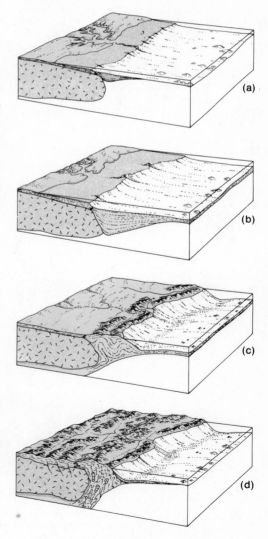

(a)

(b)

(c)

(d)

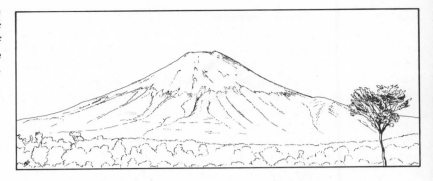

Figure 17.11 Mt. Fuji, Japan, is a typical composite volcanic cone built of andesitic lava and ash as a result of igneous activity along convergent plate margins.

Figure 17.12 Batholithic igneous intrusions associated with mountain belts. (a) A schematic diagram showing the form of a granitic batholith at Rattlesnake Mountain, California. A batholith appears to rise from the deeper parts of the crust because a magma is less dense than solid rock. (b) The block diagram shows the relations of the Sierras. The east front of the Sierras is cut by a large fault and the entire batholithic complex is tilted westward. The mountain range is composed of a number of individual intrusions which appear to pinch out at depth.

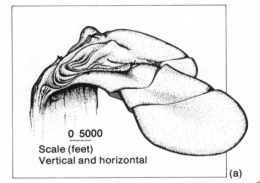

0 5000
Scale (feet)
Vertical and horizontal

(a)

of many mountain chains, metamorphism may become intense enough to produce granitic **migmatite** complexes (migmatite is a term applied to a complex mixture of thin layers of granitic material injected between sheets of schist or gneiss). The migmatite develops largely from partial melting of the preexisting rocks in the immediate vicinity and apparently does not migrate far. This provides important insight into the development of granitic magmas as they evolve in deformed mountain chains.

Igneous activity is also intimately associated with mountain-building and is manifested in two ways. The most obvious is the extrusion of andesitic lavas which are typically more viscous than basaltic lava and tend to build steep, high volcanic cones (figure 17.11). The high peaks of the Andes Mountains are excellent examples, as are the Cascades in Washington and Oregon. We have seen in the previous chapter that andesitic vulcanism is also typical of the island arcs associated with the trench systems.

Many silica-rich magmas do not reach the surface but cool at considerable depths within a mountain range to form huge batholiths. Most batholiths are roughly equivalent in composition to andesites except they may be richer in silica. They cool

(b)

The Earth's Dynamic Systems

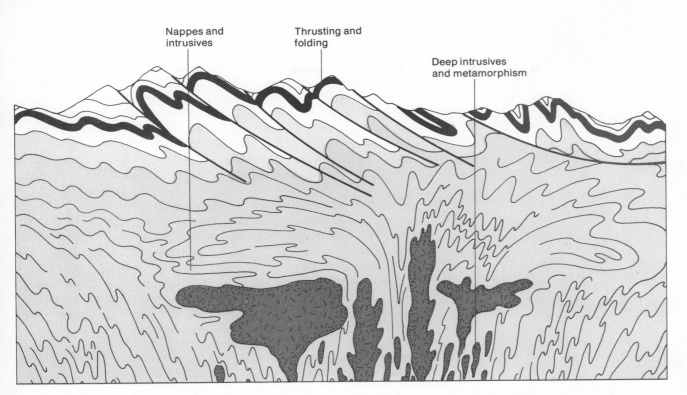

Nappes and
intrusives

Thrusting and
folding

Deep intrusives
and metamorphism

slowly at great depths and are exposed only after the mountain range is considerably eroded. Batholiths are typically elongate parallel to the axis of the fold belt and, although we know little about their appearance at depth, it appears as though they are tabular bodies such as are shown in *figure 17.12*.

The general concepts of metamorphism and igneous activity in a mountain belt and their relation to folding and faulting are shown in *figure 17.13*. Folding and thrusting are relatively shallow structures; metamorphism occurs deeper, and partial melting at a still greater depth. Granitic magma initially forms deep within the crust at a large number of points where the rock begins to melt. The melted rock is injected into the foliation of the adjacent metamorphic rock to form a migmatite. As the melted rock migrates upward (because it is less dense than the solid rock), it collects and rises along zones of weakness. Here, the boundaries of the body of magma are largely parallel with the broad zones of foliated metamorphic rock. As the granitic magma rises, it cuts across the upper folded strata which are not metamorphosed but deformed only by folding and faulting. The magma may rise high in the crust and cool within a few kilometers of the surface to form a batholith or may be extruded as andesitic or rhyolitic volcanic material.

Andesitic and other silica-rich magmas are relatively cool and, because of their high viscosity, they commonly erupt in a violent manner. As the magma works its way toward the surface, pressure is released, and the contained gas rapidly expands. The magma may then literally explode. The ejected ma-

Figure 17.13 *Diagram showing the hypothetical relationship between types of structures in a mountain belt and the depth at which they are formed. Thrust faulting occurs near the surface where rocks are not under high, confining pressure. At greater depth, tight folds develop and increase in complexity in the deeper part of the mountain belt. Large batholiths solidify at a still greater depth. In the deepest part of the mountain roots, rocks are metamorphosed and develop vertical foliation perpendicular to the compressive forces. Partial melting in this zone generates small magma bodies which rise and collect to form the batholiths which solidify at a higher level.*

terial consists of droplets of lava, bits of solid rock, vast quantities of ash, fragments of **pumice**, and early-formed crystals. This material is very hot, sometimes incandescent, and being denser than air, moves or flows across the surface as a thick, dense cloud of hot ash. French geologists described this phenomenon as a **nuée ardente** (burning cloud).

Nuées ardentes are best thought of as **ash flows**, the movement of ash particles en masse. It is not a lava flow nor an ash fall in which particles settle out independently; it is a flow phenomenon of a mass of small fragments. Ash flows may reach velocities in excess of 100 km (60 mi) per hour because gas is continuously emitted between cooling lava particles and tends to force them apart. The resulting rock deposit (**welded tuff**) consists of fused particles of crystal fragments, pumice, glass, and bits of rock.

Ash flows may be very large and form a flow unit several hundred meters thick extending over an area of thousands of square kilometers. As they cool, they contract and often develop well-defined columnar jointing. Ash flows tend to have an even upper surface; when erupted upon uneven topography, they show evidence of flowing around obstacles and down drainage channels. The violent explosive nature of ash-flow eruptions tends to develop huge explosive calderas 20 to 40 km (12 to 25 mi) in diameter.

Although eruptions of ash flows are rare and catastrophic events, a few geologists have had the good fortune to witness them and make direct observations of this type of extrusion. One of the most violent eruptions of ash flow in historic times was the 1951 eruption of Mt. Lamington in New Guinea. Until 1951, Mt. Lamington was considered extinct. It had never been examined by geologists and was not even considered by the local inhabitants to be a volcano. The Mt. Lamington eruptions are important not only because they were outstanding examples of ash-flow extrusions but also because they provided exceptional opportunities for on-the-spot studies. The main eruption was observed and photographed at close quarters from a passing aircraft, and a qualified vulcanologist began recording events on the spot within 24 hours after the main eruption. Sensitive seismographs were soon installed near the crater to monitor earth movements, and daily aerial photographic records were made.

Reactivation of the volcano began with six days of such preliminary volcanic phenomena as gas and ash emissions, earthquakes, and landslides near the crater. At 10:40 a.m. on Sunday, January 21, 1951, a catastrophic explosion burst from the crater, producing an ash flow which completely devastated the surrounding area. Nearly 3000 people perished.

The January eruption descended radially from the crater, the direction of the movement being controlled by the topography. As the ash flow moved down the slopes, it scoured and eroded the surface. Velocities of about 470 km (290 mi) per hour

The Earth's Dynamic Systems

were estimated by calculating the force required to overturn certain objects. Entire buildings were ripped from their foundations and automobiles were picked up and deposited in the tops of trees (figure 17.14).

Figure 17.14 Devastation caused by the 1951 ash flow of Mt. Lamington, New Guinea.

Orogenesis (Mountain-Building)

Statement

It is a well-known fact of geologic history that the origin of mountains begins with geosynclinal sedimentation and is terminated by intense deformation of the crust known as an **orogeny** (Gr. *oros*, "mountain"; *genesis*, "origin"). We now know, however, that mountain-building is more than simple crustal deformation; it is part of the fundamental process by which the earth's materials are differentiated and segregated according to density.

We have seen in the previous chapter how the dense mantle material is differentiated by partial melting of peridotite in the mantle to form the lighter basaltic oceanic crust. In

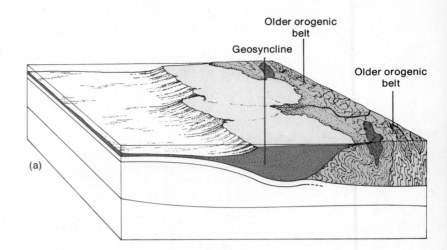

Older orogenic belt

Geosyncline

Older orogenic belt

(a)

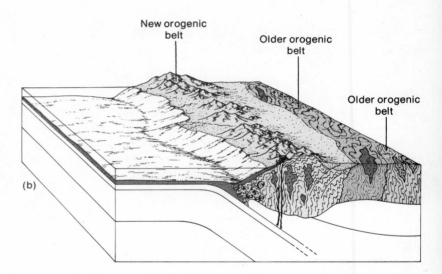

New orogenic belt

Older orogenic belt

Older orogenic belt

(b)

Figure 17.15 *Stages in the evolution of a mountain belt (orogenesis). (a) Geosynclinal stage. Sediments accumulate in a thick wedge along the passive continental margins. Sedimentary differentiation concentrates minerals with low density (quartz, clay, and calcite). (b) Deformational stage. The continental margin collides with another crustal plate to form a subduction zone, and the geosyncline is deformed. Metamorphic rocks and magma are generated in the deep zone, whereas folds and thrust faults occur at shallower depth. (c) Erosion and isostatic rebound of the deformed mountain belt results in the new orogenic belt of rocks "welded" onto the shield. The cycle begins again with geosynclinal sedimentation along the continental margins.*

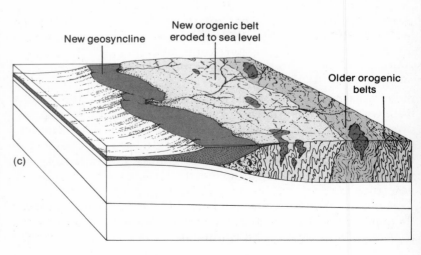

New geosyncline

New orogenic belt eroded to sea level

Older orogenic belts

(c)

mountain-building, two processes combine to further separate the lighter elements and concentrate them in the continental crust. One is partial melting of basaltic crust to form andesitic or granitic magmas. The other is sedimentary differentiation in which low-density, silica-rich minerals are concentrated by weathering and sedimentary processes.

Discussion

The major steps in orogenesis are illustrated in figure 17.15. Weathering and erosion of the continental landmass produces a supply of sediment which is carried to the continental margin by the major river systems. Minerals of low density such as quartz, clay, and calcite are concentrated by this process in sand, mud, and limestone. The transfer of sediment to the continental margin causes an isostatic imbalance so the continental margin subsides under the weight of the newly deposited sediment. As sedimentation continues, isostatic adjustment causes further subsidence, and ultimately a thick sequence of shallow-water sediments accumulates along the continental margin to form a geosyncline. This occurs on the passive margins of the continent, such as the Atlantic coast of North and South America.

Ultimately, convection patterns within the asthenosphere shift and change and a subduction zone develops adjacent to the geosynclinal deposits. The collision of lithospheric plates along the continental margin deforms the geosyncline into a folded mountain belt. Folds and thrust faults are the major structural features to form in the upper part of the mobile belt, whereas, at depth, metamorphism and partial melting of the sediments and oceanic crust occur. Partial melting also occurs as the lithosphere containing oceanic crust descends along the Benioff zone. Heat is produced in the mobile belt and along the Benioff zone by pressure from plate collision, friction of the descending plate, and radioactivity in the asthenosphere. Partial melting of the crust and associated sediment produces a magma richer in silica than the original basalt because the first minerals to melt in sediment and basalt would be those which are richest in silica. (You will recall that quartz, potassium feldspar, and sodium plagioclase have lower melting points than calcium plagioclase, amphibole, pyroxene, and olivine.) The newly generated magma rises to form granitic batholiths or andesitic extrusions.

The history of a mountain belt does not end with deformation and igneous activity. After the orogenic activity is terminated, presumably by shifting convection cells, deformation ceases, but erosion and isostatic adjustment combine to modify the orogenic belt. As illustrated in figure 17.16, the entire crust is deformed by the orogeny so that a mountain root composed of the most intensely deformed rocks extends down into the mantle. The high mountain range and its deep roots are in isostatic balance. As erosion vigorously attacts the high peaks and the sediment is carried to the continental margins, this balance is

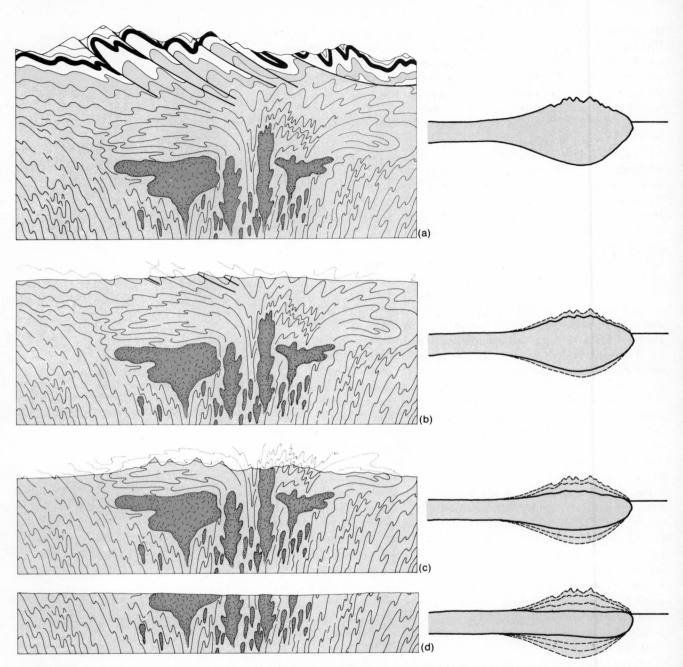

Figure 17.16 Erosion and isostatic adjustment combine to transform a newly deformed mountain belt into a segment of the stable shield. (a) After deformation, a mountain belt is in isostatic equilibrium. A root extends down into the lithosphere to compensate for the mountainous topography. Thrust faulting and folds are the dominant structures exposed at the surface. (b) As erosion removes the rock from the mountain belt, isostatic equilibrium is upset, and the deep mountain root rebounds so that some topographic relief is maintained. Tight folds, formed in the deeper parts of the mountain belt, now become the dominant structures exposed at the surface. (c) Continued erosion and isostatic adjustment reduce the topography and the root of the mountain belt below. Complex folds and igneous intrusions, originally formed deep in the mountain, are the dominant structures exposed at the surface. Ultimately isostatic equilibrium is reached again. The mountain root containing metamorphic and igneous rocks is exposed to the surface, and the topography is eroded to sea level. At this stage, the mountain belt becomes a segment of the shield.

destroyed and vertical uplift of the mountain belt occurs to re-establish an isostatic balance (figure 17.16). Uplift will continue as long as erosion removes material from the mountain range. Ultimately, a balance is reached where the mountainous topography is eroded to sea level and the mountain root is adjusted to a near-perfect balance. An important point to note here is that the rocks exposed at the surface when isostatic balance is established will be the metamorphic and igneous rock formed at great depth in the orogenic belt. These rocks are then tectonically stable and become part of the shield. Let us now consider the shield in greater detail and see how the rocks and structure of this fundamental segment of the continent fit the theory of mountain-building.

Ancient Mountains of North America

Statement

Although the Canadian Shield is extremely complex, it has been considered for many years to be a series of ancient mountain belts which have been eroded down to sea level. Evidence for this conclusion is seen in the composition and structural trends of the rocks and the radiometric age of the granitic bodies which form intrusives in them. The metamorphic rocks occur in distinct provinces, each with a definite structural trend and distinct igneous intrusions. Radiometric dates show that the metamorphic belts are not the same age; each province appears to have been formed during a separate period of deformation. In a general way, the structure of the shield consists of central areas of oldest rocks surrounded or flanked by succeedingly younger rocks deformed during later periods of deformation. Thus, it appears that the structure of the continent consists of a series of deformed mountain belts which have been eroded down to sea level to form a flat, stable platform. Other shields of the world have many of the characteristics of the Canadian Shield and are considered to have originated in a similar manner.

Discussion

In order to recognize the shield as roots of ancient mountains we must first become better acquainted with the characteristic features of the shield. Perhaps the best way to do this is to carefully study figures 17.17 and 17.18 and note some of the facts which have been observed in the field. The most obvious characteristic seen in these photographs is the vast expanse of the low surface of the shield. Throughout an area of hundreds of thousands of square kilometers, the surface of the shield is within a few hundred meters of sea level. The only features which stand out in relief are the resistant rock formations which rise 30 to 120 m (100 to 400 ft) above the adjacent surface. When we consider that the rocks which make up the shield are granitic

Figure 17.17 The complex metamorphic rocks (dark tone) in this part of the Canadian Shield are intruded by granite (light tone). The surface is eroded to within a few meters of sea level.

and metamorphic, the fact that the surface is eroded down to such low relief is even more remarkable. The nature of the rocks in the shields implies two things: (1) their highly deformed nature clearly indicates that the shields have been subjected to intense horizontal stresses, and (2) the exposure of metamorphic and igneous rocks, which form only under high temperatures and pressure, indicates that a vast amount of erosion has occurred since the rocks were formed in order for them to now be exposed. Estimates are that erosion has removed at least 8 km (5 mi) of rock from the shields in order for the deep-seated metamorphic complex to be exposed.

The structural complexities of the shield are shown by the complex patterns of erosion, alignment of lakes, and differences in the tone of the photographs. Throughout the shields, igneous

Figure 17.18 *Linear lakes emphasize the structural trends of the metamorphic rocks in this part of the Canadian Shield. The eroded metamorphic surface is extremely flat.*

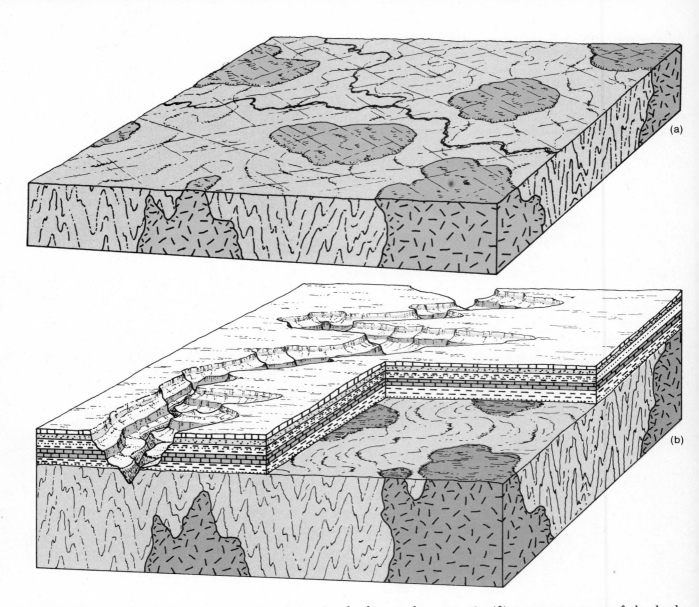

Figure 17.19 *General characteristics of a shield. (a) The schematic representation of the shield shows the complex metamorphic rocks and a variety of igneous intrusions. The upper surface is nearly flat and eroded down to sea level. (b) Throughout most of the interior of the United States, the shield is covered with a veneer of horizontal sedimentary rock, thousands of meters thick. In some places, such as the Grand Canyon, erosion has cut through the sedimentary veneer and has exposed the shield below.*

intrusive bodies make up a significant percentage of the bedrock. Faults are common but, without good reference horizons, displacement is not always obvious. Typically, faults are expressed topographically as linear depressions and, in some areas, can be traced for hundreds of kilometers. The diagrams in figure 17.19, constructed on the basis of these facts, represent our present concept of the nature of this most basic element of the continental crust. Evidence shows that the shields have been subjected to many periods of deformation in which sedimentary and volcanic rocks were converted to metamorphic complexes. Subsequently, erosion removed the upper part of the metamorphic terrain, exposing the deep roots of the mountains and the associated igneous intrusions.

The major structural elements of the Canadian Shield have

The Earth's Dynamic Systems

been plotted. The map in *figure 17.20* is a summary of a much more detailed compilation representing the results of many years of field mapping, geophysical studies, and radiometric age determination; from it we can analyze the basic structural trends of the continent. In a general way, North America is quite symmetrical. The oldest rocks, recording the first clearly recognizable geological events, are in the Superior-Wyoming and Slave provinces. They consist of volcanic flows and volcanic-derived sediments similar to those forming today in the island arc areas of the Pacific. These sediments are metamorphosed and engulfed by granitic intrusions. The granite-forming event, as determined by radioactive dating, was 2.5 to 3 billion years ago. Since the earth has been estimated to be a little more than 4.5 billion years old, this segment of crust apparently formed 1.5 to 2.5 billion years after the origin of the earth. Generally speaking, the composition of this ancient continental crust prior to the granitic intrusion was closer to that of basalt than granite. The granitic intrusions now constitute three-fourths of the area. The metamorphosed sediments and volcanics have structural trends in a northward direction as shown in *figure 17.20*. This segment

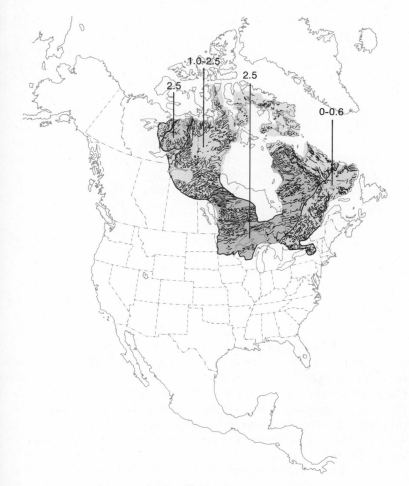

Figure 17.20 *The major structural trends and radiometric dates of the rocks in the Canadian Shield show that the shield consists of a number of geologic provinces, each representing major mountain-building events. The numbers refer to ages of major granites in billions of years, and the lines represent trends of folds. The shield appears to have grown by accretion as new mountain belts were formed along its margins.*

of crust is significant because it is thought to represent the development of the first stable and resistant granitic crust in North America. Since the time of formation, the upper surface has been eroded to sea level and has persisted as a stable platform, at times partly submerged beneath a shallow sea and at times slightly above sea level but never the site of mountain-building processes. In all probability, this block of crust was as much as 20 to 60% larger than it is presently because the trends of the metamorphic structures are abruptly terminated by younger provinces. We can assume that the original continental mass was split and fragmented by plate movement, with each segment acting as a center for future continental growth.

Surrounding the Superior-Wyoming Province to the south and west is a vast area of gneiss and granite from 1.0 to 1.4 billion years old. The rocks in this younger province are also metamorphosed, but the original sediment was quite different from that now seen in the Superior Province. These younger rocks contain less lava and show a distinct increase in quartz-rich sandstone and limestone formed on the continental shelf. In addition, the structural trends are oriented in a different direction.

Beyond this province the rocks are still younger and the granitic intrusions range in age from 1.1 to 0.8 billion years. They are mostly quartzites, marbles, and schists derived from well-sorted sand, shale, limestone, and volcanics richer in silica than the older volcanic rocks of the Superior Province.

The youngest deformed mountain belts of the continent are younger than Precambrian. The Appalachian Mountain belt is about 225 million years old and the Rockies 65 million years old.

The concentric pattern of the provinces in North America is considered to be strong evidence that the continent grew by accretion of material around the margins during an orogenic event. Each province represents a mountain-building event during which sediments, originally deposited in a geosyncline, were deformed by compressive forces into a mountain range. Subsequently, erosion removed the upper part of the deformed belt to expose only the deeper, highly deformed metamorphic rocks and associated granitic intrusions.

Rift Systems

Statement

In addition to the major mountain systems described above, several continents are broken by faults in which displacement has been predominantly in a vertical direction. These areas, in contrast to mountain belts, are under tensional stresses and typically develop long, narrow, downdropped blocks, called rift valleys. Two areas in which rift systems are well known are (1) the great rift valleys of East Africa and (2) the Basin and Range Province of the western United States. Both areas are a continental

The Earth's Dynamic Systems

extension of the mid-oceanic ridge and are considered to be places where continental blocks are being split apart.

Discussion

Rift Valleys of East Africa. The African rift valleys extend as a major structural feature from the Zambezi River to northern Ethiopia, a distance of 2900 km (1800 mi). The system then continues another 2000 km (1200 mi) northward as the rift system of the Red Sea and Jordan Valley. The rift valleys of Africa consist of large uparched segments of the crust in which long linear blocks have been dropped down near the crest (figure 17.21). Many of the rift valleys in Africa are closed depressions and are partly filled with water to form the noted lakes of East Africa, the Red Sea, the Gulf of 'Aqaba, the Dead Sea, and the Sea of Galilee. As can be seen on the physiographic map, the rift valleys are remarkably uniform in width, ranging from 40 to 60 km (25 to 37 mi) wide.

The Basin and Range. Normal faulting on a grand scale occurs throughout the Basin and Range of western North America and has produced a series of **horsts** (upthrown blocks) and **grabens** (downthrown blocks). The horsts form mountain ranges 2100 to 3000 m (7000 to 10,000 ft) high which are considerably dissected by erosion. The grabens form topographic basins partly filled with erosional debris derived from the adjacent ranges (figure 17.22). The basins are largely closed and many contain evaporite (salt) lakes.

Strike-Slip Faults. Strike-slip faults are fractures in which displacement has been horizontal rather than up or down and one block has moved past the other parallel to the strike or trend of the fault plane. Since there is little vertical movement, high cliffs do not form from strike-slip faults; instead, the fracture is expressed at the surface by a low ridge or offset drainage (figure 17.23). The San Andreas fault in California (figure 17.24) is one of the best known strike-slip faults because it is very active and has a profound influence on the environment of California. Studies of this fault show that horizontal movement results from both slow creep of less than a centimeter per year to rapid displacement of possibly more than 10 m (33 ft) during an earthquake. Total displacement may be several hundred kilometers.

Other strike-slip faults cut across the Basin and Range area and the Jordan Valley. They may, therefore, be associated with the tensional stress which produces the rift valleys.

Figure 17.21 *The East African rift valleys show where the continent is being uparched and pulled apart. If the spreading processes continue, the rift system may evolve to an elongate sea, like the Red Sea to the north. (a) Map showing location of rift valleys. (b) Cross sections across the rift valleys showing the uparched surface and the central downdropped block which forms the valleys.*

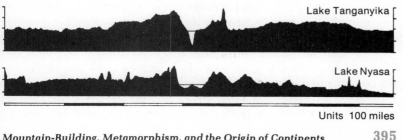

Figure 17.22 Schematic diagram showing the fault blocks of the Basin and Range Province of the western United States.

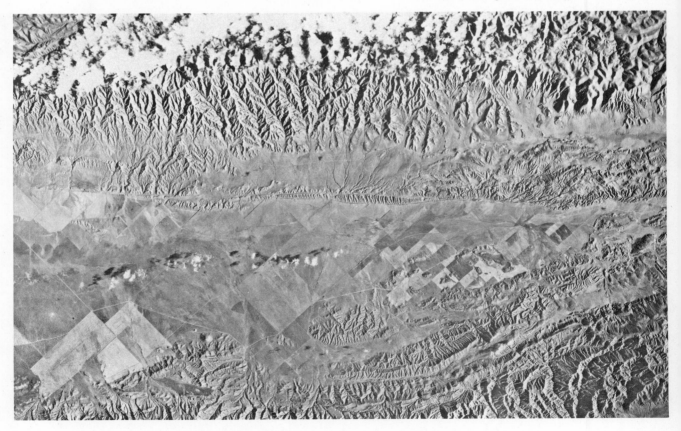

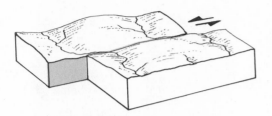

Figure 17.23 Diagram showing displacement on a strike-slip fault. The movement is predominantly horizontal, so high cliffs do not develop along the fault line. Drainage patterns, however, are offset, and rock types and landforms of one area are displaced against those of another.

Figure 17.24 High-altitude photograph of the San Andreas fault in California. The fault line trends through the central part of the picture and is marked by distinct changes in drainage and landforms.

Origin and Evolution of the Continents

Statement

Continents are believed to be the result of plate tectonics and represent the fundamental process by which materials of the earth are differentiated and segregated according to density. Erosion, sedimentation, metamorphism, igneous activity, isostasy, and sea level all play an important role in their development.

Continents are believed to originate as volcanic islands produced where crustal plates collide. Magmatic differentiation occurs by partial melting of the descending plate. The silica-rich minerals melt first and produce a silica-rich magma which erupts as andesitic volcanics or forms granitic intrusions. Sediment derived from erosion of this material is further enriched in silica by differential weathering and sedimentary differentiation. The sediment accumulating along the margins of the embryo continent is later deformed into a mountain belt when plate movement causes the embryo continent to collide with another plate. The silica-rich sediment is then metamorphosed and "welded" onto the volcanic islands to form a small continental mass rich in silica. Such a mass would be of relatively low density compared to the oceanic crust and would resist being dragged back down into the mantle along a subduction zone. As erosion levels the mountain range, isostatic adjustment would eventually raise the mountain roots to sea level. We emphasize that, throughout continental growth, the processes of weathering, transportation, and sedimentation, as well as igneous activity, combine to produce the low-density, silica-rich material of the continental crust which resists being dragged down into the mantle by a descending plate. Thus, continents are considered to grow by accretion of material along their margins as sediment is metamorphosed during a mountain-building event.

In addition to the basic process of continental growth by accretion, continents may be split and drift apart during the process of sea-floor spreading and plate fragmentation. Each continental fragment would then act as a new nucleus for future growth as an independent continental mass. Also, as continents drift on a moving lithospheric plate, they may collide and be sutured together to form a larger mass of continental crust.

Discussion

The basic elements of the theory of the origin and growth of continents are illustrated in the series of diagrams in *figure 17.25*. The original crust of the earth is assumed to have been basaltic, with new crust created along rift systems and destroyed along trenches where the plates descend into the mantle (*block a*). This assumption is based on the absence of granitic rock in the present oceanic crust. Only basaltic crust is created at the rift system above the rising convection cells in the mantle.

The initial step in development of a continent was possibly

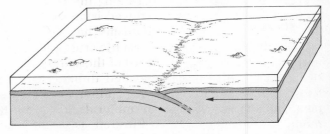

(a) Convection in the mantle initiated the tectonic system in the earth soon after it grew to its present size. Oceanic crust was created at rifts and destroyed at subduction zones.

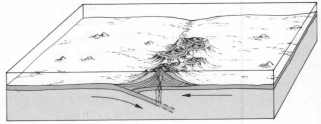

(b) Partial melting in the subduction zone produced island arcs composed of andesitic volcanic material which was too light to be pulled down into the mantle. Erosion of the islands further concentrated light material as sediment was deposited on island flanks.

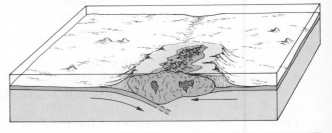

(c) The sediment and andesitic volcanoes were deformed by compression at converging plates to produce mountain belts and metamorphic rocks which formed an embryonic continent.

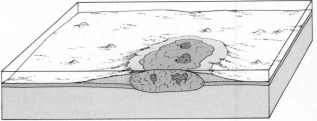

(d) Erosion of mountains concentrated light minerals (quartz, clay, and calcite) as geosynclinal sediments along margins of the small continent.

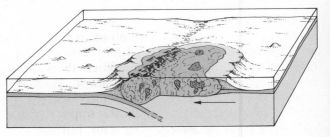

(e) The geosynclinal and orogenic cycle is repeated, with new material added to the continental mass as granitic batholiths and andesitic melting of the oceanic crust.

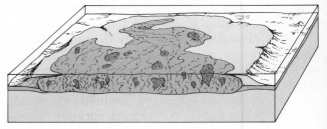

(f) The continent continues to grow by accretion of geosynclinal sediments along the margins.

a flexing of the basaltic crust along the plate junction which, together with vulcanism, produced island arcs adjacent to the trenches (*block b*). Two mechanisms then combined to separate silica-rich material from the basaltic crust and concentrate it as a "scum" of lighter granitic crust of the continents: (1) Weathering and transportation of eroded debris from the volcanic islands separate silica and clay minerals. Feldspar weathers and decomposes to form clay minerals. Any free quartz from andesitic vulcanism is concentrated as quartz sand. (Quartz plus clay is chemically similar to granite.) (2) Partial melting of basaltic crust, as one plate descends down into the mantle, yields a magma richer in silica. Through these processes, granitic intrusions and andesitic flows form small patches of rock with low density.

The volcanic material and sediment produced at the plate junction (shown in *block b*) are higher in silica and less dense than new basaltic crust produced at the rift system. As a result, they resist assimilation back into the mantle at the subduction zone. This material is then subjected to horizontal stresses where the plates collide and are folded into mountain ranges. Metamorphism occurs in the mountain roots, and granitic intrusions result from partial melting of the descending basaltic plate. This mass of light silica-rich rock floats on the denser crust and forms a nucleus for future continental growth (*block c*).

The mountains produced where the plates collide are soon eroded down to sea level, and the resulting sediment accumulates along the continental margins in a geosyncline (*block d*). These sediments may then be deformed into a new mountain range and welded back onto the mass as metamorphosed sediment intruded by granitic rock (*block e*). The structural fabric and regional trends of metamorphic fabric are roughly parallel to the axis of the geosyncline and perpendicular to the stress resulting from the converging plates which caused the deformation.

Repetition of this process develops a series of stabilized metamorphic belts welded together to form a more extensive block of continental crust or shield (*block f*). Isostatic equilibrium is eventually established when the continental crust is about 25 to 35 km (15 to 20 mi) thick. This block of granitic crust rises above the denser oceanic crust to a height of some 5 km (3 mi). Erosion planes off the upper surface of the continental platform in the stable interior, but along the continental margins, mountain ranges or geosynclines are in various stages of development.

We would like to emphasize that several forces are active upon the continental crust as it moves: (1) deformation and mountain-building occurring where plates collide, (2) erosion tending to reduce the continental surface to sea level, (3) isostatic adjustment in adding or removing weight on the crust, (4) sedimentation in the shallow seas spreading partially over the continent, and (5) movement of the continental crust, shifting horizontally with the lithosphere. All of these factors are involved in continental growth. The continental lowland is the shield and covered shield, both of which are very close to sea level. Even though sedimentary rock accumulates in river flood plains on the shield and covered shield, the process of stream flow tends to move sediment to the sea, depositing most of it on the continental shelf, while some is washed across the shelf to the deep ocean. Because erosion and isostatic adjustment tend to level the continental block to sea level, the surface of the continent (shield and covered shield) will be in a delicate isostatic balance. Broad upwarps of the interior develop as mountain-building occurs along the continental margins, and movement of the plate carrying the continent may cause the continent to "vibrate" slightly, permitting the sea to expand over the shield and covered shield. Shallow seas then advance and retreat over the lowland of the continent, and sedimentary rocks accumulate with each cycle of advance and retreat. Parts of the sedimentary cover are periodically exposed to erosion, followed by submergence and deposition of another series of sedimentary rock. In places the strata are warped into large basins and domes. In areas where a great deal of sediment is brought to the shallow sea, the weight of the new sediment is sufficient to cause subsidence and formation of a geosyncline.

Repetition of these cycles of upwarp and subsidence, followed by advance and retreat of the sea over the continental

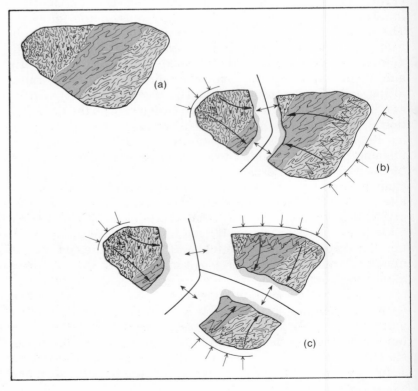

Figure 17.26 The fragmentation of continents may occur at any time during the evolution of a continental mass. (a) Original continental mass composed of orogenic belts welded together into a stable shield. (b) A spreading center splits the continent, and the fragments drift apart. A subduction zone will ultimately develop along the leading edge of the plate, and each fragment may grow as an independent continental mass. (c) Further fragmentation may occur, or the continents may collide and become sutured together.

The Earth's Dynamic Systems

platforms, results in a large stable continental mass consisting of an exposed shield, a segment of the shield covered with a veneer of sedimentary strata, and a mountain range along the **active margin** of the continental plate.

At some stage, rifts may form within the stable continental block as the spreading centers of the plate shift, presumably as a result of shifting convection currents in the mantle. The continental blocks then split and drift apart as separate continental fragments (*figure 17.26*). These continental fragments then act as separate centers for future continental growth as geosynclines and mountain chains continue to evolve. Since the earth is a sphere, the moving plates carrying the continental block eventually collide, and continental blocks may be compressed into large super-continents.

Thus, continental crust is considered to be the result of plate motion in which metamorphism and mountain-building occur where moving plates collide. By means of this process, a light granitic crust evolves as the aluminosilicate minerals are selectively concentrated by sedimentary differentiation and by partial melting of the descending plate to produce silica-rich volcanic products. It is part of the process by which dense material is concentrated near the core of the earth and light material near the surface.

Admittedly, this view of geologic history is speculative, but it seems to be the best explanation of continental history based on the facts presently available.

Summary

Continents grow by accretion as silica-rich rocks which have a low density are separated by both magmatic and sedimentary differentiation. The process involves the growth and development of a geosyncline along passive continental margins, deformation of the geosyncline into a folded mountain belt, and erosion and isostatic adjustment of the deformed belt until ultimately the mountains' roots are exposed and eroded down to sea level. The geosyncline is deformed when convection currents change direction and a passive continental margin collides with another lithospheric plate. Partial melting of the oceanic crust produces granitic or andesitic magma which is involved with the mountain-building process. Continents are too light to be pulled back into the mantle by descending plates. They float on the denser upper lithosphere and may be split apart where spreading of the lithosphere occurs beneath the continental crust. As the continents split and drift apart, a new ocean basin is produced and each continental fragment acts as a separate center for further accretion.

The differentiation of materials in the earth has produced the ocean basins and the continental platforms, plus the surface fluids which move to form the hydrologic system. We see in this fundamental process the origin of all surface features and the internal structure of the earth. In the next chapter, we shall consider how the processes operating in and on the earth affect our environment and create our natural resources.

Additional Reading

Engel, A. E. J., and C. G. Engel. 1964. "Continental Accretion and the Evolution of North America," In A. P. Subramanian and S. Balakrishna, eds. Advancing Frontiers in Geology and Geophysics: A Volume in Honour of M. S. Krishman, pp. 17-37e. Hyderabed: Indian Geophysical Union. (Reprinted in P. E. Cloud, ed. 1970. Adventures in Earth History, pp. 293-312. San Francisco: W. H. Freeman and Company.)

18 Environment and Natural Resources

In the preceding chapters we have studied the earth as a dynamic system which produces a constantly changing environment. Some rapid changes such as earthquakes, volcanoes, and floods produce catastrophic effects, but most changes are so slow by human standards that they are barely noticed, even during a lifetime. Perhaps there is comfort in realizing that most geologic processes are extremely slow and are of little concern to us as individuals, but in the last 10,000 to 20,000 years many changes, significantly affecting man's history, have occurred. For example, the present sites of many northern cities such as Chicago, Detroit, Montreal, and Toronto were buried beneath thousands of feet of glacial ice as recently as 15,000 to 20,000 years ago while, at the same time, the site of most major port cities such as New York were located many miles inland. The Missouri and Ohio rivers were not completely established at that time, and the valley where Salt Lake City is located contained a large fresh-water lake 300 m (1000 ft) deep. These changes were largely the result of worldwide climatic changes associated with the Ice Age. However, man is capable of simulating drastic geologic changes within a few years and can radically alter the environment in which he lives. For example, to irrigate an arid land is to simulate an instantaneous climatic change equivalent to more than 2.5 m of precipitation annually. When this occurs, the balance between ground-water movement and the arid surface is destroyed. New springs occur, landslides develop, sediments compact, and the surface may subside.

Another very important concern of man today is in the field of natural resources. Our utilization of natural resources in the past seemed to be based on the assumption that the earth's mineral supplies (or substitutes) were inexhaustible. Are they? How are mineral resources made? Can they be replenished? Can we continue to afford the consumption of energy and raw materials at the present rate? Can our economy continue to expand, or are there limits to growth?

Major Concepts

1. Each geologic process is intimately related to many others and is part of a major dynamic system. An attempt to change or modify in a significant way any geologic system may produce severe and often unsuspected side effects.
2. Natural resources are created in specific ways by the various geologic systems and can be replenished only by those systems. In terms of human activity they are finite, exhaustible, and nonrenewable. Man must, therefore, learn to budget the natural resources presently available.

404

Environmental Geology

Statement

The Problem. Man as a part of nature is one of the many species of animals adapted to the present natural environment. He is so well adapted that he dominates all other species, and his numbers are so great that he has become an effective agent of physical, chemical, and biological change. Presently, he is capable of modifying significantly the natural systems which operate on the earth. Many of his modifications, unfortunately, conflict with the normal evolution of the earth's environment. For example, since man has developed an agrarian culture, he has drastically changed large areas of the earth's surface from natural "wilderness" to controlled, cultivated land. This has put a great strain on the food supply of many species and animal populations which had previously established a degree of balance. As a result many species have become extinct. The advancements of science and technology have simulated the conversion of thousands if not millions of years of normal evolutionary change into days. Drainage systems are diverted and modified; the quality and flow of water both in the surface and subsurface are altered; the atmosphere is changed, and large parts of the ocean are affected. The changes instituted by man are, almost without exception, intended to produce improvements and advantages for his society, but frequently the results are quite the opposite. At best some are detrimental on a short-term basis; others are catastrophic and irreversible in the long run. Therefore, man is forced to adapt to a rapidly changing environment, one which he in part created. Changes are necessary and, at times, the undesirable side effects, anticipated or unanticipated, must be tolerated as the "necessary price." Since man is capable of changing his environment, the decision to do so should be made carefully and only after enough geological and biological facts are available to enable him to see how the natural systems operate and how they will be affected. If we decide to dam a river, we should understand the side effects of altering part of this natural system. We should question and know how the damming will affect erosion and sedimentation downstream, how the ground-water system will respond, how lakes will be altered, how the beaches along the shore far removed from the dam will be affected, and how marine and terrestrial life will be changed. If we develop a great volume of waste, we should know how it will be assimilated into the earth's system. How will it affect the quality of surface and ground water, the atmosphere and oceans? If we build metropolitan areas, we should have full understanding of how this development will affect the terrain conditions, the atmosphere, and the hydrosphere. In too many cases, in the process of spending money to correct or alter a natural condition to better suit our needs, we upset a natural balance, which requires that we spend more money to reestablish the balance.

Environment and Natural Resources

Our ability to change and alter the environment is largely limited to surface processes. Some natural earth processes operate regardless of our activities. For example, earthquakes and vulcanism are the results of internal processes and cannot be effectively stopped or altered. Indeed their occurrence cannot even be predicted with any confidence. Thus, there exist a number of geologic hazards which should be understood in our attempt to direct our growth and development.

Man exists by geologic consent because the basis of our entire environment is the geologic system. Therefore, it is incumbent upon man to understand the system and to live within it.

Discussion

Modification of Natural Systems. As population has grown, man has become increasingly involved in changing the natural water system of the continents largely through the construction of dams and canals and by utilization of ground water. Dams are built to store water for industrial and irrigation use, to control floods, and to produce electric power. Yet, when a dam is built, the balance of the river system, established over thousands or millions of years, is instantly upset. The new structure causes many unforeseen long-term effects both upstream and downstream. Some of the more important consequences of dam construction are: (1) The balance between discharge, sediment load, and stream gradient is destroyed. As a result, sediment is deposited as a delta in the reservoir and over the valley floor upstream. Downstream water released from the reservoir is practically free from sediment and may cause erosion. (2) The creation of a large artificial dam produces a completely new environment. The water table rises, evaporation increases, and new plant growth causes an increase in transpiration.

The effects of modifying a drainage system can best be understood in terms of the concept of equilibrium in river systems and the principles of sediment transportation and deposition—discussed in previous chapters. Inasmuch as the volume of water and sediment flowing in a river approaches equilibrium with the gradient, a change in any of these factors will cause a corresponding change in others. We have already seen in Chapter 7 some of the consequences of modifying the Nile River by the construction of the Aswan Dam. Without the sediment delivered to the delta, marine erosion is eating away the coastline. The aquatic food chain was seriously upset because of lack of nutrients brought down the river with the sediment. Erosion occurred downstream from the dam, and evaporation and vegetation have been problems in the lake behind the dam.

The Colorado River, as well as many others in the United States, has been altered by a series of dams resulting in environmental alterations similar to those experienced by the Nile. Our economy and population are not completely tied to one river, however, so many changes resulting from modifying the river systems go unnoticed. In the case of the Colorado River, erosion

The Earth's Dynamic Systems

downstream from Hoover Dam has extended all the way to Yuma, 560 km (350 mi) below the dam, where the channel has been lowered by approximately 3 m (10 ft). This has eliminated the gravity flow irrigation system and has necessitated pumping to lift the water to the level of irrigation intakes.

Another example of complete modification of a river resulted from the construction of the Arkansas-Verdigris navigation system. This project consists of eighteen dams and locks on the Arkansas River between its confluence with the Mississippi River and the town of Catoosa, Oklahoma. Requiring twenty years of planning and construction, it is one of the most expensive waterways ever built ($1.2 billion). The series of dams has virtually transformed this segment of the river into a series of lakes nearly 700 km (440 mi) long. Although it is too soon to recognize what effects will ultimately occur from such an extensive environmental alteration of a drainage system, conditions of ground water, sediment transportation, and surface ecology are completely altered, and many unforeseen consequences will certainly result.

Of course, there is another side to the story. In terms of property damage and loss of life, floods are considered to be one of the most disastrous of natural phenomena. Floods are caused primarily by high precipitation during a short period of time in a given drainage basin. Rapid thawing of snow and ice may also furnish flood waters during the spring thaw. In some areas like the lower Nile, flooding occurs once a year; in other drainage systems, flooding may occur at irregular intervals, with years passing between floods. However, destructive as they seem to be, floods are natural stream processes which have operated throughout geologic time and are important in maintaining the balance in a river system.

Many tragic floods destroying entire towns have been recorded, especially during the early years of our country, but the construction of dams, natural levees, and other flood prevention systems now has reduced the hazards of this natural process along most of the major rivers.

We cannot escape the long-range problem of flood control measures because dams and levees upset the balance of a river system. The gradient is modified, sediment is impounded behind dams, and levees are built higher. Ultimately, the reservoirs will be filled with sediment, and meandering rivers will change their courses regardless of levees.

Another way in which man has modified river systems is simply by urbanization. The construction of cities may at first seem unrelated to modifying a river system, but a city significantly changes the surface runoff, and the resulting changes in river dynamics are becoming very serious and costly. Modification of a river by urbanization is largely concerned with manner of surface runoff. Water that falls to the earth as precipitation normally follows several paths in the hydrologic system. Generally 54 to 97% returns into the air directly by evaporation and transpiration; 2 to 27% collects in a stream system as surface runoff; and 1 to 20% infiltrates into the ground and moves slowly in the subsurface toward the sea. Urbanization disrupts the normal hydrologic system by changing the nature of the terrain, which in turn affects rates and percent of runoff and infiltration. Roads, sidewalks, and roofs of buildings render a large percent of the surface impervious. This not only increases the volume of surface runoff, but runoff is much faster because water is channeled through gutters, storm drains, and sewers. As a result, flooding increases in intensity and frequency.

The problem of urbanization and its effect upon natural runoff can readily be understood by considering the characteristics of stream flow as shown in *figure 18.1*. This graph shows variations in stream flow with time. The shape of the curve depends upon the nature of the stream system, which is influenced by such factors as relief, vegetation cover, permeability of ground, and number of tributaries. The low, flat portion of the curve represents the portion of stream flow attributable to ground water and is referred to as base flow. The shaded part of the graph represents a period of rainfall, and the high part of the curve shows the resulting increase in discharge. An important factor shown in *figure 18.1* is the time lag or interval of time between the storm and runoff.

The effect urbanization has on runoff is shown on the hydrograph in *figure 18.2*. Lag time is materially decreased because water runs off faster from streets and roofs than from naturally vegetated areas. As the runoff time decreases, the peak rate of runoff increases. The rate of runoff is also increased by shorter man-made channels (gutters, storm drains, etc). The net result is that floods increase in number and intensity while the base flow of the stream is decreased.

Geologic Problems of Waste Disposal. Man's industrialized society produces an ever-increasing variety and quantity of toxic and noxious waste material. Most of us are aware that tra-

Figure 18.1 Hydrograph showing variations in stream flow with time under normal terrain conditions. After a flood there is a considerable lag of time between the time of maximum precipitation and maximum runoff because water is held back by vegetation, seepage into the subsurface, and a slowdown of runoff due to the curves and ends in stream channels.

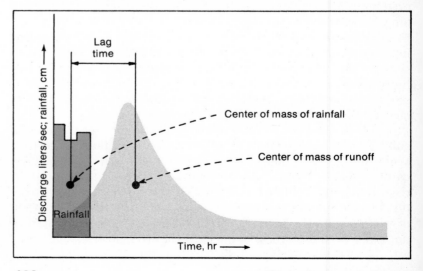

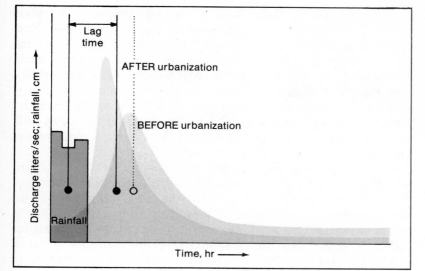

Figure 18.2 *Hydrograph showing variations in stream flow with time in an urban area. There is very little lag time between points of maximum precipitation and maximum runoff because much of the surface is made impermeable by roads, sidewalks, buildings, and parking lots. In addition, water is channeled directly through sewers for rapid runoff. As a result floods increase intensely.*

ditionally man has used fresh water to remove and dilute solid waste and the atmosphere to dilute gaseous waste products of combustion. But, until recently, we have been generally unaware of the fact that our natural systems of waste disposal have become saturated and will no longer absorb our waste products. We cannot take our waste "away" as some politicians have suggested. Our waste products are in the earth's natural systems and will remain in them. The United States in particular has developed a "wasteful" society. The problem is particularly acute because of business practices based on such things as planned obsolescence, throw-away containers, and the hard sell of new models. In addition, high labor costs make it uneconomical to repair, reclaim, or recycle, so the volume of waste grows unnecessarily at a staggering rate. This of course greatly reduces our natural resources. Unfortunately, waste is not a by-product of production; it is eventually a product itself. Our economy has reached a point where thrift is too expensive. A few examples will illustrate the magnitude of the problem.

At the present time, the average U.S. citizen produces over one ton of solid waste annually. For a city of 250,000 this amounts to 450 metric tons (500 short tons) of waste per day. Waste from New York City dumped out into the ocean far exceeds the volume of sediment contributed by local rivers. Estimates show that in the United States man produces over 1.8 billion metric tons (2 billion short tons) of waste per year as compared to 486 million metric tons (540 million short tons) of sediment carried naturally by the river systems. Clearly, man has become a major source of sediment—of a type not conducive to forming pleasant beaches, clean rivers, and healthy lakes. Since we have created a volume of waste which cannot be assimilated and diluted in the river systems, oceans, and atmosphere, we must live with it by converting it into one state or another (solid, liquid, or gas).

The problems of waste disposal have many geologic ramifications. If we bury waste, we have ground-water problems. If we dump it into streams and rivers, it is not carried away and lost; it accumulates on beaches and in estuaries, where it alters the environment of the oceans. Our previous methods of elimination have not been "waste disposal" but "waste dispersal." Any significant solution to the elimination problem must involve a consideration of what kind of waste disposal and dispersal methods the given geologic processes of an area will accommodate without critically altering the geo-biological environment.

Types of Waste Disposal and Their Geologic Problems. Few people are aware of the great variety of waste products produced by our modern culture. We commonly label waste as either municipal, agricultural, or industrial, but it is probably more important to consider waste according to composition rather than source. Carefully consider the contents of *table 18.1* as man's contribution to the hydrologic cycle.

1. Solid Wastes. Solid wastes are disposed of in many ways including landfilling, incineration, compositing, open dumping, animal feeding, fertilizing, and disposing in oceans. The geologic consequences include changes in the surface of the land where the waste is deposited and changes in the environment (rivers, lakes, oceans, ground water) where the mass of waste is concentrated. Landfilling is one of the most widely used methods of disposing of solid waste. The most acceptable form is termed sanitary landfill: refuse placed in a pit or trench and immediately covered with dirt and soil. Open dumps are also common, but they are less desirable because gases and leachates from waste can easily escape into the atmosphere and water supply. In addition, open dumps are also readily accessible to insects and animals. The major problems associated with sanitary landfills and open dumps are related to the hydrological characteristics of the site, including the porosity and permeability of the rock in which the fill is located and whether or not the waste deposit intersects the water table. The altered topography associated with dumps and landfill is also critical because it may change the drainage and ground-water conditions. Perhaps the most critical contamination problem is created by the leachate formed as circulating water passes through the landfill and dissolves the organic and inorganic compounds which enter ground-water reservoirs.

2. Liquid Wastes. Traditionally, liquid wastes are simply discharged into the surface drainage systems and diluted. Ultimately, they accumulate in lakes and the oceans, where they are stored. As the volume of liquid waste increases, the capacity for the natural water system to dilute it is overwhelmed, and the drainage system becomes a system of moving waste. Estimations are that by the year 2000 about two-thirds of the total volume of stream flow in the continental United States will be liquid waste.

The Earth's Dynamic Systems

Table 18.1. Wastes That Enter the Hydrologic Cycle

Solid Wastes

Garbage
  Waste from preparation of
    food
  Market refuse—handling,
    storage of produce and meat

Rubbish (household)
  Paper—boxes, cartons
  Plastics
  Rags
  Grass, leaves
  Metal cans
  Dirt
  Stones, brick, ceramics
  Glass

Ash
  Fly ash
  Residue from combustion of
    coal

Bulky waste
  Large auto parts
  Appliances
  Furniture
  Trees, stumps, branches

Dead animals

Construction and demolition
  Lumber and sheeting scraps
  Broken concrete, plaster, etc.
  Pipes, wire
  Asphalt paving fragments

Industrial waste
  Food processing (slaughter
    house, etc.)
  Wood, plastic scraps
  Slag

Agricultural waste
  Manures
  Crop residues
  Pesticide residues

Solid Wastes (continued)

Sewage and sewage treatment
residues
  Sludge

Mining wastes
  Concentrations of various
    minerals
  Strip mine tailings
  Placer mine tailings
  Quarries, rock piles

Liquid Wastes

Domestic sewage
  Raw sewage
  Treated residues

Industrial
  Liquid chemical residues
  Saline brines
  Paint sludge

Gaseous Wastes

Carbon monoxide
  Internal combustion of
    automobiles
Sulfur dioxide and sulfuric acid
  Sulfur-bearing fuels—power
    plants, smelters
Hydrogen fluoride
  Production of fluorine ores
Particulate matter
  Dust, soot, ash
  Metals
  Insecticides, herbicides
  Lead
  Jet exhaust

Radioactive Wastes

Explosions of nuclear devices
  Radioactive fallout
Atomic energy plants
  Radioactive waste

From Gazda and Malina, 1969.

Other methods of dispersal include barging or piping waste out into the ocean, but the sea is already showing signs of serious pollution, and large inland lakes such as Lake Erie are already highly polluted. Some liquid waste is pumped into wells and injected into the pores and natural openings of rock—a potentially catastrophic situation for the ground-water setting unless the geology and hydrodynamics of the subsurface are clearly understood. Some unforeseen disturbance of the subsurface by waste disposal may occur as illustrated by the earthquake effects produced at the Rocky Mountain Arsenal in Denver. Pressure from the injected fluids reduced the frictional resistance along fracture planes and resulted in man-made earthquakes.

One very subtle type of liquid pollutant is hot water created by cooling systems in power and industrial plants. Although the water itself is not contaminated, the temperature alone is enough to alter the biological conditions of the streams and lakes into which it flows. Such pollution is called thermal pollution.

3. Gaseous Wastes (Air Pollution). The average U.S. citizen is well aware of air pollution because the problem is heavily publicized by the news media. At sea level, our natural gaseous environment, the atmosphere, is composed of approximately 78% nitrogen, 21% oxygen, 1% argon, and about .03% carbon dioxide. Water vapor content varies from nearly zero to as much as 4%.

The population explosion with its consequent industrial expansion has produced pollutants in the form of minute liquid and solid particles and a variety of gases which are suspended in the atmosphere. In the past pollutants were expelled into the air, with the reasonable assurance that normal atmospheric processes would disperse and dilute the waste to a non-harmful, unnoticeable level. In many areas of heavy industrialization, the atmosphere's capacity for absorption and dispersal has been exceeded, and the composition of air has been severely altered. If the troposphere (that part of the atmosphere involved in most human activities) extended indefinitely into space, there would be little air pollution problem. However, few pollutants move out of the troposphere, the lower 10 to 15 km of the atmosphere, into the overlying stratosphere for any great length of time; thus, a steadily increasing volume of pollutants is concentrated mostly in the lower part of the troposphere.

Carbon dioxide is a minor but important constituent of the atmosphere used in various biological and chemical processes. Under normal conditions its concentration in the atmosphere is maintained at approximately .03%. Through the process of photosynthesis, plants extract carbon dioxide from the air and expel oxygen as waste. Through the process of respiration, animals extract oxygen from and return carbon dioxide to the atmosphere. Weathering processes "fix" carbon dioxide in new minerals. A considerable amount of carbon dioxide is extracted from sea water by chemical and biological processes to form limestone, $CaCO_3$, and great quantities have been stored in deposits of coal and peat. The respective extraction and expulsion of oxygen and carbon dioxide by these agents constitute a delicate chemical balance which makes terrestrial life possible.

With the combustion of oil, gas, and coal for industry, heating, and transportation, great quantities of carbon dioxide are suddenly released back into the atmosphere and threaten to upset the natural balance. Estimates are that in the last 80 years carbon dioxide in the atmosphere has increased 10%. The principal concern over the increasing concentration of CO_2 in the atmosphere is that it will alter the earth's temperature and climate. Carbon dioxide, along with water vapor and ozone, prevent heat loss by absorbing heat radiated from the earth toward space. The effect of this absorption produces a "greenhouse effect" in which the gas and heat are held in by the gas roof. The progressive increase in carbon dioxide will cause the earth's temperature to rise. Some predict that by doubling the carbon dioxide, the average surface temperatures will increase 3.8° C. Such a change would produce secondary changes in weather, in moisture content of the air, and possibly a more rapid melting of the glaciers.

Carbon monoxide is a major pollutant in the atmosphere produced chiefly by incomplete combustion in automobile engines. Approximately 1 kg (3 lb) are produced by each gallon of fuel burned. A definite health hazard occurs when the atmosphere contains as much as 120 ppm. Potentially this is a major health problem in urban areas.

Sulfur dioxide and sulfuric acid are produced in power plants using sulfur-bearing fuels and in smelters treating sulfide ores, as large quantities of sulfur dioxide are ejected into the atmosphere, and the sulfur dioxide converts to sulfuric acid in the presence of oxygen, sunlight, and water vapor. Sulfur pollutants are commonly the most harmful agents in smogs and are also very damaging to vegetation and crops. Around some smelters vegetation has been completely destroyed. In the U.S. alone, an estimated 22.5 million metric tons (25 million short tons) of sulfur dioxide are spewed into the atmosphere annually.

4. Radioactive Wastes. All industries are faced with problems of disposing waste products, but none are greater than the problem of disposing waste from nuclear energy plants. As nuclear energy is generated, a variety of radioactive isotopes are created, some with a short half-life and others with a very long half-life. Nuclear waste is extremely hazardous and generates large amounts of heat. Thus any disposal system must be capable of removing heat and, at the same time, completely isolating the waste from the biological environment. In addition, containment must be maintained for exceptionally long periods of time. Compared to that produced by many other industries, the volume of radioactive waste is not large, but the hazards and heat generated are considerable. In 1980 an estimated 9 million liters (2.5 million gal-

lons) of radioactive waste will be produced, and by the year 2000 the amount will increase to 144 million liters (39 million gallons).

In many respects the problems of radioactive waste disposal are similar to those of other pollutants produced by our society. What long-range effect will it have on the environment? Unlike other pollutants, radioactive waste can induce mutations in living species and, since we are looking more and more to nuclear energy to solve many of our resource problems, the long-term effect of radioactive waste disposal could be catastrophic.

One of the more promising methods of radioactive waste disposal involves the storage of waste in thick salt formations. Salt deposits are desirable because they are essentially impermeable and are isolated from circulating ground water. In addition salt yields to plastic flow so that it is unlikely to fracture and make contact with leaching solutions over extended periods of time. Salt also has a high thermal conductivity for removing heat from waste and has approximately the same shielding properties as concrete.

In theory the wastes would be solidified and sealed in containers 15 to 60 cm (6 to 24 in.) in diameter and up to 3 m (10 ft) in length. The containers would then be shipped to salt mines in the stable interior of the continent where seismic activity is minimal. There, they would be placed into holes drilled into a salt formation in a deep salt mine. When filled with waste, the hole would be closed and the room itself would be back-filled with crushed salt. Since plastic flow of salt causes consolidation and recrystallization, the salt formation would return to its original state within a few decades.

5. Mining Wastes. The waste products from mining operations include (1) tailings and dumps, (2) altered terrain (open pit mining), (3) changes in the composition of the surface, and (4) solid, liquid, and gaseous wastes produced by refining.

In the United States approximately 2.7 billion metric tons (3 billion short tons) of rock are mined each year. Approximately 85% comes from open pits or strip mining, requiring the removal of an additional 5.4 billion metric tons (6 billion short tons) of rock as overburden. In these operations about 12,000 km² (5000 mi²) of land in the United States have been affected by surface mining. The principal geologic problem arises from the alteration of the terrain by the creation of open pits and from the creation of artificial mounds and hills from tailings.

The Appalachian coal fields serve as an example of some of the undesirable results of mining. Strip mining produces large tailing piles and exposes large quantities of rock containing many sulfides to the atmosphere and drainage system. As this material weathers, sulfuric acid is produced, thus increasing the acidity of the surface waters. Estimates are that the Monongahela River, which drains parts of Pennsylvania and West Virginia, discharges the equivalent of 180,000 metric tons (200,000 short tons) of sulfuric acid each year into the Ohio River. Other mining areas inject into the drainage system lead and other heavy minerals which can affect the ecosystem of the river and make its waters unfit for human consumption.

Mining dumps from large-scale subsurface operations can also alter the environment. For example, in South Africa the gold-mining operations produce huge quantities of fine quartz sand, which is piled in mounds near the mine. This material is difficult to stabilize: thus, when the wind velocities are sufficiently high, sand storms occur. Where cyanide processes have been used to recover gold, the tailings are toxic.

Most mine dumps are unstable and highly susceptible to mass movement unless they accumulate under proper engineering supervision. In 1966, the mudflow from coal mine dumps in Aberfan, Wales, completely destroyed a school and killed or injured many children.

An additional problem arises if the mine tailings enter the drainage system, since they can readily choke a stream channel and increase the flood hazards. Alteration of a stream system can also be produced from placer mining where the movement of large quantities of sediment upsets the balance of the stream.

Resources

Statement

The important minerals upon which our civilization depends constitute almost an infinitesimally small part of the earth's crust. Whereas feldspar, quartz, and clay minerals are abundant and widely distributed, copper, tin, gold, etc., occur in such small quantities that their concentrations are measured in parts per million, and in most cases a very few parts per million. How are these very small quantities of important minerals concentrated into an economic deposit? The answer is by the various geologic processes operating in the earth's system. Essentially every geologic process plays a part in concentrating some valuable mineral deposit, including igneous processes, weathering, stream action, sedimentation, metamorphism, and deformation of the crust. It is critical to understand that these processes operate so slowly by human standards that the rates of replenishment are infinitesimally small compared to rates of consumption. A mineral deposit has finite dimensions and is therefore finite, exhaustible, and nonrenewable. If we know the approximate extent of a given resource we can predict how long it will last.

The consumption of natural resources is proceeding at a

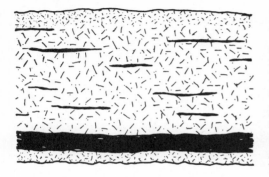

Figure 18.3 *Concentration of chromite as early-formed crystals in the magma settles to form layers near the base of an igneous body.*

phenomenal rate. Only a moment's reflection is necessary to understand that the period of rapid population and industrial growth that has prevailed during the last few hundred years is not normal but is one of the most abnormal phases of human history. The impact of our present growth and associated consumption of natural resources presents one of the most serious problems facing man today. The problem is basically one of changing from a period of growth to nongrowth and will entail a fundamental revision of those aspects of our current economic and social thinking—based on the assumption that growth can be permanent.

Discussion

Formation of Mineral Deposits. Perhaps the best way to appreciate the fact that our natural resources are finite and nonrenewable is to consider how mineral deposits are formed. Igneous processes concentrate minerals in several important ways. One is by direct segregation in which early-crystallized heavy mineral grains sink down through the fluid magma and are concentrated as a layer near the base of an igneous body. Chromite, nickel, and magnetite are good examples (*figure 18.3*).

Another way in which igneous activity concentrates minerals is by crystallization of minerals from the volatile-rich fluids during the late stages of crystallization of a magma. The early-formed minerals remove from the magma much silicon, oxygen, aluminum, iron, etc., so the remaining fluids may have much higher than normal percentages of copper, lead, zinc, gold, silver, and other rare elements. These fluids commonly move from the main magma chamber into fractures in the surrounding rock, and as they cool they precipitate various minerals to form veins. The high-temperature solutions contain much water, so the deposits are known as hydrothermal deposits (*figure 18.4*). Hydrothermal solutions may permeate a very large

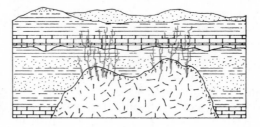

Figure 18.4 *Schematic cross section through an intrusive body showing the location of hydrothermal vein deposits.*

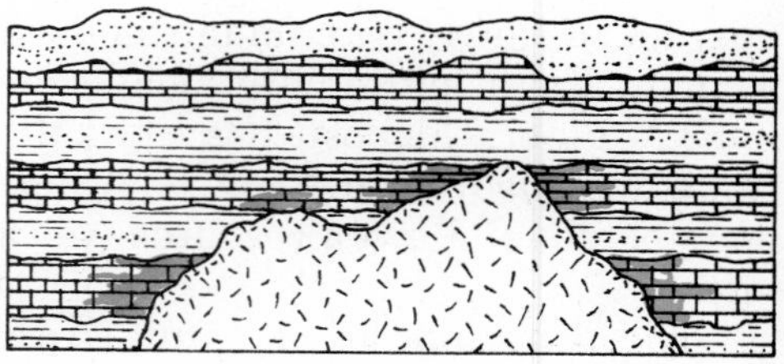

Figure 18.5 *Schematic diagram showing the concentration of ore deposits by contact metamorphism.*

The Earth's Dynamic Systems

mass of rock and disseminate minerals throughout the body. These may produce low-grade ore, such as some important reserves of copper. Hydrothermal solutions also may ultimately emerge as hot springs and deposit valuable ore in a very shallow zone near the surface.

An important process associated with igneous intrusions is **contact metamorphism.** In this process the highly active fluids of the magma react with the surrounding rock and alter it chemically by adding or replacing certain components. Limestone, for example, may be replaced by iron ore (*figure 18.5*).

Weathering of original mineral deposits may serve to concentrate and enrich ore originally formed by igneous processes. In this process weathering removes the more soluble material and concentrates the resistant insoluble minerals as a residuum (*figure 18.6*). This type of deposit is commonly referred to as a secondary enrichment. Weathering also plays a dominant role in forming bauxite, the principal ore of aluminum. Bauxite is a mixture of hydrous oxides of aluminum and forms as an insoluble product in the weathering of granite.

Sedimentary processes concentrate minerals in a variety of ways. One is by chemical precipitation in lakes and restricted embayments of the sea. Dissolved salts enter the lake or bay with the inflow of water and are concentrated as the water evaporates.

Another way in which sedimentary processes concentrate minerals is through the fluid action of streams and currents. In the process of transportation currents tend to sort out and separate the material they carry according to size and density. The heavier material will be deposited where the current action is weak, such as on the inside of meander bends in a river or on protected beaches and bars (*figure 18.7*). Through this process, minerals of high density are concentrated as layers and lenses in gravel and coarse sand and form what is known as placer deposits. Gold, tin, and diamonds all form important placer concentrations in both modern and ancient sand and gravel deposits.

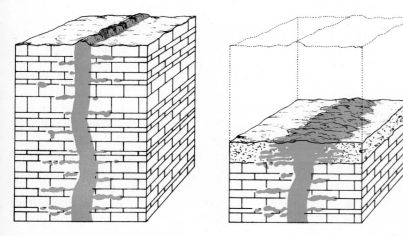

Figure 18.6 *Concentrations of ore by solution and weathering of surrounding rock. The insoluble ore accumulates as a residual deposit.*

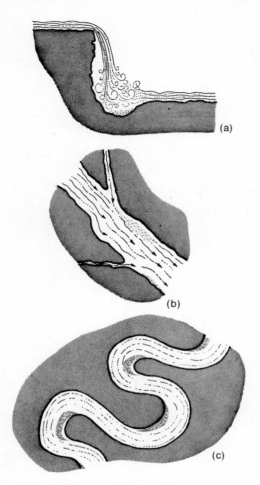

Figure 18.7 *Schematic diagrams showing several ways in which ore may be concentrated into placer deposits.*

In addition sedimentary processes are important in forming many nonmetallic deposits such as clay for bricks, limestone for cement, and sand and gravel for building material.

The main sources of energy also come from sedimentary processes. Coal and petroleum are commonly referred to as fossil fuels because they originate from once-living material which ultimately became incorporated into a sedimentary rock. **Coal** originates from plant material which was buried and preserved from decay. Only under special sedimentary environments such as coastal swamps, can vegetation accumulate to a significant thickness and be subsequently buried by sediment (*figure 18.8*). Under heat and pressure from the overlying sediment the vegetation is compressed and transformed into coal.

Petroleum and natural gas also originate in a sedimentary environment and are intimately involved with sedimentary rocks in their origin, migration, and ultimate entrapment. Petroleum originates from microscopic organisms which accumulate with dark mud in shallow seas or large lakes. After the organisms and mud are deposited, the petroleum and natural gas may be squeezed out of the shale in which it was formed and migrate under directed pressures and the pull of gravity. The petroleum and gas, being lighter than water, separate and move upward where they may be trapped under a variety of geologic conditions (*figure 18.9*). In some instances the petroleum may remain in the shale and form important reservoirs of oil shale such as those in the Rocky Mountain area.

In spite of recent discoveries our geologic resources are finite in amount and are nonrenewable during time periods of less than millions of years. If we exhaust a given mineral deposit, we cannot expect to find another one just anywhere. Areas such as shields, which do not have sedimentary rock, cannot be expected to produce petroleum or coal, and areas with no igneous intrusions cannot be expected to someday produce minerals originating from magmatic processes. Indeed there are only a limited number of places left to look for mineral deposits before a final inventory of our resources can be made.

Nonrenewable Resources and Limits to Growth. We have emphasized in the previous section that minerals and energy resources are formed by geologic processes in a systematic manner. They do not occur in a haphazard or random way, nor are they distributed evenly throughout the continents. This creates continents and nations that have resources and those that do not. Iceland and Hawaii for example, being built up exclusively of basaltic lava, cannot be considered as a good potential for petroleum and natural gas no matter how long we explore the area.

In addition, mineral deposits are a nonrenewable resource. Once a deposit is mined out and used, it is gone and cannot be replaced. It is important, therefore, to consider how our resources are used and the rates at which they are being depleted. Some of the more important mineral and energy resources used

The Earth's Dynamic Systems

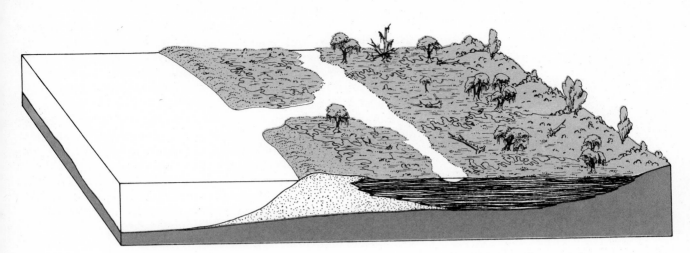

Figure 18.8 Schematic diagram showing how a swamp may be buried to form a coal deposit.

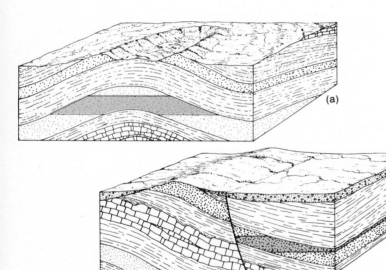

(a)

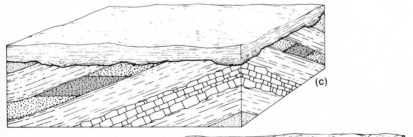

(b)

(c)

(d)

Figure 18.9 Cross sections showing various geologic conditions which may produce an oil trap. (a) Anticline—Oil, being lighter than water, migrates up the dip of permeable beds and may be trapped beneath an impermeable shale bed in the crust of an anticline. (b) Fault trap— Impermeable beds may be displaced against a permeable stratum and trap the oil as it migrates updip. (c) Unconformity— An impermeable layer caps inclined strata to form a seal which traps the upward-migrating oil. (d) Shale surrounding a sandstone lens forms an impermeable barrier and prevents the oil from escaping.

in today's major industries are listed in *table 18.2*. The number following each resource in column 3 is the number of years the present known reserves will last at the current rate of consumption. This figure assumes no growth and that the usage rate will remain constant. It is called the *static index* and is normally used to express future available resources.

The rate of resource consumption is not static but is increasing exponentially. This is a result of the growth rate in population plus the fact that the average consumption per person per year is increasing each year. In other words, the exponential increase in resource consumption is driven by growth in both population and capital. A more accurate estimate of the lifetime of ‚a given mineral is the *exponential index* or the number

Table 18.2   Nonrenewable Natural Resources

1	2	3	4	5
Resource	Known Global Reserves	Static Index (years)	Exponential Index (years)	Exponential Index Calculated Using Five Times Known Reserves (years)
Aluminum	1.17×10^9 tons	100	31	55
Chromium	7.75×10^8 tons	420	95	154
Coal	5×10^{12} tons	2300	111	150
Cobalt	4.8×10^9 lbs	110	60	148
Copper	308×10^6 tons	36	21	48
Gold	353×10^6 troy oz	11	9	29
Iron	1×10^{11} tons	240	93	173
Lead	91×10^6 tons	26	21	64
Manganese	8×10^8 tons	97	46	94
Mercury	3.34×10^6 flasks	13	13	41
Molybdenum	10.8×10^9 lbs	79	34	65
Natural Gas	1.14×10^{15} ft^3	38	22	49
Nickel	147×10^9 lbs	150	53	96
Petroleum	455×10^9 bbls	31	20	50
Platinum Group	429×10^6 troy oz	130	47	85
Silver	5.5×10^9 troy oz	16	13	42
Tin	4.3×10^6 lg tons	17	15	61
Tungsten	2.9×10^9 lbs	40	28	72
Zinc	123×10^6 tons	23	18	50

The Earth's Dynamic Systems

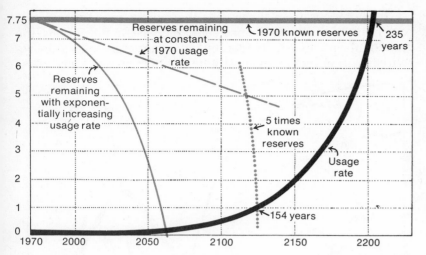

Figure 18.10 *The lifetime of known chromium reserves depends on the future usage rate of the mineral. If usage remains constant, reserves will be depleted linearly (dashed line) and will last 420 years. If usage increases exponentially at its present growth rate of 2.6% per year, reserves will be depleted in just 95 years. If actual reserves are five times present proven reserves, chromium ore will be available for 154 years (dotted line), assuming exponential growth in usage. Even if all chromium is perfectly recycled, starting in 1970, exponentially growing demand will exceed the supply after 235 years (horizontal line).*

of years our known reserves will last at an exponential consumption rate. This estimate is shown in column 4, and for many minerals the usage rate is growing even faster. You will note that exponential rates of consumption drastically reduce the estimated lifetime of a reserve. Coal will be depleted in 111 years rather than 2300. Chromium will be gone in 95 years rather than 420, and so on.

Now let us assume that our known reserves can be expanded five times the present known reserves by exploration and new discoveries. This will *not* extend the lifetime of a deposit five times because the effect of exponential growth is to consume resources at an ever faster rate. Column 5 shows the number of years that five times our known global reserves will last with consumption growing exponentially at the average annual rate of growth.

Figure 18.10 illustrates the effect of exponentially increasing consumption of a nonrenewable resource. Chromium is taken as an example because it has one of the longest-lasting static reserves of all the resources listed in *table 18.2*. At the current rate of usage the known reserves would last about 420 years. The actual reserves of chromium, however, are being consumed at an increasing rate of 2.6% annually. If the population and industrial growth rates continue, known resources will be consumed not in 420 years but in 95 years. If we assume that reserves yet undiscovered could increase our known reserves five times, this five-fold increase would extend the lifetime of the reserves only from 95 to 154 years. Even with 100% recycling of chromium so that none of the initial reserves were lost, the demand would exceed supply in 235 years.

We could draw a similar graph for each resource listed in *table 18.2*. The time scale would vary, but the general shape of the curves would be the same. The exponential index shown in column 5 is the critical figure. At our present growth in consumption rates copper will be depleted in 21 years, and in only 48 years if reserves are multiplied by five. Gold has a projected lifetime of 9 years, natural gas 22, tin 15, etc.

It is obvious from these data that, at the present resource consumption rates and the projected increase in these rates, the great majority of the currently important nonrenewable resources will be extremely costly 100 years from now, regardless

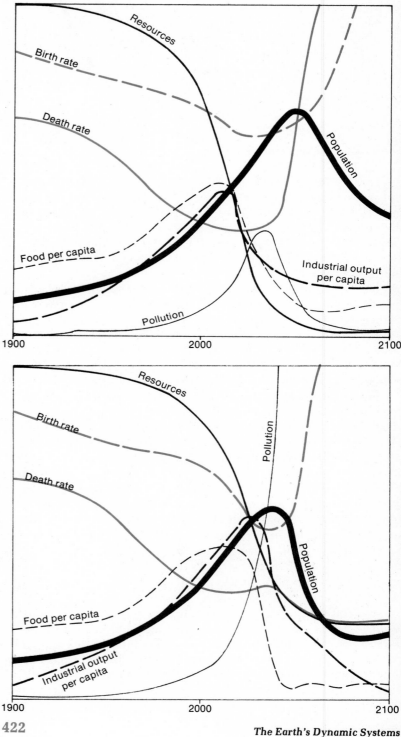

Figure 18.11 *The computer model of resource consumption and its influence on other variables assumes no major change in the physical, economic, or social relationships that have historically governed the development of the world system. All variables plotted here follow historical values from 1900 to 1970. Food, industrial output, and population grow exponentially until the rapidly diminishing resource base forces a slowing down in industrial growth. Because of natural delays in the system, both population and pollution continue to increase for some time after the peak of industrialization. Population growth is finally halted by a rise in the death rate due to decreased food and medical services. Lower diagram represents the same assumptions but double resource reserves. Industrialization can reach a higher level, but pollution rises much more rapidly, causing immediate increase in death rate and decline in food production.*

of the most optimistic assumptions about undiscovered reserves. The limit of growth in the world system will probably not be pollution but depletion of natural resources. The interaction of some of the major variables involved as the world system grows to its ultimate limits is shown in *figure 18.11*. Assuming no major change in physical, economic, or social relationships that have governed the world system historically, depletion of natural resources will be the main factor in limiting the growth of our industrial complex and population. According to *figure 18.11*, food, industrial output, and population will continue to grow exponentially until the rapidly diminishing resources force a sharp decline in industrial growth.

Summary

The natural systems which operate upon the earth constitute the basis of our ecology and the source of our natural resources. The present environment in which we live is the product of a complex of physical, chemical, and biological processes which have operated for many thousands, if not millions, of years, and it has changed and evolved very slowly by human standards. The present state of the earth's surface features—rivers, ground water, shorelines, and atmosphere—represents a balance of forces attempting to establish equilibrium, a balance which man is capable of altering almost instantaneously. When part of a natural system is altered, other parts of the system adjust to the change, which may have long-range and unforeseen side effects. When change is necessary, one must consider the impact on the entire system if a desirable future is to be achieved.

Our mineral resources are also the result of geologic processes and, for all practical human purposes, they are finite and nonrenewable. Growth in industrial output and consumption cannot be permanent.

Additional Readings

Cloud, P., chairman. 1969. Resources and Man. (A study by the Committee on Resources and Man of the National Academy of Sciences-National Research Council.) San Francisco: W. H. Freeman and Company.

Flawn, P. 1970. Environmental Geology. New York: Harper and Row, Publishers.

Hubbert, M. K. 1962. Energy Resources: A Report to the Committee on Natural Resources. Washington, D. C.: National Academy of Sciences-National Research Council.

———. 1971. "The Energy Resources of the Earth." Sci. Amer. 224(3):60-84 (Offprint No. 663).

Matthews, W. H., ed. 1970. Man's Impact on the Global Environment: Assessments and Recommendations for Action (Report of the Study of Critical Environmental Problems (SCEP). Cambridge, Mass.: The MIT Press.

Meadows, D. H., D. L. Meadows, J. Randers, and W. Behrens, III. 1972. The Limits to Growth. New York: Universe.

Park, C., Jr. 1968. Affluence in Jeopardy. San Francisco: Freeman, Cooper and Company.

Skinner, B. 1969. Earth Resources. Englewood Cliffs, N.J.: Prentice-Hall.

19 Geology of the Moon

B efore man's first landing on the moon, many scientists thought of the moon as a Rosetta stone: an airless, waterless body untouched by erosion, containing clues to events which occurred in the early years of our solar system that would help reveal details of its origin and history and provide new insights about the evolution of the earth. Now, with the Apollo missions completed, many of their fondest hopes have been realized. The thousands of photographs brought back from the moon, together with over 850 kg (1800 lb) of rock specimens and a flood of data from experiment packages, have changed some of our basic concepts of the moon. From what has been learned to date, it has become clear that the moon is one of the most exciting places in the solar system. The rocks and surface features of the moon not only contain a history of its first billion years but provide important insight into other events which transpired during the formation of other planets and during the early evolution of the sun.

The importance of the moon in studying principles of physical geology is that it provides an insight into the basic mechanics of planetary evolution, a subject which cannot be understood from studies of the earth alone. On the earth the geologic record of the very early history and processes has been entirely erased by later episodes of erosion and crustal deformation. But on the moon, there remain many features which were probably formed during the last stages of accretion of material from which the moon was built. It is from studies of the moon that we have learned most of how planets are born and events which transpired during the early, early part of their history.

Major Concepts

1. The surface of the moon can be divided into two major regions: (a) the relatively flat, smooth, dark areas called **maria** (Latin, "seas") and (b) the densely cratered, rugged highlands, originally called **terrae** (Latin, "lands").

2. Most of the craters of the moon resulted from impact, a process fundamental in planetary development.

3. **Rilles, wrinkles,** and some volcanic mounds constitute the remaining major surface features of the moon.

4. The lunar maria are large "seas" of basaltic lava extruded 3 to 4 billion years ago and represent the last major thermal event in lunar history.

5. Lunar rocks and surface material are of igneous and impact origin. The major types include (a) basalt, (b) anorthosite, (c) glass, (d) breccia, and (e) regolith.

6. The moon has a layered, structured crust about 65 km (40 mi) thick and a "mantle" rigid enough to support considerable stress to depths of 800 km (500 mi). It apparently does not have a heavy metallic core.

7. The geologic time scale for events on the moon has been established using the principles of superposition and cross-cutting relations; these provide a framework of relative time into which the major events of lunar history can be arranged in chronologic order.

Major Physiographic Divisions

Statement

When Galileo first observed the moon through a telescope, he discovered that the dark areas are fairly smooth and the bright areas are rugged and densely pockmarked with craters. He called the dark areas maria and the bright areas terrae. These terms are still used today although we know the maria are not seas of water and the terrae are not similar to our continents. The maria and terrae do represent major provinces of the lunar surface, each with different structure, landform, and history.

Discussion

Maria and terrae are readily apparent on the physiographic map of the moon. The highlands constitute about two-thirds of

Geology of the Moon

Figure 19.1 *The far side of the moon is composed almost entirely of densely cratered terrain, with only a few craters flooded with dark mare material.*

427

the visible surface; they are densely cratered and contain the highest and most rugged topography on the moon. In contrast, the maria are relatively level and have far fewer craters on them. Some maria, such as Imbrium, Serenitatis, and Crisium are confined within the well-defined walls of large circular basins, whereas Oceanus Procellarum occupies a much larger and irregular depression.

The maria and highlands not only represent different types of terrain but broadly represent two different periods in the history of the moon. The highlands, which occupy about 80% of the entire lunar surface, for the most part are an old surface which became extensively pockmarked with craters early in the moon's history. The maria basins were then formed as large impact structures and subsequently filled with lava, which in places overflowed the basins and spread over parts of the lunar highland. The maria are thus relatively young features of the lunar surface, although they were formed 3 to 4 billion years ago.

The far side of the moon was totally unknown until photos were first taken by a Russian spacecraft in 1965. Although details were poorly defined, it was a surprise to learn that the far side of the moon was composed almost entirely of densely cratered highlands. Later, American orbiting satellites completely photographed the far side of the moon with definition sufficient to map the surface in considerable detail (figure 19.1). These photographs confirmed that the far side of the moon is densely cratered with only a few large craters containing mare material. The reason that mare material is characteristic of only the near side of the moon remains a fundamental problem of lunar geology.

Craters

Statement

Craters are the most prominent landforms on the lunar surface. Early maps of the moon show more than 300 major craters but, with the aid of modern telescopes, over 300,000 craters more than 1 km (.6 mi) in diameter have been identified. With orbiting satellite photography, a vast number of additional smaller craters are seen across the lunar surface, some as small as 10 cm in diameter. Thus, whereas the major process that sculptures the earth's surface is running water, on the moon the main process is impact, and this process operates on all scales to produce craters ranging in size from minute pits to gigantic basins.

The importance of impact goes far beyond the formation of craters on the moon, for most scientists now subscribe to the accretionary process for the formation of solar planets. Impact is thus considered by many to be a fundamental and universal process in planetary development. In all probability, the earth experienced impact and shock effects on a grand scale during

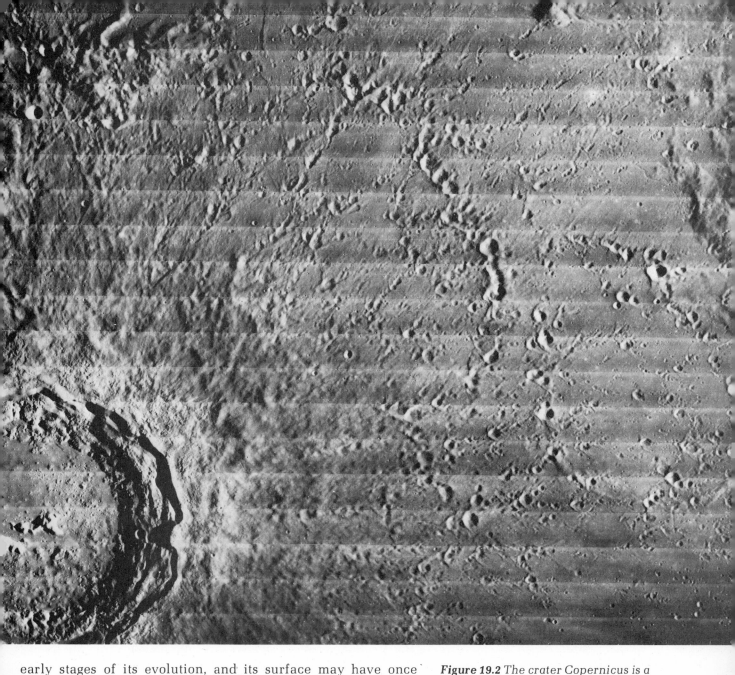

early stages of its evolution, and its surface may have once appeared much like that of the moon.

Discussion

Rayed Craters. The craters Tycho, Kepler, and Copernicus are examples of a group of large fresh craters which are distinguishable by bright streaks or rays which fan out from the crater rim like huge splash marks. Careful study of this type of crater will reveal much about the nature of impact structures.

Rayed craters are typically saucer-shaped with a rugged peak near the center (*figure 19.2*). The rim of the crater con-

Figure 19.2 The crater Copernicus is a classic rayed crater. Numerous slump blocks form concentric terraces inside the crater rim, and a bright, hummocky ejecta blanket extends out from the crater rim to a distance approximately equal to the crater's diameter. A central peak rises from the crater floor, and rays composed of glass beads and pulverized rock can be traced hundreds of kilometers as great splash marks. (See also figure 19.16.)

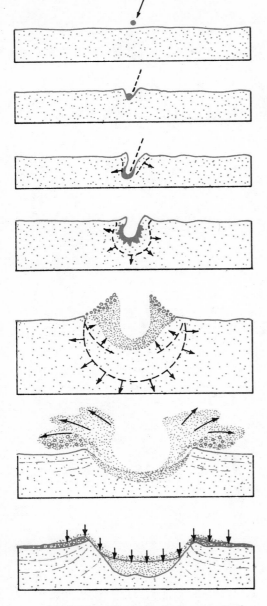

Figure 19.3 *Hypothetical stages in the formation of a meteorite impact crater. The kinetic energy of the meteorite is almost instantly transferred to the ground as a shock wave, which moves out, compressing the rock. The rock, at the point of impact, is intensely fractured, fused, and partly vaporized by shock metamorphism. The shock wave is reflected back as a rarefaction wave which throws out large amounts of fragmental debris. The solid bedrock is also forced upward to create the crater rim. A large amount of fragmental material falls back into the crater.*

sists of numerous slump blocks which form a series of concentric terraces inside the lip of the crater. Surrounding the crater rim is a hilly or hummocky terrain composed of relatively bright material which gradually diminishes away from the crater and grades into the rays. In most craters, the hummocky terrain extends out from the crater rim a distance roughly equivalent to the diameter of the crater. The rays, like the hummocky terrain, also consist of highly reflective material and may extend out from the crater hundreds of kilometers.

These observations, together with many other lines of evidence, indicate that the craters are the result of meteoritic impact. The bright, hummocky terrain surrounding the crater is believed to be a deposit of material ejected from the crater by the force of impact and is called the **ejecta blanket**. The secondary craters are formed by the impact of large rock fragments of the ejecta, and the rays are far-thrown pulverized rock and glass beads mixed with local material splashed out by the secondary craters.

Mechanism of Crater Formation. Impact processes are nearly instantaneous but can be studied in the laboratory using high-speed motion pictures. The series of diagrams in *figure 19.3* illustrates the basic processes and the resulting products. As a meteorite strikes, the kinetic energy is almost instantaneously transferred to the ground as a shock wave which moves downward and outward away from the point of impact and subjects the rock initially to extreme compression. The compression alone, however, does not form a crater. The key mechanism of cratering is the interaction between compressive waves and the "free" ground surface beyond the point of impact. When the compressive shock wave encounters the ground surface a reflected rarefaction (to make less dense) wave is produced, which is propagated in approximately the reverse direction. The material formerly compressed now expands, causing material to be ejected from the surface and be thrown out along ballistic trajectories to form the ejecta blanket and rays. The central peak on the crater floor is also formed by the rebound after initial compression, and the crater rims may be overturned. Impact cratering is similar in many respects to dropping a pebble in a pool of water. There is an initial compression followed by a rebound which ejects material out away from the point of impact (*figure 19.4*).

It is important to emphasize that the impact of meteorites produces a landform (the crater) and a new rock body (the ejecta blanket) and is, therefore, similar to other rock-forming processes which operate at the surface of a planet, such as vulcanism and sedimentation. In each case, there is a transfer of energy, material is transported and deposited to form a new rock body, and a new landform (crater, volcanic cone or flow, or delta) is formed in the process. The rock-forming processes associated with impact include fragmentation and deposition of rock particles and shock metamorphism, resulting from the

The Earth's Dynamic Systems

force of impact in which new high-pressure minerals are created and partial melting and vaporization take place. The impact process, in addition to producing a crater, produces a new rock body of fragmental material, a breccia composed of fused rock fragments, glass, plus some high-temperature minerals.

Rayless Craters. Some craters exhibit all of the morphological characteristics of rayed craters, but their rims and ejecta blankets are relatively dark, and they lack a system of rays (*figure 19.5*). It is inferred that these rayless craters initially had a bright ejecta blanket and a system of rays, but the fine dust which constitutes the rays became darkened in the course of time, possibly from bombardment of high-energy radiation and

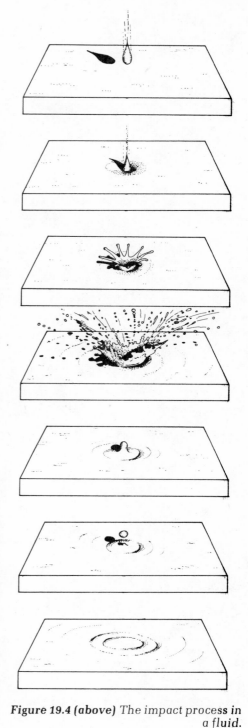

Figure 19.4 (above) *The impact process in a fluid.*

Figure 19.5 (left) *Rayless craters exhibit most of the features of rayed craters, but their ejecta blankets are dark, and they lack the system of rays.*

Figure 19.6 Crater flooded with maria basalt.

mixing with underlying darker material by the churning action of subsequent impact of meteorites and ejected rock fragments.

Other craters have been flooded with maria basalt (*figure 19.6*), which has obscured the central peak of the crater and a large part of the ejecta blanket. In some cases (*figure 19.7*) basalt flows completely fill the crater and surrounding areas so that only a "ghost" of the high crater rim is visible.

Multi-Ringed Basins and Lunar Maria

Statement

No sharp distinction can be made between large craters filled with mare material such as Ptolemaeus and Grimaldi and the great circular maria basins such as Imbrium, Serenitatis,

Figure 19.7 *Many craters, near the margins of the maria, are almost completely covered, so that only parts of their rims can be detected.*

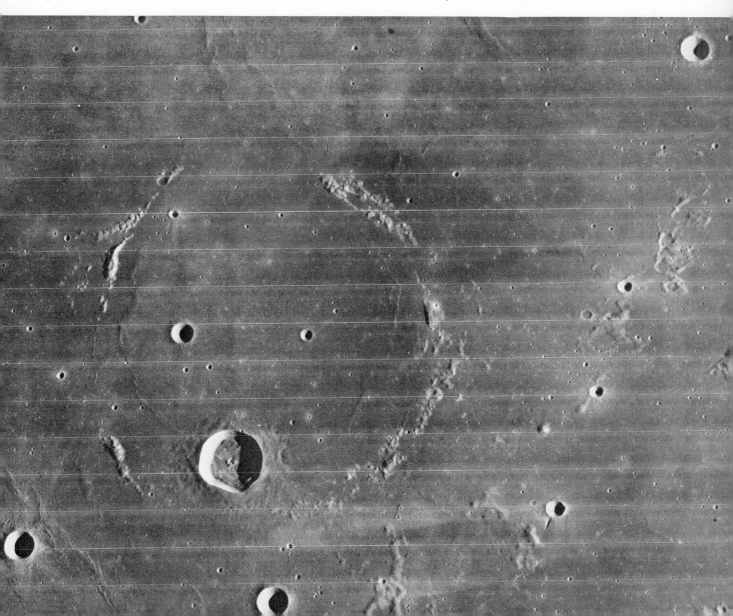

Figure 19.8 *The Orientale Basin is a spectacular multi-ringed crater resembling a gigantic bull's-eye. The rings are believed to be large waves generated in the crust by impact; these waves remain because the crust was permanently deformed. A huge ejecta blanket extends out from the crater and forms a series of radiating linear landforms.*

and Crisium. This fact suggests that the maria basins are actually very large impact craters, a theory advocated in 1893 by G. K. Gilbert, a noted American geologist. Many geologists believe that the large maria basins were originally multi-ringed craters similar to Mare Orientale, a crater near the west margin of the visible side of the moon. These spectacular craters were subsequently flooded with lava, sometimes to overflowing, to form the present extensive dark lava plains. The formation of the maria basins and subsequent extrusions of lava constitute one of the most significant events in lunar history.

Discussion

The youngest and best-preserved multi-ringed basin is Orientale (figure 19.8). The basin is largely hidden from telescopic view but, with orbiting satellite photography, it stands out as one of the most spectacular features on the lunar surface. The basin resembles a gigantic "bull's-eye" consisting of three concentric ridges and intervening lowlands. All three rings contain two types of topographic features. One is concentric to the basin, whereas the other is radial. The spacing of the rings increases outward by a regular ratio, a feature that is apparently common to all multi-ringed basins. Beyond the outermost ring most topographic features are oriented radially to the basin. Maria basalt covers much of the lowland in the center of the basin, but only isolated parts of the outer lowlands are flooded with lava.

The Orientale Basin is believed to have been produced by impact of an asteroid-sized body. The impact of such a large body would generate shock waves which could be considered analogous to the ripples formed when a stone is thrown into a pool of water. The waves were large enough to exceed the elastic limits of the lunar crust and became "frozen" (the crust remained permanently deformed) (figure 19.9). The radial ridges and valleys are believed to be ejecta with secondary impact craters. As can be seen in figure 19.8, the ejecta blanket from Orientale extends out a distance roughly equal to the diameter of the basin.

The multi-ringed structure of Mare Nectaris, located near the eastern margin of the near side of the moon just south of the equator, is somewhat more subtle, but a definite series of arcuate ridges and intervening lowlands can be seen. This is obviously a much older basin than Orientale because the concentric rings and ejecta blanket are considerably modified by subsequent impact craters. Like Orientale, only the central part of Nectaris is flooded with basalt.

Another maria basin that is useful in studying multi-ringed structures is the Imbrium Basin, the largest lunar basin (figure 19.10). It is almost completely flooded with mare material, but the concentric structure can be identified. The innermost ring is exposed only as a ring of islands 675 km (400 mi) in diameter. The second ring is exposed mainly as Montes Alpes and the

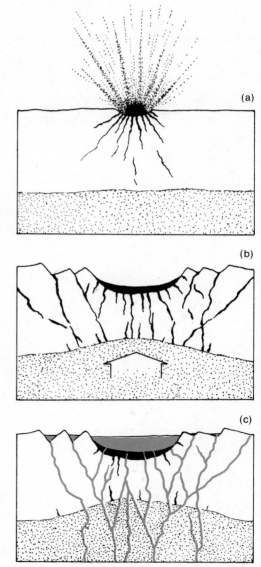

Figure 19.9 The origin of lunar maria. (a) Impact of an asteroid-size body blasts out debris. (b) The shock wave creates a multi-ringed basin, and the mantle adjusts isostatically to the loss of mass in the crater area. This causes additional fracture systems. (c) Interior heat from radioactivity causes partial melting, and basaltic magma rises, filling the basin with numerous thin layers of basalt.

Figure 19.10 The Imbrium Basin, a huge multi-ringed crater now largely flooded with basalt. The innermost ring is a series of islands. The second and third rings form famous "mountain ranges" of the moon. Rugged linear ridges can be seen extending southeastward from the southern margin.

Mare Imbrium

rugged terrain near Archimedes. The third ring is the largest and most conspicuous and corresponds to Montes Carpatus, Montes Apenninus, and Montes Caucasus. This ring marks the beginning of the ejecta blanket characterized by rugged linear ridges best developed at the southeast margin of the basin. Suggestions of still another ring can be seen a considerable distance southeast of Montes Apenninus. Part of it bounds Sinus Medii on the south.

From the evidence described above, it appears that the large maria basins of the moon are impact structures which were subsequently filled or partly flooded with lava. In some craters, such as Nectaris and Orientale, only the central basins and parts of the surrounding lowland were filled. In others such as Imbrium, Serenitatis, and Tranquillitatis, the basins were filled to the outer rim and beyond. Mare Procellarum may be an example where flooding extended beyond the outer rims, completely obscuring the basin structure.

Lava samples brought back from Apollo missions, plus orbiting satellite photos which show lobate margins of individual flow units (figure 19.11) clearly indicate that the material filling the maria basins originated as extensive basaltic lava flows. In addition, vertical exposures in the Hadley Rille, described on the Apollo 15 mission, reveal stratification in which a number of flow units were superposed and separated by breccia, ash, and dust. Perhaps the large regions of plateau basalt, such as the Columbia River Plateau in the northwestern United States and the Deccan Plateau of India, resemble the widespread floods of basalt on the maria as close as anything on earth.

Several fundamental facts about the lunar maria are generally accepted. Judging from the number of craters which are superposed upon the maria, it is apparent that the mare material is all approximately the same age. This is confirmed by radiometric dating of the samples brought back from the moon, which show the maria to be 3.5 billion years old—older than any rocks on the earth's crust even though the extrusion of maria lava appears to be a very recent event in lunar history.

The extrusion of maria material was apparently not triggered by the impact which created the circular maria basins but occurred much later. After the maria basins were formed, subsequent impact made modifications in the crater rims such as those found on the crater Sinus Iridum on the rim of Mare Imbrium. Moreover, the maria were not filled with one flow but with many; however, all occurred within a relatively short time span.

It is interesting to note that some rims of old craters project above the mare material and show that the lava is relatively thin, ranging from 300 to 1200 m (1000 to 4000 ft) thick. These "ghost" craters are most abundant near the shore and in "marshy" areas where the lava overlapped a rugged surface and produced many low islands. Excellent expressions of flooding are found in the southern maria, Nubium and Humorum, where numerous "ghost" craters barely protrude above the sur-

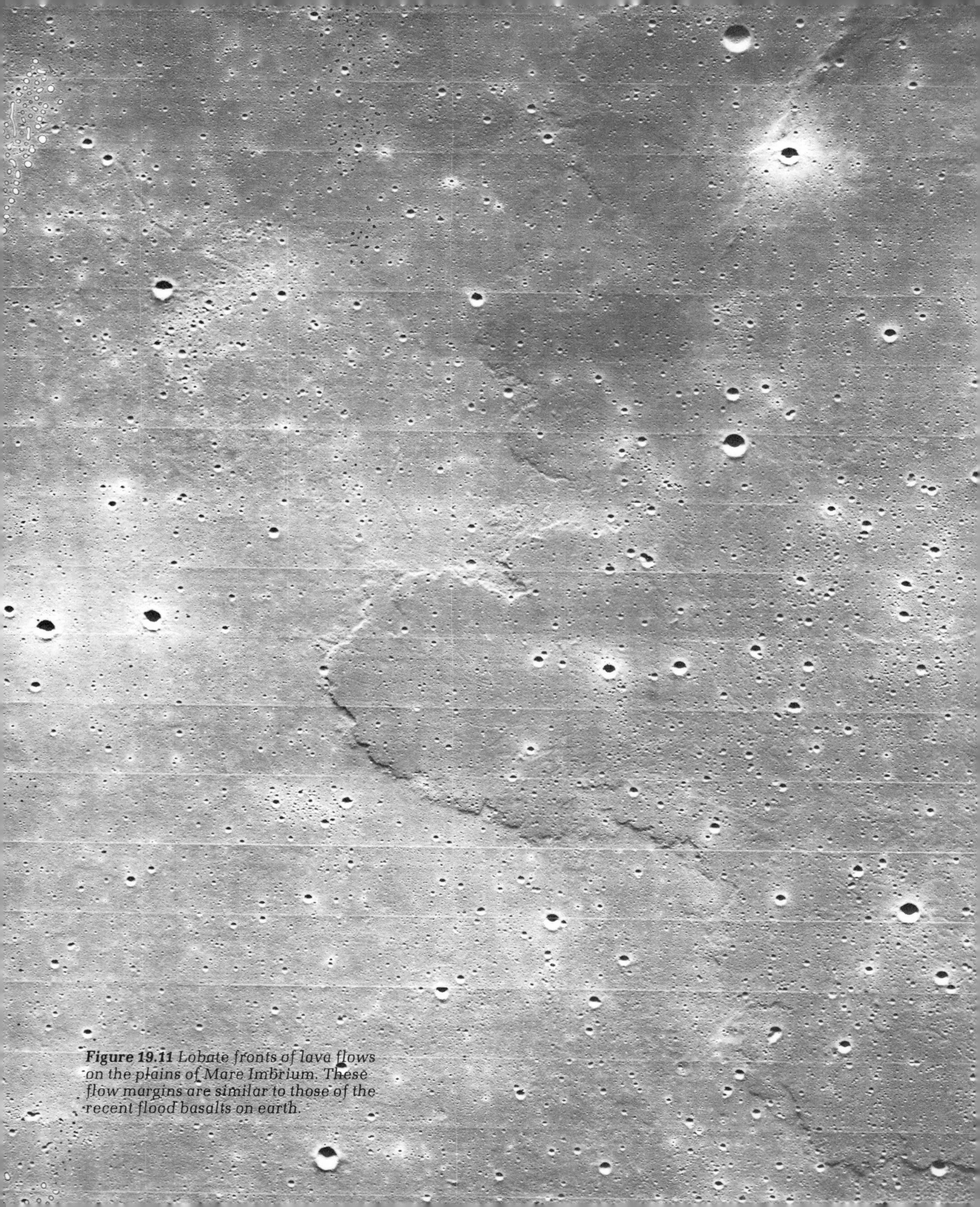

Figure 19.11 Lobate fronts of lava flows on the plains of Mare Imbrium. These flow margins are similar to those of the recent flood basalts on earth.

face. In other sections deeper in the mare basins, the lava may be much thicker because no older crater rims are visible.

The lunar maria represent a very significant event in lunar history, an event which completely modified a large part of the surface. The vast amount of lava indicates that at one time widespread melting occurred beneath the lunar surface. This thermal event occurred after the great maria basins were formed. The extrusion of lava appears to have been completed at about the same time, as the distribution of large primary impact craters such as Copernicus and Eratosthenes are relatively uniformly distributed on the mare material. There are no significantly large areas of mare lava so recent as to escape modification from subsequent impact. Photographs of the far side of the moon indicate that the large maria are primarily restricted to the near side and that the far side of the moon is largely lunar highland.

Other Surface Features

Statement

The lunar surface contains a variety of other surface features which are not nearly as numerous as craters but are important because they show the type of processes active within the moon system. Foremost among these are the rilles, wrinkles, and volcanic features.

Discussion

Rilles. Lunar rilles are long narrow trenches or valleys on the moon surface (*figure 19.12*). They are most common in the relatively smooth regions of the highlands and appear to be absent in the central part of the maria and in the most rugged mountainous regions. The majority are linear, although some are arcuate and a few are sinuous. They vary in length from tens to hundreds of kilometers, and in width from a few hundred meters to 25 km (15 mi). Generally, their depth is less than their width.

The majority of lunar rilles are sharp linear depressions; these are relatively deep compared to other types. Valley walls are straight or arcuate and stand at the same elevation, as though they had been pulled apart and the floor had subsided as a graben. Straight rilles are parallel or arranged in an echelon pattern. Some intersect, while others form a zigzag pattern similar to normal faults on earth.

There is little doubt that straight rilles are grabens resulting from normal faulting, because they so closely resemble grabens and normal fault scarps on earth. However, the stresses responsible for their formation are undoubtedly quite different from those which produced the great rift system on earth. Rilles on the moon do not form a regional global pattern but represent local subsidence.

Sinuous rilles have received considerable attention be-

Figure 19.12 Arcuate rilles in the highlands east of the Humorum Basin. These rilles are most likely the result of block faulting.

cause of their many unusual features; yet, their origin is still unclear. They differ from straight rilles in that they have a snakelike pattern (*figure 19.13*). Some meander in smooth arcs, whereas others have roughly linear segments. Most occur along the shallow edges of maria and along the flat floor of larger mare-filled craters. Many sinuous rilles begin at a crater and, when traced downslope, become progressively smaller until they disappear. Some have V-shaped profiles, but flat floors complicated by inner channels, craters, and irregular hummocks are more typical.

The Earth's Dynamic Systems

At first glance, some sinuous rilles appear to be very similar to terrestrial stream valleys, but they lack many features characteristic of stream-cut channels, such as tributary systems, increase in channel size downslope, and such associated depositional features as deltas, flood plains, and alluvial fans.

In all probability they originate in a variety of ways; some are likely collapsed lava tubes or lava channels; others may be channels created by rolling boulders. Some observers have proposed that sinuous rilles were formed when gas vented from fractures; in experimental situations, this process has produced

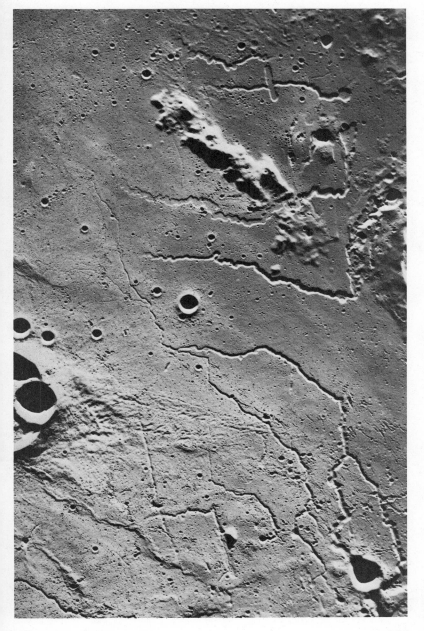

Figure 19.13 *Sinuous rilles east of the Aristarchus Plateau.*

Figure 19.14 Wrinkle ridges in the lunar maria material.

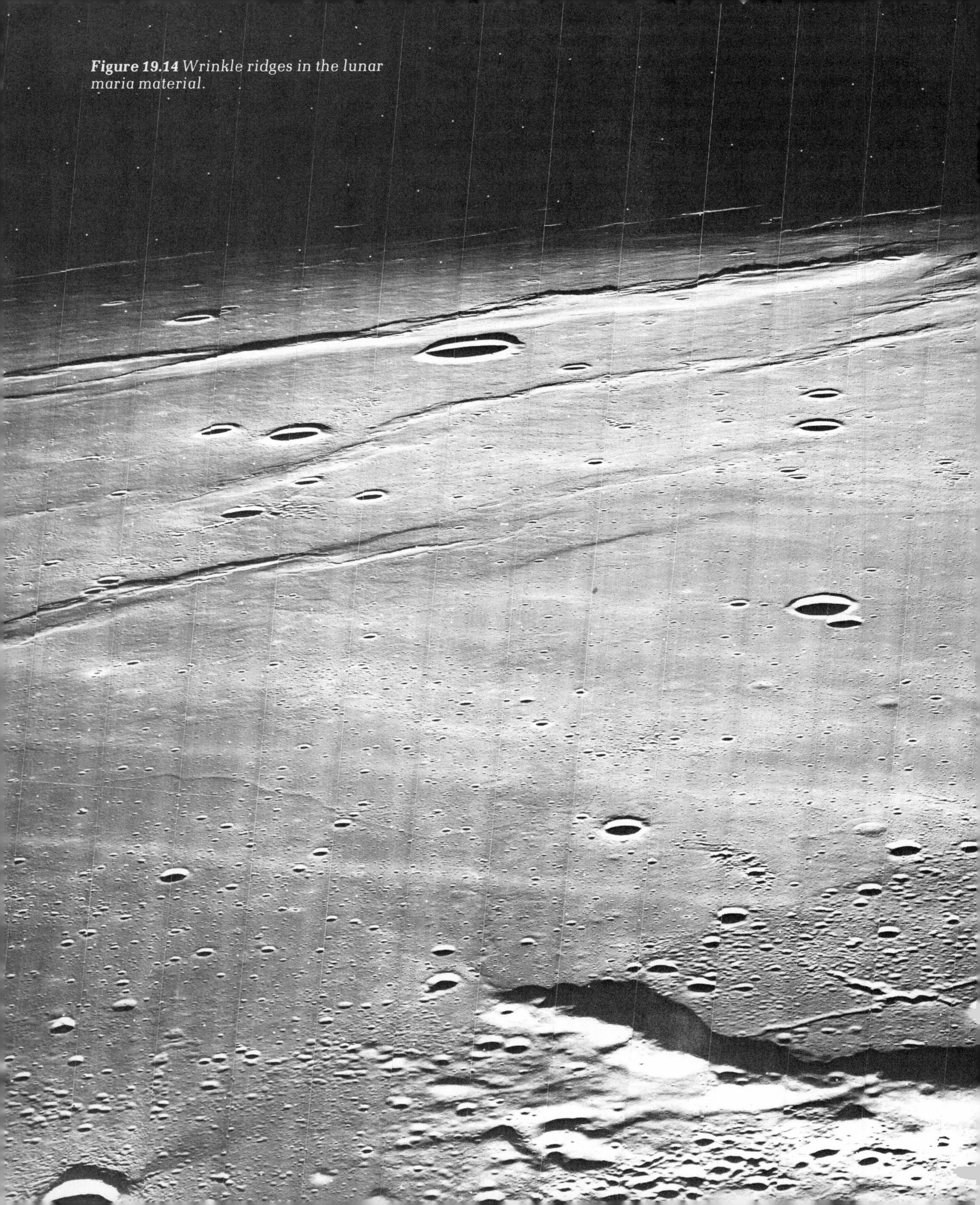

rillelike features. It may account for some of the sinuous rilles which are marked by numerous small craters strung out along much of the rille length. In contrast to straight rilles, sinuous rilles are frequently deflected by local topography and may have been formed by erosion by ash flows, which are known to erode channels on earth.

Wrinkles. Some of the most conspicuous features on the mare surface are long, narrow anticlines, sometimes referred to as "wrinkle ridges" (*figure 19.14*). Typically, they have a sinuous outline and extend discontinuously for great distances. Segments may be several kilometers wide, a few hundred meters high, and hundreds of kilometers long. The wrinkle system commonly parallels the margins of the major maria basins, although some parallel structure trends in the highlands.

A number of explanations for wrinkles have been proposed and, as is the case with rilles, they may represent several origins. Some geologists consider them to be compressional features produced as the maria subsided. Others feel that they represent differential compaction over buried ridges or are the result of intrusive structures where the crust was buckled up by the injection of shallow intrusions.

Volcanic Features. With the extensive lava fields of the maria, we might expect to find extensive volcanic craters and

Figure 19.15 *An oblique view of the Marius Hills. The domes are believed to be of volcanic origin.*

other features resulting from extrusion of lava. Some scientists have argued that most craters on the moon are of volcanic origin, but this is highly unlikely. The impact origin for most lunar craters is indicated by (1) an ejecta blanket, (2) rays and radial ridges, (3) secondary craters, and (4) a central peak on the crater floor. Still we would expect to find some volcanic craters, but it is not as easy as it may appear. The overprint of impact craters tends to obscure those of volcanic origin, but there are a number of definite volcanic features. Foremost among these are the Marius Hills on the western side of the earth-side disc just north of the equator. Here, an array of domes and cones are similar in many respects to volcanic features on earth (figure 19.15). Other cones are commonly up to 10 km (6 mi) across and a few hundred meters high; some have a summit crater. These are similar in many respects to small shield volcanoes on earth.

Sequence of Events in Lunar History

Statement

The deposits of ejecta from the craters, together with lava flows and other volcanic deposits, form a complex sequence of overlapping strata that cover most of the lunar surface. The individual deposits can be recognized by their distinctive topographic characteristics and by their physical properties such as color, tone, and thermal and electrical properties determined from measurements made with optical and radio telescopes.

A framework of relative time into which major events of lunar history can be arranged was developed by the U.S. Geological Survey in 1962. The basic principles used by the Survey to interpret lunar history are essentially the same as those used to study the history of terrestrial events, the most important of which are the law of superposition and the law of cross-cutting relations. These principles of determining relative ages are, of course, valid on the moon or Mars as they are on earth. Developed from regional studies of superposition and cross-cutting relations, the lunar time scale, like the terrestrial geologic time scale, is based on records of major events preserved in the rocks.

Discussion

Although it was recognized long ago that craters and other lunar surface features showed evidence of having been formed at different times, prior to the space program most observers studied features without relating them to their surroundings. Craters were classified according to their dimensions, statistics were calculated on crater density, but little effort was made to establish a sequence of events in lunar history.

The first lunar chronology was developed in 1962 by Shoemaker and Hackman, who interpreted the sequence of events in the vicinity of the crater Copernicus. They selected ejecta from the craters Copernicus, Eratosthenes, Archimedes, and several

The Earth's Dynamic Systems

other craters to demonstrate superposition of lunar material. From this they established a lunar geologic column.

In many ways what Shoemaker and Hackman did in providing a rationale for interpreting moon history is comparable to what Smith, Lyell, and their contemporaries did in establishing the geologic time scale on earth during the early 1800s. The planet is different, and the nomenclature is different, but the logic remains the same.

To understand the basis for establishing the lunar time scale and the meaning of the major events in lunar history, let us carefully consider the ejecta from the major craters studied by Shoemaker and Hackman and their reasons for recognizing the sequence of events they represent. As you read the following discussion, study the physiographic map of the moon and the selected illustrations, for it is only by recognizing the physical relations between the features discussed that an appreciation for the relative time involved in their formation can be gained.

Copernicus. The most outstanding feature of Copernicus is its spectacular and bright ray system which extends outward in all directions, in some cases for a distance of more than 300 km (180 mi). Within the rays, predominantly near the crater, are elongate secondary craters. The rays and ejecta blanket surrounding Copernicus are superposed on essentially every feature in their path (*figure 19.16*). Under full moon, when the rays are best observed, it is found that rays extend not only across the Mare Procellarum (Sea of Storms) and Mare Imbrium (Sea of Rains) but also up the rim and across the floor of the crater Eratosthenes, 190 km (120 mi) to the northeast. *From this superposition it is clear that Copernicus and its associated system of ejecta and ray material are younger than Eratosthenes.* Other systems of rayed craters similar to Copernicus are the slightly smaller craters to the west—Kepler and Aristarchus. In the southern hemisphere, material from similar craters such as Tycho overlap all features and may vary in age, but, as a group, they are younger than other features on the moon inasmuch as they are superposed upon them. The ejecta from rayed craters forms the uppermost and youngest system of strata on the moon and is called the *Copernican system.* The period of time during which rayed craters and their associated rim deposits were formed has been called the *Copernican period.*

Eratosthenes. About half of the craters larger than 10 km (6 mi) in diameter that occur on the maria are rayed craters. Most other craters of this size found on the maria are similar to the rayed craters, but their ejecta blankets are dark, and they lack rays. Consider, for example, the crater Eratosthenes just northeast of Copernicus (*figure 19.16*). It has terraced walls, a roughly circular floor with a central peak, a hummocky rim, and a distinctive pattern of secondary craters. However, unlike Copernicus, it does not have a visible ray system, and the secondary craters are noticeably more subdued than those

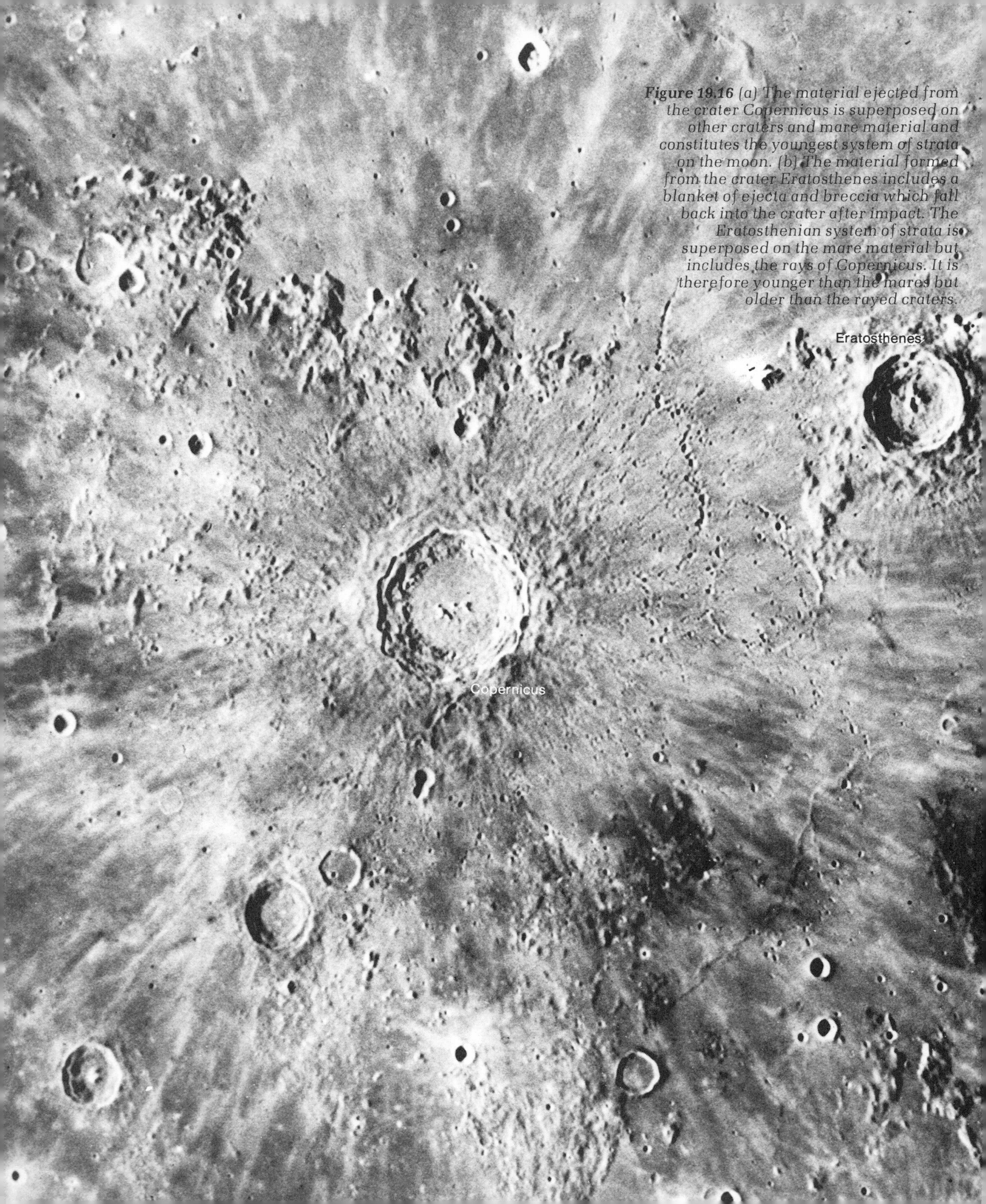

Figure 19.16 (a) The material ejected from the crater Copernicus is superposed on other craters and mare material and constitutes the youngest system of strata on the moon. (b) The material formed from the crater Eratosthenes includes a blanket of ejecta and breccia which fall back into the crater after impact. The Eratosthenian system of strata is superposed on the mare material but includes the rays of Copernicus. It is therefore younger than the mares but older than the rayed craters.

Eratosthenes

Copernicus

around Copernicus. Eratosthenes and similar craters, together with their ejecta, are superposed on the maria and are, therefore, younger than the maria lava flows on which they are formed. However, they are older than the rayed craters, as is shown by the fact that the rayed craters, ray material, secondary craters, and ejecta deposits of rayed craters are superposed on the dark crater rims. The deposit of the dark-rimmed craters is called the *Eratosthenian system* of strata, and the period of time during which they were formed is referred to as the *Eratosthenian period*. Deposits of the Copernican and Eratosthenian systems are easily recognized on the maria, but it is difficult in some cases to discriminate between Eratosthenian and older crater deposits on the lunar highlands.

The Imbrium Basin. In the northwest part of the near side of the moon is an enormous crater now largely filled with lava flows called the Imbrium Basin. We have seen in previous discussion that the Imbrium Basin is the largest multi-ringed crater on the moon and, like other craters, was formed by impact. The Imbrium Basin is surrounded by ejecta deposits similar to those formed by smaller craters—the best exposures being the Apennine Mountains extending outward from the southwest rim. These deposits are called the Fra Mauro Formation and can be traced as far as 1000 km from the mountains surrounding the basin.

The ejecta from the Imbrium Basin is partly covered with lava, as is most of the interior of the basin (Mare Imbrium). The lava plus the ejecta deposits constitute the *Imbrium system* of strata. It is apparent that some impacts occurred after the Imbrium Basin and ejecta were formed—but before the extrusion of the lava flows. Some of these deposits around larger craters such as Archimedes, Plato, and Sinus Iridum are very extensive. It is calculated that about four times as many craters larger than 10 km (6 mi) in diameter were formed on the moon during the period of deposition of the Imbrium system as during all the time since the last lava flows in Mare Imbrium were extruded.

The Ancient Terrae. The Fra Mauro Formation from the Imbrium ejecta partly overlies a complex sequence of ejecta formed in the lunar highlands. These deposits include giant sheets of ejecta around large basins such as Serenitatis, Crisium, Nectaris, and Humorum, plus ejecta blankets of a great number of other craters. These strata are collectively called Pre-Imbrium deposits and constitute the oldest layers of material on the lunar surface. The geologic structure of these deposits is very complex. Large craters are closely spaced and modified by impact. Apparently the surface of the terrae has been churned by repeated formation of large craters early in lunar history.

In light of these observations, the U.S. Geological Survey has been able to map most of the near side of the moon and outline some of the major events in lunar history. These are summarized in the time chart (*table 19.1*).

Table 19.1. Lunar Time Chart

Period	Event
Copernican	Formation of rayed craters and ejecta material (see figure 19.29)
Eratosthenian	Formation of craters and ejecta of which rays are no longer visible (see figure 19.29)
Imbrium	Formation of mare basins, ejecta material, and extrusive deposition of mare material (see figures 19.27 and 19.28)
Pre-Imbrium	Early development of cratered surface now exposed in lunar highlands (see figure 19.26)

The Structure of the Moon

Statement

It is clear from studies of lunar rocks, surface features, and lunar seismicity that the interior of the moon is layered, but the internal structure is quite different from that of the earth. The moon has a crust or "lithosphere" (outer layer) that is some 65 km (40 mi) thick, which is much thicker than the earth's crust. Below the lithosphere is a "mantle" rigid enough to support considerable stress down to a depth of 800 km (500 mi). Temperatures are possibly as high as 1650° C (3000° F). Seismic soundings suggest a partly molten outer zone in the core. The core material, however, is not iron-nickel, because the moon's magnetic field is negligible, and the density of the surface rock is not appreciably less than that of the moon as a whole.

Discussion

Our understanding of the internal structure of the moon is based upon physical observations such as density and seismicity. Much remains uncertain, but several facts place significant constraint on what the internal structure may or may not be. First, the bulk density of the moon is 3.34 g/cm³, whereas the mean density of lunar surface rocks is about 3.3. Thus, the density of the surface material, at least of the maria basalts, is only slightly less than that of the moon as a whole, and there is little possibility for a significant increase in density with depth as on earth. It is also clear that the composition of the moon's interior is different from the surface material. Our present understanding of the structure of the moon is shown in figure 19.17.

Data from the Apollo seismic stations show that the moon's crust is approximately 65 km (40 mi) thick and is layered (figure 19.18). The low velocities near the surface most likely indicate a layer of fine breccia and broken rock fragments 1 to 2 km (.6 to 1.2 mi) thick. Below this, to a depth of 25 km (15 mi), the seismic velocities have values similar to basaltic rock. The second

448

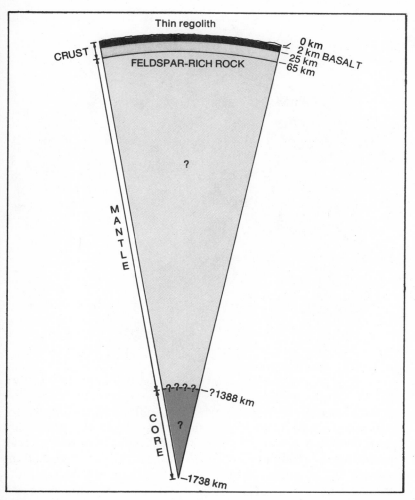

Thin regolith

CRUST

FELDSPAR-RICH ROCK

0 km
2 km BASALT
25 km
65 km

?

MANTLE

?

????
?1388 km

CORE

?

1738 km

Figure 19.17 *Generalized diagram showing our present understanding of the internal structure of the moon.*

Figure 19.18 *Graph showing variations in seismic velocities and interpreted structure of the lunar crust. Three major discontinuities in seismic velocities indicate that the moon has a layered crust. Seismic velocities increase very rapidly at shallow depths to about 10 kilometers. A very sharp increase in velocities occurs at about a depth of 25 km, and between 25 and 65 km the seismic velocity is nearly constant. A significant discontinuity for increase in velocity occurs at 65 km and is interpreted as marking the base of the crust. Comparing these seismic velocities with those of major rock types, it is interpreted that near the surface to a depth of about 1 to 2 km the composition of the rock is that of fine breccia and broken rock fragment. Below this, to a depth of about 25 km, is a layer composed largely of basalt. A second layer of lunar crust occurs between 25 and 65 km and appears to be composed of gabbro and anorthosite. Below 65 km the mantle of the moon is tentatively interpreted to be composed of magnesium-rich olivines.*

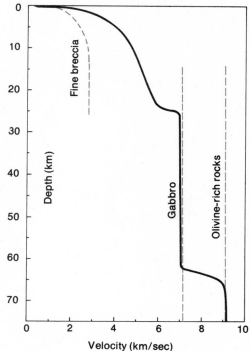

layer of the crust is distinctly different from lunar basalts sampled at the surface and is probably similar to terrestrial gabbro or anorthosite (rocks rich in plagioclase). At a depth of 65 km, another sharp change in velocity occurs, marking the contact between the lunar crust and mantle. The mantle rocks show a higher seismic velocity than most earth rocks and are believed to be rich in olivine.

It is clear from the extensive basaltic flows in the maria that the moon experienced a thermal event early in its history, and it was partly differentiated by igneous processes. This would require a source of magma at depth which would probably produce a differentiated layered structure such as that shown in figure 19.18.

The seismic energy of the moon is 1 million times less than that of the earth. This would imply that internal convection is not occurring. Approximately 80% of the energy released comes from a zone less than 10 km (6 mi) thick and located approximately 800 km (500 mi) below the surface. This indicates that the interior of the moon is rigid.

Geology of the Moon

449

Figure 19.19 *Regional gravitational fields on the near side of the moon. Note that the areas of high gravitational attractions correspond to the large circular maria basins.*

Mascons. Studies of the orbits of Apollo and Lunar Orbiter spacecraft close to the moon show that there are significant irregularities in the lunar gravitational field. The areas of high gravitational attraction are attributed to mass concentrations and are called **mascons**. Most of the mascons are found in the circular maria basins (*figure 19.19*) and do not occur with any prominence in the highland regions.

The origin of mascons is quite controversial. Some of the explanations proposed thus far are:

1. They represent a large buried meteorite of nickel and iron.
2. They represent a disk-shaped body of high-density lava filling the mare basins.
3. They represent accumulations of heavy minerals at the base of a lava lake.

More refined gravity mapping is needed before this problem can be solved.

Lunar Rocks

Statement

The samples brought back from the moon (*figure 19.20*) have been distributed to hundreds of scientists from nine nations and have been subjected to thorough and sophisticated analyses. This material consists of:

1. Basaltic igneous rock.
2. Anorthositic igneous rock derived from the terrae.
3. Glass—material fused by shock melting and rapid cooling.
4. Regolith or soil—a mixture of crystalline and glassy fragments formed by repeated fragmentation of surface material by meteorite impact.
5. Breccia—angular fragmental particles formed by fragmentation of surface material by meteorite impact and subse-

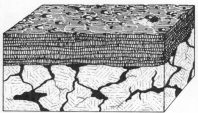

(a) Apollo 11 landed on the flood basalts of Mare Tranquillitatis. The lunar maria consists of numerous thin layers of basalt flows with an aggregate thickness of several thousand meters extruded 3.7 billion years ago. The maria basalts rest upon older rocks of the cratered highlands.

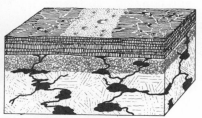

(b) Apollo 12 landed on a ray of Copernicus approximately 400 km (250 mi) south of the crater. The ray material rests upon a sequence of basalts which form Oceanus Procellarum. Below the basalts is a layer of ejecta formed during the early stages of impact and development of the lunar highlands.

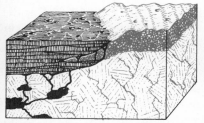

(c) Apollo 14 landed on the Fra Mauro Formation, material ejected by the Imbrium impact event. Basalts from Oceanus Procellarum lap up against the ejecta, proving that the maria lavas are younger than the highland. The rocks of the Fra Mauro Formation were thrown out of the Imbrium Basin about 4 billion years ago.

(d) Apollo 15 landed near Hadley Rille at the base of the Apennine Mountains. The maria basalts lap up against the ancient rocks of the lunar highlands, which have been dated as being more than 4 billion years old.

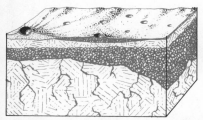

(e) Apollo 16 landed in the highlands of the Descartes region. The surface material is composed of debris churned up by North Ray crater and South Ray crater and overlies layers of breccia formed by more ancient meteorite impact.

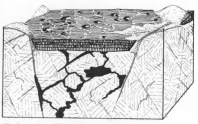

(f) Apollo 17 landed in the Taurus-Littrow valley formed by ejecta from the Serenitates Basin.

Figure 19.20 *Schematic block diagrams showing the structure of the strata at the Apollo landing sites.*

Geology of the Moon

quently compacted into a coherent rock by shock compression during cratering.

Discussion

Lunar Basalt. Most of the igneous rocks collected from the moon are very similar to terrestrial basalt, the most common rock in the earth's crust. These rocks were once totally molten, as is indicated by their vesicles, interlocking crystalline texture, and composition (*figure 19.21*). The principal minerals found in lunar basalts are plagioclase, pyroxene, ilmenite, and olivine, all found in terrestrial basalt. Only minor amounts of a few minerals unknown on earth were found. The most significant difference between lunar and terrestrial basalts is that the former contain a greater amount of heat-resistant elements (titanium, zirconium, and chromium). Lunar basalts are devoid of water and have much lower amounts of relatively volatile elements such as sodium and potassium than do terrestrial basalts. These chemical characteristics are significant in that they suggest that the material which forms the moon was, at one time, heated to higher temperatures than the material from which the earth was formed. This would explain the concentration of refractory elements and the loss of volatiles. The texture of the basalts and melting experiments which suggest that most of the basalts crystallized at 1201° C to 1060° C indicate that they were more fluid than their earthly counterparts and that they cooled fairly rapidly. The high fluidity of lunar basalts is also reflected in the broad sheets of basalt which fill the maria basins.

When subjected to high temperature and pressure, the lunar basalts have a density far in excess of the average density of the moon and, therefore, cannot represent the bulk composition of the moon.

Anorthosite. In addition to the lunar basalts, small frag-

Figure 19.21 *Photomicrograph of a sample of lunar basalt with the field approximately 4 mm.*

The Earth's Dynamic Systems

Figure 19.22 Photomicrograph of a lunar anorthosite. The width of field of view is approximately 4 mm. This rock consists of a meshwork of plagioclase crystals which are characteristically lath-shaped, with some pyroxene occupying the interstitial space between the plagioclase crystals. Olivine occurs in amounts up to 1%, with small traces of opaque minerals and glass.

ments of rock known as anorthosite were collected. Anorthosite is a rock composed almost entirely of the mineral plagioclase (*figure 19.22*). The so-called "genesis" rock collected by Scott and Irwin at the foothills of the Apennine Mountains is coarsely crystalline anorthosite with an age of 4.15 billion years. The composition of anorthosite corresponds well with the analysis of the highland region carried out by Surveyor 7 in the vicinity of Tycho. This suggests that the highlands have an anorthosite composition different from that of the maria, which is basalt.

Glass. Approximately 50% of the fine-grained material of the surface samples collected at the Apollo landing sites consists of glass formed by shock melting of rock debris during the process of impact. In addition, many rock samples contain fragments or beads of glass or spotted splashes of glass (*figure 19.23*). Glass is even more abundant in ray material, as the samples of Copernicus rays consist of 70 to 90% glass. Glass is, therefore, considered to be a significant product of shock melting, a fact which can be demonstrated from studies of impact craters on earth and also of craters made by nuclear explosions.

Regolith. The surface of the moon is mantled in most places by a thin layer of relatively loose, unconsolidated fragments of rock, crystals, and glass, similar in many respects to a fine powder. This layer is called the lunar regolith. The average thickness of the regolith depends upon the age of the stratum on which it has been formed. On ejecta from very young craters such as Tycho, the average thickness is about 10 cm. On the maria the thickness ranges from 1 to 4 m (3 to 13 ft). On the Fra Mauro Formation, the regolith is 8.5 m (28 ft) thick. As a general rule, then, the older the stratum, the thicker the regolith.

Regolith on the moon consists of debris thrown out of craters

Figure 19.23 Photo of a sample of lunar glass. This specimen is approximately 3 cm in diameter and consists of dark grey to black glass with small vesicles and cracks.

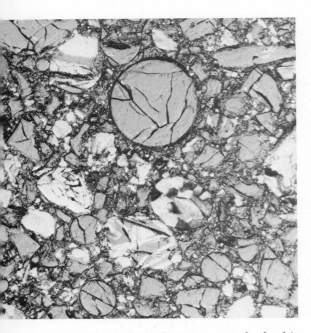

Figure 19.24 *Photomicrograph of a thin section of lunar breccia. The lunar breccia constitutes a major rock type and consists of angular fragments of broken rocks from a variety of sources. Some lunar breccia contain large amounts of glass, with particles being remarkably spherical. The fragments are typically angular and show essentially no evidence of modification by abrasion. Glass particles are the dark black material. Careful inspection will show that some of these are nearly spherical.*

and, at any given place, most of the debris has been derived from the local underlying substratum. As a result, the composition and texture varies considerably from place to place and reflects the history and the processes of the area where it is formed. For example, samples of regolith from the maria contain considerable amounts of basalt fragments and generally less than 50% glass, whereas ray material from Copernicus contains 70 to 90% glass.

We can estimate the rate at which regolith forms on the moon by measuring its thickness and determining the age of the bedrock beneath. At Tranquillitatis, the lunar basalt is about 4 billion years old, and the regolith is about 4 m (13 ft) thick. This would give an average rate of regolith formation of 1 mm per million years.

Breccias. Breccias are a mechanical mixture of rock and glass fragments compacted into a solid mass (*figure 19.24*). Samples of breccias (some are composed of very small grains and are called microbreccias) appear to be the consolidated equivalent of lunar regolith. The mechanism by which regolith is consolidated into a coherent mass of breccia presents a problem. Two possibilities are immediately apparent: (1) shock lithification—compression of the grains together as a strong shock wave passes and (2) welding of the deposits as they accumulate in a hot state after being ejected from impact craters.

Isotopic Ages of Lunar Rocks and Major Events in Lunar History

Statement

Some of the most critical information about the geology of the moon has been obtained from isotopic age determinations of lunar rocks. Four important ages have been established: (1) the age of the moon as a body, (2) the age of the formation of the Imbrium Basin, (3) the age of the maria basalts, and (4) the age of Copernicus. These dates have been integrated into the time scale determined by superposition of systems of strata so that a framework of lunar history and the dates of major events can be determined.

Discussion

The lower limit for the age of the moon has been established by uranium and lead isotopic analysis of fine regolith material derived from the terrae. The isotopic dates show that the moon is at least 4.65 billion years old. Independent studies show that the earth and the oldest known meteorites were formed about 4.6 billion years ago, so it is likely that the moon was formed at about this time. The date of 4.6 billion years, for the present, stands out as the most likely age of our solar system.

A key to the age of the Imbrium Basin is found in an unusual rock specimen collected at the Apollo 12 landing site. The rock is a breccia associated with ray material believed to have been derived from Copernicus. It is composed of light and dark components. The light material is mainly potassium feldspar and quartz and resembles terrestrial gravels in composition. Isotopic analysis indicates that the rock was first formed about 4.5 billion years ago. If the rock was thrown out of Copernicus, it may have been excavated from the Fra Mauro Formation, the substrata beneath Copernicus. If this is true, then the 4.0 billion-year heating event would correspond to the formation of the Imbrium Basin and deposition of the ejecta to form the Fra Mauro Formation. The Imbrium Basin itself would appear to have been excavated shortly after the birth of the moon. Analysis of rubidium and strontium isotopes of the lunar basalts from Mare Tranquillitatis indicates that the lava crystallized 3.65 billion years ago. A somewhat greater age is indicated by uranium and lead isotopes. The reason for the discrepancy is not known, but inasmuch as the rubidium-strontium method requires fewer assumptions, this date is generally used as the best age for the extrusion of the maria lavas.

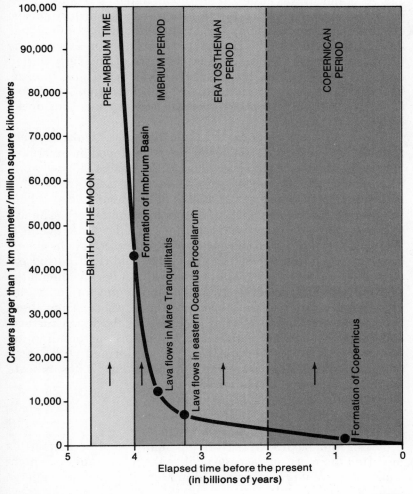

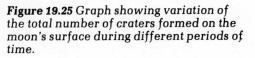

Figure 19.25 Graph showing variation of the total number of craters formed on the moon's surface during different periods of time.

Figure 19.26 (opposite) *Formation of the densely cratered highland from rapid and intense meteorite bombardment. The terrain may be in part a product of an early period of accretion, more than 4 billion years ago.*

The approximate age of Copernicus was determined from ray material collected from the Apollo 12 landing site. The date of this event is 0.8 to 0.9 billion years ago.

An additional insight into ages of lunar events can be gained from studies of numbers of craters on a given rock body. By counting the number of craters formed in different strata whose ages have been determined isotopically, it is possible to relate the number of craters to the time elapsed since the strata were deposited. *Figure 19.25* shows the variations in numbers of craters formed on surfaces of different ages. The curve shows that the rate of cratering dropped very rapidly during the first billion years of lunar history, but from 3 billion years ago to the present the rate of cratering has remained nearly constant. This rapid drop and subsequent leveling of the rate of cratering suggests two different groups of objects producing the craters on the moon.

Outline of Lunar History. The moon, like other planets of the solar system, is believed to have been formed by accretion of planetesimals. This event occurred about 4.6 billion years ago. As material was added to the moon, impact structures were produced. The earliest recorded events in lunar history consist of the highly cratered, complex topographic surface of the lunar terrae (*figure 19.26*). The craters and Pre-Imbrium strata of the terrae appear to be at least a partial record of this early period of accretion, in which bombardment was rapid and intense. Several series of events during Pre-Imbrium time are recognized by the U.S. Geological Survey but are not considered here.

Two major events occurred after the lunar highlands were formed: (1) The huge maria basins were developed by exceptionally large meteorite impacts, and (2) the basins subsequently filled and in places overflowed with lava. These events greatly modified the surface of the near side of the moon and stand out as possibly the most significant events in moon history. The use of the term Pre-Imbrium with reference to older lunar events is analogous to the use of "Precambrian" in reference to events in the geologic history of the earth. The Imbrium period extended from about 4 to 2.5 billion years ago. It began when large asteroid-size bodies collided with the moon and formed multiring basins similar to the Orientale Basin. Ejecta from these basins was distributed over a large part of the moon surface, and remnants now form the high mountains surrounding the basins (*figure 19.27*). The basins were not formed at the same time, for superposition of ejecta and radiometric dates indicate that the sequence of formation was as follows:

1. Imbrium Basin—4.0 billion years ago;
2. Tranquillitatis—3.65 billion years ago;
3. Procellarum—3.26 billion years ago.

After the basins were formed, many smaller meteorites modified the mare surface, as well as that of the adjacent highlands. This event is best expressed in Mare Imbrium, as the large

Figure 19.27 (opposite) *The formation of high multi-ringed basins from exceptionally large meteorite impact. The basins were subsequently modified by impact of smaller meteorites. These events covered a period from 4 billion to 3.26 billion years ago.*

The Earth's Dynamic Systems

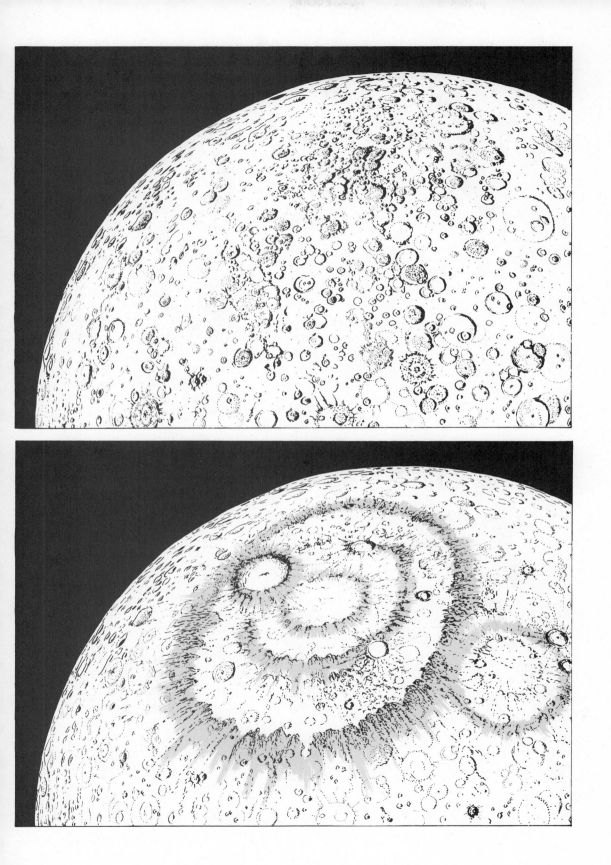

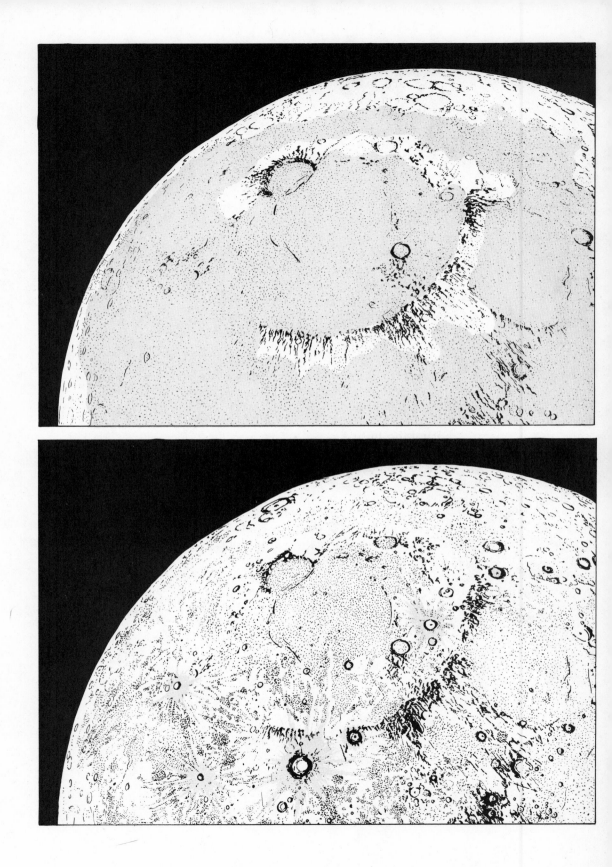

craters of Plato, Archimedes, and the large basin of Sinus Iridum greatly modify the rim of the Mare Imbrium but have themselves been flooded with mare lavas. During the Imbrium period, all but a few of the remaining planetesimals in the vicinity of the orbit of the earth and moon appear to have been swept up and, since then, relatively few impact craters have been formed.

The next major event in lunar geologic history was a thermal event, the widespread extrusion of basaltic lava which filled in the low-lying areas, including preexisting depressions or floors of small craters such as Archimedes and Ptolemaeus (figure 19.28). Most of the lava was apparently extruded during a limited interval of time and is roughly the same age throughout the entire lunar surface. This assumption is based on the relatively uniform distribution of younger primary craters on the mare material. If some maria were older than others, they would probably be pockmarked with a greater number of craters. The extrusion of lava during the Imbrium period constitutes one of the most significant events in lunar history because there is no evidence of anything like the widespread extrusion of lava occurring either before or after.

Following the widespread flooding of lava during the late Imbrium Period, large impact craters were formed in the maria and adjacent areas (figure 19.29). Their ejecta blanket and secondary craters overlie the surface of mare material and other features formed during the Imbrium time. These craters, however, have been modified and lack visible rays. They do have sharp crater walls and hummocky rim debris, which testify to their relatively young age.

The most recent events in lunar geologic history are the formation of rayed craters such as Copernicus, Tycho, and Kepler. Rays from these craters overlie all features with which they are in physical contact. Moreover, wherever a crater with faint rays occurs in an area of one with bright rays, the brighter rays are invariably superposed on the darker and dimmer rays—a fact which indicates that subdivision of the most recent lunar events is possible. The formation of the crater and ejecta blanket of Copernicus has been dated as 0.8 to 0.9 billion years ago. The age of Tycho, as determined by the age of Australian tektites thought to be associated with the formation of Tycho, is only 0.7 million years.

Summary

The moon is not a dynamic planet like the earth, but it does have a history involving both meteoritic impact and volcanic events, most of which occurred very early in its history, more than 4 billion years ago. The lunar maria are vast lava plains which record a major thermal event in lunar history in which basaltic magma was generated and extruded to fill the large

Figure 19.28 (opposite) *The extrusion of lava which filled many of the multi-ringed basins and formed the maria constitutes a major thermal event in the moon's history. This occurred approximately 3.65 billion years ago in the Tranquillitatis Basin.*

Figure 19.29 (opposite) *Since the extrusion of the maria lava, the moon has experienced relatively little change, being modified only by impact of a few meteorites. Copernicus, for example, is 0.8 to 0.9 billion years old but appears as fresh as though it was formed yesterday.*

maria basins. Since then, the lunar surface has been modified by relatively few impact craters.

The moon apparently lacks sufficient mass to become differentiated like the earth. It does not have a significant source of internal energy and does not have a tectonic system like the earth. It has no atmosphere or surface fluids, so it lacks a hydraulic system to modify its surface. As a result, it has no continents nor ocean basins and no deformed rocks resulting from mountain-building.

Additional Readings

Allen, J. P. 1972. "Apollo 15: Scientific Journey to Hadley-Apennine." Amer. Scientist 60:162-74.

Bowker, D. C., and J. K. Hughes. 1971. Lunar Orbiter Photographic Atlas of the Moon. Washington, D.C.: National Aeronautics and Space Administration.

Hinners, N. W. 1971. "The New Moon: A View." Reviews of Geophysics and Space Physics 9:447-522.

Mason, B. 1971. "The Lunar Rocks." Sci. Amer. 225(5):48-58. (A popular review of the first several groups of lunar samples.)

Mutch, T. A. 1972. Geology of the Moon. Princeton, N.J.: Princeton University Press.

Shoemaker, E. M. 1964. "The Geology of the Moon." Sci. Amer. 221(6): 38-47.

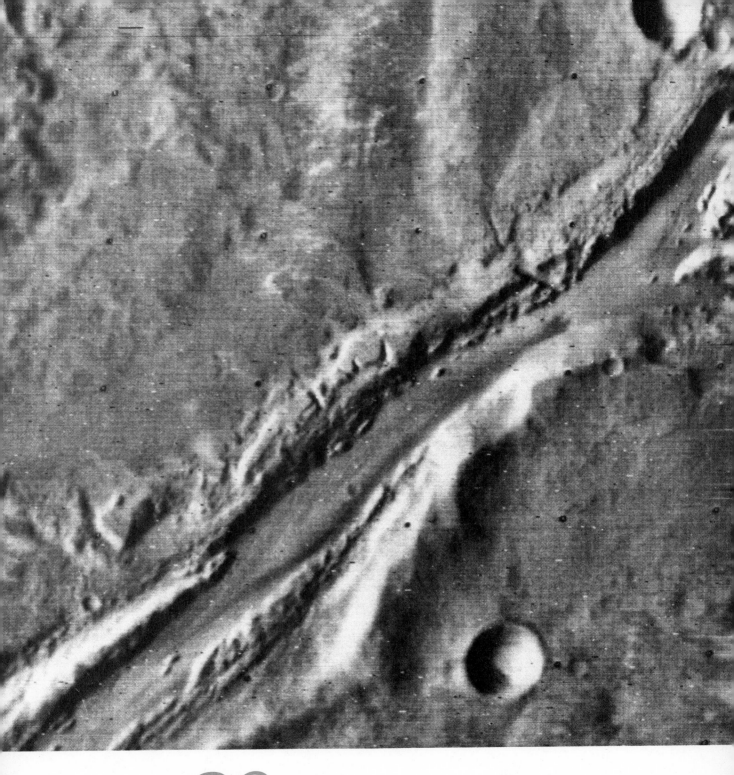

20 Geology of Mars

In November, 1971, Mariner 9 obtained an orbit around Mars and became the first man-made satellite of another planet. During its period of operation, lasting nearly a year, it provided some 7300 spectacular photographs plus a wealth of other scientific information which constitute one of the most significant scientific advancements of the space program. From these data, photomosaics of the entire planet have been made, shaded relief maps of the surface published, and geologic maps compiled in a variety of scales. This newly acquired information has drastically changed man's understanding of the planet which generations of astronomers have thought to be most like the earth.

Mars is geologically much more varied and dynamic than previously inferred. About half of the surface has been shaped by impact, the largest crater being almost twice the size of the largest crater on the moon. Other areas have been formed from volcanic activity, with some cones rising as high as 17 km (10 mi) above the surrounding surface. Tectonic activity has produced a great canyon system which has been enlarged and modified by erosion. Fluvial channels, which may have been formed by melting of permafrost, have been found. Patches of broken and jumbled blocks form a chaotic terrain, apparently the result of breakup and slumping of older geologic units. Wind action also appears to be an important surficial process and has modified most features of the planet. The polar regions are covered with glacial, eolian-layered sediments, and details of the advancing and retreating polar caps have been photographed with a type of time-lapse photography. It is also apparent that the geologic agents operating on Mars have not only varied from place to place but have also varied throughout the planet's history.

The importance of Mars is that in a study of physical geology it provides a third reference point, in addition to the earth and moon, for understanding planetary dynamics. This additional reference is a great help in emphasizing the fundamental principles of geodynamics.

Major Concepts

1. Cratering is the dominant geologic process on Mars and probably records an early period of ancient bombardment, though not the primordial accretion.
2. Vulcanism is expressed by two major features: (a) huge shield volcanoes and (b) volcanic plains.
3. Grabens and rift systems constitute the major tectonic features. No compressive structures have been found. These, together with the types of volcanic features, indicate that the planet is thermally active.
4. Fluvial features, including braided stream systems and dendritic channels, strongly suggest a period in the history of Mars when running water eroded restricted parts of the surface.
5. Areas of chaotic, jumbled blocks suggest regions of collapse and breakup.
6. Eolian processes have been observed in action, and many surface features of Mars have been modified by wind erosion or deposition.
7. The polar regions are covered with sediments of probable glacial-eolian origin.
8. Mars has experienced a varied and complex geologic history and has evolved beyond a primitive sphere dominated by impact like the moon, but the planet has not developed a system of plate tectonics as the earth has.

Impact Features

Statement

Although Mariner 9 has revealed a variety of landforms on the surface of Mars, cratering still appears to have been the dominant geologic process. The craters of Mars are similar to the types found on the moon, although the craters on Mars show evidence of much greater erosion and degradation.

As on the moon, most cratering on Mars probably occurred early in the planet's history. The oldest areas on Mars are densely cratered, with craters ranging in size from 50 to 100 km (30 to 60 mi). Most of the craters probably record an early period of intensive bombardment and represent the most ancient, though not primordial, surface features, dating back to possibly the late accretional period of Mars.

Discussion

Impact Craters. A study of the physiographic map of Mars reveals that approximately half of the planet's surface may be described as densely cratered. Most craters are greater than 20 km (12 mi) in diameter and are characteristically flat-floored. Fresh craters with well-developed ejecta blankets are rare, and the bright-rayed craters, so conspicuous on the moon, are nearly absent. Most crater rims have suffered considerable modification and appear to be worn down (*figure 20.1*).

Studies of lunar craters indicate that the first impact feature to be eroded consists of the ray material which originally extended outward a distance equal to many crater diameters. The ejecta blanket, being thicker and composed of coarser material, is much more resistant, while the crater rims are the last features to be eroded. On the moon, crater erosion occurs largely by the churning action of impact ("impact gardening"). On Mars, there is the additional process of wind action. The fine-grained ray material would apparently be transported by winds. Thus, the paucity of rayed craters is probably an indication of the effectiveness of wind erosional and depositional processes on Mars.

Multi-Ringed Circular Basins. Six large circular basins have been described on the Martian surface and are similar to the well-known multi-ringed lunar basins such as Orientale and Imbrium. These are readily apparent on the physiographic map and, although they have been modified by both erosion and deposition, they are in many ways similar to their lunar counterparts (*figure 20.2*).

The largest Martian basin is Hellas, about 2000 km (1200 mi) in diameter from rim crest to rim crest. Much of the basin is covered by plains material so the inner rings, if they exist, are concealed. The northern and eastern rims are composed of arcuate belts of rugged, multi-peaked mountains which resemble the Montes Apenninus chain forming the rim of the Imbrium Basin on the moon.

The Isidis Basin is the second largest but is not complete, be-

Figure 20.1 *Impact craters on Mars have been eroded and modified much more than those on the moon. Note the subdued and poorly defined crater in the upper left of the photograph and the large eroded and broken-floored crater left of center. Small scattered craters have sharper rims and are relatively unmodified. Well-developed ejecta blankets and bright rays are absent. The distance across the top of the photograph is about 400 km.*

cause its interior and much of its northeastern margins are buried by plains deposits. However, its identity as a multi-ringed basin is well established by the arcuate belt of rugged peaks bordering the plains on the south and along the concentric scarp in the terrae south of the basin. A segment of a third outer ring is apparent to the southwest, so the basin is tentatively interpreted as having three rings.

The Argyre Basin, located west of Hellas at 40° is slightly smaller but has the most complete rugged rim of all Martian basins. It has some features similar to Orientale on the moon—three concentric inner rings and two or three partial outer rings. In addition, radial lineations are clearly visible outside the third ring.

The Earth's Dynamic Systems

Schiaparelli, Huygens, and Lowell are smaller and might be compared to the Crisium Basin of the moon. By far the freshest is the double-ringed basin, Lowell. The outer ring is a complete concentric structure typical of a crater rim, and the inner ring consists of a well-defined line of rugged peaks much like the lunar Schrodinger.

The similarities between the multi-ringed basins on the moon and Mars are great enough to support the tentative conclusion that they were formed by the same process—impact of large meteorites.

Conclusions. The oldest areas of the Martian surface are densely cratered and probably record a very early period of intensive bombardment, though not the primordial accretionary features. This heavily cratered region falls below the saturation line found on the moon, a fact which is considered to be a sure indication that substantial erosional and depositional processes have destroyed the earliest traces of accretion. Preservation of crater features with their original circular outline indicates that the Martian crust has not been deformed by compressive forces similar to those which produce mountain ranges on earth. However, the craters are considerably modified by erosion and deposition of sand, dust, and volcanic products, suggesting that the crustal development of Mars is probably intermediate between that of the moon and the earth.

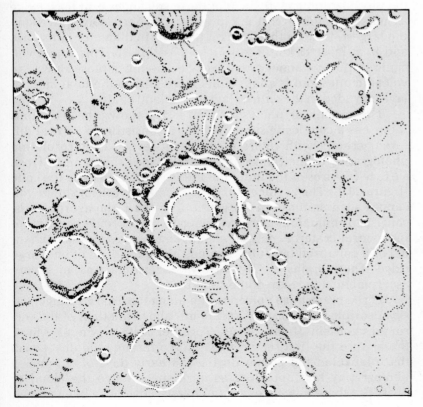

Figure 20.2 *Sketch of the multi-ringed basin Lowell located near 80° W and 50° S. Note that the outer ring completely surrounds the entire basin with a well-defined line, but the radiating ejecta blanket is poorly defined, probably due to modification by eolian deposition. Compare with the Orientale Basin on the moon, figure 19.8.*

Volcanic Features

Statement

One of the most significant results of the Mariner 9 mission was the discovery of extensive volcanic features on Mars. Previous "fly by" missions of Mariner 4, 6, and 7 sent back photographs of only limited areas of Mars in which the surface appeared to be highly cratered and moonlike, with little indication of volcanic activity. Yet, when the rest of the planet was photographed by Mariner 9, it became obvious that volcanic features may be found on nearly half of the planet.

Two types of volcanic features are found on Mars. The most spectacular are the large shield volcanoes, domes, and craters, concentrated primarily in the northwest. The four great shield volcanoes in the Tharsis region are much larger than any volcanic feature found on earth. Less spectacular but highly significant are the volcanic plains forming most of the sparsely cratered regions of Mars. These resemble the flows which form the lunar maria. Volcanic features which characterize subduction zones on earth have not been found.

Discussion

Shield Volcanoes. The most striking volcanic features on Mars are the four enormous shield volcanoes of the Tharsis-Amazonis-Elysium region near the western equatorial region of the physiographic map. Each of these volcanoes is at least twice as massive as the largest shield volcano on earth. Olympus Mons, the largest, is more than 500 km (300 mi) in diameter at its base and is at least 15 km (9 mi) high—possibly 30. This single volcanic edifice is about twice as wide as the largest Hawaiian volcanic piles and is about equal in volume to the total mass of lava extruded in the entire Hawaiian Island chain.

The mosaic of photos in *figure 20.3* shows that Olympus Mons has a composite summit caldera consisting of several different craters with floors at different levels. High-resolution photography of the flanks of Olympus Mons shows many fresh surface features of the individual flows which make up the cone (*figure 20.4*). The surface, similar to some terrestrial shield volcanoes such as those in Hawaii and in the Galapagos Islands, consists of numerous narrow, elongate ridges roughly radial to the central crater. A narrow channel suggestive of a lava channel or a collapsed lava tube extends along the crest of one large ridge. The margins of Olympus Mons terminate as an abrupt escarpment at the base of the shield.

Three other huge volcanoes lie along the Tharsis Ridge and rise possibly more than 17 km (10 mi) above the floor of the surrounding basin (*figure 20.5*). Together with Olympus Mons, these were the first features to emerge through the dust storm that blanketed the planet during the early days of Mariner 9 photography and appeared as dark spots above the cloud of dust. The three volcanoes along Tharsis ridge have been named

Figure 20.3 The volcanic shield of Olympus Mons. Olympus Mons is the largest known shield volcano in the solar system. It is approximately 500 km (310 mi) across the base and is surrounded by steep cliffs. The summit of the volcano consists of a complex of volcanic vents and caldera, 65 km (40 mi) in diameter. The mountain is more than twice as broad as the most massive volcanic pile on earth (the mountain formed by the Hawaiian Island is 225 km (140 mi) across and rises 9 km (30,000 ft) from the floor of the Pacific to the summit). The volcano lies upon a smooth plain with very low crater density, suggesting that it formed as recently as the last 100 to 300 million years.

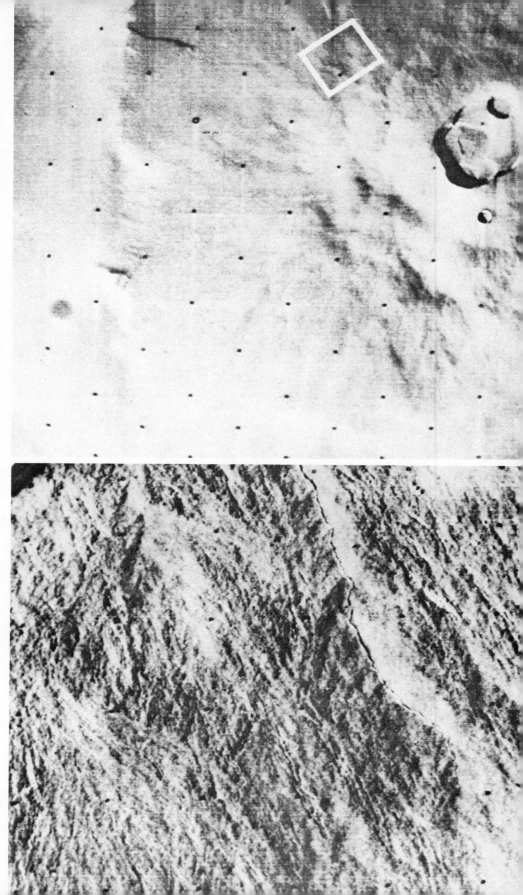

Figure 20.4 Wide-angle photo and telephoto of Olympus Mons. The wide-angle photograph covers an area of 436 km (271 mi) by 560 km (346 mi). The telephoto picture covers an area of 43 km (27 mi) by 56 km (34.5 mi), and can be located within the inscribed rectangle in the wide-angle photograph. The telephoto frame shows an intricate surface detail resulting from intersecting elongate lobes of lava flow that has moved down the slope of the volcano, radiating away from the summit crater. A raised ridge with an irregular crack trending along its crest can be seen extending from the top center of the picture diagonally to the right. This feature has been tentatively interpreted as a collapsed lava tube.

Arisa Mons, Pavonis Mons, and Ascraeus Mons. The lava flows of Arisa Mons are not long and thin like those of Olympus Mons but, in contrast, are short and stubby, indicating that the lava was more viscous, possibly andesitic in composition. Similar short, stubby flows composed of andesite are common in many continental volcanoes such as Mt. Hood and Mt. Rainier in the Cascade Mountains. As indicated by the fresh surface features of the lava flows, the sharp rims of their caldera, and the relative scarcity of impact craters, Olympus Mons and the volcanic structures in the Tharsis Ridge area are geologically recent. Some scientists estimate the age of Olympus Mons to be 100 million years and the volcanoes of the Tharsis Ridge, 300 million years. Scientists conclude that some of the volcanoes on Mars are currently active since clouds appear around the calderas of the high volcanoes. They believe that the clouds could be formed simply by water vapor condensing as it rises above the volcanoes.

Figure 20.5 *Volcanoes in the Tharsis region.*

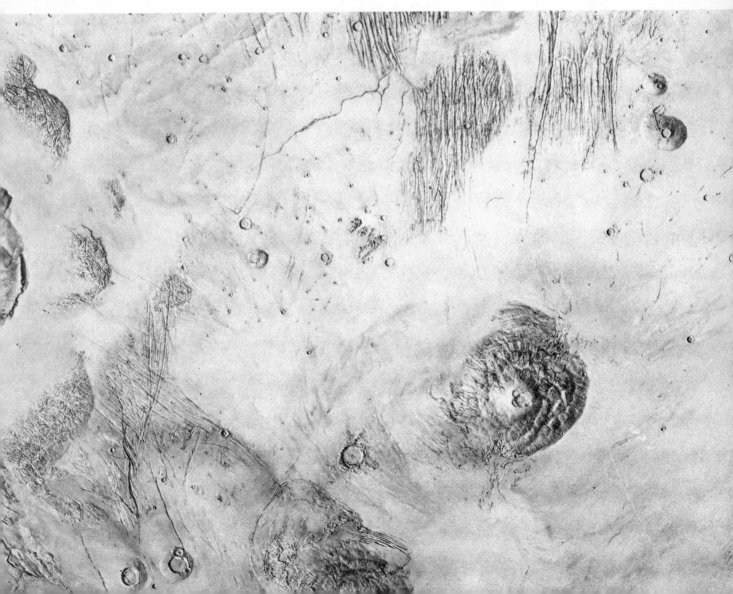

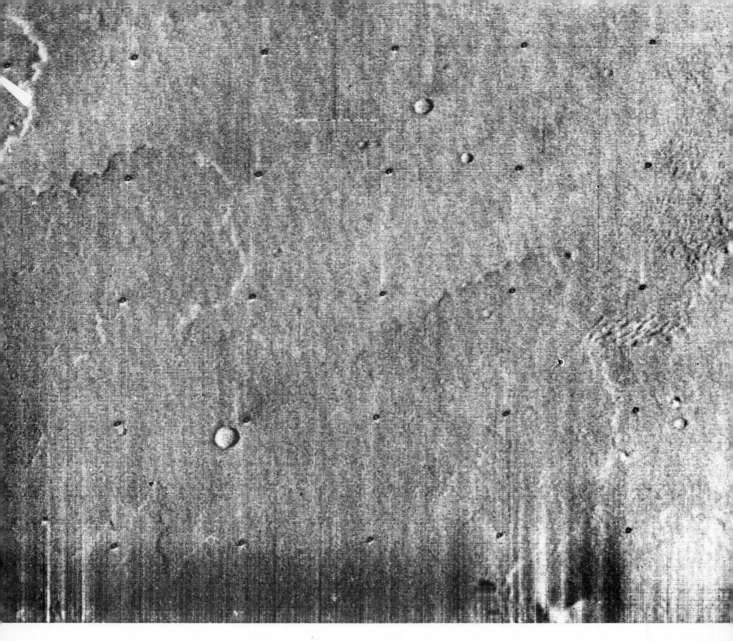

Figure 20.6 *Lobate fronts of lava flows on the sparsely cratered terrain. The lava flows on Mars are very similar to the basalt flows on earth and the maria flows on the moon. Compare with figure 19.11.*

Smaller and older volcanic cones and domes are also found on the Martian surface, indicating that volcanic activity on Mars is not restricted to its recent history. Lines of volcanic vents along fracture systems have been mapped and are similar to fissure eruptions on earth. Some craters appear to be floored with lava segmented into polygonal blocks in what appear to be frozen lava lakes similar to those near the summit of Kilauea, Hawaii.

Volcanic Plains. Although the great shield volcanoes present the most spectacular evidence of vulcanism, the flows which form the sparsely cratered plains may be volumetrically more significant. The plains commonly appear to be featureless on regional photographs but are seen to contain a succession of relatively thin lobate front lava flows on the detailed photos (*figure 20.6*). These flow margins resemble the lunar flows in Mare Imbrium and are particularly common in the south near the volcanic shield, suggesting that much of the smooth plains in these areas is formed by extensive lava flows. In addition, it is possible that the grooved terrain in the vicinity of Olympus Mons is deeply eroded lava flows. Some of these older volcanic features are highly cratered and may date as far back as 3 billion years.

The volcanic activity on Mars clearly indicates that the planet is thermally active and that some planetary differentiation has occurred. However, the rate at which a planet heats up depends upon many factors, such as amount of radioactive material in its original mass and the total mass which determines the pressure on the interior. If Mars had the same original composition as the earth, it would heat up more slowly because it has only one-tenth the mass of the earth. Perhaps Mars is just in the initial stages of developing a convecting interior—just beginning to "boil," so to speak. This process may be well advanced in the Olympus Mons-Tharsis Ridge area but has not yet developed throughout the planet as a whole. Indeed we may be witnessing on Mars an early phase in planetary evolution like the one the earth went through early in its history because of its greater mass. The huge shield volcanoes on Mars may be the result of a fixed crust in relation to the mantle which would allow greater piles of lava to build up. On earth, the same process produces a chain of volcanic islands and seamounts as the spreading sea floor moves across the hot spot.

Tectonic Features

Statement

The abundance of undeformed craters on the surface of Mars clearly indicates that the crust has not been deformed by horizontal stresses like the continental crust of the earth. Yet, there are tectonic features on Mars which, together with volcanic features, provide convincing evidence that extensive tectonic activity has occurred and that the interior of the planet is active. Grabens are the most abundant structural feature and occur both as closely spaced sets and as isolated features several thousand kilometers long. They are most abundant in the vicinity of the great shield volcanoes in the west.

The most spectacular tectonic feature on Mars is the great rift system in the Coprates region. It consists of a series of troughs which collectively form the "Grand Canyon of Mars." Tensional structures are, therefore, the dominant tectonic features on the planet, and most can be ascribed to circumferential tension in the upper part of the lithosphere and to local doming.

Mars has apparently made the transition from a relatively

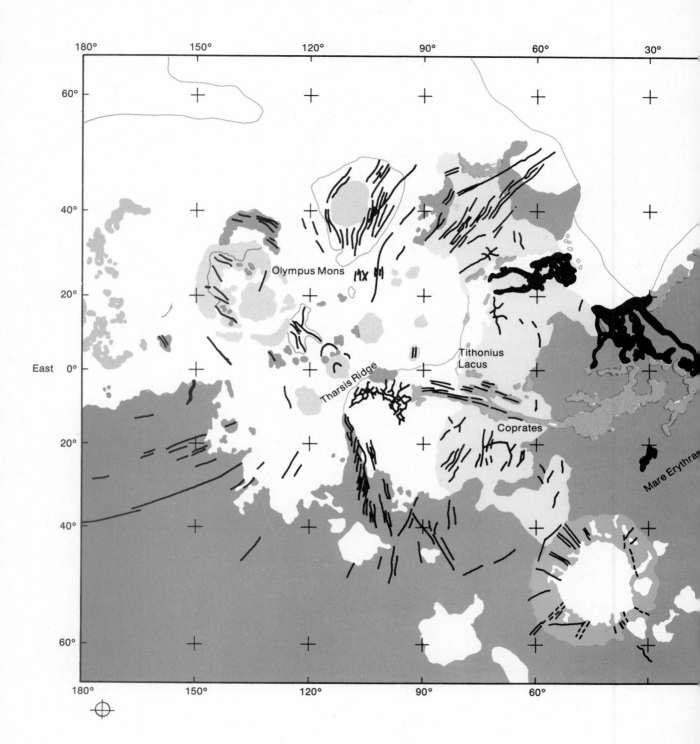

Figure 20.7 *Generalized geologic map of Mars showing the spatial relationship between the major types of deposits.*

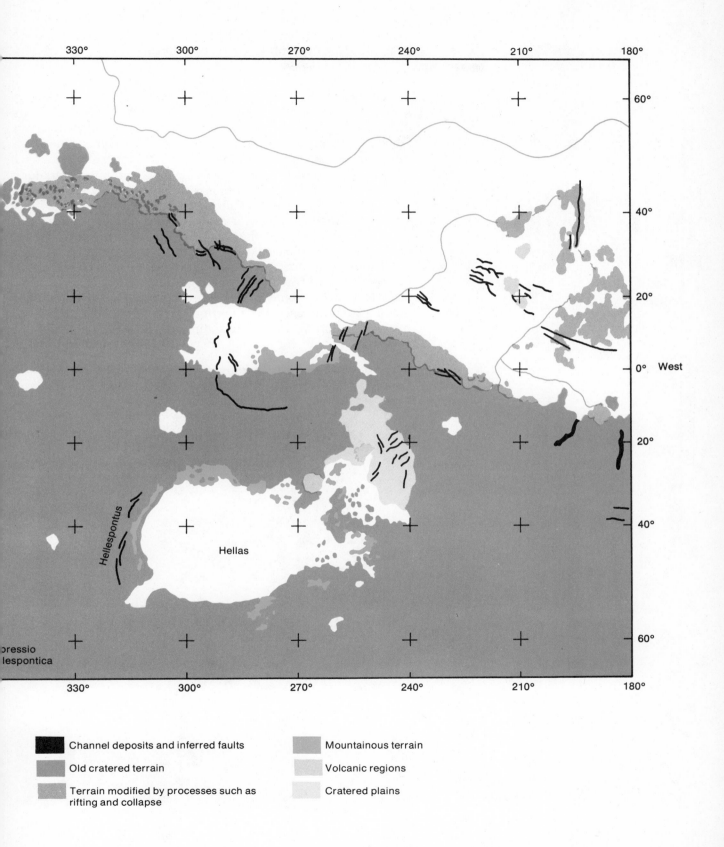

330°   300°   270°   240°   210°   180°

60°

40°

20°

0° West

20°

40°

60°

Hellespontus

Hellas

pressio
lespontica

Channel deposits and inferred faults

Old cratered terrain

Terrain modified by processes such as rifting and collapse

Mountainous terrain

Volcanic regions

Cratered plains

primitive sphere dominated by impact structures like the moon, to an orogenically mobile planet like the earth.

Discussion

Grabens. The distribution of the major faults on Mars is shown on the geologic map in *figure 20.7.* Most of the grabens are parts of a system of faults which forms a radial pattern out from the Tharsis Ridge, with the focal point being near the equator at approximately 110°. The grabens are typically 1 to 5 km (.6 to 3 mi) wide and may be several thousand kilometers long (*figure 20.8*). This pattern appears to have formed from extension associated with a broad upwarp of the Tharsis region.

Other fractures occur as concentric or radial systems associated with large impact basins. They also occur locally around the large shield volcanoes.

The Rift System. The great rift system of Mars is part of the system of radial fractures extended out from the Tharsis Ridge area. Here the grabens coalesce to form a huge canyon extending in a general east-west direction just south of the equator between 45° and 90°. The huge dimensions of this feature are vividly shown in *figure 20.9,* in which an outline map of the

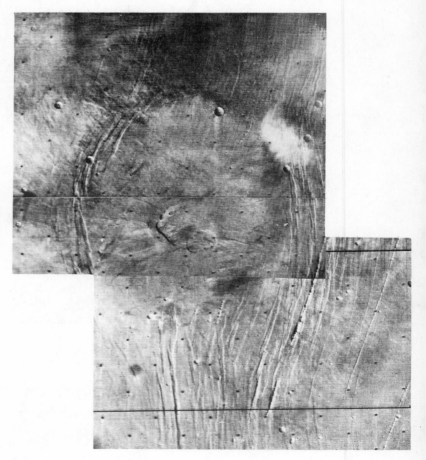

Figure 20.8 *Mosaic of wide-angle picture showing the complex faults associated with volcanic structures in the Alba region.*

The Earth's Dynamic Systems

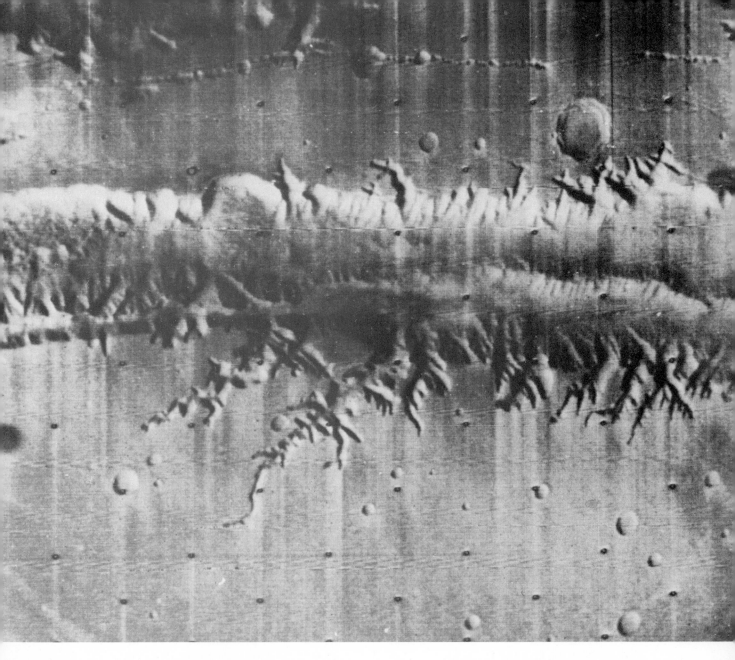

United States is superposed over the area. The rift system is far larger than anything known on earth. It is nearly four times as deep as the Grand Canyon and would extend almost across the entire United States. The rift system is expressed topographically by a series of steep-walled troughs, highly irregular in detail, with sharp indentations, arcuate alcoves, dendritic tributaries, and large open embayments (figure 20.10). Many gullies and ravines resemble, in form, those cut by running water on earth. The plateau or upland surface is separated from the canyon walls by a sharp brink which is also similar to stream-cut canyons in the desert regions of the American Southwest.

Figure 20.9 *A portion of the great Coprates Canyon (276 by 480 km [235 by 300 mi]) which has been eroded in the older smooth surface sprinkled with large craters. This vast chasm with branching canyons may represent a landform unique to Mars. The resemblance of the tributaries to terrestrial stream systems is probably superficial for many of the tributary canyons or closed depressions. Subsidence along lines of weakness in the crust and deflation by Mars winds has probably sculptured this unique pattern.*

Figure 20.10 Ravines and branching narrow divides and alcoves on the north wall of Coprates Canyon. Note the resistant ledge along the top of the canyon along with the grooving or fluting of the steep upper walls, similar to terrestrial canyons.

The configuration of the canyon walls clearly indicates that considerable erosion has occurred to widen and modify the original rift valleys. The dominant processes were probably landsliding, debris flows, and possibly running water. However, this presents a particularly vexing problem: How was the eroded material removed from the canyon system? Some debris may have been transported along the canyon and out the east end, but there are no obvious associated depositional features there such as alluvial fans or deltas which are typically formed at the mouths of river systems on earth. Moreover, many individual troughs are closed basins so that sediment could not be transported through them except possibly by the wind. It is clear that erosion has modified the canyon walls, but it is highly unlikely that processes of erosion cut the canyon system.

In considering the origin of the rift system it is important to note that on the uplands, in the vicinity of the troughs, are linear chains of pits and complexes of shallow grabens, both parallel to the regional trend of the rift. These support the contention that the troughs or canyons were initially formed by tension and collapse. Several possible mechanisms have been suggested. The proximity of the rift system to the largest young volcanic field on Mars suggests that the rift system resulted from collapse, due to withdrawal of magma. The volume of lava extruded is certainly adequate, because Olympus Mons alone represents a greater volume than is missing from the canyons. However, it is difficult to explain why the magma extruded in the Olympus Mons-Tharsis Ridge area would cause subsidence in a trough located so far to the east. Another possibility is that the rift system resulted from spreading of crustal plates. This is an exciting hypothesis, especially in light of the plate tectonic theory developed for the earth. The character and configuration of the rift system can adequately be explained by a spreading mechanism, but there is no obvious subduction zone on Mars. Moreover, there is no evidence of compressional stresses in the Martian crust because all of the craters retain their original shape. Perhaps the rift system represents only the initial stage of crustal spreading, a stage which occurred early in the earth's history because of its larger mass and greater tendency for planetary differentiation, but a stage never attained by the much smaller moon. In any event, the rift system on Mars, like the volcanoes, represents a significant aspect of the evolution of the planet and the nature of its internal processes. Evidence points to crustal updoming of a type recognized on earth, reflecting incipient convection but falling short of full-fledged plate tectonics as developed on the earth.

Fluvial Features

Statement

Perhaps the most startling finding of the Mariner project was the discovery of channels on the Martian surface which closely resemble certain types of stream channels on earth. Unlike the river systems on earth, which consist of a network of branching tributaries that funnel surface runoff to a main stream, the channels on Mars are restricted to certain areas and appear to have been only intermittently active. The fluvial landscape on Mars consists of only one or two periods of erosion, not a universal process operating like the hydrologic system on earth, and there are large areas of essentially undisturbed primitive terrain untouched by stream erosion. Although at the present time the surface temperature and pressure and water content of the Martian atmosphere make bodies of liquid water at the surface impossible, or at least confined to a few small areas during certain seasons, many features on Mars are most easily explained by assuming the presence of running water at some time in the past. Indeed, there are some channels in which running water appears to be the only explanation.

Discussion

Some of the most convincing evidence for an aqueous origin of the Martian channels is found in the region near 6° S 150° W (*figure 20.11*). This channel complex is similar in many ways to a braided stream system on earth and, were it not for the craters, it would be easy to believe that these pictures were of the earth. In the upper section, the Mangala Vallis channel consists of many broad branches which merge into a sinuous main channel with a smooth floor. Downstream, the channels broaden to form a braided complex.

Another braided channel system is in the Lunae Palus region (*figure 20.12*) and the Oxia Palus region (*figure 20.13*). In these areas the braiding is particularly obvious in the downstream portions and terraces, and remnants of islands can be seen along parts of the course.

Figure 20.11 *Section of a braided channel on the surface of Mars. Channels such as this provide convincing signs of fluvial activity on Mars inasmuch as they show features typical of terrestrial braided streams. This photograph covers an area of about 35 by 40 km (22 by 32 mi).*

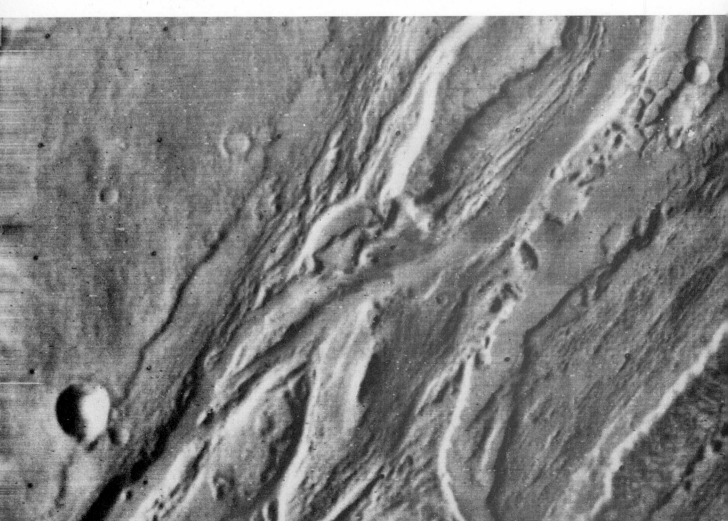

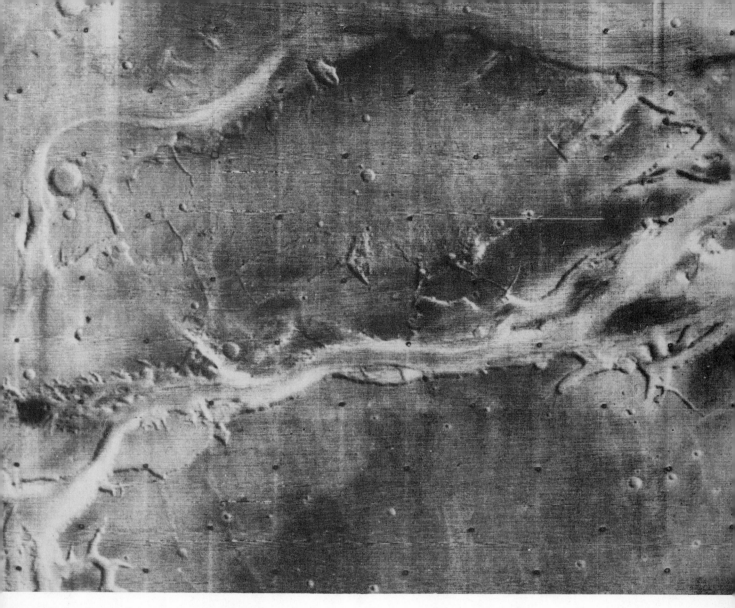

These channels are quite unlike the typical integrated drainage systems found on earth with branching tributaries which join to form channels of higher order culminating in a master stream. Rather they are similar to the Channelled Scablands of Washington, which were produced by catastrophic flooding. It appears that the braided channels on Mars did not form from regional precipitation of rain but represent flow in which there was considerable variation in velocity. The stream during the waning stages of flood was unable to transport the sediment during peak flow and deposited it to form the system of braided bars, islands, and terraces. Many scientists believe that the history of the Amazonis and related channels may have involved a succession of catastrophic floods punctuating periods of much less steady flow.

Figure 20.12 *Erosional valley and channel deposits. The valley shown in this photograph shows several characteristics typical of an eroded valley and associated channel deposits. Note the terraces along parts of the southernmost channel and remnants of islands near the right-hand margin. Fractures locally control the course of the channel, especially the small tributaries cut into the sides, which appear to represent the most recent episodes of erosion.*

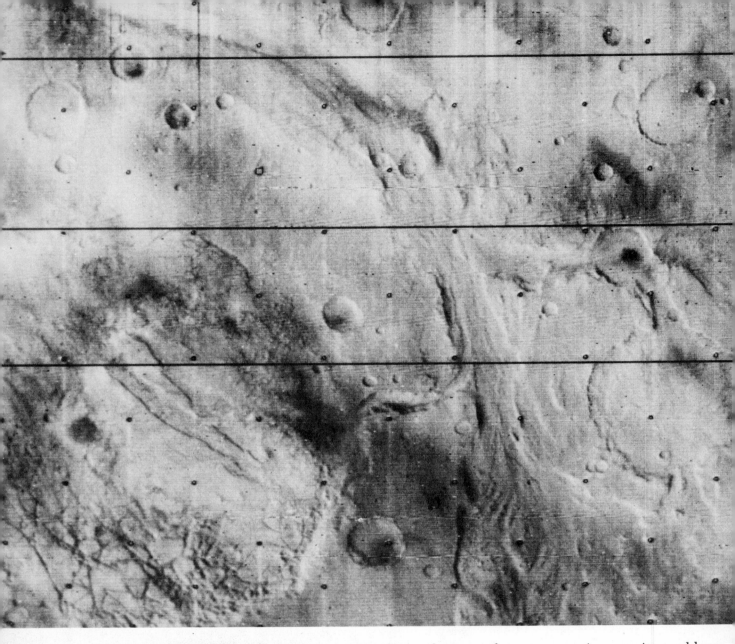

Figure 20.13 *Braided channel in the southern part of the Oxia Palus region. This is a section of a channel over 1500 km (1000 mi) long and contains what appears to be channel bars near the lower center part of the picture. Chaotic terrain is present in the lower left. Note several levels of terraces in the southern, lower part of the photograph similar to the terraces of the braided stream on earth.*

The source of water, of course, remains a major problem. Most geologists believe that the features described above are best explained by the melting of ground ice. There are other indications that ground ice may be widespread beneath the surface of Mars (see section on *collapse features*) and that melting produced by climatic changes or volcanic activity would produce liquification and runoff. Another question is whether it has rained or snowed on Mars.

Several other types of channels found in local areas on Mars probably originated by a different process. A channel pattern, entirely different from the braided channels described above, is found in the Mare Erythraeum region near 29° S 40° W (*figure*

The Earth's Dynamic Systems

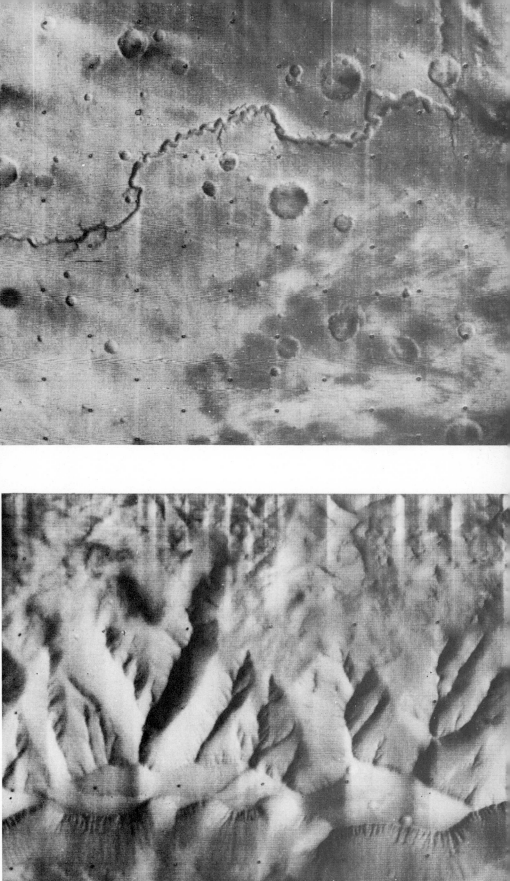

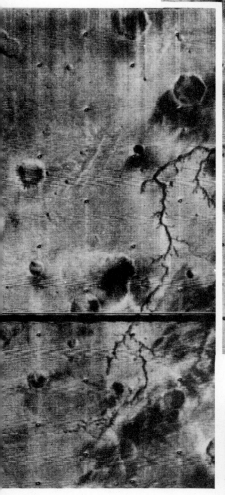

Figure 20.14 *Channels in the Mare Erythraeum region. The upper regions of the channel show branching tributaries similar to the dry river beds in the desert regions of the western United States. This channel is one of the most convincing pieces of evidence that a fluid once flowed on the surface of Mars. The channel is some 575 km (550 mi) long and 5 to 6 km (3 to 3.5 mi) wide.*

Figure 20.15 *Detailed view of ridges and erosional channels in the Coprates Canyon. Note the sharp ridges or spurs reminiscent of stream-cut topography on earth.*

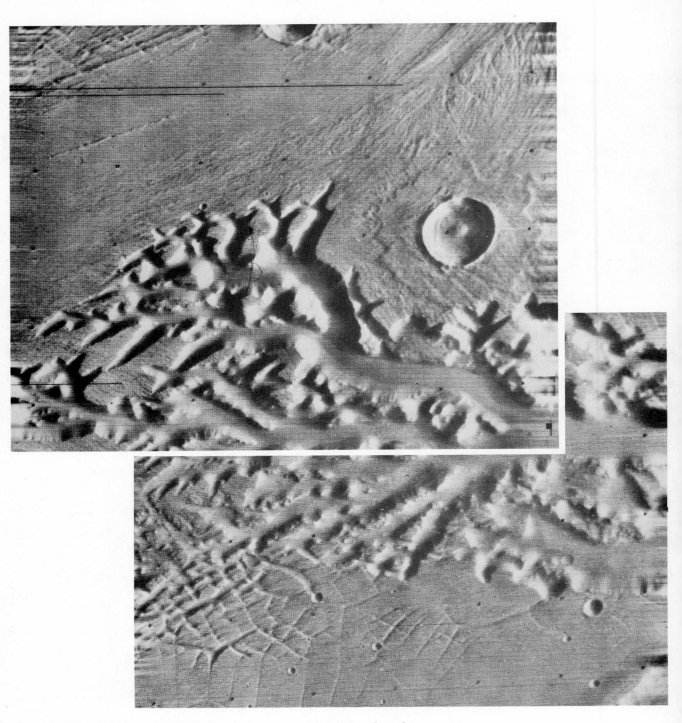

Figure 20.16 *A two-picture mosaic showing a canyon system developed by selective erosion along angular sets of fractures. Wind scour has apparently modified the surface of the upper plateau level, and the ejecta blanket of the crater in the central part of the area has been more resistant to erosion. The system of canyons forms a dendritic pattern much like the dendritic river system on earth.*

20.14). The upper segment of the channel contains branching tributaries and closely resembles dry washes in desert regions. Similar branching channels constitute erosional tributaries to the large canyons (*figure 20.15*). In addition, there are systems of dendritic canyons (*figure 20.16*) and gullies on the slopes of volcanoes (*figure 20.17*), facts which argue for some atmospheric source for water—at least in local areas.

In summary, it is apparent that the surface of Mars has been modified by flowing liquid (presumably water) and that at some earlier stage fluvial processes operated on Mars. The change from ancient fluvial periods to the present dry period may be similar to the change from glacial to interglacial periods on earth.

Figure 20.17 *Gullies and erosional channels on the slopes of the volcano Alba near 45° N 116° W. The width of this frame is about 60 km (40 mi).*

Eolian Features

Statement

When Mariner 9 entered an orbit around Mars, a planet-wide dust storm was raging, completely obscuring the surface features of the planet during the early months of the mission. This great dust storm was a boon to those studying geologic processes on Mars because it clearly indicated the lateral transport of large quantities of surface material. After the dust storm subsided and detailed analysis of Mariner 9 photography was made, it became apparent that eolian activity has produced important surface features on Mars. These include: (1) dune fields, (2) systems of parallel streaks, (3) streamlined ridges and linear grooves, (4) modified crater rims, and (5) subdued sand-mantled

Figure 20.18 Eolian features associated with craters in the Hesperia region (23° S 242° W). The streaks are believed to be fine, bright dust transported into the craters in the waning stages of a dust storm and subsequently blown out by high-velocity winds having a constant direction.

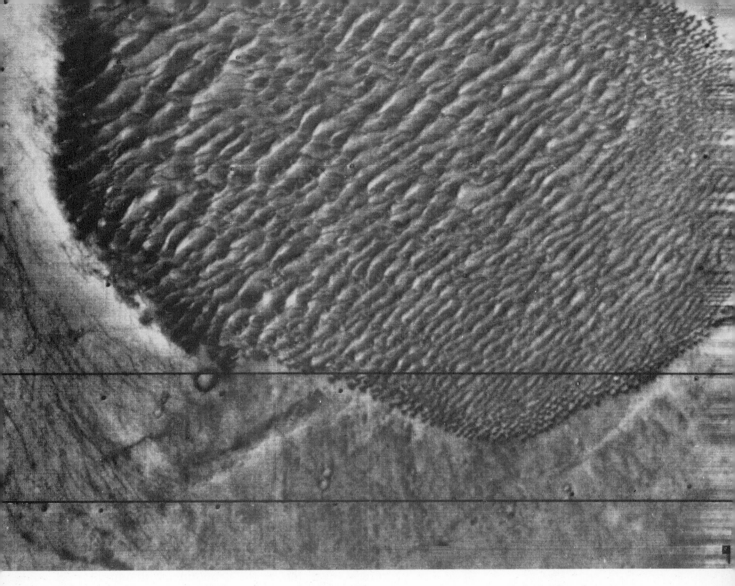

terrain. These features collectively indicate that the eolian regime on Mars is probably more powerful than on earth and that
the processes of deflation, saltation, suspension, and transport
are more significant on a planet-wide scale.

Discussion

Dunes. The most obvious eolian features are the systems of
parallel plumes and streaks which originate at a crater, ridge, or
cliff and extend hundreds of kilometers across the surface of the
planet (*figure 20.18*). These features are believed to be the result of wind erosion and deposition in which fine dust is deposited more or less uniformly over the surface during the
waning stages of a storm and subsequently eroded by high wind
velocities of later storms. There are both light- and dark-colored
plumes and streaks, and in places a single crater or topographic
form which acts as a wind shadow may have both types. Pictures

Figure 20.19 *Dune field in the
Hellespontus region near 75° S 331° W.
The wavelike structure is highly
suggestive of transverse sand dunes
found on earth. Note the tendency for the
size of the dunes to decrease towards the
margins of the field.*

taken of the same area at different times during the mission show that the configuration of the dark markings changes, whereas the light plumes remain unaltered.

Typical dune fields similar to the great desert regions on earth are not as conspicuous as the streaks and plumes. The most convincing evidence of dune formation is found in the Hellespontus region (*figure 20.19*), where a field of transverse dunes covers an area of more than 2000 km² (1500 mi²) on the floor of a large crater at 48° S, 330°. The spacing between dune crests is 1 to 2 km (.6 to 1.2 mi), becoming closer near the edges of the field as is the case in terrestrial dunes. The identification of dunes is significant in considering the eolian regime of Mars because dunes indicate that saltation occurs in spite of the tenuous atmosphere. With sand moving in saltation, numerous erosional features of eolian origin, which would not occur if the Martian winds moved only fine dust in suspension, can be expected. Dunes on Mars indicate that the surface is likely modified by wind abrasion.

The patterns of dune migration and the orientation of wind-formed streaks have been mapped, and the prevailing wind directions they indicate agree well with theoretical models of Martian atmospheric circulation deduced from rotation of the planet, solar heat, etc. Erosion appears to be most intense in the equatorial regions where sediment is picked up and carried towards the polar regions and gradually distributed over the planet.

Erosional Features. Although the craters of Mars closely resemble those of the moon in size and general appearance, they are distinctive in that they have been considerably modified. A careful study of *figure 20.20* will show that crater rims appear to be abraded and worn down. There is no sharp upturned lip of the crater margin so typical of the fresh craters on the moon, and the rays and ejecta blankets have been eroded or buried beneath windblown sediment. Elsewhere, wind erosion is strongly suggested by the pronounced alignment of elongate ridges which are commonly parallel to numerous fine-textured surface grooves (*figure 20.21*). The crests of the ridges are sharp and keellike, and the ends taper sharply. Similar features have been observed in desert regions on earth and, although the primary origin of the ridges is a matter of conjecture, the morphology of these features strongly suggests considerable modification by wind erosion. The degree to which wind erosion has modified the Martian surface cannot be determined with existing photography, but many geologists believe that it has been intense enough to completely obliterate preexisting volcanoes. Some believe that wind erosion is cutting away at the base of Olympus Mons and is responsible for developing the cliff 1 to 2 km (.6 to 1.2 mi) high, which surrounds its base.

The Martian Eolian Regime. The surface features described above, together with studies of the Martian atmosphere, permit scientists to draw some preliminary conclusions concerning the

Martian eolian regime as a whole and compare and contrast it to that of the earth and moon. The Martian atmosphere is only one-hundredth as dense as the earth's so that extremely high velocities of wind are necessary to transport sand and dust. Mariner 9 television data indicate that dust storms or clouds exceed velocities of 200 km (120 mi) /hr so that gust velocities two to three times the transverse velocities do not seem unreasonable. This far exceeds wind velocities on earth. In addition, there are a number of other important differences resulting from atmospheric differences. For example, the threshold velocity (velocity at which grains begin to move) is estimated to be ten times greater on Mars than on earth. But then the saltating grains should have ten times greater momentum and a hundred times more kinetic energy. Moreover, the thin atmosphere on Mars would produce practically no cushioning effect, permitting very small particles to act as very effective instruments of erosion.

Figure 20.20 *Mariner 6 photograph of the portion of the equatorial region of Mars showing the characteristic craters which have been modified by erosion and deposition. Note that most of the craters are flat-floored, have smooth rims, and lack a well-defined ejecta blanket.*

Geology of Mars

Figure 20.21 *Fine parallel grooves in the relatively smooth, uncratered plains, presumably formed by wind erosion.*

The relatively weak gravitational field plus the thin atmosphere on Mars combine to permit saltating grains to reach heights three to four times greater than on earth so that wind erosion would be more effective.

Fine fragmental material, including glass beads and rock fragments, would undoubtedly be produced from impact of meteorites similar to that observed on the moon, so that loose fragmental material would be readily available to the strong winds. In addition, sand and dust would be produced from the fluvial channels, mass wasting, and volcanic activity. The factor of time should also be mentioned, because wind action on Mars may have been going on without significant interruptions since the origin of its atmosphere. Only rarely have volcanic activity, tectonics, and running water interrupted the process. On earth, plate tectonics continually create and destroy the crust, and water has dominated the surface processes almost from the very beginning. Wind action has been actively limited to the restricted desert regions and has often been masked by or intimately mixed with the effects of the more universal process of running water.

The Earth's Dynamic Systems

All of these factors collectively suggest that wind action is far more intense and probably proceeds at a much faster rate on Mars than on earth. Indeed, wind action is considered by many geologists to be the principal surface process, influencing the Martian surface as universally as running water affects the surface of the earth.

Collapse Features

Statement

Isolated areas of the Martian surface are characterized by jumbled masses of large, angular blocks. These areas are referred to as chaotic terrain and are believed to result from the withdrawal of material from the subsurface and subsequent collapse of the overlying rock.

Figure 20.22 Chaotic terrain. This terrain consists of a complex of broken slabs and blocks which generally lie below the regional surface of the surrounding area. The blocks are locally so small that they can barely be seen on the limits of the resolution of this photograph. Note the braided channel in the upper right part of the photograph, which originates in a patch of chaotic terrain.

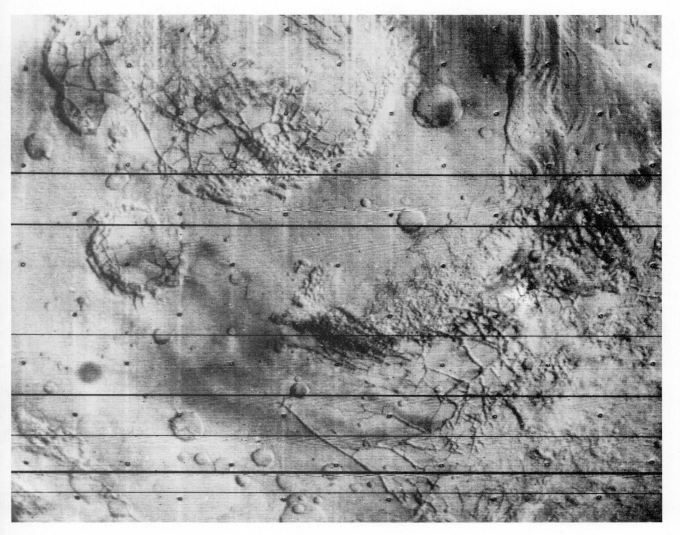

Chaotic terrain is most abundant near the eastern margins of the great canyon between 15°-30° W longitude. As can be seen in *figure 20.22*, the surface consists of a complex mosaic of broken slabs and angular blocks, generally lying below the surface level of the surrounding upland from which they were formed. Large arc-shaped slump blocks are common near the margins of the adjoining upland, which is generally cut by arcuate fractures. These appear to break up into numerous small blocks, many of which approach the limits of TV resolution. It is clear from these features that fracturing, collapse, and slumping have played an important part in the development of chaotic terrain, but the processes which produced collapse are not readily apparent. On earth collapse features are most widespread in limestone terrain and result from solution activity, but it is doubtful that soluble rocks are widespread on Mars, and there is not an abundant, suitable solvent in the Martian atmosphere. Collapse could result from the deterioration of ground ice, followed by slope retreat and slumping. Another possibility is the subsurface movement of magma associated with vulcanism.

The Polar Regions

Statement

When Mariner 9 reached Mars, it was late spring in the southern hemisphere—an ideal time to study the retreat of the polar cap and the unusual surface features beneath it. The polar regions consist of saucer-shaped depressions, possibly depressed by isostatic adjustment caused by the weight of formerly thicker polar ice. Beneath the ice cap are two sedimentary units. The older is moderately cratered and has many closed depressions that appear to be deflation hollows. This is referred to as the "etch-pitted" terrain, which is, in turn, overlain by a sequence of layered rocks originally termed "laminated terrain." These deposits are restricted to the polar areas and apparently resulted from processes associated with the ice caps.

Discussion

Polar Caps. Mars rotates once every 24.5 hours, and its axis is inclined almost exactly the same amount as the earth's. Thus, seasonal changes occur in response to the amount of solar radiation received, and the white polar caps of Mars expand and contract and alternately grow in one hemisphere and then the other every Martian year (*figure 20.23*).

The polar caps were once thought to be composed of water, but Mariner 6 and 7 confirmed that they are composed of very pure, solid carbon dioxide, "dry ice," rather than water ice. Photos of Mariner 9 show that the ice caps are relatively thin, probably no more than a few meters. Thus, they probably lack the great erosive power of continental glaciers on earth; never-

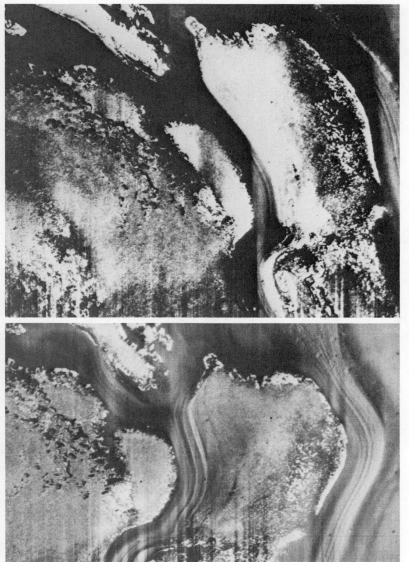

Figure 20.23 *Two telescopic views of a portion of the Martian ice caps showing the manner in which the frost cover disappears with time. The second picture was taken at a solar elevation angle of 4° and vividly displays the topography on the sinuous slopes facing the sun.*

theless, they constitute an important geologic process as they expand and contract with the seasons.

Sedimentary Blankets. The complex sequence of layered deposits was unanticipated before the Mariner 9 mission but has now been observed in both the north and south polar regions. Although these deposits are not distributed symmetrically about the geographic poles, they occupy most of the area poleward of the 80° parallels and appear to be confined to the polar regions. Throughout parts of the area, the sedimentary blanket forms a smooth, nearly featureless plain marked only by shallow linear grooves and a few fresh bowl-shaped craters. Elsewhere, the deposits are extensively eroded so that the internal stratification is well exposed (*figure 20.24*). These

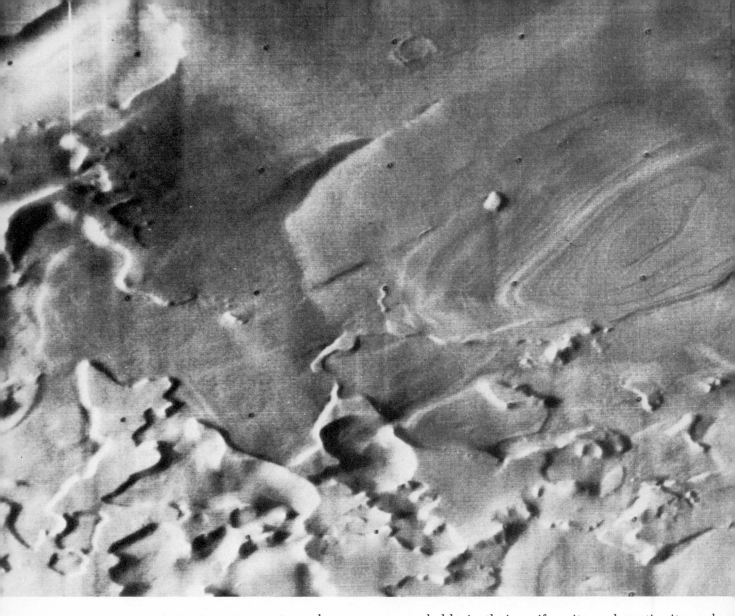

Figure 20.24 *Laminated terrain near the south pole of Mars. The oval tableland near the right part of the picture is an eroded remnant of a series of layers of light and dark sediments which may represent layers containing dust or volcanic ash alternating possibly with carbon dioxide and ice and water ice. The area covers about 47 by 60 km (29 by 37 mi).*

layers are remarkable in their uniformity and continuity and may contain a record of physical processes on Mars in much the same way as sedimentary rocks record the geologic history of continents on the earth. The sedimentary deposits consist of a series of alternating light and dark layers which are essentially horizontal so they typically resemble contours on an eroded surface. As many as 50 beds have been observed in a single exposure with a total thickness of possibly more than a kilometer. Beneath the blanket of sediment, old craters which are apparently being exhumed are partly exposed.

The sedimentary blanket is thought to result from glaciation and wind activity. Dust particles act as nuclei around which snow crystals of carbon dioxide and perhaps some water form and fall to the surface, accumulating as horizontal layers. A similar process occurs on earth when rain and snow sweep the atmosphere clean of dust particles. The alternating light and dark layers represent alternating beds of high dust and high ice content. The cyclic appearance of the laminae is likely an expression of periodic alternations in the climate of Mars.

The etch-pitted terrain is moderately cratered, has many closed depressions, and is composed of many alternating resistant and nonresistant rock units that produce numerous topographic benches and slopes. It essentially surrounds the polar regions and was probably formed by high wind velocities that form deflation basins.

Martian Satellites

Statement

A number of photographs have been made of the two little moons of Mars, Phobos and Deimos, and provide important new information concerning their size, shape, surface morphology, age, and origin. Both satellites are irregular in shape, as they are too small for their gravitational fields to impose a spherical shape. They are both highly cratered, almost to the point of saturation. Their form and shape seem to have been determined from impact fragmentation and spalling. On the basis of crater population, the satellite surfaces are estimated to be at least several billion years old.

Discussion

Figure 20.25 shows typical close-up photographs of Phobos and Deimos. Both are potato-shaped, with jagged cratered surfaces. Phobos is 25±5 km (15±3 mi) long and 21±1 km (13±0.6 mi) wide. Deimos is 13.5±2 km (8±2 mi) long and 12.0±0.5 km (7±0.3 mi) wide. The craters are typical of those produced by impact. Particularly striking is the large crater on Phobos, 5.3 km (3.2 mi) long, located near the bottom in *figure 18.25a*. The impact that produced this crater is close to the largest impact Phobos could have sustained without fracture and fragmentation.

Several pieces of evidence imply that the satellites possess a regolith, probably resulting from repeated impact fragmentation. In view of their small gravitational force, this may at first seem surprising since almost all ejecta arising from impact would escape from these bodies. However, the material would go into orbit around Mars and could eventually be recaptured by the satellites.

Phobos and Deimos are both regarded as old relative to any surface features on Mars and are possibly as old as the solar system. They appear to be remnants of a larger body (or bodies) and have evolved through a complex of collisions. At the present time, it is impossible to determine if the satellites were captured by Mars or represent material left over from the time Mars was formed. They are significant, however, in that they are

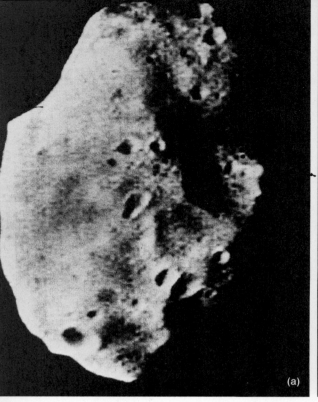

(a)

(b)

Figure 20.25 *The Martian moons are small, potato-shaped, and marked with numerous impact craters. (a) Phobos, the larger of the two moons, is approximately 25 km long and 21 km wide. (b) Demos is nearly equidimensional, being 13 km long and 12 km wide.*

probably typical of asteroids or planetesimals from which the planets were formed.

Geologic History of Mars

Statement

A generalized geologic history of Mars can be determined by using the principles of superposition and cross-cutting relations and the degree of erosion in much the same way they have been used in studying the moon and earth. Although there are many uncertainties in details of age relationships, especially in younger units which result from the more dynamic nature of Mars, a preliminary framework of the geologic history of Mars has been established.

The major events in the history of Mars are as follows:
1. Formation of densely cratered terrain.
2. Development of sparsely cratered terrain in the northern hemisphere by extrusions of flood basalt.
3. Extrusion of lava to form the great shield volcanoes, domes, and craters.
4. Development of grabens, rifts, and erosional features such as channel deposits and chaotic terrain.

Discussion

If you will study the physiographic map of Mars, you will immediately note that roughly one-half of the planet—largely concentrated in the southern hemisphere—is covered with cra-

The Earth's Dynamic Systems

ters over 20 km (12 mi) in diameter, whereas the northern part of the planet is relatively featureless and sparsely cratered. The boundary between these two major regions is quite irregular in detail and is somewhat gradational. Studies of the sparsely cratered regions in the north indicate that the surface material is largely volcanic material and windblown dust. The floods of lava and eolian deposits fill regional depressions, obscuring earlier basins and crater deposits, and resemble the lunar maria in that they are generally darker than the cratered highlands and have markedly fewer craters. On both planets the lava flooded regionally depressed areas, obscuring earlier basins and crater deposits. The lava flows and windblown deposits in the north are superposed upon the densely cratered terrain to the south and, thus, are younger. As on the moon, these relationships permit us to recognize two major events: (1) pre-lava cratering and (2) regional volcanic extrusions in which lava flooded the topographic depressions. The highly cratered terrain probably represents an early—though not primordial—record of intense bombardment. The lava floods of the plains probably range in age up to the most recent periods of geologic activity on Mars. The sparsely cratered terrain in the west is probably one of the most recent deposits.

The age relations among the younger units are not entirely clear because contact relations are commonly obscured. Several events, however, are clearly younger than the sparsely cratered lava plains. The volcanic shields in the Tharsis Ridge area lie upon the sparsely cratered plains and record a significant "recent" thermal event. The chaotic terrain is made up of both the densely cratered and sparsely cratered terrain and must be very young. The channel deposits are either younger or contemporaneous with the chaotic terrain, but both probably were formed over a considerable period of time. The great canyon of the Coprates region also cuts both the densely and sparsely cratered units and is, therefore, younger, but the canyon itself is modified by erosion and deposition. Faulting around the shield volcanoes is another recent feature, as it displaces the youngest sparsely cratered plains.

Conclusions. Although the age relationships of many younger geologic features on Mars cannot be determined in detail, it is clear that Mars has had a long and eventful geologic history. Like the moon, it shows evidence of extensive, early bombardment and the development of a primitive accretionary surface, followed by extensive floods of lava. However, Mars has been more active geologically than the moon and, subsequent to the lava floods, has experienced significant tectonic, volcanic, and erosional events.

Summary

The Mariner 9 mission to Mars has been one of the highlights of the space program and has revolutionized our knowledge of the "Red Planet." Nothing was seen of the famous "canals," but a variety of unanticipated surface features were photographed which indicate that Mars is geologically active with both internal and external processes. Among the more significant features photographed are the following:

Impact Structures. Much of the Martian surface is cratered, some of it so intense that it must date back to a time close to the original accretion of the planet. The craters on Mars are similar to those on the moon but show evidence of greater modification by erosion and sedimentation.

Vulcanism. Four great volcanic mountains in the Tharsis region and a number of smaller features elsewhere clearly indicate that Mars has volcanic activity on a scale comparable to that on earth. The great volcanic mountain, Olympus Mons, undoubtably ranks among the largest volcanic edifices in the solar system.

Tectonic Features. Fracture patterns, faults, and dikelike features are expressed in topographic relief and are eroded by wind. They are among the more prominent tectonic features seen on Mars and are probably related to volcanic activity. The great canyon in the Coprates region is probably the most impressive single feature on Mars and may be a great rift valley where the Martian crust is being pulled apart. Undeformed craters throughout the surface of Mars indicate that the crust has not been deformed by compressional forces.

Fluvial Features. Networks of erosional channels which form sinuous, branching, and braided patterns are similar to those formed by running water on earth.

Wind Action. The planet-wide dust storm that was raging at the time Mariner 9 arrived shows that winds of more than 160 km (100 mi) per hour must have swept the entire planet, transporting and depositing loose sediment. When the storm subsided, photographs revealed numerous surface features produced by wind, including dune fields, parallel streaks emanating from craters, and parallel ridges and grooves.

Collapse Features. Numerous lowlands contain a complex of closed depressions with steep cliffs, landslide blocks, and chaotic debris. The depressions have been progressively enlarged by slope retreat, probably initiated by melting of ground ice.

Polar Regions. Mariner 9 mapped the waning of the south polar cap. The ice shrank rapidly at first, as would be expected for a thin layer of CO_2 frost, but later stabilized with a residual cap 400 km (250 mi) in diameter. The size and stability of the risidual cap suggests that it is composed of water ice below the CO_2 frost cover.

Martian Satellites. The two tiny moons of Mars, Phobos and Deimos, are irregularly shaped bodies with cratered surfaces. They are geologically dead and are possibly as old as the solar system.

Geologic History of Mars. Superposition and cross-cutting

relationships show that Mars has had an eventful geologic history involving (1) formation of densely cratered terrain, (2) probable extrusion of flood basalts, (3) formation of great shield volcanoes, (4) development of a rift valley, and (5) modification by surface erosion and deposition.

Additional Readings

Journal of Geophysical Research 78(20):4009-4439, 1973. (Entire issue devoted to the geology of Mars.)

McCauley, J. F., et al. 1972. "Preliminary Mariner 9 Report on the Geology of Mars." Icarus 17(2):289-327.

Murray, B. C. 1973. "Mars from Mariner 9." Sci. Amer. 228(1):49-69.

21 The Planet Mercury

Mariner 10 was launched November 3, 1973, with the objective of obtaining scientific information about the planets Venus and Mercury. After a 94-day voyage inward through the solar system the space probe passed about 3600 miles from the planet Venus on February 5, 1974, just as planned, and obtained approximately 3000 pictures of Venus.

The primary goal of the Mariner 10 mission, however, was to obtain data about Mercury, which had never before been visited by a spacecraft. Twin high-resolution television cameras and six non-imaging experiments constituted its 150-lb scientific payload, which was to provide new insight into the nature of the closest planet to the sun and how it fits into the overall picture of the solar system. On March 23, 1974, Mariner 10 began photographing Mercury and, by April 3, it had collected an unprecedented store of scientific data, including more than 2000 high-resolution television pictures. The Mariner 10 spacecraft passed within about 725 km (460 mi) of Mercury's surface at the point of closest approach. A nearly complete photographic coverage of the illuminated half of the planet was obtained, with some photographs showing features as small as 150 m (500 ft) in diameter. Although these pictures will require many months and years of study before an integrated study of the geology of Mercury can be published, a brief description and preliminary interpretation of the images initially available reveals much about the general nature of the planet. This chapter is based largely on preliminary reports and photographs of the Mariner 10 project published by the Jet Propulsion Laboratory.

Major Concepts

1. The surface features of Mercury are remarkably similar to those of the moon with the major landforms being: (a) craters ranging from multi-ringed basins to small depressions; (b) sparsely cratered plains presumably flooded with maria basalt.
2. Craters range in age from old, highly eroded features to young rayed craters surrounded with a halo of bright ejects.
3. Prominent fault scarps extend across the surface of Mercury, but sinuous rilles and grabens were not observed.
4. Large shield volcanoes, rift valleys, and eolian features similar to those on Mars were not seen on the photos of Mercury.
5. A preliminary interpretation of the major events in the history of Mercury is as follows: (a) early period of intense bombardment which formed highly cratered terrain; (b) impact of large meteorites to form multi-ringed basins; (c) formation of plains material—presumably by flooding of basalt; (d) subsequent impact at a much lower frequency.

Surface Observations

Statement

The surface of Mercury as observed by Mariner 10 photography is strikingly similar to that of the moon. Indeed, at first glance it would be difficult for many nonspecialists to tell the two apart and one member of the Mariner 10 team was heard to remark that the surface of Mercury looked so much like the moon that the photography could be previously unreleased Lunar Orbiter photos. The Mariner 10 photographs reveal that the major landforms on Mercury are basins, scarps, ridges, and plains. The entire surface of Mercury, like that of the moon, is pockmarked with craters ranging in size from large multi-ringed basins 1300 km (780 mi) in diameter downward to small depressions which are barely discernible on the highest-resolution photographs. The craters represent a wide range in age with the older craters overlapping to form a rugged terrain similar to the lunar highland. Many are greatly modified by subsequent impact. The younger craters are typically surrounded by a halo of bright ejecta and an extensive system of rays. Many of the large basins and lowland areas of the planet are flooded with plains material which has many characteristics similar to those of the lunar maria. The plains of Mercury are relatively smooth, contain wrinkle ridges, and are cratered to approximately the same degree as the lunar maria. This would suggest a volcanic origin for much of this material. Prominent scarps up to 1 km (.6 mi) high can be traced across the landscape but grabens and sinuous rilles are lacking on photographs examined to date; neither are there large shield volcanoes similar to those in the Tharsis region of Mars. Pictures from the satellite search phase revealed no Mercurian satellites, but Mercury does have a thin atmosphere consisting in part of helium. Mercury has a significant magnetic field tilted less than 10° away from the pole.

Discussion

Description of Surface Features. The striking similarity between Mercury and the moon is immediately obvious from the photomosaics of Mercury shown in *figure 21.1*. The mosaics were made from a series of computer-enhanced pictures taken from a distance of approximately 230,000 km (140,000 mi). The entire surface is heavily cratered, much like the far side of the moon, and as on the moon, the individual craters range greatly in size and age. Numerous bright-eyed craters comparable to Copernicus and Tycho on the moon are apparently the youngest features on the surface of Mercury, as the bright halo of ejecta and the long splashlike rays appear to be superposed on everything they come into contact with. The rayed craters have an average diameter of approximately 40 km (25 mi). Some of the rays can be traced for over 1000 km (600 mi). There are also two large curvilinear features that resemble light-colored rays, trending east-west and north, but these are solitary, not radial,

501

Figure 21.1 (a) Photomosaic of Mercury made from 9 computer-enhanced photographs taken at a distance of 234,000 km (140,000 mi) 6 hours before the closest approach to the planet. The evening terminator (twilight zone) is at about 20° W and the bright limb is at 110° W. The first surface feature of Mercury to be named is the large rayed multiple crater (arrow) slightly above center. This is named after the late Gerard Kuiper, principal investigator on Ranger 7, which returned the first lunar photographs. (b) Photomosaic of computer-enhanced pictures taken at a distance of 210,000 km (126,000 mi) 5½ hours after the closest approach. North is at the top and the equator extends from left to right about two-thirds of the way down the photograph. The bright limb is at 100° W and the terminator is at 200° W. This picture and picture a show the entire luminated portion of the planet. Adjoining areas lie along the bright limb of both pictures. This part of Mercury is heavily cratered, with craters ranging in size up to almost 200 km. The bright rayed craters are generally less than 40 km in diameter. (c) Telephoto view of the crater Kuiper.

clusters extending out from a crater. Similar features have not been found on the moon or Mars. The bright ejecta and systems of rays on the moon are relatively new features formed after the initial heavy bombardment which formed the densely cratered surface similar to the lunar highlands.

Smaller fresh, new craters similar in size and shape to the small cup-shaped craters which occur on the maria regions of the moon are common throughout the surface of Mercury (figure 21.2). These typically have a well-defined, sharp rim and a relatively light ejecta blanket. The rims and ejecta from the older craters have been denuded and greatly modified by impact of smaller craters. In figure 21.3 craters as small as 150 m (500 ft) across can be seen. This photograph was taken only minutes after Mariner 10 made its closest approach to Mercury and is one of the highest-resolution pictures obtained during the

Figure 21.2 Craters on Mercury range in age from relatively new, fresh craters with sharp rims to older craters which have been considered modified by subsequent impact. The fresh crater in the center of the older crater basin near the center of the photograph is about 12 km (7.5 mi) across. Note that the rims of the older craters are considerably eroded and modified by smaller cratering. This picture covers an area 130 by 170 km (80 by 105 mi) and was taken from a distance of about 20,700 km (12,400 mi) a half hour before the closest approach to the planet.

Figure 21.3 *High resolution photograph taken from 7342 km (4400 mi) only minutes after Mariner 10 made its closest approach to the surface of Mercury. Craters as small as 150 m (500 ft) can be seen on this relatively smooth area of plains material similar to the lunar maria. The ridge at the left resembles a flow front of lunar flood basalts. Both primary and secondary craters can be recognized; the secondary craters are elongate or elliptical and typically occur in clusters or chains. The area covered by this photograph is located at 25° N latitude and 161° W longitude and covers an area of 40 by 50 km (25 by 30 mi).*

mission. Larger craters which have been modified and have lost clear definition of their ejecta deposits and secondary crater fields have no prominent raised rims and are typically shallow, flat-floored, and sometimes filled with plains material. The similarity of crater size, density, and degree of denudation suggest that craters on Mercury, like those on the moon, probably range in size downward to small pits of microscopic size. The similarity in size and form between lunar craters and those found on Mercury, plus evidence that both planets have the same degree of degradation and modification, indicates that similar formations and erosion processes have been active on the two planets. There are some significant differences between the lunar and Mercurian craters. On Mercury, the ejecta deposits of the larger craters do not extend as far from the crater rim as for similar-sized lunar craters. Similarly, the maximum density of secondary craters is closer to the rim of Mercurian craters. Also, preliminary depth measurements suggest that the craters on Mercury are significantly shallower than similar-sized lunar craters. These three differences are to be expected because of Mercury's greater gravitational acceleration, which would reduce the ballister range of ejecta and create a greater degree of collapse through slumping of the crater rim and walls.

It is apparent from *figure 21.1* that the heavily cratered terrain on Mercury is similar in general appearance to the lunar highlands. Statistical studies of crater populations on the two planets confirm this observation as the crater frequency distribution on the heavily cratered terrain of Mercury is essentially identical to that of the lunar highlands. Both surfaces have attained a steady-state condition and are saturated. Craters as large as at least 100 km (60 mi) in diameter have apparently survived since the end of the period of intense bombardment by small planetesimals.

Judging from the surface optical properties, Mercury, like the moon, is apparently covered with a fine-grained regolith of dust and rock fragments.

Large multi-ringed basins strikingly similar to the large basins on the moon and Mars also occur on Mercury. The largest basin is approximately 1300 km (780 mi) in diameter and is located near the central part of the photomosaic in *figure 21.1*. Ejecta from the main ring form hills and valleys which radiate outward, similar to the multi-ringed craters which form the Orientale and Imbrium basins on the moon (compare *figure 21.4* to *figures 19.8* and *20.2*). This is possibly some of the most rugged terrain on Mercury. Like the large basins on the moon and Mars, the multi-ringed basins on Mercury are relatively old features which were formed early in the development of the planet by impact of asteroid-size material. Smaller basins commonly have two well-preserved rings, with the outer ring having close to twice the diameter of the inner ring. On the moon, 24 well-developed basins larger than 300 km (185 mi) have been counted, whereas 8 basins larger than 300 km have been observed on the approximately one-third of the surface of Mercury. This would suggest about the same total of 24 for a body with a surface area about twice that of the moon.

The inner portions and parts of the adjacent areas of the large basins are covered with plains material, similar in most respects to the lunar maria (*figure 21.5*). The plains material on Mercury clearly formed after the large basins and after a subsequent interval of cratering which modified and postdates the basins. Wrinkle ridges are very common in the plains material, more so than on the maria of the moon. In *figure 21.5* large areas of the plains material within the main ring of the basins are covered with polygonal patterns of wrinkle ridges, some of which appear to form depressions rather than ridges. The

Figure 21.4 The largest multi-ringed basin observed on Mercury closely resembles those on the moon and Mars. Hills and valleys formed from ejecta in a radial fashion outward from the main ring. The interior of the basin is completely flooded with plains material which spreads out over the surrounding lowlands.

pattern vaguely resembles the hexagonal patterns formed by columnar jointing in basalt.

The plains material covers large areas beyond the multi-ringed basins, and covers the floor of many moderate- to large-size craters (figures 21.5 and 21.7). It not only fills and embays large areas, but it covers the floors of many isolated craters. This suggests a local source of fluid, presumably highly fluid basalt. Throughout large areas, the plains material is probably relatively thin, much thinner than the maria basalts which fill the lunar basins. This conclusion is based on the fact that parts of the rim of numerous craters protrude through the cover of plains material so that on a regional scale the cover appears to be incomplete and discontinuous (figure 21.6).

It is important to note that the plains material on Mercury, like the lunar maria, has a much lower density of cratering than

Figure 21.5 Close-up view of the large basin shown in figure 21.4. Half of the basin is hidden in the dark side of the planet beyond the terminator.

Figure 21.6 *An area of dark, relatively uncratered plains material on Mercury photographed two hours after the closest approach to the planet at a range of 86,800 km (52,000 mi). Most of the area is covered with marelike material, although parts of the rugged, densely cratered highlands can be seen in the upper left. The plains material is apparently not deep in this area because many of the older craters are not completely covered and parts of their rims protrude through the younger material. A history of heavy cratering followed by flooding of volcanic material seems to be similar to the early history of the moon. The prominent, sharp, fresh crater with a central peak (center) is 30 km (19 mi) across. It is located on the upper left edge of a very bright ejecta blanket from the smaller adjacent crater.*

on the surrounding highlands. This clearly indicates that it formed after the early intense bombardment.

The multi-ringed basins and plains material are highly significant in deciphering the history of Mercury because they provide important and reliable reference surfaces on a regional basis which permit one to discriminate the chronological sequence of events. Although it is premature to attempt to establish a firm framework of Mercury's geologic history, it appears from preliminary analysis of Mariner 10 photographs that the major events in the history of the moon and Mercury are similar and can be tentatively arranged as follows:

1. An early period of intense bombardment which formed the highly cratered terrain.
2. Impact of large meteorites to form the multi-ringed basins.
3. Formation of plains material, probably by flooding of basalt.
4. Subsequent impact, but at a much lower frequency.
4. Subsequent impact, but at a much lower frequency.

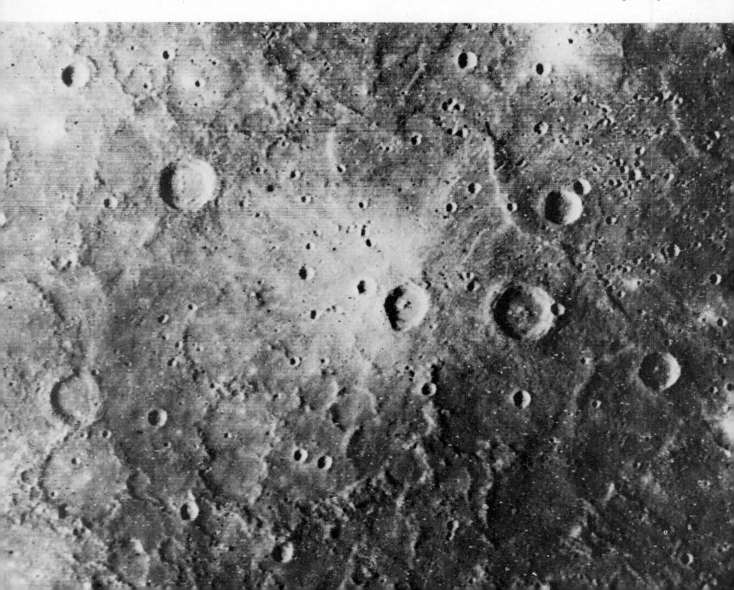

In addition, the plains material is significant in that it provides several important implications concerning the interior of the planets. The large horizontal extent of such features implies a silicate composition with a density of approximately 3 g/cm³ for the entire outer region of the planet, not just the upper meters, as indicated by remote optical, infrared, and radio measurements. Mercury, however, has a planetary density of 5.5 g/cm³, so a very much denser material must occur at depth. This would suggest that Mercury is a chemically differentiated planet, probably containing a large iron core.

The photographs sent back by Mariner 10 show that Mercury does not have large shield volcanoes, rift valleys, windblown deposits, nor channels which are so prominent on Mars, nor are there numerous rilles like those found on the moon. Prominent scarps, however, do occur and appear to be the result of faulting. The large scarps of great linear extent commonly cut both the highly cratered terrain and plains material. Several are indicated on *figure 21.9*. Some of the scarps are cut by large craters, indicating that at least some of the scarps were formed during the final stages of intense bombardment. The form and

Figure 21.7 *Most of the area shown in this photograph is densely cratered and is apparently part of the older surface of Mercury. In the lower left corner a portion of a crater 61 km (35 mi) in diameter shows the margins of a flood front extending across the crater floor and filling more than half of the crater. Craters as small as 1 km are visible on the young plains material.*

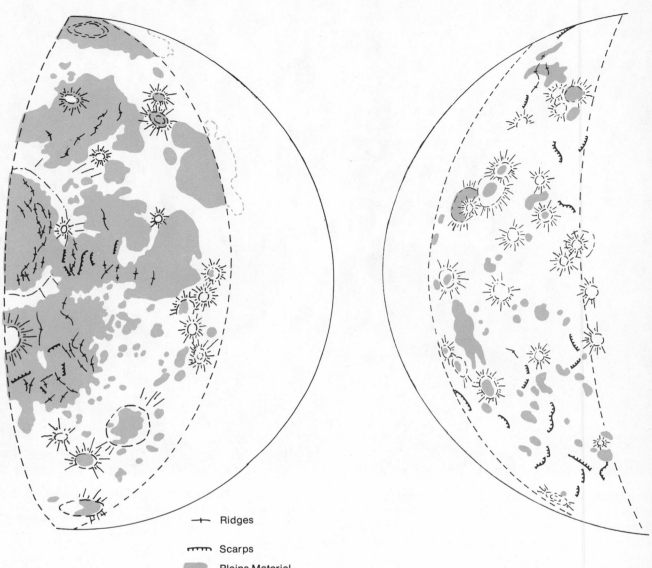

+—+ Ridges

ᴍᴍᴍ Scarps

⬭ Plains Material

Figure 21.8 *Sketch map showing major physiographic provinces of Mercury as viewed by Mariner 10 photography. The rims of basins larger than 200 km in diameter are shown with dash lines, and ejecta and secondary craters surrounding the major craters are shown by radial lines.*

orientation of the scarps suggest that they may be thrust or reverse faults caused by compressive stresses. If this interpretation is correct, Mercury is the first planet other than the earth to show evidence of compressive stress on a global scale. One such scarp is seen in *figure 21.9* near the horizon. The cliff is estimated to be up to 1 km in height and extends for hundreds of kilometers across the surface, cutting across large craters and intercrater areas alike. Similar features are absent on the moon.

Although the cratered surface of Mercury shows many similarities with the moon, there are some obvious differences. One is the "wormy" texture of parts of the surface seen in *figure 21.10*. This texture is possibly the result of extensive secondary

The Earth's Dynamic Systems

cratering, which forms elliptical craters, commonly grouped or aligned so that the ridges between craters appear braided.

Summary

Preliminary Conclusions. The wealth of data sent back by Mariner 10 will be studied for years, but initial studies of the photographs combined with our knowledge of the earth, moon, and Mars suggest the following preliminary conclusions:

1. The densely cratered surface of Mercury is similar to that on the moon and probably records the last periods of heavy bombardment.
2. The relatively flat plains material similar to the lunar maria basalts suggests extensive flooding, probably basaltic lava.

Figure 21.9 *A view of Mercury's northern region showing prominent east-facing scarps extending from the horizon near the middle of the photograph southward for hundreds of kilometers. Most of the area consists of highly cratered terrain similar to the lunar highlands. The "tear" on the horizon near the top of the picture was caused by loss of data. This picture was taken at a distance of 77,800 km (47,000 mi). The area shown at the base of the photo is about 580 km (360 mi) long.*

The extensive horizontal extent of this material implies a silicate composition. This suggests that Mercury is a differentiated planet with an iron-rich core.

3. A striking feature of Mercury (and of the moon, as well) is that the primitive, highly cratered terrain has been preserved throughout extensive regions without being modified by internal processes. The survival of this ancient terrain places limits on the time when materials were differentiated. Differentiation must have occurred very early in the planet's history, before the period of intense bombardment. The density of Mercury is 5.5, similar to that of the earth. It must have an iron-rich core which would extend outward 75 to 80% of the radius of the planet. The silicate outer layers would be approximately 500 to 600 km (300 to 370 mi) in thickness.

4. The surprising similarity in the marialike plains material of the moon, Mars, and Mercury suggests that a thermal event has occurred early in the history of all three planets. Similarities in densities of craters on the plains material vs. the densely cratered highlands suggests that the impact histories of the moon, Mars, and Mercury are similar. All appear to have experienced an early period of intense bombardment

The Earth's Dynamic Systems

with a high rate of influx, impact of large bodies to form multi-ringed basins, extensive flooding of plains material, presumably basalt, and subsequent cratering which decreased exponentially.

5. In the half of the planet observed by Mariner 10, there are no signs of planet-wide crustal shifts on Mercury similar to the plate tectonics on earth, nor are there indications of large volcanic cones similar to those on Mars. Some crustal adjustments are indicated by the long scarps, but the planet has not been tectonically active. Perhaps some of the original accretionary surface still remains.

6. It is clear from studies of lunar geology that extrusion of maria basalts occurred after the end of the period of intensive bombardment. Very little activity, external or internal, has subsequently modified the planet. The history of Mercury is remarkably similar to that of the moon; if the relationship between incident impact flux and time also proves to be similar, then absolute time scales are comparable as well.

   caution, however, in generalizing planetary history from only limited samples. We cannot exclude the possibility of volcanic or tectonic processes on the unexplored parts of Mercury.

7. The development of large basins on the moon, Mars, and Mercury, followed by the formation of plains material (presumably basaltic flows) appears to be a common aspect of terrestrial planetary development. These events may have occurred early in the earth's history as well.

   With the exploration of the moon, Mars, and Mercury, we have reached across space and have viewed new worlds and have seen for the first time the processes in the very formation of terrestrial planets. In the next chapter we will compare and contrast in greater detail the earth, the moon, Mars, and Mercury.

Additional Readings

Fink, D. E. 1974. "Mariner Surpasses Mercury Flyby Goals." Aviation Week and Space Technology, April 8, pp. 14-17.

Murry, D. C., et al. 1974. "Mariner Venus/Mercury." Status Bull. No. 29. Pasadena, Calif.: Jet Propulsion Lab. C. I. T.

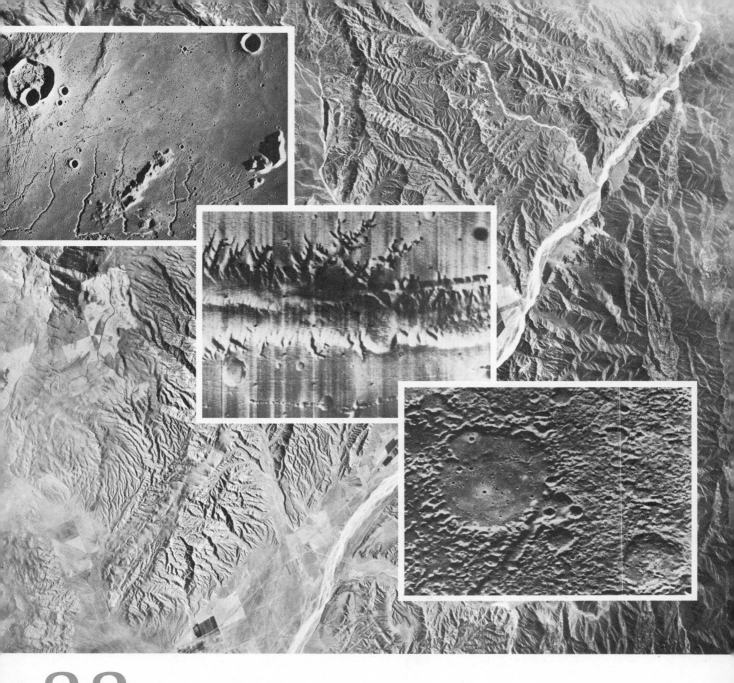

22 Geological Contrasts of the Earth, the Moon, Mars, and Mercury

One of the most fascinating results of the space program for geologists is that it has provided firsthand knowledge of the rocks and structure of other planets. This does much more than merely satisfy scientific curiosity, for, as we are able to compare details of the geologic nature and history of planets, we can better recognize those principles and processes fundamental to the geology of the earth and those which are of secondary importance.

Scientists no longer consider the moon to be a dead clinker orbiting the earth, cold and inert, and after the flights of Mariners 4, 6, and 7 Mars appears much more similar to the earth than it was previously believed to be.

Mercury has surface features remarkably similar to those of the moon, yet the density and internal structure of Mercury are more like that of the earth. All of the planets have had major thermal events and are differentiated into layers, and all appear to share the same birthdate of 4.6 billion years ago.

Yet, there are many important differences, such as size, density, and type of geologic processes, which have operated throughout the planets' history. We wonder why the moon has no atmosphere, no folded mountain belts, and no recent vulcanism. Why does Mars have such large shield volcanoes and the great canyon? Did Mars ever have an abundance of surface water? If so, why is it so dry at the present time? Why is the surface of Mercury so much like that of the moon, when its density is like that of the earth? Why did Mercury, the moon, and Mars all have a major thermal event and outpouring of basalt early in their history? How important is the earth's atmosphere to its surface features? If the earth had no surface fluids, would its surface resemble that of the moon or Mars, or would it have a different style of surface features? To gain an insight into these and other questions about the four planets, let us venture into a brief study of comparative planetology in an effort to understand their similarities and differences.

Major Concepts

1. The earth, the moon, Mars, and Mercury form a family of related planets in which size, density, and distance from the sun determine many individual characteristics of each planet.
2. The earth is the most dynamic of the terrestrial planets and has been modified throughout most of its history by its tectonic and hydrologic systems.
3. The moon and Mercury are primitive bodies and their surfaces have not experienced change from atmospheric erosion nor continuing convection in its interior.
4. Mars appears to represent an intermediate phase in planetary development, transitional between an impact-dominated body like the moon and Mercury and a tectonically active, water-dominated planet like the earth.
5. The early histories of the moon and Mercury are similar in that early intense bombardment developed highly cratered terrain, followed by development of large multi-ringed basins and floods of basalt and a subsequent decline in cratering. These events probably occurred early in the earth's history, but evidence of them has been erased by changes resulting from the hydrologic and tectonic systems.

Contrasts and Comparisons

Statement

Space photography from the Lunar Orbiter, Apollo, Mariner 9 and Mariner 10 missions provides an unprecedented store of scientific data concerning the origin and evolution of the terrestrial planets in our solar system. The overwhelming impression gained from studying the surface features and physical characteristics of the planets is that the inner solar system forms a family of planets in which the size and distance from the sun determines many individual characteristics within the family. The presence of a clear-cut gradient in densities outward from the sun supports the idea that the more volatile elements condensed further out in the solar system.

The early history of the moon, Mars, and Mercury involves widespread cratering, probably dating back to the final stages of accretion. The formation of large multi-ringed basins and flooding of marelike material followed as the rate of cratering decreased exponentially.

Similar events were probably present in the early history of the earth, but the tectonic system and surface fluids have recycled and obliterated any evidence of this early history of our planet. The tectonic mobility and hydrologic system make the earth unique and are probably the result of its size, mass, and greater concentration of volatile elements due to its distance from the sun.

Discussion

Size, Volume, and Density. The most obvious differences between the earth, the moon, Mars, and Mercury are the size and density of the four spheres (*figure 22.1*).

The earth is the largest, with a diameter of 12,682 km (7926 mi) and a mean density of 5.5. The latter figure is very important, for the density of the rocks on the earth's crust is only 2.7. We have known from this fact alone, long before seismology, that the interior of the earth is much denser than the crust and that the earth is differentiated into layers. The estimated density of the core is between 9.6 and 10.7. It is believed to be composed of the heavy elements such as iron and nickel. Heavy elements are concentrated near the core—and lighter elements are concentrated in the crust.

In contrast, the moon is not only much smaller than the earth, with a diameter of 3456 km (2160 mi), but it has a mean density of only 3.3. The density of the entire moon is not much greater than the density of the rocks exposed on its surface. However, seismic soundings on the moon suggest some differentiation into a crust, mantle, and core. The core is probably partly molten because S seismic waves do not pass through the interior. The material in the lunar core, however, does not resemble the earth's iron-nickel core.

517

Mars is a little more than one-half the size of the earth, with a diameter of 6800 km (4200 mi). Its density is 4.0—intermediate between the moon (3.3) and the earth (5.5). Mars must be differentiated to some degree, probably more than the moon but

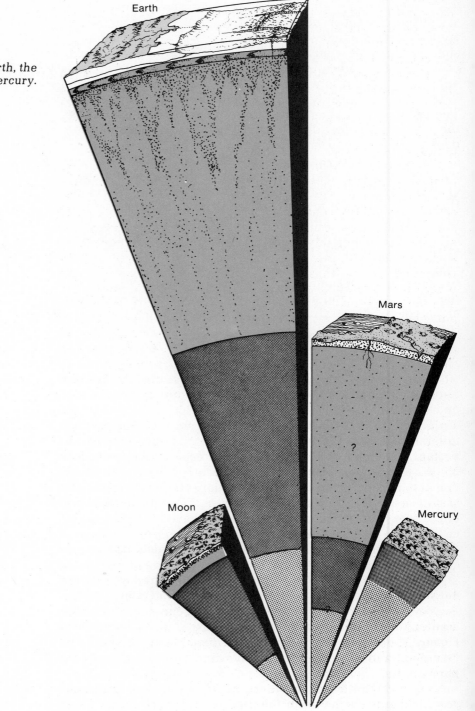

Figure 22.1 *A comparison of the earth, the moon, Mars, and Mercury.*

The Earth's Dynamic Systems

less than the earth. It could contain a small nickel-iron core some 1000 km (600 mi) in radius, but this is still uncertain.

Mercury is the nearest planet to the sun and has an eccentric orbit. It is a small planet, having a diameter of 4848 km (3030 mi), less than half that of the earth. The mean density of Mercury is 5.5, about the same as the mean density of the earth. Temperatures on the planet measured by an infrared radiometer on board Mariner 10 ranged from 188° C (370° F) on the planet's day side to −138° C (−280° F) on the night side.

These physical statistics may at first seem unimportant, but they are among the most significant facts underlying the geologic contrasts among the four planets. Most geologic differences among these planets may be explained by the differences in their size and density, which would govern their internal heat, atmosphere, and type and degree of tectonic activity.

Atmospheric Differences. The remarkable thing about terrestrial geologic processes is the gradual but profound changes which they incur. The entire surface of the earth appears to remain essentially constant from day to day and from year to year, yet changes occur in almost infinitely small increments and, in time, completely alter the landscape. Rocks at the earth's surface react with the elements in the atmosphere and are decomposed. Water transports the weathered debris to the sea, where it is deposited to ultimately form parts of a new crust. The landscape evolves through a predictable series of stages, and high mountain ranges are ultimately worn down and may be covered with a blanket of younger sediment. This pattern of gradual change has resulted in a remarkably detailed record of the earth's history, in which segments of each interval of relatively recent geologic time are preserved. In contrast, the oldest features of the earth appear to be completely destroyed.

The moon and Mercury, however, do not have surface fluids, and without water there is no erosion on these planets comparable to that on earth. Their surface features, though, do change with time as a result of natural processes operating on those spheres. The most significant process appears to be churning action and fragmentation by impact. However, some processes are similar to those operating on the earth, such as surface creep, faulting, and volcanic eruptions. Disintegration also operates on the moon, as is evidenced by the upper surfaces of boulders being more rounded than their bases. Rays from craters are obliterated with time, and crater rims are greatly modified by impact from later meteorites. Impact from meteorites, however, represents a rapid change and a sharp pulse of geologic activity. Meteorite impact does not occur as a continuing, gradual process such as weathering or erosion by running water. Therefore, it appears that forces working gradually on the moon and Mercury play a subordinate role in shaping the topography.

On both the moon and Mercury, the density of craters on the highlands compared to the density of craters on the maria provides good reason for thinking that meteoritic bombardment was concentrated during an early phase of their history and that volcanic activity was concentrated during one or several limited periods. In contrast to the earth, therefore, there is little evidence for gradualism in the development of their surface features. Craters appear to have been produced instantaneously and then modified largely by additional impact or flooding by volcanic material. They have not been deformed by folding. If maria are indeed 3.5 billion years old, as radioactive dates on lunar samples indicate (and a similar date is inferred for the plains material on Mercury), these planets suffered very little change since mare time.

In terms of atmosphere, Mars appears to be midway between the earth and the moon. The presence of an atmosphere around Mars has been known for many years, as evidenced by the seasonal variations in polar caps. The principal constituents of the Martian atmosphere are carbon dioxide, a minute quantity of carbon monoxide, and a trace of molecular nitrogen. Small, seasonally variable amounts of water vapor have also been detected but are very much less than those found in the earth's deserts. The atmospheric pressure on Mars is only 0.6% of the earth's. Clouds have been seen over the polar regions and near the summit of the shield volcanoes, but Mars does not have a hydrologic system—anything comparable to that of the earth. With an atmosphere eolian activity is possible and appears to be a major surface process eroding, transporting, and depositing surface material.

Impact Craters. Impact craters are the dominant surface features on the moon and Mercury and appear to be nearly as prominent on Mars. The bombardment of planets by meteorites, asteroids, and comet heads appears to be a fundamental process in the solar system. Many thousands of meteorites have struck the moon, Mercury, and Mars, whereas only a few impact structures are found on the earth.

One explanation for this is that they burn up in the earth's atmosphere before they reach the surface, but there are other important reasons. Although meteoritic impact has been observed on both the moon and the earth, extensive bombardment by meteorites appears to have been an event largely concentrated in the early history of the planets. The highly cratered lunar highlands were partly flooded with basalt 3 billion years ago when the maria were formed, and since then relatively few impact structures have been produced. A similar record appears on Mercury and Mars, where the highly cratered surfaces are the oldest terrain and relatively few impact structures are found on the younger volcanic fields and eolian deposits. As on the moon, the extensively cratered surfaces on Mercury and Mars were probably formed more than 3 billion years ago.

In contrast, there are essentially no rocks remaining on earth from this early period. Tectonic activity on earth has deformed this material, and much of it has been recycled into

The Earth's Dynamic Systems

sedimentary, metamorphic, and igneous rocks—not once but several times. The process of impact which so dominates the surface features of the other planets may have modified the earth to a much greater extent than we first imagined, but this was during its early history; plate tectonics, continental drift, and erosion have obliterated the effects of impact.

Continents and Ocean Basins. One of the most fundamental characteristics of the earth's crust is the distinction between continental masses and ocean basins. Continents are composed of relatively light, granitic rocks which float on the denser basalt forming the oceanic crust. The two types of crust form two dominant levels—the continental level, which averages 500 m (1640 ft) above sea level, and the abyssal floor, which averages 4120 m (13,200 ft) below sea level. The continents are high because they are composed of lighter rock and float like icebergs in a state of isostatic equilibrium. In contrast, the crust of the moon, Mars, and Mercury is fairly homogeneous to a considerable depth. If it were segregated into heavy and lighter rock types, this would almost certainly be reflected in two preferred levels on the surface of these planets. Since the time of Aristotle, 2000 years ago, the dark areas of the moon have been compared with the earth's oceans and, until recently, some observers have maintained that the maria may be analogous to the earth's ocean basins but not filled with water. Some similarities do exist. Both are extensive lowlands or depressions in the planetary crust, both are covered with basalt, and both have large, smooth expanses as compared to the rugged terrain of the upland. However, these similarities are superficial, and the fundamental nature of the maria on the moon and the plains material on Mars and Mercury compared to the ocean basins on earth is a study of contrasts.

The ocean basins on earth are interconnected and form, in reality, a worldwide ocean above which the continental blocks rise as isolated islands. The ocean floor covering 60% of the earth's surface is the typical surface of our planet—not the continents. If all the water were drained from the oceans, the continental slopes would stand out as one of the most significant topographic features.

In contrast, the maria of the moon are not interconnected but are usually restricted to large basins produced by impact. The maria are simply depressions of the lunar highland filled with lava. This is clearly evident from the outline of numerous craters beneath the mare floors. The maria are not a special type of lunar crust but simply ponds of lava. It is apparent, therefore, that although the maria and the earth's ocean basins may have some superficial similarities, they originated in completely different ways and are really worlds apart.

Tectonic Activity. The differences between the surface features of the earth, the moon, Mars, and Mercury go beyond the erosional and depositional features produced by running water and wind. If the earth had no atmosphere and were free from the effects of stream erosion, it would still not resemble these other planets. The earth's crust is extremely mobile and shows the effects of deformation by compressive forces.

The most striking deformational features are the great folded mountain ranges in which crustal shortening on the order of 120 to 160 km (75 to 100 mi) can be demonstrated. Perhaps the best record of horizontal stresses is found in the shield areas of the continents. Here all of the rocks have suffered extreme deformation, as is evident in the small-scale contortions in the metamorphic rocks, such as foliation in gneiss and schist. The great bulk of metamorphic rock has nearly vertical foliation, a structure produced only by horizontal stresses.

The crust of the ocean floor is very young. The North Atlantic basin is only 70 million years old, and the Antarctic Ocean between Australia and Antarctica began only 40 million years ago. New ocean basins such as the Red Sea are just being born. Others have undoubtedly come and gone. Our earth's crust has been mobile throughout its history and has continually been shaped and remolded by internal processes. The mantle is in motion, giving rise to sea-floor spreading, folded mountain ranges, metamorphic rocks, oceanic trenches, and a worldwide rift system.

Evidence of similar compression on the crust of other planets is completely absent. There are no linear mountain ranges and no suggestion of thrust faults and folds. The only deformation has been impact of meteorites, local wrinkles, and normal faulting along the basin margins. Considerable stability of the crust of the moon, Mars, and Mercury is also clearly indicated by the circular shape of craters. Regardless of origin, their circular form indicates that they were formed by *vertical forces*. Therefore, the circular craters provide an excellent reference system that has evidently been fixed throughout time, not deformed by compression. The preservation of these structures shows that the lunar crust has never been subjected to compressive forces similar to those which formed the folded mountains on earth. Compressional structures such as folds, thrust faults, and strike-slip faults, if present, should be clearly expressed on these planets, because they would not be masked by sedimentary cover, water, or vegetation.

Vulcanism. Although there are many things about the nature of terrestrial vulcanism which are not completely understood, the concept of convection within the earth's mantle can explain most aspects of volcanic activity on earth. Vulcanism in linear zones, related to mountain-building, island arcs, and oceanic trenches, can all be explained by plate tectonics.

The rock record makes it abundantly clear that vulcanism has occurred throughout geologic time. Some of the greatest outpourings of lava are the plateau basalts, now situated near the margins of continents, which appear to be related to the initial phases of continental drift. These flood basalts have occurred from the Precambrian to the present.

There is little convincing evidence for similar volcanic activity on the moon and Mercury. Blankets of ejecta material and systems of rays show that most craters are impact features rather than collapse caldera, and the flood basalts which form the lunar maria and presumably the plains material of Mercury represent a major period of volcanic activity early in the history of the planets but relatively little activity since this initial outpouring of lava. Certainly volcanic activity is not restricted to linear belts or mobile crust but is related more intimately to large impact craters.

Vulcanism on Mars might also be considered intermediate between that of the earth and moon. Early flood basalts probably exist, and ancient volcanoes remain in the Hellas region. The great shield volcanoes of the Tharsis region are much larger than anything on the earth and moon.

Summary

We have seen in previous chapters that the earth is extremely dynamic, as it is continually molded and reshaped by internal processes—processes which have produced folded mountains, metamorphic rocks, a worldwide rift system, and deep oceanic trenches. In addition, the earth's atmosphere and water continually erode and sculpt the surface features produced by the mobile crust. Gradual but profound changes occur because of the tectonic and hydrologic systems so that a record of the early history is completely obliterated.

In contrast, the moon and Mercury appear to be primitive bodies. They have no atmosphere to provide a mechanism for erosion, no convection in the mantle capable of generating new crust. As a result, planetary dynamics have not proceeded very far because there is no mountain-building or differentiation of the crust into continents and ocean basins. The surface features of the moon and Mercury are largely the result of impact from external material, plus early volcanic flows, probably triggered by the larger meteorites. The surface of the moon has changed very little during the last 3.5 billion years, whereas the surface of the earth is continually being built up and eroded down.

Mars apparently represents an intermediate step in planetary evolution, transitional between a relatively primitive, impact-dominated body like the moon and Mercury and a tectonically mobile and volcanically active, water-dominated planet like the earth. Mars is similar to the moon in that it has extensive highly cratered regions as well as large multi-ringed basins. However, over much of the planet are spectacular tectonic and volcanic features, quite unlike anything found on the moon and Mercury and only partly like features on earth. Extensive tectonic activity has occurred on Mars, but no compressional features have yet been recognized. Vulcanism has played an important role in Martian history and has contributed greatly to its atmosphere by expelling gases. It has occurred over a much larger span of the planet's history than has lunar vulcanism and is dramatically more varied.

Erosion and sedimentation, independent of impact cratering, have occurred on Mars on a planet-wide scale. Transport of sediment through channels by moving fluids has occurred but has probably been in episodic surges. Wind action appears to be the dominant surface process and has modified most topographic features by erosion and deposition.

The surface of Mercury appears so much like that of the moon that we may forget that it is quite different on the inside. The high density of Mercury suggests that it has a large iron core similar to that of the earth. The silica-base minerals apparently are concentrated near the surface. There is evidence from the cratered patterns on Mercury that the core was developed very early in the planet's history, before the final stages of accretion.

The overwhelming impressions gained from studying the surface features and physical characteristics of the planets is that the inner solar system forms a family of related planets in which the size and distance from the sun determines many individual characteristics within the family. The presence of a clear-cut gradient in the densities outwards from the sun obviously ties in neatly with the idea that the more volatile elements condensed further out in the collapsing proto solar system. The more we learn about the planets, the more we realize the universality of the laws of nature. The processes which operate in the evolution of the planets are fundamentally the same, with the major differences resulting from different proportions and intensities of surface-forming processes. Meteorite impact, tectonism, vulcanism, and various erosion and sedimentation processes are not distinctive and unique to any one planet but appear to operate in different styles and to different degrees on all of them.

The early history of the moon, Mars, and Mercury have many similarities. Intense bombardment in the very early development formed the densely cratered surface but, as the material in the vicinity of their orbits was swept up, the intensity of cratering declined exponentially. Large multi-ringed basins then formed and were followed by extrusion of flood basalts and a continued decline in cratering. These events are known to be the dominant elements in the early history of the moon, Mars, and Mercury and were probably present in the early development of the earth.

The unique features of the earth may be largely the result of its size and mass and the greater concentration of volatile elements due to its position farther from the sun.

The size and density of the earth and its position in the solar system may also account for the atmosphere and abundance of water on the earth's surface; inasmuch as water and atmospheric gas are created by vulcanism, they are held close to the earth by the strong gravitational attraction of this moderately large, dense planet. Heat from the sun provides the energy for the hydrologic cycle, so that the water on the earth's surface is continually washing and eroding the surface of the continents at such a pace that the folded mountains are worn down almost as fast as they are formed.

In contrast, the moon, being smaller, lacks the gravitational force to become differentiated, at least to the degree that the earth is. The moon could not hold an atmosphere if one were produced. The moon has had a thermal history but is not a thermally active body like the earth. It is rigid and, lacking convection overturn, has retained a rigid base. Surface features are neither strained by internal movements nor destroyed by the powerful erosional forces of water.

Mars is approximately twice the size of the moon but only half the size of the earth, so it is not surprising that Mars is a planet intermediate between the moon and earth in its evolutionary sequence. It is large enough to develop a greater degree of differentiation than the moon and has, therefore, developed a more advanced style of tectonism and vulcanism and a thin atmosphere. However, it has not developed a convecting mantle capable of forming folded mountains and continental masses as the earth has. All features of Mars indicate it is a planet that has partly made the transition from a primitive, impact-dominated body like the moon to a tectonically mobile and volcanically active planet like the earth. The most primitive bodies of the solar system studied to date appear to be the small moons of Mars, Phobos and Deimos, which are likely remnant planetesimals from which the planets formed by aggregation. Elucidation of these relationships will certainly be one of the major scientific products of future space research and should contribute significantly to a better understanding of the earth.

Mercury has many surface features like those of the moon but an interior differentiated like that of the earth. It is, however, too small to develop a tectonic system and lacks an atmosphere and ocean.

Igneous Rocks

The rate at which a magma cools greatly influences the size of the resulting crystals. If cooling is extremely rapid, such as on the surface of lava flows, there is insufficient time for crystals to grow. This results in a **glassy** texture. Slightly slower cooling on the interior of a flow will permit crystals to grow, but their size is microscopic. This fine-grained texture is termed **aphanitic.** Slow cooling of magma beneath the surface will develop large crystals which form a coarse-grained mass of interlocking minerals called **phaneritic** texture. If the magma cools at two different rates, first slowly and then more rapidly, large crystals called **phenocrysts** develop during the slow cooling, and finer-grained crystals form from the remaining liquid during the period of more rapid cooling. A texture in which phenocrysts are present is termed **porphyritic.** Texture, therefore, provides an important insight into the cooling history of igneous rock and is one of the characteristics upon which the classification of igneous rocks is based. Composition is also important and is used in the system of classification. A simple chart showing variations in composition horizontally and textural variations vertically, provides an effective system of classifying and naming igneous rocks (*table A.1*). The following is a brief description of the most common igneous rocks.

Rocks with Phaneritic Textures

Granite. Granite is a coarse-grained igneous rock composed predominantly of feldspar, quartz, and mica (*figure A.1*). Potassium feldspar is the most abundant mineral and is usually easily recognized by its pink color. Sodium plagioclase is present in moderate amounts and is usually distinguished by its white color and porcelainlike appearance. Mica is very conspicuous as black or bronzelike flakes usually distributed evenly throughout the rock. The texture of the rock, together with lab-

Appendix
Classification
and
Description
of Rocks

Table A.1. Classification of Igneous Rocks

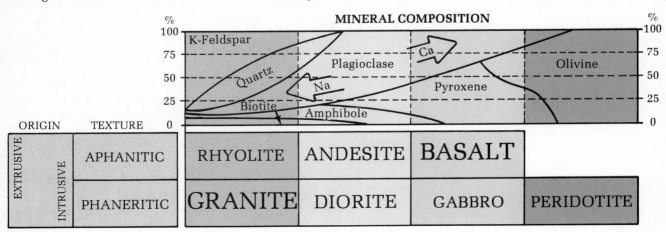

From W. Kenneth Hamblin and James D. Howard, 1971, *Physical Geology Laboratory Manual*, 3rd ed., Burgess Publishing Company, Minneapolis.

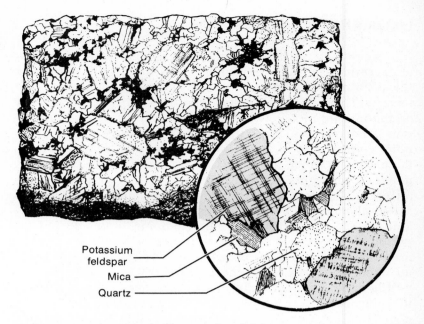

Figure A.1 Granite—A phaneritic rock composed predominantly of K-feldspar, plagioclase, and quartz.

Potassium feldspar

Mica

Quartz

oratory experiments, indicates that feldspars and mica are the first minerals to crystallize from the magma.

Granite is a light-colored rock that has an overall color of gray or pink, depending upon the variety of potassium feldspar present. In many granites, the interlocking crystals, all approximately the same size and uniformly distributed throughout the rock, produce a remarkably homogeneous texture. In others, the early-formed feldspar crystals form phenocrysts embedded in a matrix of smaller grains of quartz, feldspar, and mica. This texture is technically referred to as porphyritic, and the rock is called a porphyritic granite.

A seemingly insignificant but very important property of granite is its relatively low specific gravity of about 2.7. In contrast to basalt and related mafic rocks, which have a specific gravity of 3.2, granite is a light rock. This fact is very important in considering the nature of continents and the contrast between continental and oceanic crust. Granite and related rocks comprise the great bulk of continental crust, whereas the oceanic crust is composed of basalt.

Diorite. Diorite is similar to granite in texture but differs in composition. Plagioclase feldspar is the dominant mineral, and quartz and potassium feldspar are very minor constituents. Amphibole is an important constituent, and some pyroxene may be present. The composition of diorite is intermediate between granite and basalt.

Gabbro. Gabbro is not an abundant rock, but it is important in some older intrusive bodies. It has a coarse-grained texture similar to granite but is composed almost entirely of pyroxene and calcium-rich plagioclase, with minor amounts of olivine (*figure A.2*). Gabbro is dark green, dark gray, or almost black because of the dominance of dark-colored minerals.

Appendix

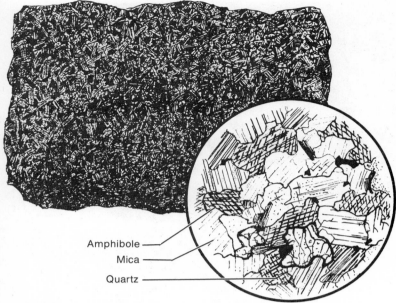

Figure A.2 *Gabbro—A phaneritic rock composed predominantly of Ca-plagioclase, pyroxene, and olivine.*

Amphibole

Mica

Quartz

Rocks with Aphanitic Textures

Basalt. Basalt is the most common aphanitic rock. It is a very fine-grained, dark-colored rock and occurs widespread as lava flows. The mineral grains are so small that they can rarely be seen without the aid of a microscope. When a thin section (a thin, transparent slice of rock) is viewed through a microscope, the individual minerals can be seen and studied (figure A.3).

Basalt is composed predominantly of calcium-rich plagioclase, olivine, and pyroxene. The plagioclase occurs as a mesh of elongate lathlike crystals surrounding the more equidimensional pyroxene and olivine crystals. In some cases, large crystals of olivine or pyroxene form large phenocrysts, resulting in a porphyritic texture. Many basalts have varying amounts of glass, especially near the tops of flows.

Peridotites and Other Ultramafic Rocks. **Ultramafic** rocks are composed almost entirely of two minerals, olivine and pyroxene. Peridotite, a rock composed of 70 to 95% olivine, has a coarse-grained texture in which the minerals are visible to the unaided eye. Other ultramafic rocks are composed almost exclusively of pyroxene (pyroxenite) or calcium plagioclase (anorthosite). These rock types are not common on the earth's surface nor, as far as we can tell, within the continental crust, but they are very important rock types in studies of the subcrustal part of the earth (mantle). The high specific gravity of these rocks (3.3), together with other physical properties, indicates that the great bulk of the earth, the mantle, is most likely composed of peridotite and closely related rock types. The peridotites of the Alps and of St. Paul's Rocks (islets in the Atlantic Ocean) are two areas where rocks from the mantle appear to have been pushed through the crust to the earth's surface.

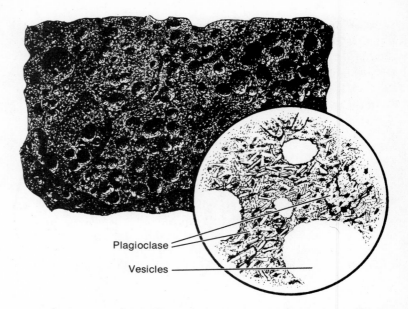

Plagioclase

Vesicles

Andesites. Andesite is an aphanitic rock composed of plagioclase, pyroxene, and amphibole but usually no quartz. It is generally porphyritic, with phenocrysts of feldspar and ferromagnesian minerals. The term comes from the Andes Mountains where lava flows have produced the rocks in great abundance. Probably the next most abundant lava type after basalt, it occurs most frequently along the continental margins but is not found in the ocean basins, nor is it abundant in the continental interior. The origin of andesite along the continental margins is probably related to continental drift.

Tuff (Ash Falls). Volcanic eruptions of basaltic and andesitic lavas commonly produce large volumes of ash. The fragments range from dust size to large blocks more than a meter in diameter and accumulate as cinder cones at volcanic vents. Fine volcanic ash may be carried a considerable distance downwind and deposited as a blanket over the countryside. The rock resulting from the accumulation of ash falls is referred to as a **tuff** and, although of volcanic origin, it has many characteristics of a sedimentary rock.

Ash Flow Tuff. Ash flow tuff is a rock composed of fragments of volcanic glass, broken fragments of crystals, rock fragments, and pieces of solidified lava fused or "welded" together into a tight, coherent mass. Many fragments are typically flattened or bent out of shape. These unique textures indicate that the rock mass, although composed of ash fragments, was at the time of extrusion very hot. *Figure A.4* shows these features as they appear under the microscope.

Ash flow tuffs occur in thick layers ranging from several tens of meters to nearly 300 meters thick, and they may cover areas of thousands of square kilometers. Columnar jointing is common; this feature further substantiates the fact that the rock unit cooled from relatively high temperatures.

Figure A.4 Ash flow tuff—An igneous rock composed of fragmental material fused or welded together into a compact mass.

Ash flow tuffs originate from granitic magmas which are highly viscous and resistant to flow. The lava is highly charged with gas which separates into bubbles and causes the lava to expand into a froth. As the magma is extruded onto the surface, it explodes violently, blowing fragments of pumice and ash into the air. The ash, being relatively heavy, moves or "flows" as a hot mass over the surface and accumulates as a single layer of "welded" tuff.

Sedimentary Rocks

Sedimentary rocks are classified on the basis of size, shape, and composition of their constituent particles. Two main groups are recognized: (1) *clastic* rocks, consisting of deposits made from fragments of other rocks, and (2) *chemical* or *biochemical*, rocks formed by chemical and biological processes.

Clastic Rocks

Clastic rocks are generally subdivided on the basis of grain size of the component material. These include the accumulation of familiar sedimentary material such as gravel, sand, silt, and mud, which, when consolidated into hard rock, are referred to as conglomerate, sandstone, and shale.

Conglomerates. A conglomerate consists of consolidated deposits of gravel, with variable amounts of sand and mud deposited in the space between the larger grains (figure A.5). The cobbles and pebbles are usually well-rounded fragments over 2 mm in diameter. Most conglomerates show a crude stratification and include beds and lenses of sandstone. Conglomerates are

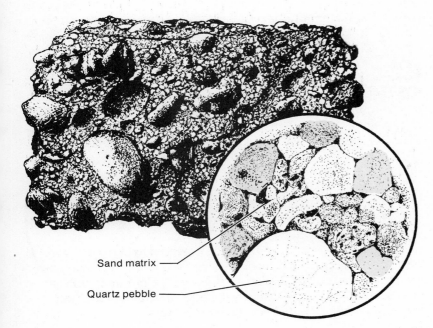

Sand matrix

Quartz pebble

Figure A.5 Conglomerate—A coarse-grained clastic rock in which the fragments are greater than 2 mm in diameter.

accumulating today at the base of mountain ranges, in stream channels, and on beaches.

Sandstones. Sandstone is probably the most familiar sedimentary rock, since it is well exposed, easily recognized, and is generally resistant to weathering. Sand may be composed of almost any material, but quartz grains are usually most abundant because quartz is a common constituent in many other rock types and does not break down easily by abrasion or chemical action (*figure A.6*). The particles of sand in most sandstones are cemented by calcite, silica, or iron oxide.

The composition of a sandstone provides an important clue to its history. During prolonged transportation, unstable minerals such as olivine, feldspar, mica, and small rock fragments are broken down to finer particles and winnowed out, leaving only the ultra-stable quartz. Clean, well-sorted sandstone composed of well-rounded quartz grains indicates prolonged transportation or even several cycles of erosion and deposition.

Shales (Mudstone). Deposits of fine, solidified mud and clay are known as shale (*figure A.7*). The particles which make up the rock are less than 1/16 mm in diameter and in many cases are too small to be clearly seen and identified under the microscope. Shale is the most abundant sedimentary rock, but it is usually soft and weathers into a slope, so that relatively few fresh, unweathered exposures are found. It is usually well stratified with thin laminae. Black shales are rich in organic material and accumulate in a variety of quiet-water environments such as lagoons, restricted shallow seas, or tidal flats. Red shales are colored with iron oxide and indicate oxidizing conditions in the environments in which they accumulate such as stream channels, flood plains, or tidal flats.

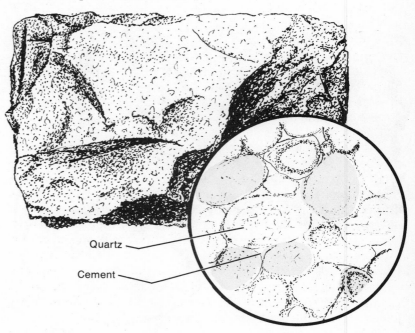

Quartz

Cement

Figure A.6 *Sandstone—A fine-grained clastic rock composed of fragments which may range from 1/16 to 2 mm in diameter.*

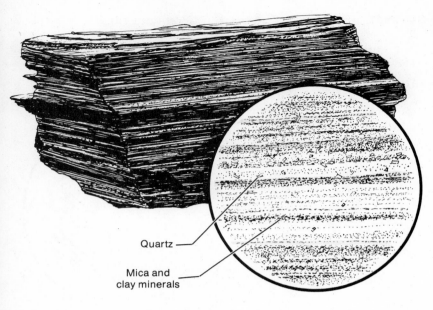

Quartz

Mica and clay minerals

Chemical and Organic Rock

Limestone. Limestone is by far the most abundant nonclastic rock. It is composed principally of calcium carbonate ($CaCO_3$) and originates by both chemical and organic processes. Limestones have a great variety of rock textures, and there are many different limestone types. Some of the major groups are skeletal limestone, oolitic limestone, and microcrystalline limestone.

Many plants and invertebrate animals extract calcium carbonate from water and use it to construct their shells and hard parts. When the organisms die, the shells accumulate on the sea floor and, over a long period of time, build up a deposit of limestone in which the texture consists of shells and shell fragments (*figure A.8*). This type of limestone, composed mostly of skeletal debris, may be several hundred meters thick and extend over thousands of square kilometers. Chalk, for example, is a skeletal limestone in which the skeletal fragments are remains of microscopic plants and animals.

Some limestones are composed of small, spherical concretions of calcium carbonate called oolites. The individual grains are about the size of a grain of sand and can be observed forming at the present time in the shallow waters of the Bahamas where currents and waves are active. Evaporation and increase in temperature of the sea water increases the concentration of $CaCO_3$ to a point at which it can be precipitated. Small fragments of shells or tiny grains of $CaCO_3$ moved by waves or currents become coated with successive layers of $CaCO_3$ as they roll along the sea floor. The sorting action of the waves and currents keeps the grain size of oolites quite uniform. Naturally, fragments of shells and oolites accumulate together in many places, forming an oolitic skeletal texture.

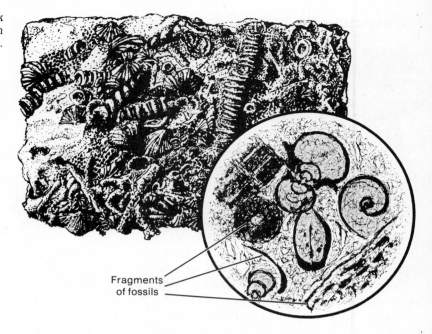

Figure A.8 Fossiliferous limestone—A rock composed predominantly of calcium carbonate and containing many fossils.

Fragments
of fossils

Oolitic limestone is a special type of non-clastic sedimentary rock in which moving currents rather than running water or wind carry the grains to the places where they are deposited.

In quiet water, calcium carbonate is precipitated to form tiny, needlelike crystals which settle to the bottom and accumulate as limy mud. Soon after deposition, the grains are commonly modified by compaction and recrystallization. This modification produces a rock with a dense, very fine-grained texture known as microcrystalline limestone. In such a rock, individual crystals can be seen only under high magnification. Microcrystalline limestone is also precipitated from springs and dripping water in caves, but the total amount of "dripstone" compared to marine limestone is negligible.

Other Sedimentary Rocks

Dolostone. Dolostone is a rock composed of the mineral dolomite, a calcium magnesium carbonate ($CaMg[CO_3]_2$) which is similar to limestone ($CaCO_3$) in most textural and structural features and general appearance. It may develop by direct precipitation from sea water, but most dolomites appear to originate by a substitution of the mineral dolomite for calcite in limestones.

Rock Salt and Gypsum. Rock salt is composed of the mineral halite and forms by evaporation in saline lakes such as the Dead Sea or in restricted bays along the shore of the ocean. Gypsum, composed of $CaSO_4 \cdot 2H_2O$, also originates from evaporation and collects in layers as calcium sulfate is precipitated. Since evaporites accumulate only in restricted basins subjected to prolonged evaporation, they are important indicators of ancient climatic and geographic conditions.

Appendix

Metamorphic Rocks

Although metamorphic rocks are highly complex, it is possible to develop a simple classification of the major types based on texture and composition. Following this scheme, two major groups of metamorphic rocks are recognized: (1) those that are foliated and possess a definite planar structure and (2) those that lack foliation and have a granular texture. The foliated rocks are further subdivided on the basis of the type of foliation. The major rock names may be qualified by adjectives describing their chemical or mineralogical composition.

Foliated Rocks

Slate. Slate is a very fine-grained metamorphic rock generally produced by low-grade metamorphism of shale and other fine-grained rocks and is characterized by excellent foliation (slaty cleavage) which causes the rock to split into thin sheets (figure A.9). The foliation results from the growth of platy minerals (mica and chlorite) perpendicular to the applied stresses. Individual mineral grains are so small that they can rarely be distinguished, and the separation of mineral type into individual layers is not present. The parallel arrangement of perfect cleavage in these smaller mica grains develops parallel planes of weakness which cause the rock to split into smooth slabs along the mineral cleavage directions. It should be emphasized that the foliation in slates (slaty cleavage) is the result of mineral growth under directed stress and is *not* original stratification.

Schist. Schist is a coarse-grained, foliated rock. Foliation results from the parallel arrangement of large grains of platy minerals such as mica, chlorite, talc, and hematite. The mineral

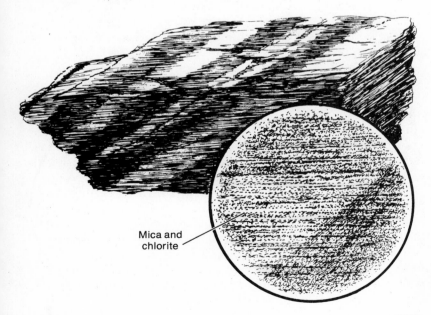

Mica and
chlorite

Figure A.9 *Slate—A fine-grained metamorphic rock with slaty cleavage; a metamorphosed shale.*

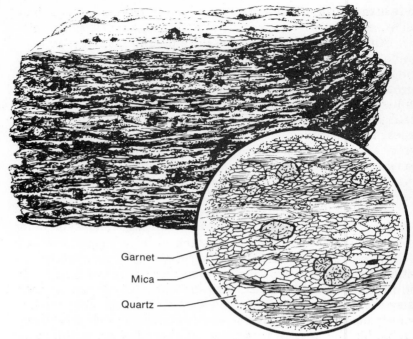

Figure A.10 *Schist—A coarse-grained metamorphic rock with schistosity.*

Garnet

Mica

Quartz

grains are readily visible and are oriented in an overlapping parallel arrangement (*figure A.10*). Thus, the foliation differs from slaty cleavage largely in the size of the crystals. In addition to the platy minerals, significant quantities of quartz, feldspar, garnet, amphibole, and other minerals may occur, providing a basis for subdividing schists into many varieties based on composition (chlorite schist, mica schist, amphibole schist, etc.).

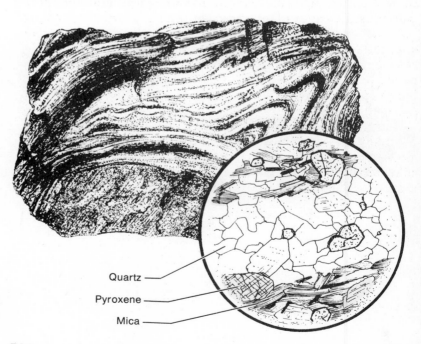

Quartz

Pyroxene

Mica

Figure A.11 *Gneiss—A coarse-grained metamorphic rock which contains gneissic banding.*

536

Schists result from a higher intensity of metamorphism than slates. They may have a variety of parent rock types (shale, tuff, ash flow tuff, etc.) and are one of the most abundant rock types.

Gneiss. Gneiss is a coarse-grained, granular metamorphic rock in which the foliation results from thin layers of alternating light and dark minerals (*figure A.11*). The composition of gneiss is very similar to that of granite; the major minerals are quartz, feldspar, and ferromagnesian minerals. Feldspar is especially abundant. Gneisses are coarse rocks in which most mineral grains are over 2 mm in diameter. The foliation in gneisses is generally less planar than in schists and slates, and the layering often shows tight folds and extreme contortions. Gneiss is formed during high-intensity, regional metamorphism.

Non-Foliated Rocks

Rocks such as sandstone and limestone are composed predominantly of one mineral which crystallizes in an equidimensional form. Metamorphism of these rocks does not result in a strong foliation, although mica grains scattered through the rock may assume a parallel orientation. The minerals may be flattened, stretched, or elongated and may show a preferred orientation, but the mass of rock does not develop a strong foliation. The resulting texture is best described as granular, or simply non-foliated.

Quartzite. Quartzite is a metamorphosed, quartz-rich sandstone. It is non-foliated because quartz grains, the principal constituent, do not form platy crystals. The individual grains are deformed and fused into a very tight mass so the rock breaks across the grains as easily as it breaks around them. Pure quartzite is white or light colored, but iron oxide and other minerals may impart various tones of red, brown, green, or other colors.

Marble. Marble is recrystallized limestone or dolomite. Calcite, the major constituent, is equidimensional; therefore, the rock is non-foliated. The crystals are commonly large and compactly interlocked, forming a dense rock. Many marbles show bands or streaks resulting from organic matter or other impurities in the original sedimentary rock.

Metaconglomerate. Metaconglomerate is not an extremely abundant metamorphic rock but illustrates the degree to which a rock may be deformed in the solid state. Under directional stress, the individual pebbles are stretched and fused into a mass which shows a very distinct linear fabric.

Glossary

Aa. Lava flow whose surface is typified by angular, jagged blocks.

Ablation. Reduction of a glacier by melting, evaporating, iceberg calving, or deflation.

Abrasion. The mechanical wearing away of a rock by friction, rubbing, scraping, or grinding.

Absolute time. Geologic time measured in a specific duration of years as contrasted to relative time, which signifies only the chronologic order of events.

Abyssal. Of or pertaining to the great depths of the sea, generally below 1000 fathoms (6000 ft).

Abyssal fan. A fan-shaped accumulation of sediment on the abyssal floor at the mouth of a submarine canyon.

Abyssal hills. That part of the ocean floor consisting of hills rising up to 1000 m (3000 ft) above the surrounding floor. Abyssal hills are found seaward of most abyssal plains and occur in profusion in basins isolated from continents by ridges, rises, or trenches.

Abyssal plains. Flat areas of the ocean floor which slope less than 1:1000. Most abyssal plains lie at the base of the continental rise and are simply areas where abyssal hills are completely covered with sediment.

Accessory mineral. Minerals which are rare or of minor importance and are not essential in classifying the rock.

Active margin (plate tectonics). The leading edge of a plate.

Aftershock. An earthquake which follows a larger earthquake. After a major earthquake, there are generally many aftershocks which may be felt for many days or even months.

Agate. A variety of cryptocrystalline quartz in which colors occur in bands. Agate is commonly deposited in cavities.

Aggradation. The process of building up a surface by deposition of sediment.

A-horizon. The topsoil layer in a soil profile.

Airy hypothesis. That hypothesis which explains isostasy by assuming equal densities and postulates that differences in elevation result from different thickness of the outer layer—in a way analogous to the underwater extension of an iceberg floating in the ocean or wooden blocks floating in water.

Alcove. A large niche or recession formed in a precipitous cliff.

Alkalic (petrology). Igneous rocks rich in sodium and/or potassium.

Alluvial fan. A fan-shaped deposit of sediment built by a stream where it emerges from an upland or mountain range into a broad valley or plain. Fans are typical of arid to semiarid climates but are not restricted to them.

Alluvium. A general term for all sedimentary accumulations which occur as the result of the comparatively recent action of rivers. Alluvium thus includes sediment laid down in river beds, flood plains, and alluvial fans.

Alpine glacier. A glacier occupying a valley. Also called mountain glaciers or valley glaciers.

Amorphous solid. A solid in which atoms or ions are not arranged in a definite crystalline structure. Example: glass, amber, obsidian.

Amphibole group. An important rock-forming mineral group of ferromagnesian silicates with a double chain of silicon-oxygen tetrahedra. Common example: hornblende.

Andesite. A fine-grained igneous rock composed mostly of plagioclase feldspar and 25 to 40% amphibole and biotite. It has no quartz or potassium feldspar. It is abundant in the mountains around the borders

Aa flow

Abyssal plains

Alluvial fan

539

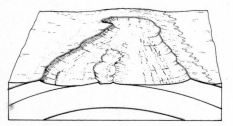
Angular unconformity

Anticline

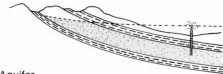
Aquifer

Asymmetric fold

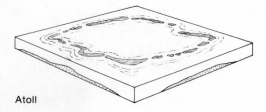
Atoll

of the Pacific Ocean. Andesitic magma is believed to originate from the fractionation of partially melted basalt.

Andesite line. The boundary in the Pacific Ocean separating volcanoes of the inner Pacific Ocean, discharging only basalts, from those near the continental margins, which discharge both andesite and basalt.

Angle of repose. Also known as critical slope. The maximum slope at which material such as loose rock fragments or sand grains remain stable (usually about 30° to 34°). When exceeded, mass movement may occur.

Angular unconformity. An unconformity in which older strata dip at a different angle (generally steeper) than the younger strata.

Anomaly. A deviation from the norm or the average.

Anorthosite. An igneous rock composed almost entirely of plagioclase.

Anticline. A fold in which the limbs dip away from the crest. When eroded, the oldest rocks are exposed in the central core.

Aphanitic texture. A rock texture in which individual crystals are so small they cannot be identified without the aid of a microscope. The rock in hand specimen appears to be dense and structureless.

Aquifer. A permeable stratum or zone below the surface through which ground water moves.

Arete. A narrow, sharp ridge separating two adjacent glacial valleys.

Arkose. A sandstone containing at least 25% feldspar.

Artesian basin. A geologic structural feature in which ground water is confined and under artesian pressure.

Artesian water. Ground water confined to an aquifer and under sufficient pressure so as to rise above the top of the aquifer when tapped by a well.

Ash. Volcanic fragments the size of dust.

Ash flow. A turbulent blend of unsorted pyroclastic material (mostly fine grained) mixed with high-temperature gas ejected explosively from fissures or craters.

Asteroid. A relatively small celestial body believed to be either the remains of an exploded planet or matter that never completed the planet-forming process.

Asthenosphere. A zone in the earth directly below the lithosphere 50 to 200 km below the surface. Seismic velocities are distinctly lower in the asthenosphere, suggesting that the material is soft and will yield to plastic flow.

Astrogeology. The application of geologic methods and knowledge to the study of extraterrestrial bodies.

Asymmetric fold. A fold (anticline or syncline) in which one limb dips more steeply than the other.

Atmosphere. The mixture of gases that surround a planet. The earth's atmosphere is chiefly oxygen and nitrogen, with some argon and carbon dioxide. Synonymous with *air*.

Atoll. A ring of low coral islands surrounding a central lagoon.

Attitude. The three-dimensional orientation of a bed, fault, dike, or other geologic structure. Determined by the combined measurements of dip and strike.

Axial plane. As applied to folds, an imaginary plane through a fold that intersects the crest or trough in such a manner as to divide the folds as symmetrically as possible.

Axis. 1. In crystallography, an imaginary line passing through a crystal around which the parts are symmetrically arranged. 2. In folds, a line where folded beds show maximum curvature; also defined as a line formed by the intersection of the axial plane with the bedding surface.

Backswamp. Marshy area of a flood plain at some distance beyond and lower than the natural levees which confine the river.

Backwash. The return sheet flow down a beach after the wave is spent.

Badlands. An area nearly devoid of vegetation and dissected by stream erosion into an intricate system of closely spaced, narrow ravines.

Bajada. A system of coalesced alluvial fans.

Bar. An offshore submerged elongate ridge of sand or gravel built on the sea floor by waves and currents.

Barchan dune. A crescent-shaped dune, the tips or horns of which point downwind. Formed in desert areas where sand is scarce.

Barrier island. An elongate island of sand or gravel built parallel to a coast.

Barrier reef. An elongate coral reef that trends parallel to the shore of islands or continents, separated from them by lagoons.

Basalt. A dark-colored aphanitic igneous rock composed of plagioclase (over 50%) and pyroxene. Olivine may or may not be present. Basalt and andesites represent 98% of all volcanic rocks.

Base level. The level below which a stream cannot effectively erode. Sea level is the ultimate base level, but lakes form temporary base levels for drainage systems inland.

Basement complex. A series of igneous and metamorphic rocks lying beneath the oldest stratified rocks of the region. The shield is the basement complex for the Paleozoic and Mesozoic rocks of the mid-continent. In some places younger metamorphosed rocks form the basement complex.

Basin. 1. Structural—a circular or elliptical downwarp with younger beds in the center. 2. Topographic—a depression into which the surrounding area drains.

Batholith. A large body of intrusive igneous rock at least 100 km² in areal extent.

Bathymetric chart. A topographic map of the bottom of a body of water (such as the sea floor).

Bathymetry. The measurement of ocean depths and the charting of the topography of the ocean floor.

Bauxite. A mixture of various amorphous or crystalline hydrous aluminum oxides and aluminum hydroxides commonly formed as a residual clay deposit in tropical or subtropical regions. Bauxite is the principal commercial source of aluminum.

Bay (coast). A wide, curving recess or inlet between two capes or headlands.

Beach. A deposit or wave-washed sediment deposited along the coast between the landward limit of wave action and the outermost breakers.

Bed. A layer of sediment 1 cm or more in thickness.

Bedding plane. A surface separating layers of sedimentary rock.

Bed load. Material transported along the bottom of a stream or river in contrast to material carried in suspension or solution.

Bedrock. The continuous solid rock that underlies the regolith everywhere and is exposed locally at the surface.

Beheaded stream. A stream that has lost a part of its drainage system as a result of stream piracy.

Benioff zone. A zone of deep earthquakes which dips beneath the continents or island arcs from the deep-sea trenches.

Barrier island

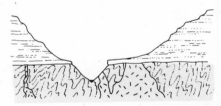

Barrier reef

Basement complex

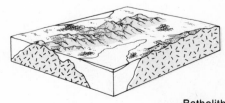

Batholith

Volcanic bomb

Berm. A nearly horizontal portion of the beach or backshore formed by storm waves. Some beaches have no berms, others have several.

B-horizon. The soil zone of accumulation where some of the material derived from leaching in the overlying A-horizon is deposited.

Biologic material. A general term for material originating from organisms. Example: fossils (shells, bones, leaves), peat, coal, etc.

Biosphere. A term which signifies the totality of living things on the earth's surface.

Biotite. "Black mica." A rock-forming ferromagnesian silicate with the tetrahedra arranged in sheets.

Birdfoot delta. A delta with distributaries extending seaward and resembling in plan the outstretched claws of a bird. The Mississippi River delta is an excellent example.

Block faulting. A type of normal faulting in which the crust is divided into fault blocks of different elevations and orientations.

Blowout. A basin excavated by the wind.

Bombs, volcanic. Hard fragments of lava that were liquid or plastic at the time of ejection and have forms and surface markings acquired during flight through the air. Fragments range from a few millimeters to more than a meter in diameter.

Boulder size. Fragments larger than 256 mm in diameter (approximately the size of a volleyball). Next size larger than a cobble.

Boulder train. A series of glacial erratics that are derived from a single locality and deposited in the shape of a fan, narrowing toward the original source.

Bracketed intrusion. An intrusive rock which has been exposed at the surface by erosion and subsequently covered by younger sediment. The stratigraphic age of the intrusive thus falls between (is bracketed by) the ages of the younger and older sedimentary deposits.

Braided stream. A complex of converging and diverging stream channels separated by bars or islands. Forms where more sediment is available than can be removed by the discharge of the stream.

Breaker. A collapsing water wave.

Breccia. A rock composed of cemented angular fragments.

Butte. A somewhat isolated hill usually capped with a resistant layer of rock and bordered by talus. Represents an erosional remnant of a former more extensive slope.

Butte

Calcite. A mineral composed of calcium carbonate ($CaCO_3$).

Caldera. A large, more or less circular depression or basin having a diameter many times greater than that of the included volcanic vents. Calderas are believed to result from subsidence or collapse and may or may not be related to explosive eruptions.

Calving. The process by which large blocks of ice break off from glaciers which terminate in a body of water.

Capacity. The potential quantity of sediment a stream, glacier, or wind can carry under a given set of conditions.

Capillary. A small, tubular opening with a diameter about the size of a human hair.

Capillary action. The action by which a fluid such as water is drawn up in small openings such as pore space in rocks as a result of surface tension.

Capillary fringe. A zone above the water table where water is lifted by surface tension in openings of capillary size.

Carbonaceous. Containing carbon (C).

Carbonate minerals. Minerals formed by the combination of the com-

Capillary fringe

542

plex ion ($[CO_3]^{2-}$) with a positive ion. Examples: Calcite ($CaCO_3$), dolomite ($CaMg[CO_3]_2$).

Carbonate rock. Rocks composed chiefly of carbonate minerals. Examples: limestone, dolomite.

Carbon-14. Radioisotope of carbon which has a half-life of 5730 years.

Catastrophism. The belief that geologic history consists of major catastrophic events far beyond any process we see today. (Contrast with *uniformitarianism.*)

Cave. A naturally formed subterranean open area or chamber or series of chambers commonly formed in limestone by solution activity.

Cement. Minerals precipitated from ground-water solution in the pore space of sedimentary rock which bind the particles together.

Chalcedony. A general term referring to fibrous cryptocrystalline quartz.

Chalk. A variety of limestone composed of the shells of microscopic oceanic organisms.

Chert. Granular cryptocrystalline silica.

C-horizon. The zone of soil consisting of partly decomposed bedrock, which lies directly beneath the B-horizon and grades downward into fresh, unweathered bedrock (see diagram under B-horizon).

Cinder cone. A cone-shaped hill which is composed of loose volcanic fragments.

Cinders. Fragments of volcanic ejecta 0.5 to 2.5 cm in diameter.

Cirque. An amphitheater-shaped depression at the head of a glacial valley; excavated mainly by ice plucking and frost wedging.

Clastic. Fragments such as mud, sand, and gravel produced by the mechanical breakdown of rocks.

Clastic texture. A texture of sedimentary rocks resulting from physical transportation and deposition of broken particles of minerals, rock fragments, and organic skeletal remains.

Clay minerals. Finely crystalline hydrous silicates formed by weathering of minerals such as feldspar, pyroxene, and amphibole.

Clay size. Particles less than 1/256 mm in diameter.

Cleavage. The tendency for a mineral to break in a preferred direction along smooth planes.

Cobble size. Fragments having a diameter between 64 mm (size of a tennis ball) and 256 mm (size of a volleyball).

Color. An observable physical property of minerals which is not considered diagnostic in making an identification.

Columnar jointing. A system of fractures that split a rock body into long prisms or columns. Characteristic of basalt flows.

Competence. The maximum potential size of particles that a stream, glacier, or wind can move at a given velocity.

Composite cone. A large volcanic cone built by extrusion of alternating layers of ash and lava. Synonymous with *strato-volcano.*

Compression. A system of stresses that tends to reduce the volume of or shorten a substance.

Conchoidal fracture. A type of fracture that produces a smooth, curved surface. A characteristic of quartz and of obsidian.

Concretions. Spherical to elliptical nodules formed by accumulation of mineral matter after deposition of the sedimentary rock.

Cone of depression. A conical depression in the water table surrounding a well which results from heavy pumping.

Conglomerate. A sedimentary rock composed of rounded fragments of pebbles, cobbles, or boulders.

Connate water. Water trapped in the sediment as it was deposited.

Glossary

Cirques

Columnar jointing

Composite cone

Conchoidal fracture

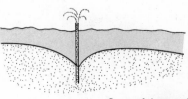

Cone of depression

Consequent stream

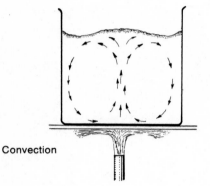

Convection

Coquina

Consequent stream. A stream whose course is a direct consequence of the original slope on which it developed.

Contact. The surface separating two different rock bodies.

Contact metamorphism. Metamorphism resulting from heat and pressure near the contact with a magma.

Continent. A large landmass 20 to 60 km thick composed mostly of granitic rock which rises abruptly above the deep-ocean floor; includes marginal areas submerged beneath sea level. Example: African Continent, Australian Continent, Asian Continent, South American Continent.

Continental accretion. The theory that continents have grown by incorporation of deformed sediments along their margins.

Continental crust. The type of the earth's crust which underlies the continents (including the continental shelves). It is sometimes referred to as *sial*. The continental crust is about 35 km thick with a maximum of 60 km beneath mountain ranges. The density of the continental crust is 2.7 g/m³, and the velocities of compressional seismic waves through it are less than 6.2 km/sec. (Contrast with *oceanic crust*.)

Continental drift. The concept that continents have moved relative to one another.

Continental glacier. A thick ice sheet that covers large parts of a continent. Examples are existing glaciers in Greenland and Antarctica.

Continental margins. The zone of transition from the continental mass to the ocean basins. Generally includes continental shelf, continental slope, and continental rise.

Continental rise. The gently sloping surface located at the base of the continental slope (see diagram for abyssal).

Continental shelf. Submerged margins of the continental mass extending from the shores to the first prominent break in slope at about 120 m depth.

Continental slope. The slope that extends from the continental shelf down to the ocean deep. In some areas such as off eastern North America, the continental slope grades into a more gentle slope of the continental rise (see diagram for continental shelf).

Convection. A movement of portions of any fluid medium resulting from density differences produced by heating.

Convection cell. Space occupied by a single convection current (see diagram).

Convection current. The transfer of material within a closed system as a result of thermal convection. Convection currents are characteristic of the atmosphere and bodies of water and are believed to be generalized within the interior of the earth. Convection within the mantle is thought to be the process responsible for plate tectonics (see diagram).

Coquina. A limestone composed of an aggregation of shells and shell fragments.

Coral. A bottom-dwelling marine organism.

Core. The central part of the earth, 3380 to 3540 km in diameter, which is surrounded by the mantle (see diagram for asthenosphere).

Coriolis effect. The effect produced by the *Coriolis force*, viz., the tendency of all particles of matter in motion on the earth's surface to be deflected to the right in the northern hemisphere and to the left in the southern hemisphere. This results from the centrifugal force produced by rotation of the earth.

Coriolis force. The force caused by the earth's rotation which serves to deflect a moving body on the earth's surface to the east in the northern hemisphere and to the west in the southern hemisphere.

Country rock. A general term for rock surrounding an igneous intrusion.

Crater. 1. Volcanic—a circular depression at the summit of a volcano with a diameter less than three times its depth. Compare with caldera. 2. Lunar—a circular depression characteristic of the moon's surface. Most lunar craters are believed to have been caused by impact from meteorites.

Creep. 1. The imperceptibly slow movement of material downslope. 2. The slow permanent yielding to stresses that are less than the yield point if applied only for a short time.

Crevasse. 1. Glaciers—a deep crack in the upper surface of a glacier. 2. Natural levees—a break in a natural levee.

Cross-bedding. Stratification inclined to the original horizontal surface upon which the sediment accumulated. Produced by deposition on the slope of a dune or sand wave.

Cross-cutting relations. A rock is younger than any rock across which it cuts.

Crust (earth's structure). The outermost layer or shell of the earth—generally defined as that part of the earth above the Mohorovičić discontinuity. It represents less than 1% of the earth's total volume. See also: *continental crust; oceanic crust.*

Crustal warping. Gentle upward or downward bending of sedimentary rocks in the continental interior.

Cryptocrystalline. A texture in which crystals are too small to be identified with an ordinary microscope.

Crystal. A solid polyhedral form bounded by natural plane surfaces resulting from growth of a crystal lattice.

Crystal face. A natural smooth plane surface of a crystal.

Crystal lattice. A systematic symmetrical network of atoms within a crystal.

Crystalline texture. The texture of a rock resulting from the simultaneous growth of crystals.

Crystallization. The process of crystal growth. May occur because of condensation from a gaseous state, precipitation from solution, or cooling of a melt.

Crystal structure. The orderly arrangement of atoms in a crystal.

Cuesta. An elongate ridge formed on the upturned and eroded edges of gently dipping strata.

Daughter isotope. The isotope created by the radioactive decay of a parent isotope. The amount of a daughter isotope continually increases with time.

Debris flow. The rapid downslope movement of debris—rock, soil, and mud.

Debris slide. A type of landslide involving slow-to-rapid downslope movement of comparatively dry rock fragments and soil. The mass of debris does not show backward rotation (as in a slump) but slides and rolls forward.

Declination, magnetic. The horizontal angle between true north and magnetic north.

Decomposition. Weathering by chemical processes.

Deep-focus earthquakes. Earthquakes which originate at depths greater than 300 km.

Deep-sea fan. A cone or fan-shaped deposit of land-derived sediment

Creep

Cross-bedding

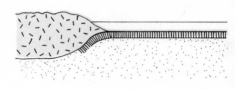

Crust

Crystal form

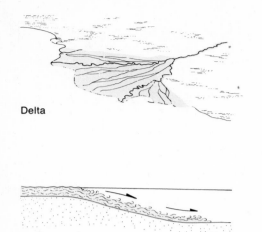

Delta

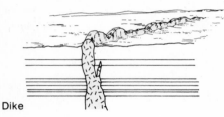

Density current

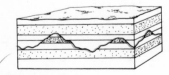

Dike

Disconformity

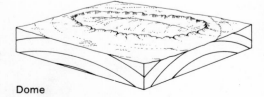

Dome

located seaward of large rivers or submarine canyons. Synonymous with *submarine cone*, *abyssal cone*, *abyssal fan*.

Deep-sea trenches. See *trench*.

Deflation. Erosion of loose rock particles by the wind.

Degradation. The general lowering of the surface of the land by processes of erosion.

Delta. A deposit of sediment, roughly triangular, at the mouth of a river.

Dendritic. A branching stream pattern that resembles the branching habit of certain trees such as the oaks or maples.

Density. The measure of the concentration of matter in a substance expressed in grams per cubic centimeter (gm/cm³), weight per unit volume.

Density current. A current which flows as a result of differences in density. In the ocean, density currents may be produced by differences in temperature, salinity, or turbidity (material held in suspension).

Desert pavement. A veneer of pebbles left in place where wind has removed the finer material.

Desiccation. Drying out. Used in sedimentation in reference to loss of water from pore space through evaporation or compaction.

Detrital. Fragments, clastic.

Diastrophism. Large-scale deformation involving mountain-building and metamorphism.

Differential erosion. Variations in rates of erosion on different rock masses. As a result of differential erosion, resistant rocks form steep cliffs, whereas nonresistant rocks form gentle slopes.

Differentiated planet. A planet in which the various elements and minerals are separated according to density and concentrated at various levels. The earth, for example, is differentiated with the heavy metals, iron and nickel, concentrated in the core, lighter silicate minerals in the mantle, and still lighter material in the crust, hydrosphere, and atmosphere.

Dike. A tabular intrusive rock which occurs across strata or other structural features of the surrounding rock.

Dike swarm. A group of associated dikes.

Dip. The angle between a horizontal plane and the surface of bedding (joints, faults, foliation, etc.).

Disappearing stream. A stream that disappears into an underground channel and that does not reappear in the same or even in an adjacent drainage basin. Typical of karst regions, where streams disappear into sink holes and follow channels through caves.

Disconformity. An unconformity in which beds above and below are parallel.

Discontinuity. A sudden or rapid change in the physical properties of the earth. Discontinuities are recognized by seismic data. See also *Mohorovičić*.

Disintegration. Weathering by mechanical processes. Synonymous with mechanical weathering.

Distributaries. Branches of a stream into which a river divides when it reaches its delta.

Divide. A line separating two drainage basins.

Dolomite. 1. A mineral composed of CaMg(CO₃)₂. 2. A rock composed primarily of dolomite.

Dome. A circular or elliptical uplift whose beds dip away in all directions from a central area.

Glossary

Downwarp. A downward bend or subsidence of a part of the earth's crust.

Drainage basin. The total area that contributes water to a given river.

Drift (glacial). A general term referring to any sediment deposited directly by ice or deposited in lakes, oceans, or streams as a result of glaciation.

Drip curtain (caves). A thin sheet of dripstone hanging from the ceiling or wall of a cave.

Dripstone. A cave deposit formed by the precipitation of calcium carbonate in ground water entering an underground cavern.

Drumlin. A smooth, glacially streamlined hill elongate in the direction of ice movement. Drumlins are generally composed of till.

Dune. A low mound of fine-grained material which accumulates in response to sediment transport in a current system. A dune has a geometric form which is maintained as it migrates. Sand dunes are commonly classified according to shape. (See also *barchan, longitudinal, parabolic, seif, star,* and *transverse* dunes.)

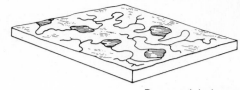

Deranged drainage

Earthquake. Groups of elastic waves propagating in the earth, initiated where elastic limits are exceeded.

Ecology. A study of the relationship between organisms and their environment.

Ejecta (crater). Rock fragments, glass, and other material thrown out of a crater during formation.

Ejecta blanket. A deposit of material thrown out of a crater during formation which accumulates in the area beyond the crater rim. In lunar craters the material is chiefly crushed rock. Typically, the ejecta blanket forms a hummocky terrain extending a distance roughly equal to the diameter of the crater. Rays are also ejecta but do not form a continuous blanket deposit.

Elastic deformation. Temporary deformation after which a rock returns to its original size and shape. Example: bending mica flakes.

Elastic limit. The maximum stress any specimen can stand without undergoing permanent deformation either by solid flow or rupture.

Elastic rebound theory. The theory that explains earthquakes as a result of energy released by faulting. Earthquake waves are caused by the sudden release of stored strain.

End moraine. A ridge of till marking the former front of a glacier.

Entrenched meander. A meander cut into the underlying rock as a result of regional uplift.

Environment of sedimentation. The physical, chemical, and biological conditions at the site where sediment accumulates.

Eolian. Pertaining to the wind.

Epicenter. The area on the surface directly above the focus of an earthquake.

Epoch. A division of geologic time; a subdivision of a period. Example: Pleistocene epoch.

Erosion. The process that loosens and moves sediment to another place on the earth's surface. Agents of erosion include water, ice, wind, and gravity.

Erratic. A large boulder carried by ice to an area far removed from its point of origin.

Escarpment. A steep cliff.

Esker. A long, narrow, sinuous ridge of stratified glacial drift deposited in a former tunnel or stream bed beneath a glacier.

Estuary. A bay at the mouth of a river formed by subsidence of land or

Drumlins

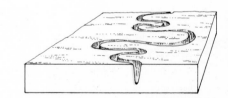

Entrenched meanders

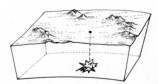

Epicenter

Exfoliation

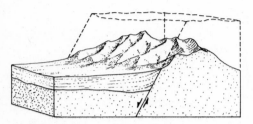

Fault block

Fissure eruption

Fold

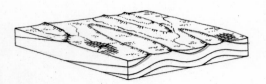

Eroded fold

rise of sea level. Fresh water from the river mixes with and dilutes the sea water.

Eustatic change of sea level. Worldwide change in sea level resulting from changes in volume of water or capacity of ocean basins.

Evaporites. Rocks that result from evaporation of mineralized water. Common examples are rock salt and gypsum.

Exfoliation. A general term referring to spalling of exposed rock.

Exposure. Bedrock not covered with soil or regolith.

Extrusive rock. Rock originating from a mass of magma that flowed out onto the surface of the earth. Example: lava.

Fabric. The orientation of particles in a rock.

Faceted spur. A spur or ridge that has been beveled or truncated by faulting, erosion, or glaciation.

Facies. A distinctive group of characteristics within a rock, such as composition, grain size, and fossils, that differ as a group from those elsewhere in the same unit. Example: conglomerate facies, shale facies, brachiopod facies, etc.

Fan. A fan-shaped deposit of sediment.

Fault. A surface along which a rock body has broken and been displaced.

Fault block. A mass of rock bounded by faults on at least two sides.

Fault scarp. A cliff produced by faulting.

Faunal succession. A law discovered by William Smith which states that fossils in a stratigraphic sequence succeed one another in a definite, recognizable order.

Feldspar. A group of silicate minerals linked together in a three-dimensional network. Examples: K-feldspar, Ca-plagioclase, Na-plagioclase.

Felsite. A general term for light-colored aphanitic igneous rocks. Example: rhyolite.

Ferromagnesian minerals. Silicate minerals containing abundant iron and magnesium. Olivine, pyroxene, and amphibole are common examples.

Fetch. The area of open ocean over which wind blows to create a wave system.

Fiord. A glaciated valley now flooded by the sea to form a long, narrow, steep-walled inlet.

Firn. An elliptical granule of recrystallized snow. Sometimes referred to as a névé.

Fissure. An open fracture in a rock.

Fissure eruption. Extrusion of lava along a fissure.

Flint. A popular name for dark-colored chert (cryptocrystalline quartz).

Flood basalt. Extensive volumes of basalt erupted largely along fissures. Synonymous with *plateau basalt*.

Flood plain. Level areas bordering a stream which are occasionally flooded.

Focus. The actual location of earthquake origin.

Fold. A bend or flexure in a rock.

Foliation. A planar element in metamorphic rocks produced by the secondary growth of minerals. Three major types are recognized: (1) slaty cleavage, (2) schistosity, and (3) gneissic banding.

Footwall. The block beneath a dipping fault surface. Example: See illustration for hanging wall.

Foreshore. The seaward part of the shore or beach lying between high tide and low tide.

Fossil. The natural preserved evidence of past life, such as bones, shells, casts, impressions, or trails.

Fossil fuel. Fuel containing solar energy preserved in chemical compounds of plants and animals of former ages. Includes petroleum, natural gas, and coal.

Fringing reef. A reef marginal to the shore of a landmass.

Frost heaving. The lifting of unconsolidated material produced by the freezing of subsurface water.

Frost wedging. The mechanism of forcing rocks apart by the growth of ice in fractures and pore space.

Fumarole. Vent discharging gas and steam. Associated with hydrothermal activity.

Fringe reef

Gabbro. A dark phaneritic rock composed of plagioclase, pyroxene, and possibly olivine, but no quartz. *Course grain*

Gas. The state of matter that has neither independent shape nor volume. It can be compressed readily and tends to expand indefinitely.

Geode. A cavity lined with crystals so that when weathered from the rock body it appears as a partly hollow, rounded rock.

Geologic column. A composite diagram showing in a single column the subdivisions of geologic time and the rock units formed during each major period.

Geologic cross section. A diagram showing the structure and arrangement of rocks as they would appear in a vertical plane below the surface.

Geologic map. A map showing the distribution of rocks at the surface.

Geologic time scale. The time interpreted from the geologic column and radiometric dates.

Geosyncline. A subsiding part of the lithosphere in which thousands of meters of sediment accumulate.

Geothermal. Pertaining to heat of the interior of the earth.

Geothermal gradient. The rate of increase of temperature with depth. The approximate average in the earth's crust is about 25° C/km.

Geyser. A thermal spring which intermittently erupts steam and boiling water.

Glacier. A mass of ice formed from recrystallized snow and thick enough to flow plastically.

Glass. A form of matter that has many properties of a solid but lacks crystalline structure.

Glassy texture. A texture of an igneous rock in which the material is not crystalline but is in the form of natural glass.

Global tectonics. A study of the characteristics and origin of structural features of the earth which have regional or global significance.

Glossopteris flora. An assemblage of late Paleozoic fossil plants named for the seed-fern *Glossopteris*, one of the plants in the assemblage. This flora is widespread in South America, Africa, Australia, India, and Antarctica and has been important in the development of concepts of continental drift.

Gneiss. A coarse-grained metamorphic rock containing a foliation consisting of alternating layers of light- and dark-colored minerals. Composition is generally similar to that of a granite.

Gondwanaland. The name of the southern continental landmass thought to have split apart in Mesozoic time to form the present-day continents of South America, Africa, India, Australia, and Antarctica.

Graben. An elongate fault block that has been lowered relative to the blocks on either side.

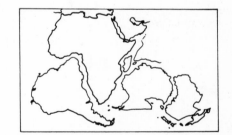

Gondwanaland

Graben

Glossary

549

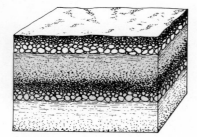

Graded bedding

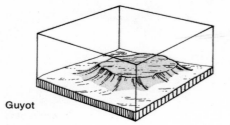

Guyot

Horn

Gradation. Leveling of the land resulting from agents of erosion such as river systems, ground water, glaciers, wind, and waves.

Graded bedding. A type of bedding in which there is a characteristic increase in grain size from bottom to top.

Graded stream. A stream which has attained a delicate adjustment or balance between erosion and deposition so that there is just the velocity necessary to transport the sediment load supplied from the drainage basin. A graded stream is in equilibrium so that neither erosion nor deposition occurs.

Gradient (stream). The slope of the stream channel measured along the course of the stream.

Grain. A particle of mineral or rock generally lacking well-developed crystal facies.

Granite. A coarse-grained igneous rock composed of potassium feldspar, plagioclase, and quartz, with minor ferromagnesian minerals.

Granitization. The formation of granite from a metamorphic rock without complete melting.

Gravity anomaly. An abnormal gravitational attraction within the earth.

Graywacke. An impure sandstone consisting of rock fragments, quartz, feldspar, and a matrix of clay-size particles.

Groundmass. The fine grains in a porphyritic rock.

Ground moraine. Glacial deposits that cover an area formerly occupied by a glacier. Ground moraine typically produces a landscape of low, gently rolling hills.

Ground water. Water below the earth's surface. Generally occurs in pore space of rock and soil.

Guyot. A seamount with a flat top.

Half-life. The amount of time required for half of a radioactive isotope to decay to another isotope of less mass.

Hanging valley. A tributary valley whose floor lies ("hangs") above the valley of the main stream or shore into which it flows. Hanging valleys are commonly created by deepening of the main valley by glaciation, but they may also be produced by faulting or rapid retreat of a sea cliff.

Hanging wall. The surface or block of rock above an inclined fault plane.

Hardness. 1. Minerals—a measure of the resistance of a mineral to scratching or abrasion. 2. Water—the amount of calcium carbonate and magnesium carbonate in solution.

Headland. An extension of land seaward out from the general trend of the coast. A promontory, cape, or peninsula.

Headward erosion. Extension of a stream headward up the regional slope of erosion.

Heat flow. The amount of heat escaping from the interior of the earth.

High-grade metamorphism. Metamorphism that is accomplished under high temperature and pressure.

Hogback. A narrow, sharp ridge formed on steeply inclined, resistant rock.

Horizon. 1. Geologic—a plane of stratification assumed to have originally been horizontal; 2. Soil—a layer of soil distinguished by characteristic physical properties generally designated by letters, e.g., A-horizon, B-horizon, etc.

Horn. A sharp peak formed by the intersecting headwalls of three or more cirques.

Hornblende. A variety of the amphibole mineral group.

Horst. An elongate fault block that has been uplifted relative to the adjacent rocks.

Hummock. A rounded or conical knoll, mound, hillock, or a surface of other small irregular shapes. A surface not equidimensional or ridge-like.

Hydraulic. Pertaining to fluids in motion.

Hydrolysis. The chemical combination of water with other substances.

Hydrosphere. The waters of the earth as distinguished from the rocks (lithosphere), air (atmosphere), or living things (biosphere).

Hydrostatic pressure. The pressure exerted by the water at any given point in a body of water.

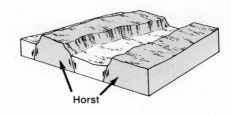

Horst

Ice sheet. Large body of ice of considerable extent, not confined to valleys. Localized ice sheets are sometimes called ice caps.

Igneous rocks. Rocks which originate from cooling and solidification of molten silicate minerals (magma). Include volcanic rocks and plutonic rocks.

Inclination (magnetic). The angle between a horizontal plane and the magnetic line of force.

Inclusion. A fragment of an older rock incorporated into an igneous rock.

Intermittent stream. A stream through which water flows only part of the time.

Internal drainage. A drainage system that does not extend to the sea.

Interstitial. Material that occurs in the pore space of a rock. Petroleum and ground water are interstitial fluids. Minerals deposited by ground water in a sandstone are interstitial minerals.

Intrusion. The process of placing or injecting a magma into a preexisting rock.

Intrusive rock. Plutonic rock which, while fluid, penetrated into or between other rocks and solidified before being exposed at the surface.

Inverted valley. A valley filled with lava or some other resistant material which is subsequently eroded into an elongate ridge.

Island. A body of land smaller than a continent completely surrounded by water.

Island arc. A chain of islands generally convex toward the open ocean. Example: Aleutians.

Isoclinal fold. A fold in which the limbs are parallel.

Isostasy. The theory that all parts of the earth's outer layers tend to establish a condition of flotational balance as they rest on the soft, denser layers beneath. Layers which have low density and layers which are thicker will rise higher than those which are denser or thinner.

Isotopes. Elements with the same atomic number but different atomic weights caused by the variations in number of neutrons in the nucleus.

Isoclinal folds

Joint. A fracture in a rock along which there has been no appreciable displacement.

Joint set. A regional pattern of groups of parallel joints.

Joint system. Two or more joint sets that intersect.

Kame. A body of stratified glacial sediment. A mount, knob, or irregular ridge deposited by a subglacial stream as an alluvial fan or delta.

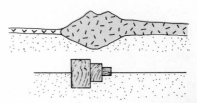

Isostasy

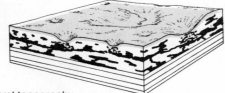

Karst topography

Laccolith

Karst topography. A landscape characterized by sinks, solution valleys, and other features resulting from ground-water activity.

Kettle. A closed depression in glacial drift created by the melting of a buried or partly buried block of ice.

Laccolith. A concordant igneous intrusion that has arched up the strata into which it was injected so it forms a pod- or lens-shaped body. The laccolith has a floor which is generally horizontal.

Lag deposits. A residual accumulation of coarse fragments remaining on the surface after the finer material has been removed.

Lagoon. A body of shallow water separated from the ocean by a barrier island or reef.

Lamina. A layer of sediment less than 1 cm thick.

Laminar flow. Flow in which the fluid moves in parallel lines. (Contrast with *turbulent flow*.)

Landform. Any feature of the earth's surface having a characteristic shape as the product of natural processes. Includes major features such as continents and ocean basins, plains, plateaus, mountain ranges, and minor features such as hills, valleys, slopes, drumlins, or dunes. Taken together, landforms make up the entire surface configuration of the earth.

Landslide. A general term applied to relatively rapid mass movement such as debris flows, slumps, rock slides, etc.

Lateral moraine. A deposit of till along the margins of a valley glacier. Accumulates as a result of mass movement of debris onto the sides of the glacier.

Laterite. A soil rich in oxides of iron and aluminum formed by deep weathering in tropical and subtropical areas.

Laurasia. The original continental landmass comprising what is now Europe, Asia, North America, and Greenland.

Lava. Magma that reaches the earth's surface.

Leach. To separate and remove by dissolving the soluble constituents of a rock.

Leachate. A solution obtained by leaching. Example: Water percolating through a waste disposal site would dissolve certain soluble substances and would contain these substances in solution.

Leading edge (plate tectonics). The oldest edge or margin of a plate which is located furthest from the spreading center. (Compare *trailing edge*, which is the plate margin at the spreading center.)

Lee slope. That part of a hill, dune, or rock that is sheltered or turned away from the wind. Also called *slip face*.

Levee (natural). A broad, low embankment built up along the sides of a river channel during floods.

Limbs. The flanks or sides of a fold.

Limestone. A sedimentary rock composed principally of calcium carbonate.

Liquid. A state of matter that flows freely and lacks a crystalline structure. Unlike a gas, a liquid retains its independent volume.

Lithification. The processes by which sediment is converted into sedimentary rock. Includes cementation and compaction.

Lithosphere. The relatively rigid outer zone of the earth. Includes the continental platform, the oceanic layers, and the part of the mantle above the softer asthenosphere.

Load. The total amount of sediment carried at a given time by a stream, glacier, or wind.

Loess. Unconsolidated, wind-deposited silt and dust.

Longitudinal dune. An elongate sand dune oriented in the direction of the prevailing wind.

Longitudinal wave. A seismic body wave in which particles oscillate along lines in the direction the wave travels. Also called *primary* or *P wave*.

Longshore current. A current in the surf zone moving parallel to the shore. Longshore currents result where waves strike the shore at an angle and push water and sediment obliquely up the beach. The backwash is straight down the beach face. The water and sediment thus follow a zigzag pattern, with a net movement that is parallel to the shore.

Low-grade metamorphism. Metamorphism that is accomplished under conditions of low to moderate temperature and pressure.

L waves. Surface seismic waves distinguished by their long periods relative to P and S waves.

Maar crater. An explosive volcanic crater with little or no cone.

Mafic rock. An igneous rock containing more than 50% ferromagnesian minerals.

Magma. A mobile silicate melt which may contain liquid, suspended crystals, and dissolved gases.

Magmatic differentiation. A general term for the various processes by which early-formed crystals or early-formed liquid is separated and removed to form a rock with a composition different from that of the original magma.

Magnetic reversal. The complete 180° reversal of the polarity of the earth's magnetic field.

Mantle. The zone of the earth's interior between the Moho discontinuity and the core.

Marble. A metamorphic rock which originates from limestone or dolomite.

Maria. The relatively smooth, low, dark areas of the moon, formed by the extrusion of lava.

Mascons. Concentrations of mass in local areas beneath the maria of the moon.

Mass movement. The transfer of material downslope through the direct action of gravity. Also referred to as *mass-wasting*.

Mass-wasting. Movement of material downslope under the pull of gravity without a flowing medium such as a river or glacial ice.

Matrix. Small particles of a rock that occupy the space between larger particles.

Meander. A broad looplike bend in the course of a river.

Mechanical weathering. The breaking down of rock by physical processes such as frost wedging. Synonymous with *disintegration*.

Medial moraine. A ridge of till in the middle of a valley glacier formed by the junction of two lateral moraines where two valley glaciers meet.

Melt. A substance altered from a solid to a liquid state.

Metaconglomerate. A metamorphosed conglomerate.

Metamorphic rock. A major class of rocks formed within the earth's crust by changes in the texture and/or composition as a result of high temperatures, pressure, and fluids.

Meteoric water. Ground water derived principally from the atmosphere.

Meteorites. Particles of solid matter that have fallen to the earth, moon, or other planets from space.

Meander

Medial moraine

Monocline

Mudflow

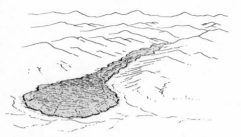

Nonconformity

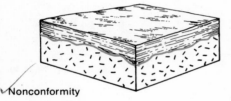

Normal fault

Nuée ardente

Micas. A group of silicate minerals exhibiting perfect cleavage in one direction.

Microcontinent. A relatively small, isolated fragment of continental crust. Example: Madagascar.

Mid-Atlantic ridge. The mountain range extending down the central part of the Atlantic Ocean.

Mid-oceanic ridges. The system of ridges on the ocean floor extending from the Arctic Ocean through the central part of the Atlantic and Indian oceans and along the eastern part of the Pacific.

Migmatite. A mixture of igneous and metamorphic rocks in which thin dikes and stringers of granitic material interfinger with metamorphic rock.

Mineral. A naturally occurring inorganic solid having a definite internal structure and a definite chemical composition which varies only within strict limits. The chemical composition and internal structure produce specific physical properties, including the tendency to assume a specific geometric form (crystal).

Mobile belts. Long, narrow belts of the continents which have been subjected to mountain-building processes.

Mohorovičić discontinuity. The first global seismic discontinuity below the surface. Commonly referred to as the "Moho." Depth varies from about 5 to 10 km beneath the ocean floor to about 35 km below the continents.

Monadnock. An erosional remnant rising above a peneplain.

Monocline. A type of anticline in which the beds on either side of the fold dip uniformly at low angles.

Moraine. A general term for landforms composed of till.

Mountain. A general term for any landmass that stands above its surroundings. In the stricter geologic sense, a mountain belt is a highly deformed part of the earth's crust which has been injected with igneous intrusions and which has had deeper parts metamorphosed. The topography of young mountains is high, but old mountains may be eroded down to a flat lowland.

Mud crack. A crack in a deposit of mud or silt resulting from contraction which accompanies drying.

Mudflow. A flow of a mixture of mud and water.

Nappe. Faulted and overturned folds.

Névé. Granular ice formed by recrystallization of snow. Synonymous with *firn*.

Nodule. An irregular knobby or rounded body that is generally harder than the surrounding rock.

Nonconformity. An unconformity in which stratified rocks rest upon granitic or metamorphic rocks.

Normal fault. A steeply inclined fault in which the hanging wall has moved down relative to the footwall. Also referred to as *gravity fault*.

Nuée ardente. A hot cloud of volcanic fragments and superheated gases which flows as a mass because it is denser than air. Upon cooling it forms a rock called an ash flow tuff or welded tuff.

Obsidian. A glassy igneous rock with a composition equivalent to that of granite.

Ocean basin. The low part of the lithosphere that lies between continental masses. The rocks are dominantly basalt with a veneer of oceanic sediment.

Oceanic crust. The type of crust that underlies the ocean basin. It is

554

Glossary

about 5 km thick, composed predominantly of basalt. It has a density of 3.0 g/cm³, and compressional seismic-wave velocities traveling through it exceed 6.2 km/sec. (Compare with *continental crust*.)

Oceanic rise (ridge). The continuous ridge or broad fractured topographic swell which extends through the central part of the Arctic, Atlantic, Indian, and South Pacific oceans. It is several hundred kilometers wide and has a relief of 600 m or more. It is thus a major structural and topographic feature of our planet.

Offshore. The area from low tide seaward.

Oil shale. Shale rich in hydrocarbon derivatives. In the U.S., the chief oil shale is the Green River Formation of the Rocky Mountain region.

Oolite. A limestone consisting largely of spherical grains of calcium carbonate having concentric spherical layers.

Ooze. Marine sediment consisting of more than 30% shell fragments of microscopic organisms.

Orogenic. Pertaining to deformation of the continental margins to the extent that mountain ranges are formed.

Orogenic belt. A mountain belt.

Orogeny. The processes of mountain-building.

Outcrop. An exposure of bedrock.

Outwash. Stratified sediment "washed out" from a glacier by meltwater streams and deposited in front of the terminal moraine.

Outwash plain. The area beyond the margin of a glacier where melt water deposits glacier-derived sand, gravel, and mud.

Overturned fold. A fold in which at least one limb has been rotated through more than 90°.

Oxbow lake. A lake formed in the channel of an abandoned meander.

Oxidation. The chemical combination of oxygen with a mineral.

Pahoehoe. Lava with a billowy or ropy surface. (Contrast with *aa*.)

Paleocurrent. An ancient current that existed in the geologic past and whose direction can be inferred from cross-bedding, ripple marks, and other sedimentary structures.

Paleogeography. The study of geography in the geologic past, including the patterns of the earth's surface, distribution of land and sea, ancient mountains, etc.

Paleomagnetism. The study of the earth's magnetic field during geologic time.

Paleontology. The study of ancient life.

Paleowind. An ancient wind in the geologic past. Its direction can be inferred from patterns of ancient ash falls, orientation of crossbedding, and growth rates of colonial corals.

Pangaea. A hypothetical continent from which the present continents originated through processes of drifting from the Mesozoic era to the present.

Parabolic dune. A dune shaped like a parabola with the concave side toward the wind.

Passive margins (tectonics). Margins of lithospheric plates in which crust is neither created nor destroyed; generally marked by transform faults.

Peat. An accumulation of partly reduced plant material containing approximately 60% carbon and 30% oxygen. Peat is the intermediate material in the process of coal formation.

Pebble size. Sediment particles with a diameter ranging from 2 mm to 64 mm (about the size of a match head to the size of a tennis ball).

Pediment. A gently sloping erosional surface developed as a mountain

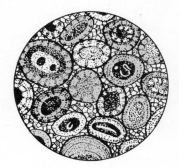

Oolite

Overturned fold

Pediment

Peneplain

Plateau basalt

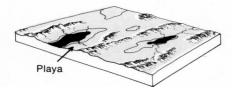

Playa

Point bar

front or cliff recedes. A pediment surface cuts across bedrock and may be covered with a veneer of sediment. Forms in arid to semiarid climates.

Peneplain. An extensive erosional surface worn down almost to sea level.

Peninsula. An elongate body of land extending out into the water.

Perched water table. A local zone of saturation above the regional water table.

Peridotite. A dark, coarse-grained igneous rock composed of olivine, pyroxene, and some other ferromagnesian minerals, but essentially no feldspar and no quartz.

Permafrost. Permanently frozen ground.

Permeability. The ability of a material to transmit fluid.

Phaneritic. A texture of igneous rock in which the interlocking crystals are large enough to be seen without aid of magnification.

Phenocryst. A crystal that is significantly larger than those that surround it. It forms during an early slow-cooling stage of the magma.

Physiographic map. A map showing surface features.

Physiography. A study of the surface features and landforms of the earth.

Pillar. A landform shaped like a pillar.

Pillow lava. An ellipsoidal mass of igneous rock formed by extrusion of lava under water.

Pinnacle. A tall tower or spire-shaped pillar of rock.

Placer. A mineral deposit formed by the sorting or washing action of water. Usually a heavy mineral such as gold.

Plagioclase. A group of feldspar minerals having a composition ranging from $NaAlSi_3O_8$ to $CaAl_2Si_2O_8$.

Plastic deformation. Permanent change in the shape or volume that does not involve failure by rupture.

Plateau. An extensive upland region.

Plateau basalt. Extensive layers of nearly horizontal basalt that tend to erode into great plateaus. Synonymous with *flood basalt*.

Plates (tectonics). Broad segments of the lithosphere (may include the rigid upper mantle, continental crust, and/or oceanic crust) which "float" on the underlying asthenosphere and move independently of other plates.

Plate tectonics. The theory of global dynamics in which the lithosphere is believed to be broken into individual plates which move in response to convection in the upper mantle. The margins of the plates are the sites of considerable activity.

Playa. A depression in the center of a desert basin, the site of occasional playa lakes.

Playa lake. A shallow temporary lake formed in a desert basin after a rain.

Plucking (glacial). The process of glacial erosion by which large rock fragments are loosened, detached, and transported by freezing of the melt water along fractures and bedding planes.

Plunging folds. Folds whose axes are inclined.

Plutonic rock. Igneous rock formed beneath the surface.

Pluvial lake. A lake that formed during a former climate when rainfall in the region was higher than at present.

Point bar. A crescent-shaped accumulation of sand and gravel deposited on the inside of a meander bend.

Polar wandering. Movement of the magnetic pole relative to the continents.

Glossary

Polarity epoch. A period of time in which the earth's magnetic field has been oriented in either a normal or a reverse direction.

Polarity event. A shorter interval of opposite polarity within a polarity epoch.

Pore fluid. Fluids in the pore space of a rock; can be ground water or liquid rock material resulting from partial melting.

Pore space. The space in a rock body unoccupied by solid material. May be space between grains, fractures, or voids formed by solution.

Porosity. The percentage of pore space within a rock or sediment.

Porphyritic. A texture of igneous rock in which some crystals are distinctly larger than others; phenocrysts.

Pothole. A hole in a stream bed formed by sand and gravel swirled around in one spot by eddy action.

Pratt hypothesis. A hypothesis which explains isostasy by assuming greater crustal density under mountains than under oceans.

Primary coast. Coasts shaped by subaerial erosion, deposition, vulcanism, or tectonic activity.

Pumice. A rock consisting of frothy natural glass.

P waves. Primary seismic waves. Waves in the earth which are propagated like sound waves. The material involved in the wave motion is alternately compressed and expanded.

Pyroclastic. Fragments of volcanic debris.

Pyroxene. A group of silicate minerals with a single chain of silica tetrahedra. Compare with amphibole which has a double chain.

Quartz. An important rock-forming silicate mineral composed of silica tetrahedra joined in a three-dimensional network. Distinguished by its hardness, glassy luster, and conchoidal fracture.

Quartzite. A metamorphosed sandstone.

Radar imagery. An image produced on photographic material by radar energy.

Radial drainage. A stream pattern in which the streams radiate outward from a central zone.

Radioactive dating. Calculating the age in years for minerals by measuring the ratios of the original material to the decayed product. Synonymous with *radiometric dating.*

Radioactivity. The spontaneous disintegration of an atomic nucleus with the emission of radiant energy.

Radiocarbon. The radioactive isotope of carbon, C^{14}, which is formed in the atmosphere and is circulated throughout living matter.

Radiogenic heat. Heat generated by the process of radioactive decay.

Ray craters. Lunar craters which have a system of rays extending like splash marks from the crater rim.

Recessional moraine. A ridge of till deposited at the end of a glacier during a period of temporary stability in its general recession.

Recharge. Addition of water to the ground-water reservoir.

Recrystallization. Reorganization of elements of the original minerals in a rock as a result of heat, pressure, and pore fluid.

Recumbent fold. A fold in which the axial plane is essentially horizontal.

Reef. A solid structure built by shells and other secretions of marine organisms.

Regolith. Soil and loose rock fragments which overlie the bedrock.

Rejuvenated. A change in a regimen in which more active erosion occurs.

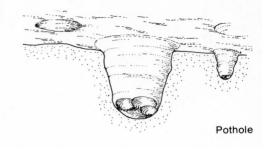

Pothole

Rayed crater

Recumbent fold

Reverse fault

Roche moutonnée

Saltation

Sea arch

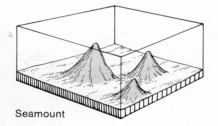

Seamount

Relative age. The age of a rock or event as compared to some other event.

Relative time. Dating by means of arranging events in their proper chronologic order. (Compare with *absolute time.*)

Relief. The difference in altitude between the high and low parts of an area.

Reverse fault. A fault in which the hanging wall has moved up relative to the footwall. A high-angle thrust fault.

Rift system. A system of fractures and faults in the earth resulting from extension.

Rift valley. A valley formed by block faulting in which tensional stresses tend to pull the crust apart. A graben.

Rill. A very small stream of trickling water.

Rille. An elongate trench or cracklike valley on the moon's surface. Rilles may be extremely irregular and meandering or relatively straight structural depressions.

Ripple mark. Small waves produced in sand or mud by the effects of drag of moving wind or water.

River system. A river with all its tributaries.

Roche moutonnée. An abraded knob of bedrock formed by an over-riding glacier. They are typically striated and have a gentle slope facing the upstream direction of ice movement.

Rock. An aggregate of minerals which form an appreciable part of the lithosphere.

Rock fall. The most rapid type of landslide, ranging from large masses of rock to small fragments loosened from the face of a cliff.

Rock glacier. A mass of poorly sorted angular boulders cemented with interstitial ice. It moves slowly through the action of gravity.

Rock slide. A landslide involving a sudden and rapid movement of a newly detached segment of bedrock sliding over an inclined surface of weakness such as a joint or bedding plane.

Runoff. Water that flows over the surface.

Saltation. The transportation of particles in a current of wind or water by movement through a series of bounces.

Salt dome. A dome in sedimentary rock produced by the upward movement of a body of salt.

Sand. Fragments ranging in size from 0.0625 mm to 2 mm in diameter. Much sand is composed of quartz grains because quartz is abundant and resists chemical and mechanical disintegration. Other material, such as shell fragments and rock fragments, may form sand grains.

Sandstone. A sedimentary rock composed mostly of sand-size particles cemented usually by calcite, silica, or iron oxide.

Sand wave. A wave of sand created by the effects of drag of air or water moving over the surface. Includes dunes and ripple marks.

Scarp. A cliff produced by faulting or erosion.

Schist. A metamorphic rock containing strong foliation as a result of parallel orientation of platy minerals.

Scoria. An igneous rock containing abundant vesicles.

Sea arch. An arch cut by wave erosion through a headland.

Sea cave. A cave formed by wave erosion.

Sea cliff. A cliff produced by wave erosion.

Sea-floor spreading. The theory that the sea floors spread laterally away from the mid-oceanic ridges as new lithosphere is created along the crest of the ridges by igneous activity.

Seamount. An isolated conical mound rising more than 1000 m above

Glossary

the floor of the ocean. Seamounts are probably submerged shield volcanoes.

Sea stack

Sea stack. An isolated pillarlike rocky island near a cliffy shore, detached from the mainland by wave erosion.

Secondary coast. A coast formed by marine processes or the growth of organisms.

Sediment. Material that has been transported and deposited by wind, water, ice, or gravity. Also includes materials precipitated from solution and deposits of organic origin such as coal and coral reefs.

Sedimentary rock. Rocks formed from the accumulation of sediment.

Seep. A spot where water oozes from the earth.

Seif dune. A longitudinal dune of great height and length.

Seismic. Pertaining to waves produced by natural or artificial earthquakes.

Seismic discontinuity. A physical interface within the earth, detected by seismic studies.

Seismograph. An instrument which records seismic waves.

Shale. A fine-grained clastic sedimentary rock formed by the consolidation of clay and mud.

Sheeting. A set of joints essentially parallel to the surface which allows layers to spall off when the rock is exposed by erosion; especially well developed in granitic rock.

Shield. Large areas where igneous and metamorphic rocks are exposed and have approached equilibrium with respect to erosion and isostasy.

Shield volcano. A large volcano built up almost entirely of lava. Slopes seldom exceed 10° so the profile resembles a shield or broad dome.

Shore. The zone between high and low tide; a narrow strip of land immediately bordering any body of water, especially lakes or seas.

Sial. A general term for the silica-rich rocks which form the continental masses.

Sill. A tabular body of intrusive rock injected between layers of the enclosing rock.

Siltstone. A fine-grained clastic rock in which particles range from 1/16 to 1/256 mm in diameter.

Sima. A term referring to magnesium-rich igneous rocks (basalt, gabbro, and peridotite) of the ocean basins.

Sink hole. A depression formed by collapse of a cavern roof.

Slate. A fine-grained metamorphic rock with foliation resulting from the parallel arrangement of microscopic platy minerals such as mica.

Slip face. See *lee slope.*

Slope retreat. The progressive recession of a scarp or side of a hill or mountain by mass movement and stream erosion.

Slump. A type of mass movement in which material moves along a curved surface or rupture.

Soil. The surface material of the continents, produced by rock disintegration and decomposition as well as organic processes. Regolith that has undergone chemical weathering in place.

Soil profile. A vertical section of soil which displays all of its horizons and parent material.

Solid. Matter with a definite shape and volume and some fundamental strength.

Solifluction. Mass movement in which material moves slowly downslope in areas where the soil is saturated with water. Common in permafrost areas.

Sink hole

Solution. Chemical weathering by dissolving soluble minerals.

Spheroidal weathering

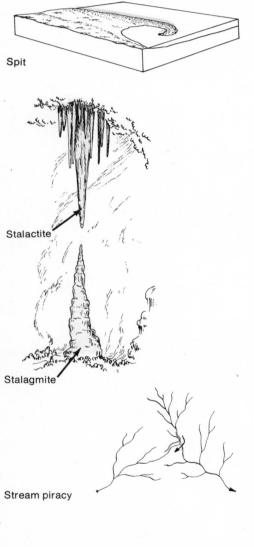

Spit

Stalactite

Stalagmite

Stream piracy

Strike-slip fault

Solution valley. A valley produced by solution activity either by dissolving surface materials or by removal of subsurface material such as limestone, gypsum, or salt.

Sorting. The separation of particles according to size, shape, or weight. Sorting occurs during transportation by running water or wind.

Specific gravity. The weight of a substance compared to the weight of an equal volume of water.

Spheroidal weathering. The tendency for a rock surface to become rounded as it weathers.

Spit. A sandy bar projecting from the mainland or promontory into open water.

Splay. A small delta formed when water and sediment are diverted through a crevasse in a levee.

Spring. A place where water seeps naturally to the surface.

Stack. A small island just offshore, formed by wave erosion of a headland.

Stalactite. An iciclelike deposit of dripstone hanging from the roof of a cave.

Stalagmite. A conical deposit built up from the cave floor.

Star dune. A mound of sand with a high central point and arms radiating in various directions.

Stock. A small, roughly circular intrusive body usually less than 100 km² in surface exposure.

Strata. Layers of sedimentary rock.

Stratification. The layered structure of sedimentary rock.

Strato-volcano. A volcano built of alternating layers of ash and lava flows. Synonymous with *composite volcano*.

Streak. The color of a powdered mineral.

Stream piracy. The diversion of the headwaters of one stream into another stream. The process is accomplished by headward erosion of a stream having greater erosional activity.

Stress. Force applied to a material that tends to change the material's dimensions or volume.

Striations. Scratches or grooves on a bedrock surface.

Strike. The bearing (compass direction) of a horizontal line on a plane (bedding plane, fault plane, etc.)

Strike-slip fault. A fault in which the movement has been parallel to the strike of the fault.

Strike valley. A valley eroded parallel to the strike of the underlying strata.

Strip mining. A method of mining in which soil and rock cover are removed to obtain the sought-after material.

Subaerial. Occurring beneath the atmosphere or in the open air. Conditions or processes such as erosion that operate on the land. (Contrast with submarine or subterranean.)

Subduction. The subsidence of the leading edge of a lithospheric plate down into the mantle.

Submarine canyon. A steep-sided, V-profile trench or valley cut into the continental shelf or slope.

Subsequent stream. A tributary stream eroded along a belt of weak, nonresistant rock.

Subsidence. A sinking of a part of the earth relative to its surrounding parts. Synonymous with *sinking*.

Succession (in landscape development). A principle of relative dating that states that landscapes develop in a definite order.

Superposed stream. A stream whose course was established on young

Glossary

rocks and subsequently cut down into whatever rocks happen to underlie its course. In this way the stream pattern is superposed, or placed
upon, buried ridges or other structural features.

Superposition, Law of. The principle which states that, in a series of
sedimentary rocks that have not been overturned, the oldest rocks are
at the base and the youngest at the top.

Swash. Rush of water up onto the beach after a wave breaks.

S waves (secondary waves). A seismic wave in which energy vibrates
at right angles to the direction the wave travels. (Contrast with
P wave.)

Symmetrical fold. A fold in which the two limbs are essentially mirror
images of each other.

Syncline. A fold in which the limbs dip toward the axis and the youngest beds are in the central part or core.

Superposed stream

Talus. Rock fragments that accumulate in a pile at the base of a ridge
or cliff.

Tectonic. A term referring to large structure features of the earth.

Tension. Stress that tends to pull material apart.

Tephra. A general term referring to all pyroclastic material ejected
from a volcano. Includes ash, dust, bombs, etc.

Terminal moraine. A ridge of glacier-deposited material that accumulates at the point of maximum advance of the glacier.

Terrace. A nearly level surface bordering a steeper slope (stream terrace, wave-cut terrace).

Terrae. The rugged, light-toned highlands of the moon.

Texture. The size, shape, and arrangement of the particles that make
up a rock.

Thin section. A slice of rock mounted on a glass slide and ground about
30 microns thick.

Thrust fault. A low-angle fault (45° or less) in which the hanging wall
has moved up relative to the footwall. Horizontal compression rather
than vertical displacement is characteristic.

Tidal bore. A violent rush of tidal water.

Till. Unsorted and unstratified glacial deposits.

Tillite. A rock formed by lithification of glacial till (unsorted, unstratified glacial sediment).

Tombolo. A beach or bar that connects an island to the mainland.

Topography. The shape and form of the earth's surface.

Transform fault. A strike-slip fault that offsets an active spreading
center or subduction zone.

Transpiration. The process by which water vapor is released into the
atmosphere by plants.

Transverse dune. An asymmetrical dune ridge formed perpendicular
to the prevailing winds.

Travertine terrace. A terrace formed from calcium carbonate that has
been deposited from water on a cave floor.

Trellis pattern. A drainage pattern in which the tributaries are arranged
in a rectilinear pattern similar to a garden trellis.

Trench (marine geology). A narrow, elongate depression of the deep-
sea floor oriented parallel to trends of continents or island areas.

Tributary. A stream flowing into or joining another larger stream.

Tsunami. A long, low wave in the ocean developed by earthquakes,
faulting, or landslides on the sea floor. Velocity may be up to 600 km
per hour. Commonly misnamed a tidal wave.

Tuff. A fine-grained rock composed of volcanic ash.

Talus

Terrace

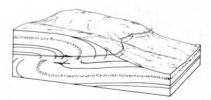

Thrust fault

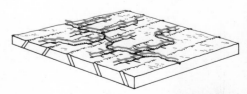

Trellis pattern

561

Glossary

Ventifact

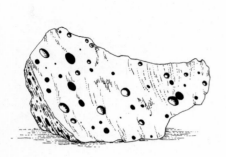

Vesicles

Volcanic neck

Water table

Turbidity current. A current generated by turbid (muddy) water that moves relative to the surrounding water because of its greater density.

Turbulent flow. Fluid flow in which the path of motion is very irregular, with eddies and swirls.

Ultimate base level. The lowest level to which a stream can erode the earth's surface. Sea level.

Ultramafic rock. An igneous rock composed entirely of ferromagnesian minerals.

Unconformity. A discontinuity in the succession of rocks in which there is a gap in the geologic record. A buried erosional surface. (See *angular unconformity*, *disconformity*, and *nonconformity*.)

Uniformitarianism. The theory that the earth is a result of slowly acting natural processes, many of which are operating at the present time.

Upwarp. An arched or uplifted segment of the crust.

Valley glacier. A glacier confined to a stream valley. Synonymous with *alpine glacier* or *mountain glacier*.

Varve. A pair of thin sedimentary layers, one relatively coarse and light colored and the other fine and dark as the result of deposition of one year in a lake; the coarse layer formed during spring runoff and the fine during the winter when the lake was frozen.

Ventifact. A pebble or cobble shaped and polished by the wind.

Vesicles. Small holes in volcanic rock formed by gas bubbles which became trapped as the lava cooled.

Viscosity. The tendency within a body to oppose flow.

Volcanic ash. Dust-size particles ejected from a volcano.

Volcanic neck. The solidified lava filling the vent or neck of an ancient volcano. Exposed by erosion.

Vulcanism. A term referring to the processes asssociated with the transfer of material from the earth's interior to its surface.

Wash. Dry stream bed.

Water gap. A pass in a ridge through which a stream flows.

Water table. The upper surface of the zone of saturation.

Wave base. The lower limit of wave transportation and erosion (equal to half the wave length).

Wave-built terrace. A terrace built by wave-washed sediments. Usually lies seaward of a wave-cut terrace.

Wave crest. The highest part of a wave.

Wave-cut cliff. A cliff along the coast formed by undercutting of waves and currents.

Wave-cut terrace or platform. A terrace cut across bedrock by wave erosion.

Wave height. The vertical distance between a wave crest and the preceding trough.

Wave length. The horizontal distance between similar points on two successive waves measured perpendicularly to the crest.

Wave period. The time a wave crest takes to travel a distance equal to one wave length; the time for two successive wave crests to pass a fixed point.

Wave refraction. The process by which a wave is turned from its original direction as it approaches shore. The part of the wave advancing in shallower water moves more slowly than the part moving in deeper water.

Wave trough. The lowest part of a wave between successive crests.

Glossary

Weathering. The chemical and mechanical breakdown of rock.

Welded tuff. A volcanic ash hot enough so that the particles became fused together.

Wind gap. A gap in a ridge through which a stream used to flow, abandoned as a result of stream piracy.

Wind shadow. The area behind an obstacle where air movement is not capable of moving material.

Wrinkle ridges. Ridges on the lunar maria.

Zone of aeration. The zone below the surface and above the water table in which pore space is normally filled with air.

Zone of saturation. The zone in the subsurface in which all openings are filled with water.

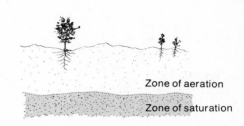

Zone of aeration

Zone of saturation

Cover photo — Skylab photograph of part of the Colorado Plateau taken from 270 miles above the surface shows the Colorado River as it flows through Lake Powell (right), circles the Painted Desert (center), and plunges southwest into Marble Canyon. Skylab Photo, NASA, U.S. Dept. of Agriculture—ASCS Western Aerial Photo Lab, Salt Lake City.

Title page photo — W. K. Hamblin.

Opening page photo, Chapter 1 — Courtesy NASA.

Figure 1.1 — Courtesy National Space Data Center, NASA.

Figure 1.2 — ERTS Photo, NASA, U.S. Dept. of Agriculture—ASCS Western Aerial Photo Lab, Salt Lake City.

Figures 1.3, 1.4, 1.5 — Skylab Photo, NASA, U.S. Dept. of Agriculture—ASCS Western Aerial Photo Lab, Salt Lake City.

Figure 1.6 — ERTS Photo, NASA, U.S. Dept. of Agriculture—ASCS Western Aerial Photo Lab, Salt Lake City.

Figure 1.7 — Courtesy NASA, Technology Application Center, University of New Mexico.

Figure 1.8 — Courtesy Amman Aerial Survey.

Figure 1.9 — Courtesy NASA, Technology Application Center, University of New Mexico.

Figure 1.10 — ERTS Photo, NASA, U.S. Dept. of Agriculture—ASCS Western Aerial Photo Lab, Salt Lake City.

Figure 1.11 — Courtesy NASA, Technology Application Center, University of New Mexico.

Figure 1.12 — ERTS Photo, NASA, U.S. Dept. of Agriculture—ASCS Western Aerial Photo Lab, Salt Lake City.

Figures 1.13, 1.14, 1.15 — Courtesy NASA, Technology Application Center, University of New Mexico.

Figure 1.16 — ERTS Photo, NASA, U.S. Dept. of Agriculture—ASCS Western Aerial Photo Lab, Salt Lake City.

Figure 1.17 — National Air Photo Library, Dept. of Energy, Mines, and Resources, Canada.

Figures 1.18, 1.19, 1.20, 1.21 — Courtesy NASA, Technology Application Center, University of New Mexico.

Figure 1.22 — After profile records of Mid-Atlantic Ridge at 44° N Lat., from A. E. Maxwell (ed.), 1970, The Sea, Vol. 4, pp. 11-36, Wiley Interscience, New York.

Figure 1.24 — After profile records of Murray Fracture zone, from A. E. Maxwell (ed.), 1970, The Sea, Vol. 4, Part II, p. 111, Wiley Interscience, New York.

Figure 1.25 — After profile records, from H. W. Menard, 1964, Marine Geology of the Pacific, McGraw-Hill, New York.

Figure 1.26 — After profile records of Kane 9, U.S. Naval Oceanographic Office, U.S. Government Printing Office, Washington, D.C.

Figure 1.28 — After profile records, from A. E. Maxwell (ed.), 1970, The Sea, Vol. 4, p. 40, Wiley Interscience, New York.

Figure 1.29 — After profile records, National Science Foundation, 1971, Deep Sea Drilling Project, Vol. VI, NSFSP-6, U.S. Government Printing Office, Washington, D.C.

Figure 1.30 — After profile records of Kane 9, U.S. Naval Oceanographic Office, U.S. Government Printing Office, Washington, D.C.

Opening page photo, Chapter 2 — W. K. Hamblin.

Figure 2.1 — Modified after J. H. Zumberge, 1972, Elements of Geology, 3rd ed., Wiley Interscience, New York.

Figure 2.2 — Modified after A. N. Strahler, 1971, The Earth Sciences, p. 386, Harper and Row, New York.

Figure 2.3 — Modified after R. H. Dott and R. L. Batten, 1971, Evolution of the Earth, McGraw-Hill, New York.

Opening page photo, Chapter 3 — W. K. Hamblin.

Figure 3.3a — After B. Isacks, J. Oliver, and L. R. Sykes, 1968, "Seismology and the New Global Tectonics," J. Geophys. Res. 73:5855-99.

Figure 3.3b — After E. Bullard, 1969, "The Origin of the Oceans," Sci. Amer. 221(3):66-75.

Figure 3.5 — After D. M. Crittenden, 1963, New Data on the Isostatic Deformation of Lake Bonneville, U.S. Geol. Survey Prof. Paper 454-E, p. 9.

Opening page photo, Chapter 4 — W. K. Hamblin.

Figure 4.1 — Courtesy Kent Dalley.

Figure 4.21 — U.S. Geological Survey.

Figures 4.27, 4.28 — National Air Photo Library, Dept. of Energy, Mines, and Resources, Canada.

Opening page photo, Chapter 5 — W. K. Hamblin.

Figure 5.3 — Modified after D. L. Eicher, 1968, Geologic Time, Prentice-Hall, Englewood Cliffs, New Jersey.

Opening page photo, Chapter 6 — U.S. Geological Survey.

Figure 6.11 — After N. M. Strahkov, 1967, Principles of Lithogenesis, Vol. 1, trans. J. P. Fitzemms, Oliver and Boyde, Edinburgh.

Opening page photo, Chapter 7 — U.S. Geological Survey.

Figure 7.1a — ERTS Photo, NASA, U.S. Dept. of Agriculture—ASCS Western Aerial Photo Lab, Salt Lake City.

Figures 7.1b and c — U.S. Geological Survey.

Figure 7.6 — U.S. Geological Survey.

Figure 7.10 — After K. Davis and L. Leopold, 1970, Water, Time-Life Books, New York.

Figure 7.12 — After Straub, L., in Meinzer, O. E. (ed.), 1942, Hydrology, p. 625, McGraw-Hill, New York.

Figure 7.13 — Courtesy Swiss Air Aerial Photo.

Figure 7.14 — After Hjulström, 1935, Studies of the Morphological Activities of Rivers as Illustrated by the River Fyris: Upsala University, Geol. Inst. Bull., V. 25, pp. 221-527.

Figure 7.16 — Courtesy U.S. Forest Service.

Opening page photo, Chapter 8 — U.S. Geological Survey.

Figure 8.6 — U.S. Geological Survey.

Figure 8.9 — Modified after Varnes, 1958, Landslide Types and Processes, Ch. 3, in E. B. Eckel (ed.), Landslides and Engineering Practices, Highway Research Board, Special Report 29, NAS-NRD 544, Washington, D.C.

Figure 8.15 — U.S. Geological Survey.

Figure 8.18 — After H. N. Pollack, 1969, A Numerical Model of Grand Canyon, in Four Corners Geol. Soc. Guidebook, p. 62.

Figure 8.19 — U.S. Geological Survey.

Figure 8.24 — U.S. Corps of Engineers, Vicksburg, Mississippi.

Figure 8.25 — U.S. Geological Survey.

Figure 8.27 — U.S. Geological Survey.

Figure 8.28 — Modified after Welder, 1959, Processes of Deltaic Sedimentation in the Lower Mississippi River, Louisiana State University Coastal Studies Inst. Tech. Report 12.

Figure 8.30 — Modified after Kolb and Lopik, 1966, Depositional environments of the Mississippi River Coastal Plan, U.S. Army Corps of Engineers Waterway Dept. Sta. Tech. Report 3-483 and 3-484.

Figure 8.33 — U.S. Geological Survey.

Opening page photo, Chapter 9 — U.S. Geological Survey.

Figure 9.2 — U.S. Geological Survey.

Figure 9.3 — National Air Photo Library, Dept. of Energy, Mines, and Resources, Canada.

Figures 9.7, 9.8 — ERTS Photo, NASA, U.S. Dept. of Agriculture—ASCS Western Aerial Photo Lab, Salt Lake City.

Figure 9.9 — After W. K. Hamblin and J. D. Howard, 1971, Physical Geology Laboratory Manual, 3rd ed., Burgess Publishing Company, Minneapolis.

Figure 9.21 — Modified after J. Shelton, 1966, Geology Illustrated, W. H. Freeman and Company, San Francisco.

Figure 9.22 — After A. N. Strahler, 1951, Physical Geography, 1st ed., John Wiley and Sons, New York.

Figure 9.25 — Modified after J. G. Vedder and R. E. Wallace, 1970, U.S. Geological Survey Map I-574.

Opening page photo, Chapter 10 — U.S. Geological Survey.

Figure 10.1 — Modified after A. N. Sayre, 1950, "Ground Water," Sci. Amer. 183(5):14-19.

Figure 10.9 — Modified after W. R. Keefer, 1972, The Geologic Story of Yellowstone National Park, U.S. Geol. Survey Bull. 1347.

Figure 10.10 — After W. W. Varnedoe and R. W. Faurbridge (eds.), in Encyclopedia of Geomorphology, 1968, Van Nostrand Reinhold Company, New York.

Figure 10.11 — After W. K. Hamblin and J. D. Howard, 1971, Physical Geology Laboratory Manual, 3rd ed., Burgess Publishing Company, Minneapolis.

Figure 10.12 — U.S. Geological Survey.

Figure 10.15 — Courtesy Onadaga Cave, Missouri.

Figure 10.16 — Modified from W. J. Schneider, 1970, Hydrologic Implications of Solid Waste Disposal, U.S. Geol. Survey Circ. 601-F.

Figure 10.18 — Modified after F. Ward, 1972, "The Imperiled Everglades," National Geographic 141(1):1.

Opening page photo, Chapter 11 — Dept. of Energy, Mines, and Resources, Canada.

Figure 11.3 — National Air Photo Library, Dept. of Energy, Mines, and Resources, Canada.

Figure 11.5 — Lake Gillian and Conn Lake topographic maps, N.W. Territories, Geological Survey of Canada.

Figure 11.6 — Courtesy J. D. Ives.

Figure 11.7 — After Atwood, 1940, Physiographic Provinces of North America, Ginn and Company, Boston.

Figure 11.11 — After R. F. Flint, 1971, Glacial and Quaternary Geology, John Wiley and Sons, New York.

Figure 11.12 — U.S. Geological Survey.

Figure 11.13 — Sketched from a photo by J. D. Ives.

Figure 11.15 — National Air Photo Library, Dept. of Energy, Mines, and Resources, Canada.

Figure 11.16 — After A. N. Strahler, 1951, Physical Geography, 1st ed., John Wiley and Sons, New York.

Figure 11.17 — National Air Photo Library, Dept. of Energy, Mines, and Resources, Canada.

Figure 11.19 — Compiled from Glacial Map of the U.S. east of the Rocky Mountains, the Geological Society of America. Base map courtesy of E. R. Raisz.

Figure 11.20 — After R. F. Flint, 1971, Glacial and Quaternary Geology, John Wiley and Sons, New York.

Figure 11.22 — Compiled from R. J. W. Douglas, 1970, Geology and Economic Minerals of Canada, p. 691, Dept. of Energy, Mines, and Resources, Geological Survey of Canada, Ottawa, and R. F. Flint, Glacial and Quaternary Geology, p. 234, John Wiley and Sons, New York.

Figure 11.24 — National Air Photo Library, Dept. of Energy, Mines, and Resources, Canada.

Figure 11.25 — After J. L. Hough, 1958, Geology of the Great Lakes, University of Illinois Press, Urbana, and R. J. W. Douglas, 1970, Geology and Economic Minerals of Canada, pp. 714-725, Dept. of Energy, Mines, and Resources, Geological Survey of Canada, Ottawa.

Figure 11.27 — After E. Fromm, 1955, Atlas Over Sverige, Suomen Geol. Seura, Vol. 4, No. 7, pp. 33-34.

Figure 11.28 — After R. F. Flint, 1971, Glacial and Quaternary Geology, p. 447, John Wiley and Sons, New York. Base map courtesy of E. R. Raisz.

Figure 11.30 — U.S. Geological Survey.

Opening page photo, Chapter 12 — W. K. Hamblin.

Figure 12.7 — U.S. Dept. of Agriculture—ASCS Western Aerial Photo Lab, Salt Lake City.

Figure 12.8 — After W. Bascom, 1960, "Beaches," Sci. Amer. 203(2):81-94.

Figure 12.13 — Courtesy Iceland Tourist Bureau.

Figure 12.15 — U.S. Dept. of Agriculture—ASCS Western Aerial Photo Lab, Salt Lake City.

Figure 12.17 — U.S. Geological Survey.

Opening page photo, Chapter 13 — Courtesy Aramco.

Figure 13.1 — After A. N. Strahler, 1969, Physical Geography, 3rd ed., p. 162, John Wiley and Sons, New York.

Figure 13.2 — U.S. Geological Survey.

Figure 13.5 — After R. A. Bagnold, 1941, The Physics of Blown Sand and Desert Dunes, p. 36, Methuen Publishing Company, London.

Opening page photo, Chapter 14 — W. K. Hamblin.

Figures 14.2a, 14.3a — U.S. Geological Survey.

Figure 14.4a — Courtesy Aramco.

Figure 14.5a — Courtesy J. D. Ives.

Figure 14.6a — Courtesy Swiss Air.

Figure 14.8a — U.S. Geological Survey.

Figure 14.10a — Courtesy J. K. Rigby.

Figures 14.18, 14.19 — W. K. Hamblin, 1961, Paleogeographic Evolution of Lake Superior Region from Late Keweenawan to Late Cambrian Time, Geol. Soc. Amer. Bull., V. 72, p. 1.

Figures 14.20, 14.21, 14.22 — W. K. Hamblin, 1969, Marine Paleocurrent Directions in Limestones of the Kansas City Group (Upper Pennsylvanian) in Eastern Kansas, State Geol. Survey of Kansas Bull. 194, Part 2.

Figure 14.24 — After P. B. King, 1959, The Evolution of North America, Princeton University Press, Princeton, New Jersey.

Opening page photo, Chapter 15 — U.S. Geological Survey.

Figure 15.1 — After Wegener, 1915, Origins of Continents and Oceans, Figure 1, Dover, New York. (Paperback, S1708, English translation of 4th ed., 1929).

Figure 15.2 — After P. M. Hurley, 1968, "The Confirmation of Continental Drift," Sci. Amer. 218(4):52-64.

Figure 15.3 — Modified after Takeuchi et al., 1970, Debate about the Earth, p. 46, Freeman, Cooper and Company, San Francisco.

Figure 15.6 — After American Association of Petroleum Geologists, 1928, Theory of Continental Drift, A Symposium, Figure 2, Tulsa, Oklahoma.

Figure 15.8 — After A. Cox, G. B. Dalrymple, and R. R. Doell, 1967, "Reversals of the Earth's Magnetic Field," Sci. Amer. 216(2):44-54.

Figure 15.10 — After J. Heirtzler et al., 1966, Deep Sea Research, Vol. 13, pp. 427-443.

Figure 15.13 — After I. G. Gass et al., Understanding the Earth, 1971, p. 308, MIT Press, Cambridge, Massachusetts.

Figure 15.17 — After B. Isacks et al., 1968, "Seismology and the New Global Tectonics," J. Geophys. Res. 73:5855-99.

Figure 15.18 — After E. Bullard, 1969, "The Origin of the Oceans," Sci. Amer. 221(3):66-75.

Figure 15.23 — After E. Bullard, 1969, "The Origin of the Oceans," Sci. Amer. 221(3):66-75.

Figures 15.24, 15.25, 15.26, 15.27, 15.28, 15.29 — After R. S. Dietz, and J. C. Holden, 1970, "The Breakup of Pangaea," Sci. Amer. 223(4):30-41.

Opening page photo, Chapter 16 — U.S. Geological Survey.

Figure 16.3 — After I. G. Gass et al., Understanding the Earth, 1971, p. 310, MIT Press, Cambridge, Massachusetts.

Figure 16.5a — From Map of World Seismicity 1961-1969, National Earthquake Information Center, Washington, D.C.

Figure 16.5b — After B. Isacks et al., 1968, "Seismology and the New Global Tectonics," J. Geophys. Res. 73:5869.

Figure 16.7 — After H. W. Menard, 1946, Marine Geology of the Pacific, p. 68, McGraw-Hill, New York.

Figure 16.10 — Compiled from data in F. P. Shepard, 1963, Marine Geology, pp. 321-22, Harper and Row, New York.

Opening page photo, Chapter 17 — National Air Photo Library, Dept. of Energy, Mines, and Resources, Canada.

Figure 17.6 — National Air Photo Library, Dept. of Energy, Mines, and Resources, Canada.

Figure 17.10 — After R. S. Deitz, 1963, "Collapsing Continental Rises," J. of Geology 13(6):324.

Figure 17.12a — After R. S. MacColl, 1964, Geochemical and Structural Studies in Batholithic Rock of Southern California, Geol. Soc. Amer. Bull., Vol. 75, p. 805.

Figures 17.17, 17.18 — National Air Photo Library, Dept. of Energy, Mines, and Resources, Canada.

Figure 17.19 — After R. S. Deitz, 1972, "Geosynclines, Mountains, and Continent Building," Sci. Amer. 226(3):30-38.

Figure 17.20 — After C. H. Stockwell, 1964, Structural Provinces, Orogenies, and Time Classification of Rocks in the Canadian Precambrian Shield, Geol. Survey of Canada Paper 61-71.

Figure 17.21 — After B. Heezen, 1962, The Deep Sea Floor, in S. K. Runcorn (ed.), Continental Drift, pp. 235-288, Academic Press, N. Y.

Figure 17.24 — U.S. Geological Survey.

Opening page photo, Chapter 18 — Courtesy Texas Highway Department, Travel and Information Division.

Figures 18.1, 18.2 — After L. B. Leopold, 1968, Hydrology for Urban Land Planning, U.S. Geol. Survey Circ. 554, Figure 1, p. 3.

Figures 18.10, 18.11 — From The Limits to Growth: A Report for The Club of Rome's Project on the Predicament of Mankind, by Donella H. Meadows, Dennis L. Meadows, Jørgen Randers, William W. Behrens, III. A Potomac Associates book published by Universe Books, New York, 1972. Graphics by Potomac Associates.

Opening page photo, Chapter 19 — Courtesy NASA.

Figures 19.1, 19.2 — Courtesy National Space Data Center, NASA.

Figure 19.3 — After E. M. Shoemaker, 1960, Penetration Mechanics of High Velocity Meteorites, Illustrated by Meteor Crater, Arizona, International Geological Congress XXI, Vol. 18, p. 418.

Figures 19.6, 19.7, 19.8 — Courtesy National Space Data Center, NASA.

Figure 19.9 — After K. F. Weaver, 1973, "Have We Solved the Mysteries of the Moon?" National Geographic 144(3):318.

Figures 19.11, 19.12, 19.13, 19.14, 19.15 — Courtesy National Space Data Center, NASA.

Figure 19.18 — After F. Toksoz et al., 1972, Lunar Science Abstracts, January 10-13, p. 670.

Figure 19.19 — After P. M. Muller and W. L. Sjogren, 1968, "Mascons: Lunar Mass Concentrations," Science 161:680-684.

Figure 19.20 — Modified after K. F. Weaver, 1973, "Have We Solved the Mysteries of the Moon?" National Geographic 144(3):318.

Figures 19.21, 19.22, 19.23, 19.24 — Courtesy NASA and Garth H. Ladle.

Figure 19.25 — After E. M. Shoemaker, Geology of the Moon and Project Apollo, Contribution No. 2037 of the Division of Geological and Planetary Sciences, Calif. Inst. Tech., Pasadena, California.

Opening page photo, Chapter 20 — Courtesy NASA.

Figures 20.1, 20.3, 20.4, 20.5, 20.6 — Courtesy National Space Data Center, NASA.

Figure 20.7 — After M. H. Car et al., 1973, "A Generalized Geologic Map of Mars," J. Geophys. Res. 78(20):4032.

Figures 20.8, 20.9, 20.10, 20.11, 20.12, 20.13, 20.14, 20.15, 20.16, 20.17, 20.18, 20.19, 20.20, 20.21, 20.22, 20.23, 20.24, 20.25 — Courtesy National Space Data Center, NASA.
Opening page photo, Chapter 21 — Courtesy NASA.
Figures 21.1, 21.2, 21.3, 21.4, 21.5, 21.6, 21.7 — Courtesy National Space Data Center, NASA.
Figure 21.8 — After D. C. Murry et al., 1974, Mariner Venus/Mercury, Status Bull. No. 29, Jet Propulsion Lab, Calif. Inst. Tech., Pasadena, California.
Figure 21.9 — Courtesy National Space Data Center, NASA.
Opening page photo, Chapter 22 — Earth photo, courtesy U.S. Geological Survey; photos of the moon, Mars, and Mercury, courtesy NASA.

Index

Yazoo tributaries, **172**
Yellow River, 306

Zagros Mountains, Iran, **9**
Zones

Benioff, **348**, 361, 362
in soil profile, 133
of aeration, 212-214
of saturation, 212-214